Intelligent Information Processing
and Web Mining

Advances in Soft Computing

Editor-in-chief
Prof. Janusz Kacprzyk
Systems Research Institute
Polish Academy of Sciences
ul. Newelska 6
01-447 Warsaw, Poland
E-mail: kacprzyk@ibspan.waw.pl
http://www.springer.de/cgi-bin/search-bock.pl?series=4240

Robert Fullér
Introduction to Neuro-Fuzzy Systems
2000. ISBN 3-7908-1256-0

Robert John and Ralph Birkenhead (Eds.)
Soft Computing Techniques and Applications
2000. ISBN 3-7908-1257-9

Mieczysław Kłopotek, Maciej Michalewicz
and Sławomir T. Wierzchoń (Eds.)
Intellligent Information Systems
2000. ISBN 3-7908-1309-5

Peter Sinčák, Ján Vaščák, Vladimír Kvasnička
and Radko Mesiar (Eds.)
The State of the Art in Computational Intelligence
2000. ISBN 3-7908-1322-2

Bernd Reusch and Karl-Heinz Temme (Eds.)
Computational Intelligence in Theory and Practice
2000. ISBN 3-7908-1357-5

Rainer Hampel, Michael Wagenknecht,
Nasredin Chaker (Eds.)
Fuzzy Control
2000. ISBN 3-7908-1327-3

Henrik Larsen, Janusz Kacprzyk,
Sławomir Zadrozny, Troels Andreasen,
Henning Christiansen (Eds.)
Flexible Query Answering Systems
2000. ISBN 3-7908-1347-8

Robert John and Ralph Birkenhead (Eds.)
Developments in Soft Computing
2001. ISBN 3-7908-1361-3

Mieczysław Kłopotek, Maciej Michalewicz
and Sławomir T. Wierzchoń (Eds.)
Intelligent Information Systems 2001
2001. ISBN 3-7908-1407-5

Antonio Di Nola and Giangiacomo Gerla (Eds.)
Lectures on Soft Computing and Fuzzy Logic
2001. ISBN 3-7908-1396-6

Tadeusz Trzaskalik and Jerzy Michnik (Eds.)
Multiple Objective and Goal Programming
2002. ISBN 3-7908-1409-1

James J. Buckley and Esfandiar Eslami
An Introduction to Fuzzy Logic and Fuzzy Sets
2002. ISBN 3-7908-1447-4

Ajith Abraham and Mario Köppen (Eds.)
Hybrid Information Systems
2002. ISBN 3-7908-1480-6

Przemysław Grzegorzewski, Olgierd Hryniewicz,
Maria A. Gil (Eds.)
*Soft Methods in Probability, Statistics
and Data Analysis*
2002. ISBN 3-7908-1526-8

Lech Polkowski
Rough Sets
2002. ISBN 3-7908-1510-1

Mieczysław Kłopotek, Maciej Michalewicz
and Sławomir T. Wierzchoń (Eds.)
Intelligent Information Systems 2002
2002. ISBN 3-7908-1509-8

Andrea Bonarini, Francesco Masulli
and Gabriella Pasi (Eds.)
Soft Computing Applications
2002. ISBN 3-7908-1544-6

Leszek Rutkowski, Janusz Kacprzyk (Eds.)
Neural Networks and Soft Computing
2003. ISBN 3-7908-0005-8

Jürgen Franke, Gholamreza Nakhaeizadeh,
Ingrid Renz (Eds.)
Text Mining
2003. ISBN 3-7908-0041-4

Tetsuzo Tanino, Tamaki Tanaka,
Masahiro Inuiguchi
*Multi-Objective Programming and Goal
Programming*
2003. ISBN 3-540-00653-2

Mieczysław A. Kłopotek
Sławomir T. Wierzchoń
Krzysztof Trojanowski (Eds.)

Intelligent Information Processing and Web Mining

Proceedings
of the International IIS:IIPWM'03 Conference
held in Zakopane, Poland, June 2-5, 2003

With 100 Figures
and 70 Tables

Springer

Professor Dr. Mieczysław A. Kłopotek
Professor Dr. Sławomir T. Wierzchoń
Dr. Krzysztof Trojanowski

Polish Academy of Sciences
Inst. Computer Science
Ul. Ordona 21
01-237 Warsaw
Poland

ISSN 16-15-3871
ISBN 3-540-00843-8 Springer-Verlag Berlin Heidelberg NewYork

Cataloging-in-Publication Data applied for.
Bibliographic information published by Die Deutsche Bibliothek. Die Deutsche Bibliothek lists this
publication in the Deutsche Nationalbibliografie; detailed bibliographic data is available in the Internet
at <http://dnb.ddb.de>.

This work is subject to copyright. All rights are reserved, whether the whole or part of the material is
concerned, specifically the rights of translation, reprinting, reuse of illustrations, recitation,
broadcasting, reproduction on microfilm or in other ways, and storage in data banks. Duplication of
this publication or parts thereof is permitted only under the provisions of the German Copyright Law of
September 9, 1965, in its current version, and permission for use must always be obtained from
Springer-Verlag. Violations are liable to prosecution under German Copyright Law.

Springer-Verlag Berlin Heidelberg New York
a member of BertelsmannSpringer Science+Business Media GmbH

http://www.springer.de

© Springer-Verlag Berlin Heidelberg 2003
Printed in Germany

The use of general descriptive names, registered names, trademarks, etc. in this publication does not
imply, even in the absence of a specific statement, that such names are exempt from the relevant
protective laws and regulations and therefore free for general use.

Typesetting: Digital data supplied by the authors
Cover-design: E. Kirchner, Heidelberg
Printed on acid-free paper   62 / 3020 hu – 5 4 3 2 1 0

Preface

This volume contains articles accepted for presentation during The Intelligent Information Processing and Web Mining Conference IIS:IIPWM'03 which was held in Zakopane, Poland, on June 2-5, 2003. This conference extends a series of 12 successful symposia on Intelligent Information Systems, organized by the Institute of Computer Science of Polish Academy of Sciences, devoted to new trends in (broadly understood) Artificial Intelligence.

The idea of organizing such meetings dates back to 1992. Our main intention guided the first, rather small-audience, workshop in the series was to resume the results gained in Polish scientific centers as well as contrast them with the research performed by Polish scientists working at the universities in Europe and USA and their foreign collaborators. This idea proved to be attractive enough that we decided to continue such meetings. As the years went by, the workshops has transformed into regular symposia devoted to such fields like Machine Learning, Knowledge Discovery, Natural Language Processing, Knowledge Based Systems and Reasoning, and Soft Computing (i.e. Fuzzy and Rough Sets, Bayesian Networks, Neural Networks and Evolutionary Algorithms). At present, about 50 papers prepared by researches from Poland and other countries are usually presented.

This year conference is an attempt to draw a much broader international audience on the one hand, and to devote much more attention to the newest developments in the area of Artificial Intelligence. Therefore special calls for contributions on artificial immune systems and search engines. In connection with these and related issues, contributions were accepted, concerning:

- immunogenetics
- recommenders and text classifiers
- natural language processing for search engines and other web applications
- data mining and machine learning technologies
- logics for artificial intelligence
- time dimension in data mining
- information extraction and web mining by machine
- web services and ontologies
- foundations of data mining
- medical and other applications of data mining

The above-mentioned topics were partially due to four invited sessions organized by F. Espozito, M. Hacid, J. Rauch and R. Swiniarski.

Out of an immense flow of submissions, the Program Committee has selected only about 40 full papers for presentation and about a dozen of posters.

On behalf of the Program Committee and of the Organizing Committee we would like to thank all participants: computer scientists mathematicians, engineers, logicians and other interested researchers who found excitement in advancing the area of intelligent systems. We hope that this volume of IIS:IIPWM:S03 Proceeding will be a valuable reference work in your further research.

We would like to thank the Programme Committee Members for their effort in evaluating contributions and in making valuable suggestions both concerning the scientific level and the organization of the Conference.

We would like to thank Mr. M. Wolinski for his immense effort in resolving technical issues connected with the preparation of this volume.

Zakopane, Poland, *Mieczysław A. Kłopotek*, Conference Co-Chair
June 2003 *Sławomir T. Wierzchoń*, Conference Co-Chair
Krzysztof Trojanowski, Organizing Committee Chair

We would like to thank to the PC Members for their great job of evaluating the submissions

- Peter J. Bentley (University College London, UK)
- Petr Berka (University of Economics, Czech Republic)
- Dipankar Dasgupta (University of Memphis, USA)
- Piotr Dembinski (Polish Academy of Sciences, Poland)
- Wlodzislaw Duch (Nicholas Copernicus University, Poland)
- Tapio Elomaa (University of Helsinki, Finland)
- Floriana Esposito (University of Bari, Italy)
- Ursula Gather (University of Dortmund, Germany)
- Jerzy W. Grzymala-Busse (University of Kansas, USA)
- Mohand-Said Hacid (Université Claude Bernard Lyon 1, France)
- Mirsad Hadzikadic (University of North Carolina at Charlotte, USA)
- Ray J. Hickey (University of Ulster, UK)
- Olgierd Hryniewicz (Polish Academy of Sciences, Poland)
- Janusz Kacprzyk (Polish Academy of Sciences, Poland)
- Samuel Kaski (Helsinki University of Technology, Finland)
- Willi Kloesgen (Frauenhofer Institute, Germany)
- Jozef Korbicz (University of Zielona Gora, Poland)
- Jacek Koronacki (Polish Academy of Sciences, Poland)
- Witold Kosinski (Polish-Japanese Institute of Information Technologies, Poland)
- Stan Matwin (University of Ottawa, Canada)
- Maciej Michalewicz (NuTech Solutions Polska, Poland)
- Zbigniew Michalewicz (NuTech Solutions, USA)
- Ryszard Michalski (George Mason University, USA)
- Fionn Murtagh (Queen's University Belfast, UK)
- Zdzislaw Pawlak (Scientific Research Committee, Poland)
- James F. Peters (University of Manitoba, Canada)
- Adam Przepiorkowski (Polish Academy of Sciences, Poland)
- Zbigniew W. Ras (University of North Carolina at Charlotte, USA)
- Jan Rauch (University of Economics, Czech Republic)
- Henryk Rybinski (Warsaw University of Technology, Poland)
- Andrzej Skowron (Warsaw University, Poland)
- Katia Sycara (Carnegie Mellon University, USA)
- Roman Swiniarski (San Diego State University, USA)
- Ryszard Tadeusiewicz (University of Mining and Metallurgy, Poland)
- Jonathan Timmis (University of Kent, UK)
- Antony Unwin (University of Augsburg, Germany)
- Alicja Wakulicz-Deja (University of Silesia, Poland)
- Jan Weglarz (Poznan University of Technology, Poland)
- Stefan Wegrzyn (Polish Academy of Sciences, Poland)
- Krzysztof Zielinski (University of Mining and Metallurgy, Poland)
- Djamel A. Zighed (Lumiere Lyon 2 University, France)

- Jana Zvarova (EuroMISE Centre, Czech Republic)

We would like also to thank to additional reviewers

- Anna Kupść (Polish Academy of Sciences, Poland)
- Agnieszka Mykowiecka (Polish Academy of Sciences, Poland)
- Stanislaw Ambroszkiewicz (Polish Academy of Sciences, Poland)
- Witold Abramowicz (The Poznan University of Economics, Poland)

Table of contents

Part III. Natural Language Processing for Search Engines and Other Web Applications

Part IV. Data Mining and Machine Learning Technologies

Part V. Logics for Artificial Intelligence

Part VI. Time Dimension in Data Mining

Part VII. Invited Session: Information Extraction and Web Mining by Machine

Part VIII. Invited Session: Web Services and Ontologies

Part IX. Invited Session: Reasoning in AI

Part X. Invited Session: AI Applications in Medicine

Part I

Immunogenetics

Model of the Immune System to Handle Constraints in Evolutionary Algorithm for Pareto Task Assignments

Jerzy Balicki and Zygmunt Kitowski

Computer Science Department, Naval University of Gdynia,
Śmidowicza 69, 81-103 Gdynia, Poland

Abstract. In this paper, an evolutionary algorithm based on an immune system activity to handle constraints is discussed for solving three-criteria optimisation problem of finding a set of Pareto-suboptimal task assignments in parallel processing systems. This approach deals with a modified genetic algorithm cooperating with a main evolutionary algorithm. An immune system activity is emulated by a modified genetic algorithm to handle constraints. Some numerical results are submitted.

1 Introduction

Evolutionary algorithms (EAs) are an unconstrained search technique and, subsequently, have to exploit a supplementary procedure to incorporate constraints into fitness function in order to conduct the search correctly. An approach based on the penalty function is the most commonly used to respect constraints, and there have been many successful applications of the penalty function, mainly exterior, for finding a sub-optimal solution to an optimisation problem with one criterion. Likewise, the penalty technique is frequently used to handle constraints in multi-criteria evolutionary algorithms to find the Pareto-suboptimal outcomes [1]. However, penalty functions have some familiar limitations, from which the most noteworthy is the complicatedness to identify appropriate penalty coefficients [11].

Koziel and Michalewicz have proposed the homomorphous mappings as the constraint-handling technique of EA to deal with parameter optimisation problems in order to avoid some impenetrability related to the penalty function [11]. Then, Coello Coello and Cortes have designed a constrained-handling scheme based on a model of the immune system to optimisation problems with one criterion [2].

In this paper, we propose an improved model of the immune system to handle constraints in multi-criteria optimisation problems. The problem that is of interest to us is the new task assignment problem for a distributed computer system. Both a workload of a bottleneck computer and the cost of machines are minimized; in contrast, a reliability of the system is maximized. Moreover, constraints related to memory limits, task assignment and computer locations are imposed on the feasible task assignment. Finally, an evolutionary algorithm based on tabu search procedure and the immune system model is proposed to provide task assignments to the distributed systems.

2   Models of immune system

The immune system can be seen, from the information processing perspectives, as a parallel and distributed adaptive system [2]. Learning, using memory, associative retrieval of information in recognition and classification, and many local interactions provide, in consequence, fault tolerance, dynamism and adaptability [5]. Some con-ceptual and mathematical models of these properties of the immune system were constructed with the purpose of understanding its nature [10,14]. A model of primary response in the presence of a trespasser was discussed by Forrest et al. [7]. Moreover, the model of secondary response related to memory was assembled by Smith [12]. Both detectors and antigens were represented as strings of symbols in a small alphabet in the first computer model of the immune system by Farmer et al. [6]. In addition, molecular bonds were represented by interactions among these strings.

The model of immune network and the negative selection algorithm are two main models in which most of the current work is based [8]. Moreover, there are others used to simulate ability of the immune system to detect patterns in a noise environment, ability to discover and maintain diverse classes of patterns and ability to learn effectively, even when not all the possible types of invaders had been previously presented to the immune system [12].

Jerne has applied differential equations to simulate the dynamics of the lymphocytes by calculation the change of the concentration of lymphocytes' clones [9]. Lymphocytes do not work in an isolated manner, but they work as an interconnected network. On the other hand, the negative selection algorithm (NSA) for detection of changes has been developed by Forrest at el. [7]. This algorithm is based on the discrimination principle that is used to know what is a part of the immune system and what is not [8]. Detectors are randomly generated to reduce those detectors that are not capable of recognising themselves. Subsequently, detector capable to identify trespassers is kept. Change detection is performed probabilistically by the NSA. It is also robust because it looks for any unknown action instead of just looking for certain explicit pattern of changes.

In this paper, the NSA is used to handle constraints by dividing the contemporary population in two groups [2]. Feasible solutions called 'antigens' create the first group, and the second group of individuals consists of 'antibodies" – infeasible solutions. Therefore, the NSA is applied to generate a set of detectors that determine the state of constraints. We assume the fitness for antibodies is equal to zero. Then, a randomly chosen antigen G^- is compared against the σ antibodies that were selected without replacement. Afterwards, the distance S between the antigen G^- and the antibody B^- is calculated due to the amount of similarity at the genotype level [2]:

$$S(G^-, B^-) = \sum_{m=1}^{M} s_m(G^-, B^-),\qquad(1)$$

where

M – the length of the string representing the antigen G (the length of the antibody B is the same),

$$s_m = \begin{cases} 1 \text{ if } G_m^- \\ 0 \text{ in the other case} \end{cases}, \quad m = \overline{1, M}.$$

The fitness of the antibody with the highest matching magnitude S is increased by adding its amount of similarity. The antibodies are returned to the current population and the process of increasing the fitness of the winner is repeated typically tree times the number of antibodies. Each time, a randomly chosen antigen is compared against the same subset of antibodies.

Afterwards, a new population is constructed by reproduction, crossover and mutation without calculations of fitness. Above process is repeated until a convergence of population or until a maximal number of iterations is exceeded. Then, the final population of the NSA is returned to the external evolutionary algorithm.

The negative selection algorithm is a modified genetic algorithm in which infeasible solutions that are similar to feasible ones are preferred in the current population. Although, almost all random selections are based on the uniform distribution, the pressure is directed to improve the fitness of appropriate infeasible solutions.

The measure of genotype similarity between antigen and antibody depends on the representation. The measure of similarity for the binary representation can be redefined for integer representation:

$$S'(G^-, B^-) = \sum_{m=1}^{M} |(G_m^- - B_m^-)|.\qquad(2)$$

3   Negative selection algorithm with ranking procedure

The fact that the fitness of the winner is increased by adding the magnitude of the similarity measure to the current value of fitness may pass over a non-feasible solution with the relatively small value of this total measure. However, some constraints may be satisfied by this solution. What is more, if one constraint is exceeded and the others are performed, the value of a similarity measure may be low for some cases. That is, the first of two similar solutions, in genotype sense, may not satisfy this constraint and the second one may satisfy it.

For example, an antigen represented by the binary vector $(1,0,0,0,0)$ satisfies constraint $x \geq 32$; however, an antibody $(0,0,0,0,1)$ with the magnitude of similarity equal to 3 does not satisfy this inequality constraint. On the other hand, the antibody $(0,1,1,1,1)$ has the amount of similarity to $(1,0,0,0,0)$ equal to 0, and this antibody with the lower similarity to $(1,0,0,0,0)$ than $(0,0,0,0,1)$ is very close to satisfy the constraint $x \geq 32$. Therefore, the antibody $(0,1,1,1,1)$ is supposed to be preferred than $(0,0,0,0,1)$, but the opposed preferences are incorporated in the above NSA version.

To avoid this limitation of the NSA, we suggest introducing some distance measures from the state of an antibody to the state of the selected antigen, according to the constraints. The constraints that are of interest to us are, as follows:

$$g_k(x) \leq 0, \quad k = \overline{1, K}, \tag{3}$$

$$h_l(x) = 0, \quad l = \overline{1, L}. \tag{4}$$

Constraints (3) and (4) are included to the general non-linear programming problem [1] as well as they are met in task assignment problem [13]. Especially, memory constraints belong to the class described by inequalities (3). On the other hand, computer allocation constraints and task assignment constraints fit in the class defined by the equalities (4).

The distance measures from the state of an antibody B to the state of the selected antigen G are defined, as below:

$$f_n(B^-, G^-) = \begin{cases} g_k(B^-) - g_k(G^-), \ k = \overline{1, K}, n = k, \\ |h_l(B^-)|, \ l = \overline{1, L}, n = K + l, \end{cases} \quad n = \overline{1, N}, N = K + L. \tag{5}$$

The distance $f_n(B^-, G^-)$ is supposed to be minimized for all constraint numbers n. If the antibody B^- is marked by the shorter distance $f_n(B^-, G^-)$ to the selected antigen than the antibody C^-, then B^- ought to be preferred than C^- due to the improvement of the nth constraint. Moreover, if the antibody B^- is characterized by the all shorter distances to the selected antigen than the antibody C^-, then B^- should be preferred than C^- due to the improvement of the all constraints. However, it is possible to occur situations when B^- is characterized by the shorter distances for some constraints and the antibody C^- is marked by the shorter distances for the others. In this case, it is difficult to select an antibody.

Therefore, we suggest introducing a ranking procedure to calculate fitness of antibodies and then to select the winners. A ranking idea for non-dominated individuals has been introduced to avoid the prejudice of the interior Pareto alternatives.

Now, we adjust this procedure to the negative selection algorithm and a subset of antibodies. Firstly, distances between antigen and antibodies are calculated. Then, the nondominated antibodies are determined according to their distances, and after that, they get the rank 1. Subsequently, they are temporary eliminated from the population. Next, the new nondominated antibodies are found from the reduced population and they get the rank 2. In this procedure, the level is increased and it is repeated until the subset of antibodies is exhausted. All non-dominated antibodies have the same reproduction fitness because of the equivalent rank.

If B^- is the antibody with the rank $r(B^-)$ and $1 \leq r(B^-) \leq r_{max}$, then the increment of the fitness function value is estimated, as below:

$$\Delta f(B^-) = r_{max} - r(B^-) + 1. \tag{6}$$

Afterwards, the fitness of the all chosen antibodies are increased by adding their increments. The antibodies are returned to the current population and the process of increasing the fitness of antibodies is repeated typically tree times the number of antibodies as it was in the previous version of the NSA. Each time, a randomly chosen antigen is compared against the same subset of antibodies. Next, the same procedure as for the NSA is carried out. Afterwards, a new population is constructed by reproduction, crossover and mutation without calculations of fitness. Above process is repeated until a convergence of population emerges or until a maximal number of iterations is exceeded. Then, the final population of the negative selection algorithm is returned to the external evolutionary algorithm.

4   Constrained multi-criterion task assignment problem

Let the negative selection algorithm with the ranking procedure be called NSA+. To test its ability to handle constraints, we consider a new multi-criteria optimisation problem for task assignment in a distributed computer system.

Finding allocations of program modules may decrease the total time of a program execution by taking a benefit of the particular properties of some workstations or an advantage of the computer load. An adaptive evolutionary algorithm and an adaptive evolution strategy have been considered for solving multiobjective optimisation problems related to task assignment that minimize Z_{max} – a workload of a bottleneck computer and F_2 – the cost of machines [1]. The total numerical performance of workstations is another criterion for assessment of task assignment and it has been involved to multicriteria task assignment problem in [1]. Moreover, a reliability R of the system is an additional criterion that is important to assess the quality of a task assignment.

In the considered problem, both a workload of a bottleneck computer and the cost of machines are minimized; in contrast, a reliability of the system is maximized. Moreover, constraints related to memory limits, task assignment and computer locations are imposed on the feasible task assignment.

A set of program modules $\{M_1, \ldots, M_m, \ldots, M_M\}$ communicated to each others is considered among the coherent computer network with computers located at the processing nodes from the set $W = \{w_1, \ldots, w_i, \ldots, w_I\}$. A program module can be activated several times during the program lifetime and with the program module runs are associated some processes (tasks). In results, a set of program modules is mapped into the set of parallel performing tasks $\{T_1, \ldots, T_v, \ldots, T_V\}$ [13].

Let the task T_v be executed on computers taken from the set of available computer sorts $\Pi = \{\pi_1, \ldots, \pi_j, \ldots, \pi_J\}$. The overhead performing time of the task T_v by the computer π_j is represented by an item t_{vj}. Let π_j be failed independently due to an exponential distribution with rate λ_j. We do not take into account of repair and recovery times for failed computer in assessing the logical correctness of an allocation. Instead, we shall allocate tasks to computers on which failures are least likely to occur during the execution of tasks. Computers can be allocated to nodes and tasks can be assigned to them in purpose to maximize the reliability function R defined, as below:

$$R(x) = \prod_{v=1}^{V} \prod_{i=1}^{I} \prod_{j=1}^{J} \exp(-\lambda_j t_{vj} x_{vi}^m x_{ij}^\pi), \tag{7}$$

where

$$x_{ij}^\pi = \begin{cases} 1 \text{ if } \pi_j \text{ is assigned to the } w_i, \\ 0 \text{ in the other case,} \end{cases}$$

$$x_{vi}^m = \begin{cases} 1 \text{ if task } T_v \text{ is assigned to } w_i, \\ 0 \text{ in the other case,} \end{cases}$$

$$x = [x_{11}^m, \ldots, x_{1I}^m, \ldots, x_{vi}^m, \ldots, x_{VI}^m,$$
$$x_{11}^\pi, \ldots, x_{1J}^\pi, \ldots, x_{ij}^\pi, \ldots, x_{I1}^\pi, \ldots, x_{Ij}^\pi, \ldots, x_{IJ}^\pi]^T.$$

A computer may be chosen several times from the set Π to be assigned to the node w_i and one computer is allocated to each node. On the other hand, each task is allocated to any node.

A computer with the heaviest task load is the bottleneck machine in the system, and its workload is a critical value that is supposed to be minimized [1]. The workload $Z_{\max}(x)$ of the bottleneck computer for the allocation x is provided by the subsequent formula:

$$Z_{\max}(x) = \max_{i \in \overline{1,I}} \left\{ \sum_{j=1}^{J} \sum_{v=1}^{V} t_{vj} x_{vi}^m x_{ij}^\pi + \sum_{v=1}^{V} \sum_{\substack{u=1 \\ u \neq v}}^{V} \sum_{i=1}^{I} \sum_{\substack{k=1 \\ k \neq i}}^{I} \tau_{vuik} x_{vi}^m x_{uk}^m \right\}, \tag{8}$$

where τ_{vuik} – the total communication time between the task T_v assigned to the ith node and the T_u assigned to the kth node.

Each computer should be equipped with necessary capacities of resources for a program execution. Let the following memories $z_1, \ldots, z_r, \ldots, z_R$ be available in an entire system and let d_{jr} be the capacity of memory z_r in the workstation p_j . We assume the task T_v reserves c_{vr} units of memory z_r and holds it during a program execution. Both values c_{vr} and d_{jr} are nonnegative and limited.

The memory limit in a machine cannot be exceeded in the ith node, what is written, as bellows:

$$\sum_{v=1}^{V} c_{vr} x_{vi}^m \leq \sum_{j=1}^{J} d_{jr} x_{ij}^\pi, \quad i = \overline{1, I}, \quad r = \overline{1, R}. \tag{9}$$

The other measure of the task assignment is a cost of computers [1]:

$$F_2(x) = \sum_{i=1}^{I} \sum_{j=1}^{J} \kappa_j x_{ij}^\pi, \tag{10}$$

where κ_j corresponds to the cost of the computer π_j.

The total computer cost is in conflict with the numerical performance of a distributed system, because the cost of a computer usually depends on the quality of its components. The faster computer or the higher reliability of it, the more expensive one. Additionally, the workload of the bottleneck computer is in conflict with the cost of the system. If the inexpensive and non-high quality components are used, the load is moved to the high quality ones and workload of the bottleneck computer increases.

In above new multiobjective optimisation problem related to task assignment, a workload of a bottleneck computer and the cost of machines are minimized [1]. On the other hand, a reliability of the system and numerical performance are maximized. Let (X, F, P) be the multi-criterion optimisation question for finding the representation of Pareto-optimal solutions. It is established, as follows:

---

1) $\mathbb{X}$ – an admissible solution set

$$\mathbb{X} = \{\, x \in \mathcal{B}^{I(V+J)} \mid \sum_{v=1}^{V} c_{vr} x_{vi}^m \leq \sum_{j=1}^{J} d_{jr} x_{ij}^\pi, \quad i = \overline{1, I}, \quad r = \overline{1, R}$$

$$\sum_{i=1}^{I} x_{vi}^m = 1, \quad v = \overline{1, V}; \quad \sum_{j=1}^{J} x_{ij}^\pi = 1, \quad i = \overline{1, I} \,\},$$

where $\mathcal{B} = \{0, 1\}$
2) F – a quality vector criterion $F : \mathbb{X} \mapsto \mathbb{R}^3$,
where
$\mathbb{R}$ – the set of real numbers,
$F(x) = [-R(x), Z_{\max}(x), F_2(x)]^T$ for $x \in \mathbb{X}$,
$R(x), Z_{\max}(x), F_2(x)$ are calculated by (7),(8) and (10), respectively
3) P – the Pareto relationship [9]

---

5 Tabu-based adaptive evolutionary algorithm using NSA

An overview of evolutionary algorithms for multiobjective optimisation problems is submitted in [3,4]. Zitzler, Deb, and Thiele have tested an elitist multi-criterion evolutionary algorithm with the concept of non-domination in their strength Pareto evolutionary algorithm SPEA [15].

An analysis of the task assignments has been carried out for two evolutionary algorithms. The first one was an adaptive evolutionary algorithm with tabu mutation AMEA+ [1]. Tabu search algorithm [13] was applied as an additional mutation operator to decrease the workload of the bottleneck computer. However, initial numerical examples indicated that obtained task assignments have not satisfied constraints in many cases. Therefore, we suggest reducing this disadvantage by introducing a negative selection algorithm with ranking procedure to improve the quality of obtained task assignments.

Better outcomes from the NSA are transformed into improving of solution quality obtained by the adaptive multicriteria evolutionary algorithm with tabu mutation AMEA*. This adaptive evolutionary algorithm with the NSA (AMEA*) gives better results than the AMEA+ (Fig. 1). After 200 generations, an average level of Pareto set obtaining is 1.4% for the AMEA*, 1.8% for the AMEA+. 30 test preliminary populations were prepared, and each algorithm starts 30 times from these populations. For integer constrained coding of chromosomes there are 12 decision variables in the test optimisation problem. The search space consists of 25 600 solutions.

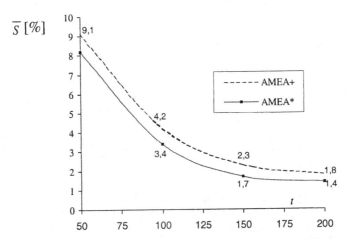

Fig. 1. Outcome convergence for the AMEA* and the AMEA+

6   Concluding remarks

The tabu-based adaptive evolutionary algorithm with the negative selection algorithm is an advanced technique for finding Pareto-optimal task allocations in a new three-objective optimisation problem with the maximisation of the system reliability. Moreover, the workload of the bottleneck computer and the cost of computers are minimized.

The negative selection algorithm can be used to handle constraints and improve a quality of the outcomes obtained by an evolutionary algorithm. Our future works will concern on a development the NSA and evolutionary algorithms for finding Pareto-optimal solutions of the other multiobjective optimisation problems.

References

1. Balicki, J., Kitowski, Z.: Multicriteria Evolutionary Algorithm with Tabu Search for Task Assignment. Lectures Notes in Computer Science, Vol. 1993 (2001) 373–384
2. Coello Coello, C. A., Cortes, N.C.: Use of Emulations of the Immune System to Handle Constraints in Evolutionary Algorithms. Knowledge and Information Systems. An International Journal, Vol. 1 (2001) 1–12
3. Coello Coello, C. A., Van Veldhuizen, D. A., Lamont, G.B.: Evolutionary Algorithms for Solving Multi-Objective Problems. Kluwer Academic Publishers, New York (2002)
4. Deb, K.: Multi-Objective Optimization using Evolutionary Algorithms, John Wiley & Sons, Chichester (2001)
5. D'haeseleer, P., et al. An Immunological Approach to Change Detection. In Proc. of IEEE Symposium on Research in Security and Privacy, Oakland (1996)
6. Farmer, J.D., Packard, N.H., Perelson, A.S.: The Immune System, Adaptation, and Machine Learning. Physica D, Vol. 22 (1986) 187–204
7. Forrest, S., Perelson, A.S.: Genetic Algorithms and the Immune System. Lecture Notes in Computer Science (1991) 320–325
8. Helman, P. and Forrest, S. An Efficient Algorithm for Generating Random Antibody Strings. Technical Report CS-94-07, The University of New Mexico, Albuquerque (1994)
9. Jerne, N.K.: The Immune System. Scientific American, Vol. 229, No. 1 (1973) 52–60
10. Kim, J. and Bentley, P. J. (2002), Immune Memory in the Dynamic Clonal Selection Algorithm. Proc. of the First Int. Conf. on Artificial Immune Systems, Can-terbury, (2002) 57–65
11. Koziel, S., Michalewicz, Z.: Evolutionary Algorithms, Homomorphous mapping, and Constrained Parameter Optimisation. Evolutionary Computation, Vol. 7 (1999) 19–44
12. Smith, D.: Towards a Model of Associative Recall in Immunological Memory. Technical Report 94-9, University of New Mexico, Albuquerque (1994)
13. Weglarz, J. (ed.): Recent Advances in Project Scheduling. Kluwer Academic Publishers, Dordrecht (1998)

14. Wierzchon, S. T.: Generating Optimal Repertoire of Antibody Strings in an Artificial Immune System. In M. Klopotek, M. Michalewicz and S. T. Wierzchon (eds.) Intelligent Information Systems. Springer Verlag, Heidelberg/New York (2000) 119–133
15. Zitzler, E., Deb, K., and Thiele, L.: Comparison of Multiobjective Evolutionary Algorithms: Empirical Results. Evolutionary Computation, Vol. 8, No. 2 (2000) 173–195

Function Optimization
with Coevolutionary Algorithms

Franciszek Seredynski[1,2], Albert Y. Zomaya[3], and Pascal Bouvry[4]

[1] Polish -Japanese Institute of Information Technologies, Koszykowa 86, 02-008
Warsaw, Poland
[2] Institute of Computer Science of Polish Academy of Sciences, Ordona 21,
01-237 Warsaw, Poland
[3] School of Information Technologies, University of Sydney, Sydney, NSW 2006,
Australia
[4] Institut Superieur de Technologie, Luxembourg University of Applied Science,
6, rue Coudenhove Kalergi, L-1359 Luxembourg-Kirchberg, Luxembourg

Abstract. The problem of parallel and distributed function optimization with coevolutionary algorithms is considered. Two coevolutionary algorithms are used for
this purpose and compared with sequential genetic algorithm (GA). The first coevolutionary algorithm called a loosely coupled genetic algorithm (LCGA) represents a
competitive coevolutionary approach to problem solving and is compared with another coevolutionary algoritm called cooperative coevolutionary genetic algorithm
(CCGA). The algorithms are applied for parallel and distributed optimization of
a number of test functions known in the area of evolutionary computation. We
show that both coevolutionary algorithms outperform a sequential GA. While both
LCGA and CCGA algorithms offer high quality solutions, they may compete to
outperform each other in some specific test optimization problems.

1 Introduction

The use of evolutionary computation (EC) techniques to evolve solutions of
both theoretical and real-life problems has seen a dramatic increase in popularity and success over last decade. The most popular and widely applied
EC technique was a *sequential GA* ([5]) which computational scheme is based
on a single population of individuals representing a single *species*. Development of parallel machines stimulated parallelization of a sequential GA and
resulted in two parallel EC techniques known respectively as *island model*
and *diffusion model* (see, e.g. [2]). These both models widely used today have
been still exploring a notion of a single species, but individuals representing
the species live in different subpopulations .

While these techniques are very effective in many applications, new more
difficult problems were set. These problems (e.g. modeling economic phenomena such as a market) are in their nature distributed, i.e. can be seen
as a number of independent interacting entities with own goals, where a
global behavior can observed as the result of interactions. To meet these new

requirements, researches in the area of EC were looking for new more powerful paradigmes of natural processing. In the result *coevolutionary algorithms* based on modelling phenomena of coexistance of several species emerged [3] as a very promising area of EC.

In this paper we present a *competitive coevolutionary algoritm* called *LCGA*[8]), which is based on a game-theoretical model. The algorithm is parallel and distributed and can be interpreted as a multi-agent system with locally expressed (where it is possible) goals of agents and a global behavior of the system. We use our algorithm to solve the problem of *optimization of function* and compare it with another known from the literature *cooperative coevolutionary algorithm CCGA* ([7]) which is partially parallel and needs a global synchronization.

The paper is organized as follows. In the next section we shortly overview coevolutionary models existing in the area of EC. In Section 3 we describe test functions used in experiments. Sections 4 and 5 contain presentation of two coevolutionary algorithms *LCGA* and *CCGA* studied in the paper. Section 6 contains results of experiments conducted with use of coevolutionary algorithms. The last section contains conclusions of the paper.

2   Coevolutionary Genetic Algorithms

The idea of coevolutionary algorithms comes from the biological observations which shows that coevolving some number of *species* defined as collections of phenotypically similar individuals is more realistic than simply evolving a population containing representatives of one species. So, instead of evolving a population (global or spatially distributed) of similar individuals representing a global solution, it is more appropriate to coevolve subpopulations of individuals representing specific parts of the global solution.

A number of coevolutionary algorithms have been presented recently. The *coevolutionary GA* [6]) described in the context of the constraint satisfaction problem and the neural network optimization problem is a low level parallel EA based on a *predator-prey* paradigm. The algorithm operates on two subpopulations: the main subpopulation $P^1()$ containing individuals $\bar{x}$ representing some species, and an additional subpopulation $P^2()$ containing individuals $\bar{y}$ (another species) coding some constraints, conditions or test points concerning a solution $\bar{x}$. Both populations evolve *in parallel* to optimize a global function $f(\bar{x}, \bar{y})$.

The *cooperative coevolutionary GA* (CCGA) [7] has been proposed in the context of a function optimization problem, and competetive coevolutionary algorithm called *loosely coupled GA* (LCGA) [8] has been described in the context of game-theoretic approach to optimization. These two coevolutionary algorithms are the subject of study presented in next sections. Another coevolutionary algorithm called *coevolutionary distributed GA* [4] was presented in the context of integrated manufacturing planning and scheduling problem. It combines features of diffusion model with coevolutionary concepts.

3 Test Functions

In the evolutionary computation literature (see, e.g. [5]) there is a number of test functions which are used as benchmarks for contemporary optimization algorithms. In this study we use some number of such functions, which will be the subject of minimization. We use the following test functions:

- sphere model: a continuous, convex, unimodal function

$$f_1(x) = \sum_{i=1}^{n} x_i^2; \qquad x \in R^n, \tag{1}$$

with $-100 \le x_i \le 100$, a minimum $x^* = (0, ..., 0)$ and $f_1(x^*) = 0$
- Rosenbrock's function: a continuous, unimodal function

$$f_2(\mathbf{x}) = \sum_{i=1}^{n} \left(100 \left(x_i^2 - x_{i+1} \right)^2 + (1 - x_i)^2 \right); x \in R^n, \tag{2}$$

with $-2.12 \le x_i \le 2.12$, a global minimum
$f_2(\mathbf{x}^*) = 0$ at $\mathbf{x}^* = (1, 1, \ldots, 1)$
- Rastrigin's function: a continuous, multimodal function

$$f_4(x) = \sum_{i=1}^{n} (x_i^2 - 10 \cdot \cos(2\pi x_i)); \qquad x \in R^n, \tag{3}$$

with $A = 10, \omega = 2 \cdot \pi - 5.12, \le x_i \le 5.12$,
$f_4(x^*) = 0$ in $x^* = (0, 0, ..., 0)$
- Schwefel's function: a continuous, unimodal function

$$f_5(x) = \sum_{i=1}^{n} (\sum_{j=1}^{i} x_j)^2; \qquad x \in R^n, \tag{4}$$

with $-100 \le x_i \le 100, f_4(x^*) = 0$ in $x^* = (0, 0, ..., 0)$
- Ackley's function: a continuous, multimodal function

$$f_6(x) = -20 e^{(-0.2 \sqrt{\frac{1}{n} \sum_{i=1}^{n} x_i^2})} - e^{(\frac{1}{n} \sum_{i=1}^{n} \cos 2\pi x_i)} + \tag{5}$$

$+20 + e$
with $-32 \le x_i \le 32, f_6(x^*) = 0$ in $x^* = (0, 0, ..., 0)$
- Griewank' function: a continuous, multimodal function

$$f_7(x) = \frac{1}{4000} \sum_{i=1}^{n} x_i^2 - \prod_{i=1}^{n} \cos(\frac{x_i}{\sqrt{i}}) + 1, \tag{6}$$

with $-600 \le x_i \le 600, f_4(x^*) = 0$ in $x^* = (0, 0, ..., 0)$.

4   Competitive Coevolutionary Approach: Loosely Coupled Genetic Algorithms

Loosely coupled genetic algorithm (LCGA) [8,9] is a medium-level parallel and distributed coevolutionary algorithm exploring a paradigm of *competitive coevolution* motivated by noncooperative models of game theory. Local chromosome structures of LCGA are defined for each variable, and local subpopulations are created for them. Contrary to known sequential and parallel EAs, the LCGA is assumed to work in a distributed environment described by locally defined functions. A problem to be solved is first analyzed in terms of its possible decomposition and relations between subcomponents, expressed by a problem defined communication graph G_{com} called a graph of interaction.

In the case of functions like e.g. the Rosenbrock's function a decomposition of the problem, designing local functions $f_2^i(x_i, x_{i+1})$, and a graph of interaction (a local function assigned to an agent associated with the node i depends on a variable x_i associated with this node, and on the node $(i + 1)$ with associated variable x_{i+1}) is straightforward (see, [1]).

Many real-life problems e.g. describing behavior of economic systems are naturally decentralized, or their models can be designed in such a way to decentralize their global criterion. When it is not possible, a communication graph G_{com} is a fully connected graph, and a global criterion becomes a local optimization criterion associated with each agent.

LCGA can be specified in the following way:

Step 1: for each agent-player create a subpopulation of his actions:
- create for each player an initial subpopulation of size *sub_pop_size* of player actions with values from the set S_k of his actions.

Step 2: play a single game:
- in a discrete moment of time each player randomly selects one action from the set of actions predefined in his subpopulation and presents it to his neighburs in the game.
- calculate the output of each game: each player evaluates his local payoff u_k in the game.

Step 3: repeat step 2 until *sub_pop_size* games are played.

Step 4: for each player create a new subpopulation of his actions:
- after playing *sub_pop_size* games each player knows the value of his payoff received for a given action from his subpopulation.
- the payoffs are considered as values of a local fitness function defined during a given generation of a GA; standard GA operators of selection, crossover and mutation are applied locally to the subpopulations of actions; these actions will be used by players in the games played in the next game horizon.

Step 5: return to step 2 until the termination condition is satisfied.

After initializing subpopulations, corresponding sequences of operations are performed *in parallel* for each subpopulation, and repeated in each generation. For each individual in a subpopulation a number of n_i (n_i-number of neighbors of subpopulation $P^i()$) of *random tags* is assigned, and copies of individuals corresponding to these tags are sent to neighbor subpopulations, according to the interaction graph. Individuals in a subpopulation are matched with ones that arrived uppon request from the neighbor subpopulations. Local fitness function of individuals from subpopulations is evaluated on the base of their values and values of arrived taggeted copies of individuals. Next, standard GA operators are applied locally in subpopulations. Coevolving this way subpopulations compete to maximize their local functions. The process of local maximization is constrained by neighbor subpopulations, sharing the same variables. As the result of this competitive coevolution one can expect the system to achieve some equilibrium, equivalent to a Nash point equilibrium in noncooperative models of game theory.

A final performance of the LCGA operated in a distributed environment is evaluated by some global criterion, usually as a sum of local function values in an equilibrium point. This global criterion is typically unknown for subpopulations (except the case when G_{com} is a fully connected graph), which evolve with their local criteria.

5   Cooperative Coevolutionary Approach: Cooperative Coevolutionary Genetic Algorithm

Cooperative coevolutionary genetic algorithm (CCGA) has been proposed [7] in the context of a function optimization problem. Each of N variables x_i of the optimization problem is considered as a species with its own chromosome structure, and subpopulations for each variable are created. A global function $f(\bar{x})$ is an optimization criterion. To evaluate the fitness of an individual from a given subpopulation, it is necessary to communicate with selected individuals from all subpopulations. Therefore, the communication graph G_{com} is fully connected.

In the initial generation of CCGA individuals from a given subpopulation are matched with randomly chosen individuals from all other subpopulations. A fitness of each individual is evaluated, and the best individual I^i_{best} in each subpopulation is found. The process of *cooperative coevolution* starts from the next generation. For this purpose, in each generation the following cycle consisting of two phases is repeated in a round-robin fashion. In the first phase only one current subpopulation is active in a cycle, while the other subpopulations are frozen. All individuals from an active subpopulation are matched with the best individuals of frozen subpopulations. A new better individual is found this way for each active subpopulation. In the second phase the best found individual from each subpopulation is matched with a single, randomly selected individual from other subpopulations. A winner individual is a better individual from these two phases. When the evolutionary process is completed a composition of the best individuals from each subpopulation represents a solution of a problem.

6   Experimental Study

Both LCGA and CCGA algorithms were tested on the set of functions presented in Section 4. Results of these experiments were compared with the results of a sequential GA. In all experiments the accuracy of x_i was not worse than 10^{-6}. Experiments were conducted with number of variables $n = 5, 10, 20, 30$, but only results for $n = 30$ are reported in the paper. All algorithms run 200 generations.

The following parameters were set for LCGA: the size of subpopulation corresponding to given species was equal 100, p_m ranged for different functions from 0.001 (Rosenbrock's, Rastrigin's function) to 0.005, p_k ranged from 0.5 (Griewank's function), 0.6 (Rosenbrock's function) to 0.9. Ranking selection for Rastrigin's function and proportional selection for Ackley's function was used. For remaining functions a tournament selection with a size of tournament equal to 4 was used.

The following parameters were set for CCGA: the size of subpopulation corresponding to given species was equal 100, p_m ranged for different functions from 0.001 (Rosenbrock's, Rastrigin's, and Ackley's function) to 0.008, p_k ranged from 0.6 (Rosenbrock's function) to 1.0. Proportional selection for Rosenbrock's and Rastrigin function was used and ranking selection for Griewank's function was used. For remaining functions tournament selection with a size of tournament equal to 4 was used.

The following parameters were set for the sequential GA: the size of a global population corresponding to given species was equal 100, p_m ranged for different functions from 0.001 (Ackley's function) to 0.0095, p_k ranged from 0.5 (Griewank's function) to 0.95. Ranking selection for Rastrigin's and Griewank's function was used and tournament selection with a size of tournament equal to 3 was used for remaining functions.

Fig. 1 and 2 show results of experiments conducted with use of LCGA, CCGA and the sequential GA applied to minimize test functions presented in Section 4. Each experiment was repeated again 25 times. The best results in each generations of 20 the best experiments were averaged and accepted as results shown in following figures as a function of a number of generations.

One can easy notice that both coevolutionary algorithms LCGA and CCGA are better than the sequential GA in the problem minimization of function for all test functions used in the experiment. However, comparison of coevolutionary algorithms is not so straigthforward. For the test problem corresponding to the sphere model both coevolutionary algorithms present almost the same behavior (speed of convergence and the average of minimal values) for the number of variables from the range $n = 5$ to $n = 20$ (not shown in the paper), with some better performance of LCGA. For $n = 30$ (see, Fig. 1a) the difference between both coevolutionary algorithms becomes more visible: LCGA maintenances its speed of convergence achieving the best (minimal) value after about 60 generations while CCGA is not able to find this value within considered number of 200 generations.

For the Rosenbrock's function CCGA shows slightly better performance that LCGA for all n, and this situation is shown in Fig. 1b for $n = 30$. One can see that this test function is really difficult for both algorithms. The opposite situation takes place for the Rastrigin's function. LCGA is distinctively better than CCGA for small values of n ($n = 5$) and slightly better for greater values of n. Fig. 1c illustrates this situation for $n = 30$.

For the Schwefel's function and $n = 5$ all three algorithms achieve the same minimal value. CCGA needs for this about 60 generations, LCGA needs about 110 generations and the sequential GA needs 200 generations. For $n = 10$ both coevolutionary algorithms are better than the sequential GA, but CCGA is slightly better than LCGA. For greater values of n CCGA outperforms LCGA (see, Fig. 2a). For the Ackley's function CCGA outperforms LCGA for all values of n (see, Fig. 2b) and the Griewank's function situation is similar Fig. 2c).

7 Conclusions

Results of ongoing research on the development of parallel and distributed evolutionary algorithms for function optimization have been presented in the paper. Coevolution - a new very promising paradigm in evolutionary computation has been chosen as an engine for effective parallel and distributed computation. Two coevolutionary algorithms based on different phenomena known as *competition* (LCGA) and *cooperation* (CCGA) were studied.

LCGA presents fully parallel and distributed coevolutionary algorithm in which subpopulations, using game-theoretic mechanism of competition, act to maximize their local goals described by some local functions. The competition between agents leads to establishing some equilibrium in which local goals cannot be more improved, and at the same time some global goal of the system is also achieved. The global state of the system (a value of a global goal) is not directly calculated, but is rather observed. To achieve this global goal no coordination of agents is required.

CCGA is partially parallel and centralized coevolutionary algorithm in which subpopulations cooperate to achieve a global goal. In a given moment of time only one subpopulation is active while the other subpopulations are frosen. The global goal of the system is at the same time a goal of each subpopulation. To evaluate a global goal a coordination center needs to communicate with each subpopulation to know a current local solution.

Results of experiments have shown that the LCGA is an effective optimization algorithm for problems where the global goal of the system is the

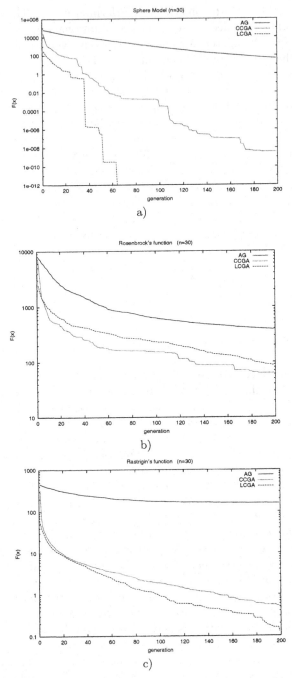

Fig. 1. Sphere model (a), Rosenbrock's function (b) and Rastrigin's function (c)

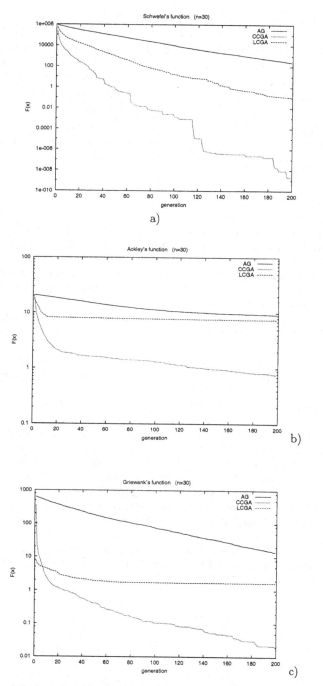

Fig. 2. Schwefel's function (a), Ackley's function (b) and Griewank's function (c)

sum of local goals. For such unimodal problems (sphere model) LCGA clearly shows its ability of fast (a number of generations) parallel and distributed optimization at low computational cost (no need for communication between agents to collect values of local functions and calculate a value of a global function) and high quality of solution, better than offered by CCGA. LCGA poseses such a ability also for highly multimodal functions (Rastrigin's function) expressed as a sum of local functions. However for problems expressed in more complex way (Schwefel's, Ackley's and Griewank's functions) CCGA with cooperation mechanism and a global coordination shows better performance than LCGA.

References

1. Bouvry, P., F. Arbab, F., Seredynski, F. (2000) Distributed evolutionary optimization, in Manifold: Rosenbrock's function case study, Information Sciences **122**, 141-159
2. Dorigo, M., Maniezzo, V. (1993) Parallel Genetic Algorithms: Introduction and Overview of Current Research, Parallel Genetic Algorithms, Stender, J. (ed.), The Netherlands: IOS Press, 5-42
3. Hillis, W. D. (1992) Co-evolving Parasites Improve Simulated Evolution as an Optimization Procedure, Artificial Life II, Langton, C. G. et al. (eds.), Addison-Wesley
4. Husbands, P. (1994) Distributed Coevolutionary Genetic Algorithms for Multi-Criteria and Multi-Constraint Optimization, Evolutionary Computing, Fogarty, T. C. (ed.), LNCS 865, Springer, 150-165
5. Michalewicz, Z. (1996) Genetic Algorithms + Data Structures = Evolution Programs, Springer
6. Paredis, J. (1996) Coevolutionary Life-Time Learning, Parallel Problem Solving from Nature – PPSN IV, Voigt, H. -M. et al. (eds.), LNCS 1141, Springer, 72-80
7. Potter, M. A., De Jong, K. A. (1994) A Cooperative Coevolutionary Approach to Function Optimization, Parallel Problem Solving from Nature – PPSN III, Davidor, Y. et al. (eds.), LNCS 866, Springer
8. Seredynski, F. (1994) Loosely Coupled Distributed Genetic Algorithms, Parallel Problem Solving from Nature – PPSN III, Davidor, Y. et al. (eds.), LNCS 866, Springer
9. Seredynski, F. (1997) Competitive Coevolutionary Multi-Agent Systems: The Application to Mapping and Scheduling Problems, Journal of Parallel and Distributed Computing, **47**, 39-57

Studying Properties
of Multipopulation Heuristic Approach
to Non–Stationary Optimisation Tasks

Krzysztof Trojanowski[1] and Sławomir T. Wierzchoń[1,2]

[1] Institute of Computer Science of Polish Academy of Sciences,
Ordona 21, 01-237 Warsaw, Poland
[2] Department of Computer Science, Białystok Technical University,
Wiejska 45a, 15-351 Białystok, Poland

Abstract. Heuristic optimisation techniques, especially evolutionary algorithms were successfully applied to non–stationary optimisation tasks. One of the most important conclusions for the evolutionary approach was a three-population architecture of the algorithm, where one population plays the role of a memory while the two others are used in the searching process. In this paper the authors' version of the three–population architecture is applied to four different heuristic algorithms. One of the algorithms is a new iterated heuristic algorithm inspired by artificial immune system and proposed by the authors. The results of experiments with a non–stationary environment showing different properties of the algorithms are presented and some general conclusions are sketched.

1 Introduction

In this paper we study properties of heuristic algorithms equipped with a memory. Common feature of the algorithms used in the study is a three–population architecture resembling that of proposed by Branke [2]. It must be stressed however, that instead of comparing the behaviour of these algorithms, we rather study their ability of exploiting information contained in the memory.

In most studies, e.g. [3] or [7], it is assumed that the algorithm knows somehow the position of current global or local optimum. That is in [7] the string representing optimum is used to compute the affinity of each individual in the population while in [3] it is assumed that the distance of an individual from the target point is known at each iteration. In our study we relax this assumption and we assume that a number of local optima move at some constant velocity through the search space. The aim is to search for an algorithm which is able both: (a) to locate the global optimum and (b) to trace its location as it moves through search space.

An iterated heuristic algorithm inspired by an artificial immune system, called Artificial Immune Iterated Algorithm – AIIA in short – is a searching engine applied in our study. It proved to be useful in solving both (a) and (b) problems stated above in case when a single optimum changes its location in the search space after every t iterations (t is a parameter) [13]. It should be stressed that the AIIA (described in details in Section 2) is a kind of evolutionary algorithm; it uses reproduction and genetic variation, affinity (i.e. fitness) evaluation and selection — consult [4] for a deeper discussion.

In this study the goal is to test a number of population management strategies in the search for one that is optimal for optimisation in non–stationary environments [11,12]. In this research we also studied the influence of the selected population management strategy on the behaviour of a few well known classic heuristic techniques to compare the results with results given by our algorithm. In 1999 in papers published by two independent authors (i.e. [2,10]) it was shown that there are two significant properties of an algorithm to be well suited to non–stationary optimisation tasks: (a) explorative skills, which make the algorithm able to adapt instantly in case of any changes in the optimisation task, and (b) memory mechanisms which can be especially useful in case of cyclically repeated changes in the task.

In our first study [11] it was shown, that the presence of memory can improve efficiency of AIIA searching for optima in non–stationary environments. The main problem however was a form of memory control. By the latter we mean a memory population structure, and rules of remembering, reminding and forgetting. Next paper [12] presented two memory management strategies, one of them, called *double population method* proved to be very efficient in our version of artificial immune optimisation. In this study we compare four algorithms — instances of different metaheuristics — equipped with a very similar memory structures and adopted to the requirements of the double population method.

In Section 2 description of AIIA and presentation of other compared algorithms are included. In Section 3 we discuss memory management strategy and rules of access to memory structures applied to all tested algorithms. Section 4 presents testing non–stationary environment and Section 5 — measures applied to obtained results and the results of experiments. Section 6 recapitulates the text and sketches some conclusions.

2   Tested algorithms

In this paper we use the same algorithm which has already been tested in our previous papers [11,12]. It is a modification of the original Gaspar and Collard algorithm [7] and it resembles the clonal selection algorithm originally proposed by de Castro and von Zuben [4]. Its pseudocode is given in Figure 1.

Step of *clonal selection* controls explorative properties of the algorithm. One of the key features of this step is a criterion of searching for the highest matching specificity of the antibody to the antigen. In our algorithm the best solution in the population (i.e. the one with the best value of the fitness function) is called antigen, so the matching of the antibody to the antigen

procedure AIIA
begin
   $t \leftarrow 0$
   initialize population of antibodies $P(t)$
   while (**not** termination–condition) **do**
   begin
      $t \leftarrow t + 1$
      evaluate $P(t)$
      select the best in $P(t)$; the best one is called *antigen*
      Clonal selection: select n antibodies with highest matching specificity to the antigen
      Somatic hypermutation: make c_i mutated clones of i-th antibody.
        if $fc_{(i)} > f_i$ **then** the clone $c_{(i)}$ with highest fitness replaces original antibody
      Apoptosis: replace d antibodies with lowest fitness value by randomly generated new antibodies
   end
 end

Fig. 1. A high level description of the structure of The Artificial Immune Iterated Algorithm.

returns the difference between fitness values of these two solutions. The fitness value of solution is a value of the fitness function for the solution modified with normalised value of its success register. The success register of antibody (called *NormSuccR*) is a counter attached to every solution in the population. The register is incremented every time the antibody is the best among all the other antibodies in the population. This kind of prize is awarded every iteration. The fitness value of antibody is evaluated according to Equation 1a:

$$f'(X) = (1 + \alpha * NormSuccR) * f(X). \tag{1a}$$

where:

  $f(X)$ — a fitness value of an antibody X,
  α — an experimentally tuned parameter. In our experiments it was set to 10,
  NormSuccR — a normalized value of the success register. The normalized value belongs to the range $[0, 1]$,
  $f(X)$ — a value of the fitness function for the antibody X.

Three other algorithms were also compared in our experiments :

1. simple genetic algorithm — a modified SGA comes from [8]. The modification was in selection method, i.e. individuals for crossover and mutation were selected with tournament selection method of size 2 instead of roulette wheel. Other components of the algorithm have been left unchanged.

2. simple genetic algorithm with Random Immigrants (RI) mechanism —
the modified SGA is described above (point no. 1). Here, it was extended
by RI mechanism coming from [6]. The RI mechanism enhances algorithm's ability to track the optimum of a changing environment and
makes the algorithm more competitive to artificial immune system. Replacement rate of RI mechanism was set to 0.4[1].

3. simulated annealing — a simulated annealing algorithm comes from Handbook of Evolutionary Computation [1], which in its general assumptions
is based on the publication of Kirkpatrick [9].

3   Memory management strategy — double–population method

There are two populations of individuals in this approach: the first one, called
population of explorers, and the second one, called population of exploiters.
There is also a memory buffer of a constant size, which represents memory
cells of the system. In the approach discussed here, the memory buffer has a
specific structure: it is an array of cells, where each cell is able to store a set
of solutions. In our case, the capacity of a single cell is large enough to keep
all the individuals from a single population there.

Applied strategy of population management assumes that there are two
search engines which start simultaneously with two different initial populations. One of the engines works on the population which is called a population of explorers because it is generated randomly and therefore uniformly
distributed over the search space. The other one builds its initial population
taking into account solutions stored in the cells of the memory buffer. This
search engine selects one of the populations stored in the cells of the memory
and builds its initial population copying all the individuals from the selected
cell. This population is much more exploitative than the first one, because
its individuals are usually already located in a relatively small region of the
search space. Their role is to exploit this region hoping that it belongs to
the basin of attraction of a global optimum. Obviously, in case of simulated
annealing algorithm, where there is only one individual subjected subsequent
modifications, each memory cell stores exactly one solution.

After every change in the optimised environment, both populations start
again their search processes. And again the difference is in starting points of

---

[1] Value 0.4 of the replacement rate was suggested as the best value for a set of
tests presented in [6]. Although there are some differences between the genetic algorithm used by Cobb and Grefenstette and our simple genetic algorithm (Cobb
and Grefenstette used GENESIS version 5.0 written by J. Grefenstette, the population size was 100, there was a two-point crossover with crossover rate of 0.6
and solution was represented as Grey coded binary string with 16 bits per dimensions) we assumed that in our case the optimum value of the replacement
rate can be quite similar.

these populations. This type of memory is slightly different than the memory described in [12], where a single cell stored exactly one individual. In that case, exploitative population took the best individual from the memory, and build its population using this individual as a "seed", i.e. all the individuals in exploitative population were created by a light mutation of the seed. In the research presented here, we decided to modify this and turned to population based memory cells.

Given fixed memory structure, we have to design rules which allow to manage this memory.

- **Rules of remembering** — at the first iteration of the search process the memory buffer is empty. Then, after every change in the testing environment, all the individuals from the better population of the two populations: population of explorers and exploitative population are written to the first free cell of the memory buffer. Thus, what is remembered (i.e. the content of the memory buffer) increases as the number of changes increases.

- **Rules of forgetting** — when the buffer is filled in, i.e. all cells of the buffer include their sets of solutions, it is necessary to release one of cells to make room for the new information. FIFO strategy of choosing the cell to be emptied is applied. In other words, the strategy selects the cell in which the oldest information is written to be emptied.

- **Rules of recalling** — the system recalls, i.e. reaches to the memory every time the change appears in the environment (it is assumed, that the system knows about the changes from any external source immediately after the change appears). The best set of the individuals among those stored in the cells is searched. Here "the best" means that all the individuals from the non–empty memory cells are re-evaluated with current evaluation function (i.e. the function after change), and then the best individuals fitness values from each cell are compared to select the best one. All the individuals from the selected cell are copied to become an initial population of the exploitative subprocess of the algorithm.

4   Testing environment

We carried out three groups of experiments with two types of environments. Our test–bed was a generator proposed by Trojanowski and Michalewicz [10]. The generator creates a convex search space, which is a multidimensional hypercube. The space is divided into a number of disjoint subspaces of the same size with defined simple unimodal functions of the same paraboloidal shape but possibly different value of optimum[2]. In case of two–dimensional

---

[2] More detailed description of testing environment can be found in [10–12] and at www address: http://www.ipipan.waw.pl/~stw/ais. Animated figures of non-stationary fitness landscapes created with the generator as well as electronic

search space which was selected in the experiments presented in this paper we simply have a patchy landscape, i.e. a chess–board with a hill in the middle of every field.

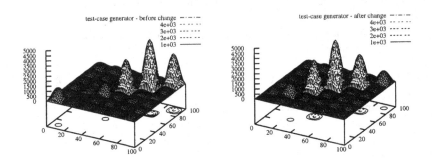

Fig. 2. Sample environment created with test–case generator: before and after a change in the fitness landscape

Hills do not move but periodically change their heights which makes the landscape varying in time. The aim of the algorithm is to find the highest hill at the specific moment. In our environments there were sequences of fields with varying hills' heights. Other fields of the search spaces were static.

All experiments were performed with two–dimensional search space and for each of the tested algorithms a BCD representation of a point coordinates with 16 bits per dimension was used. Four different environments were generated. We did experiments with the chess–boards of size 4 by 4, i.e. with 16 fields (environments 1 and 2), and of size 6 by 6, i.e. with 36 fields (environments 3 and 4). Thus, the search spaces consisted of 15 local optima and one global optimum in the first case, and of 35 local optima and one global optimum in the second one. The environments 1 and 3 were test–beds for experiments with cyclic changes, while the environments 2 and 4 were test-beds for experiments with both cyclic and acyclic changes. Each environment had a different set of fields with varying fitness landscape.

For experiments with cyclic changes, a single epoch obeyed 5 cycles of changes. In all the experiments each antigen has been presented through 10 iterations. Thus, in case of the environment 1 a single epoch took 200 iterations, in case of the environment 2 — 400 iterations, in case of the environment 3 — 300 iterations, and in case of the environment 4 — 600 iterations. Experiments with non–cyclic changes were based on the environments 2 and

---

versions of other papers concerned with artificial immune optimisation algorithms are also available at this www address.

4 and a single epoch included just one cycle of changes and took 80 and 120 iterations respectively.

For the six cases described above we carried out series of experiments with the memory buffer of different number of cells — from zero to 30. Every experiment was repeated through 200 epochs and in the later figures we always study average values of these 200 epochs.

5   Results evaluation criteria and obtained results

For estimations of non–stationary optimisation results, we used a measure called *average deficiency* (Formula 2a):

$$Ad = \frac{1}{K} \sum_{i=1}^{K} (err_{i,\tau-1}). \qquad (2a)$$

where:

τ — the number of generations between two consecutive changes,
K — the number of changes of the fitness landscape during the run,
$err_{i,j}$ — a difference between the value of the current best individual in the population of j–th generation after the last change ($j \in [0, \tau - 1]$), and the optimum value for the fitness landscape after the i–th change ($i \in [0, K - 1]$).

It is a difference between the value of the current best individual in the population of the "just before the change" generation and the optimum value averaged over the entire run. For this measure the smaller values are the better results.

Obtained values of average deficiency are presented in the figures. In every figure we compare graphs of average deficiency for the four methods (e.g. AIIA, SGA, SGA-RI, and SA) with 7 sizes of the memory buffer applied to a group of environments. The results of experiments with cyclic changes are presented in Figure 3, and the results of experiments with non–cyclic changes — in Figure 4.

6   Conclusions

The goal of these comparisons was to check if the influence of the memory strategy on the algorithms is the same for all of them. It can be seen that memory improved artificial immune optimisation algorithm and Simulated Annealing algorithm significantly for both cyclic and non–cyclic changes in the environment. The graphs of average deficiency for all the environments go down when the memory buffer is in use, i.e. average deficiency has a high value for experiments without memory relatively to the values of experiments when the memory buffer is of non–zero size.

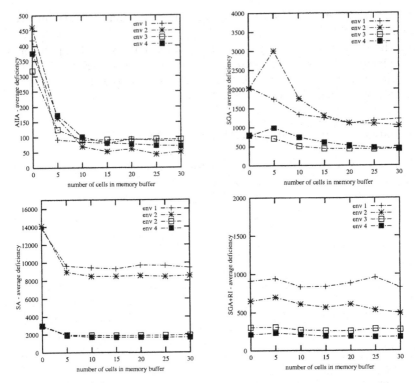

Fig. 3. Results of experiments with 7 sizes of the memory buffer (from 0 to 30) performed with AIS, SGA, SGA+RI, and SA applied to four environments: 1, 2, 3 and 4 (cyclic changes)

Behavior of SGA (with and without RI mechanism) was different than the other algorithms — when the memory was too small, the results deteriorated, i.e. values of average deficiency grew up. However, for larger sizes of the memory buffer an improvement of obtained results was eventually observed. Additionally, for environments with non–cyclic changes the memory did not improve SGA but even decreased its efficiency. This observation corresponds to conclusions presented in [2,10], that memory structures applied to evolutionary algorithms can cause an effect of deterioration in quality of obtained results for small sizes of the memory buffers. As opposed to SGA results, results obtained with SA and AIIA do not show this effect and in case of these algorithms even a very small memory buffer improves quality of the algorithm.

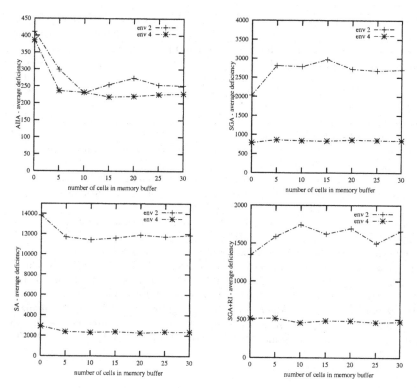

Fig. 4. Results of experiments with 7 sizes of the memory buffer (from 0 to 30) performed with AIS, SGA, SGA+RI, and SA applied to two environments: 2 and 4 (non–cyclic changes)

References

1. Back, T., Fogel, D.B., and Michalewicz, Z., Eds. (1997) *Handbook of Evolutionary Computation*, Institute of Physics Publishing and Oxford University Press, NY

2. Branke, J., (1999) Memory Enhanced Evolutionary Algorithm for Changing Optimization Problems, Proc. of the 1999 Congress on Evolutionary Computation — CEC'99, IEEE Publishing, pp. 1875-1882

3. Carlisle, A., Dozier, G., (2000) Adapting particle swarm optimization to dynamic environments, Proc. of the Int. Conference on Artificial Intelligence — ICAI 2000, pp. 429-434

4. de Castro, L. N., and Von Zuben, F. J., (2000) The Clonal Selection Algorithm with Engineering Applications, Proc. of the Genetic and Evolutionary Computation Conference — GECCO'00, Morgan Kaufmann Publishers, pp. 36-37

5. de Castro, L. N., (2002) Immune, swarm, and evolutionary algorithms. Part II: Philosophical Comparisons, Proc. of the Int. Conference on Neural Information Processing, Workshop on Artificial Immune Systems, vol. 3, pp. 1469-1473

6. Cobb, H.G., Grefenstette, J.J., (1993) Genetic Algorithms for Tracking Changing Environments, Proc. of the 5th IEEE International Conference on Genetic Algorithms — V ICGA'93, Morgan Kauffman, pp. 523-530
7. Gaspar, A., Collard, Ph., (1999) From GAs to Artificial Immune Systems: Improving Adaptation in Time Dependent Optimisation, Proc. of the 1999 Congress on Evolutionary Computation — CEC'99, IEEE Publishing, pp. 1859-1866
8. Goldberg, D.E., (1989) *Genetic Algorithms in Search, Optimisation and Machine Learning*, AddisonWesley, Reading, MA
9. Kirkpatrick, S., Gelatt, C.D., Vecchi, M.P., (1983) Optimization by simulated annealing, *Science* **220**, pp. 671-680
10. Trojanowski, K., and Michalewicz, Z., (1999) Searching for Optima in Non-Stationary Environments, Proc. of the 1999 Congress on Evolutionary Computation — CEC'99, IEEE Publishing, pp. 1843-1850
11. Trojanowski, K., Wierzchoń, S.T., (2002) Memory Management in Artificial Immune System, Proc. of International Conference on Neural Nets and Soft Computing — ICNNSC02, Physica-Verlag (Advances in soft computing)
12. Trojanowski, K., Wierzchoń, S.T., (2002) Immune Memory Control in Artificial Immune System, National Conference on Evolutionary Computation and Global Optimization — KAEiOG02, Warsaw University of Technology Press
13. Wierzchoń, S.T, (2001) Algorytmy immunologiczne w dzialaniu: optymalizacja funkcji niestacjonarnych, XII Oglnopolskie Konwersatorium nt. "Sztuczna Inteligencja — nowe wyzwania". SzI-16'2001, Akademia Podlaska, PAN, WAT, pp. 97-106

An Immune-based Approach to Document Classification

Jamie Twycross and Steve Cayzer

Biologically Inspired Complex Adaptive System (BICAS) Group,
Hewlett-Packard Laboratories, Filton Road, Stoke Gifford, Bristol U.K.
{jamie.twycross, steve.cayzer}@hp.com

Abstract. The human immune system is a complex adaptive system which has provided inspiration for a range of innovative problem solving techniques in areas such as computer security, knowledge management and information retrieval. In this paper the construction and performance of a novel immune-based learning algorithm is explored whose distributed, dynamic and adaptive nature offers many potential advantages over more traditional models. Through a process of cooperative coevolution a classifier is generated which consists of a set of detectors whose local dynamics enable the system as a whole to group positive and negative examples of a concept. The immune-based learning algorithm is first validated on a standard dataset. Then, combined with an HTML feature extractor, it is tested on a web-based document classification task and found to outperform traditional classification paradigms. Further applications in content filtering, recommendation systems and user profile generation are also directly relevant to the work presented.

1 Introduction

This paper explores the novel application of a biologically-inspired learning algorithm based on the human immune system to the problem of document classification. Using this approach, we demonstrate a novel, working system which is able to perform better than the currently available learning algorithms. In order to give substance to the claims made, we first validate our methodology using a standard dataset. The performance of our system is then compared to that of other methods in a systematic and rigorous manner.

In this, the introductory section, we give a brief overview of the concepts and themes central to the work presented here. Section 2 contains a review of related work. Details of the methodology used can be found in Section 3, while Section 4 contains the evaluation results. In Section 5, we discuss these results and offer some concluding remarks.

1.1 The Document Classification Problem

Document classification is an important technique in the field of information retrieval. Work in this field has grown steadily since the 1940's and the advent of computers, and has been driven by the need for systems which are able to quickly and accurately access the increasingly large amounts of data being produced and stored on computers. With the birth of the Internet and World Wide Web this need has become more pressing than ever, but the problem of effective retrieval still remain largely unsolved [1]. Much of the work within the field of information retrieval belongs to three main areas: content analysis, information structure, and evaluation. Content analysis is concerned with transforming documents into a form suitable for processing; information structure with improving the effectiveness and efficiency of information retrieval systems through the exploitation of relationships between documents; and evaluation with the assessment of the performance of information retrieval systems. In terms of the concept learner presented here, these areas can be equated to deciding what to feed into the learner (feature extraction, discussed in 3.4), the learning algorithm (concept learning, discussed below), and how to assess how well it works (evaluation, discussed in Sect. 3.5) respectively.

1.2   Concept Learning for Document Classification

Concept learning can be framed as the problem of acquiring the definition of a general category given a sample of positive and negative training examples of the category. In this paper, we consider the category of 'web pages relevant to my current task', which forms the target concept for which we wish to acquire a definition. Our sets of positive and negative training examples are web pages that we have already rated as 'useful' or 'not useful'. In this case, the problem of concept learning can be summarised as one of inferring a Boolean-valued function from a set of training examples.

The immune system as a concept learner The learning algorithm we implement and study is based on aspects of the dynamics of the human immune system (HIS), part of whose function in its role as protector of the body can be broadly seen as the classification of proteins in the body into two classes: self – belonging to the body; and non-self – not belonging to the body and potentially harmful. This classification is carried out by a set of detectors called antibodies.

The question then arises how we can learn such a set of discriminating detectors (or, equivalently, classification rules). The approach taken by Potter and his colleagues [9] explores the use of a cooperative coevolutionary algorithm. Here, individuals belonging to non-interbreeding subspecies represent partial solutions to the problem at hand, and are combined to form a complete solution. We adopt this approach for the generation of antibodies, where each species produces one antibody. The evolved antibodies are then combined to form a serum, which performs the classification required. In order to validate this approach, we replicate and extend the results of Potter & de Jong [10] on a standard dataset, before turning our attention to web page classification.

2   Related Work

2.1   Document Classifier Systems

Pazzani et al. [8] describe a system, instantiated as a software agent, which learns a profile of user's interests from a collection of explicitly rated web pages, and uses this profile to identify other web pages that may be relevant. The profile can be used to classify links to pages which the user has not visited, thereby providing intelligent guidance for web browsing.

To construct the profile, Pazzani et al. compare several different standard learning algorithms [7,8]. They examine a naive Bayesian classifier (NBC), a nearest neighbour algorithm, a decision tree and a neural network, finding that the NBC generally performs best. They also investigate the role of feature selection in the predictive accuracy of the classifiers, and find that appropriate feature extraction algorithms significantly reduces classification error. They go on to implement the NBC in a system, Syskill and Webert, which automatically filters search results for users.

We reimplement Pazzani's naive Bayesian classifier and in turn compare and contrast its performance with that of the immune-based concept learner.

Many other concept learning systems exist, most employing some form of inductive learning algorithm which arrives at hypotheses by considering specific examples. Many of these algorithms are surveyed in [5], and we compare several with our immune-based classifier in Section 4.

2.2   Immune-based classifier systems

Self-nonself discrimination in the HIS can be thought of as a form of 2-class classification. Indeed, a number of artificial immune systems use this metaphor. For example, in AIRS [12], a class is represented by a number of B-cells which encode prototype vectors. AIRS is an effective classifier, but we want to explore a different metaphor, where the classes are represented by explicitly meaningful concepts.

We use a coevolutionary approach for detector generation. This idea, introduced by Potter & De Jong [10], has been proven to be of use in a variety of settings. It involves the evolution of a number of non-interbreeding subspecies, individuals of which represent partial solutions to the problem at hand, and are combined to form a complete solution. There are a number of refinements and variations possible to this basic approach; we describe our implementation in detail, and other possibilities are reviewed and discussed in [11].

3   Methodology

3.1   Immune-based Classifier

The immune-based classifier is based on one described by Potter and De Jong [10], and is composed of a set of detectors, each of which is instantiated as a ternary schema of the same length as the feature vectors it will classify. Associated with each detector is a real-valued threshold, which indicates the percentage of matching bits between schema and feature vector necessary before a match is said to have occurred. The strength of the match between detector and feature vector is the percentage of matching bits in the schema and feature vector, ignoring any positions where the schema contains a #. For example, a detector '01#1##11' will match a feature vector '11100101' in 2 out of 5 non-# bits, so the binding strength between the detector and feature vector is $2/5 = 0.4$. The calculated binding strength must be greater than the threshold of the detector to consider a match to have occurred. Detectors can be of one of two types, Type 0 or Type 1, with a Type 0 detector, as in the human immune system, classifying any feature vector it matches as non-self, while a Type 1 detector contrarily classifying matching feature vectors as self.

3.2   Naive Bayes Classifier

The naive Bayesian classifier (NBC) is a probabilistic method of classification, which calculates the probabilities of a particular feature vector belonging to each possible class and then classifies the feature vector as belonging to the class for which this probability is highest. Formally, if $\mathbf{a} = [a_1, a_2, \cdots, a_n]$ is a feature vector made up of n features, a_i, and $V = \{v_1, v_2, \cdots, v_m\}$ is a set of m classes, then the class $v_{NB} \in V$ that the NBC classifies the example $\mathbf{a}$ as belonging to is given by:

$$v_{NB} = \underset{v_j \in V}{\operatorname{argmax}} \, P(v_j) \prod_{i=1}^{n} P(a_i|v_j), \ \forall \ v_j \in V$$

3.3   Co-Evolutionary algorithm

The cooperative coevolutionary algorithm consists of a number of non-interbreeding species of detectors, whose encoding will be described shortly, and initially starts with one randomly initialised species whose fitness is evaluated as described below. The initialisation of species is controlled by bias parameters, which control the *type* (class 0 or 1) and *generality* (number of #'s) of the initial detectors. The species are each allowed to evolve according to a standard genetic algorithm, except that the fitness evaluation (see below) involves a *serum*, which is the fittest detector taken from each species. The selection strategy is fitness-proportionate with balanced linear scaling, recombination is by uniform crossover, and mutation involves bit flipping. The replacement strategy is generational. If the fitness of the serum fails to increase above a certain stagnation threshold over several consecutive generations, a new species is added and any species not contributing to the fitness of the serum are removed. The parameters used in the experiments were as follows: Species size=100, crossover rate=0.6, mutation rate=2/(*genome length*), stagnation threshold=0.001, stagnation generations=2, bias parameters (both)=0.5.

Encoding Detectors are encoded as proposed by Potter & De Jong [10]. This scheme employs binary genomes, each containing 4 genes. The first gene, the threshold gene, encodes the 8 bit value for the detector's threshold. A Gray coding was used for this gene in order to reduce the probability of small changes in the genotype producing disproportionately large changes in the phenotype. The threshold value is calculated by converting the gene to base 10 and then dividing this value by 255 to get a real number in the range [0, 1]. The second and third genes, the pattern and mask genes, are combined to form the detector's schema. Each of these genes has the same number of bits as the number of bits in the feature vectors the immune-based classifier is designed to operate on. The mask gene is overlaid onto the pattern gene and any positions at which the mask gene is 1 changes the corresponding bit in the pattern gene to a #. A value of 0 in the mask gene leaves the corresponding bit of the pattern gene unchanged. In this way the schema is formed by copying the pattern gene, modified by the mask gene. The fourth gene stores the detector's type. The overall arrangement is shown in Fig. 1 for a detector recognizing 8-bit patterns.

Genome

threshold	pattern	mask	type
00110111	10100101	01101110	1

⇓

Detector

threshold	schema	type
0.215686	1##0###1	1

Fig. 1. Detector encoding scheme

We ran several experiments with alternative encodings without finding any significant performance improvements.

Fitness evaluation The fitness of each individual is assessed by adding it to a serum consisting of the fittest individual from each of the other species. Serum fitness is calculated by presenting it with each training vector in the training set in turn. The detector in the serum which matches the current training vector with the greatest binding strength is then found, and if this strength is greater than the detector's threshold, the detector is said to have matched the training vector, and assigns it to Class 1 (or Class 0 if the detector is Type 1), otherwise if no match occurs the training vector is assigned to Class 0. The assigned class is then compared with the actual class of the training vector, and if equal the serum is said to have classified the training vector correctly. The proportion of correct classifications made by the serum gives its predictive accuracy on the training set.

3.4   Data sets

Two sets of test data were used in our experiments, both taken from the UCI Repository of Machine Learning Databases [2]. The first data set, the 1984 United States Congressional Voting Records , was used to validate our algorithm. This data set contains the voting records for 267 Republican and 168 Democrat members of the U.S. House of Representatives. Each record holds the vote cast by the member on 16 different issues, and the original records have been simplified to record this vote as yea (01), nay (10) or abstain (00). Each member's voting record is represented as a Boolean feature vector, with each consecutive pair of bits encoding a vote for a particular issue. Using this encoding, it is possible to generate antibodies that generalize over votes (eg '0#' matches both 'abstain' and 'yea'). Associated with each record is the class the record belonged to: 0 for Democrat, 1 for Republican.

The second data set used was the Syskill and Webert Web Page Ratings [6]. These data consist of 4 sets: Bands, BioMedical, Goats, and Sheep; each containing HTML pages related to a particular topic (the encoding of the pages is described in the next section). A user rated each page in a set as not interesting or interesting, which allowed a page to be assigned to one of two classes: Class 0 (cold) and Class 1 (hot) respectively. The task in this problem is to make predictions about whether examples from an unseen set of web pages would be interesting or not from the information contained within a training set of ranked pages. Table 1 provides a summary of the data sets.

Table 1. HTML Documents: Data set summary

data set	total examples	number of positives
Bands	61	15
BioMedical	131	32
Goats	70	32
Sheep	65	14

Feature Extraction In applying learning algorithms to the classification of text and HTML, the documents must be presented in the form of feature vectors. Following Pazzani et al. [8], a feature extraction algorithm was used to convert a raw HTML document into a Boolean feature vector. Each bit in the feature vector represents the absence or presence (at least once) of some associated feature, in this case a word, in the document. The task of the feature extraction algorithm is to decide from which words to compose the feature vector, and this is done using an information-based approach to extract the most informative words from a collection of documents.

Initially, the feature extraction algorithm takes the complete set of pages, S, and creates a list of all the words, W, contained in the pages. If a word, considered as a sequence of upper of lower case letters [a-zA-Z] separated by nonalphabetic characters, occurs more than once on the same page or across several pages, it is only represented once on this list. All words were converted to upper case and any words occurring on a list of frequently used words (Table 2) were removed. The *expected information gain*, $E(w, S)$, that the presence or absence word $w \in W$ gives towards the classification of S is:

$$E(w, S) = I(S) - [P(w = \text{pres})I(S_{w=\text{pres}}) + P(w = \text{abs})I(S_{w=\text{abs}})]$$

with $P(w = \text{pres})$ is the probability a word is present at least once on any page, $S_{w=\text{pres}}$ the set of pages containing the word w, and,

$$I(S) = \sum_{C \in \{\text{hot,cold}\}} -P(S_C) \log_2[P(S_C)]$$

where S_C is the set of pages belonging to class C, and $P(S_C)$ is the probability of a page belonging to that class.

To create n features the extraction algorithm uses the n words with the highest values of $E(w, S)$. Each HTML document is then converted to a Boolean feature vector by assigning a 1 to the appropriate feature if the document contains the word at least once, and a 0 if the document does not contain the word.

3.5   Evaluation

For the voting data set, 10-fold crossvalidation was used. This involves randomly dividing the complete data set into 10 equally sized disjoint sets, and then using 1 subset as a test set and the other 9 as a training set. The training set is used by the learning algorithm to create a classifier, whose sample predictive accuracy is then calculated using the test set. This process is repeated for each of the 10 subsets, the mean sample predictive accuracy of these 10 trials forming an unbiased estimate of the true predictive accuracy. Averaging the 10-fold crossvalidation process over 5 trials further refined these results. A randomly constructed crossvalidation set was used in each trial and the trials were paired, meaning the same training and test sets were used to train and test the two classifiers on each iteration.

Table 2. Frequent words removed from the word list

AND	WIC	GIF	AVAILABLE	ADDRESS	HAVE	HTM
FOR	FONT	THE	INFORMATION	WHAT	COM	ALT
SIZE	THIS	GOPHER	BOTTOM	TITLE	BODY	WHO
ORG	PAGE	ABOUT	WIDTH	PLEASE	HEAD	SRC
WWW	GNN	ALIGN	HEIGHT	SERVER	ALSO	IMG
FTP	NAME	LINKS	COMMENTS	HTTP	LIST	EDU
NET	ARE	WITH	CENTER	OTHER	NEWS	WEB
CGI	YOU	STRONG	WELCOME	CAN	HREF	BIN
TOP	WILL	HTML	INTERNET	MORE	GIFS	OUR
HAS	FROM	ALL	MAILTO	THAT	HOME	GDB
NOT	MAIL	WUSTL	INDEX	HERE	YOUR	IBC
ANY	TOC	GOV				

The web page rating data sets range in size from between 61 to 131 samples, making techniques such as 10-fold crossvalidation an unreliable means of estimating the predictive accuracy due to the relatively small size of the test set produced by this method. In such cases alternative methods need to be employed, one of which, and the one used here, is to create a training set by randomly selecting n samples from the original data set without replacement. The remaining unselected samples then become the test set. This method is advantageous for our purposes in that it can be used to assess the performance of classifiers over a range of training set sizes, giving a good indication of number of pages a user would have to rate in order to get reliable results from the classifier. After the training and test sets were created, the 128 most informative words were used to transform the training and test sets into Boolean feature vectors as previously explained. This process was repeated 30 times for each training set size, and the mean predictive accuracy of the resulting classifiers measured.

4   Results

All the code was written on a 1.4GHz Athlon Linux box, originally in C and then in C++, compiled using g++ v2.95.3 with level 2 optimisation, and released under the GNU General Public License. The experiments were carried out on this Linux box and on four 1.8GHz Pentium 4's also running Linux. For the voting data set, a typical 100 generation run with 400 training examples took around 40 seconds. For the document classification problem, the immune-based classifier took around 6 seconds to train over 100 generations on a training set of size 20 with 128 features.

4.1   Classifier validation (voting data)

In order to validate our classifier, we compared the predictive accuracy of the immune-based classifier and NBC on the voting data set. The results of these experiments are shown in Fig. 2, which is a density plot of the distribution of predictive accuracies of the classifiers produced in the crossvalidation trials.

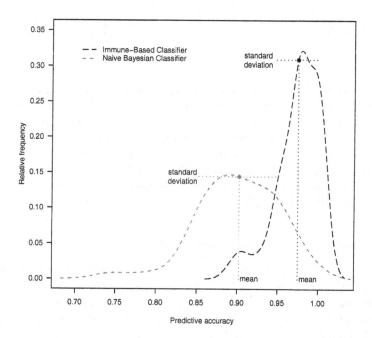

Fig. 2. 10-fold crossvalidation (voting data set). Predictive accuracy is represented along the x-axis, and the relative frequency with which a classifier with this predictive accuracy was observed during the experiments along the y-axis.

For classifiers produced by the naive Bayesian algorithm, Fig. 2 shows a right-skewed unimodal distribution with a single low, spread peak (standard deviation: 0.049) almost symmetrical about its mean predictive accuracy of 0.901. In contrast the immune system concept learner produced a less symmetric distribution with a higher mean of 0.974, rising fairly steeply and then dropping off even more steeply, giving a tighter distribution of values (standard deviation: 0.026) than the NBC. The immune-based learning algorithm thus produced classifiers which were both significantly (Wilcoxon rank sum test) more accurate and more likely to be closer to the true predictive accuracy than that of an NBC.

Table 3. Comparison of classifier performance (voting data set). The classifiers implemented in this paper are shown in **bold**.

Algorithm	predictive accuracy	standard deviation	95% confidence interval	error rate
Immune-based	0.974	0.026	0.057	0.026
Immune-based [10]	0.964		0.018	0.036
QUEST [5]	0.963			0.037
AQ15 [10]	0.956		0.023	0.044
POLYCLASS [5]	0.948			0.052
Fuzzy Classifier [4]	0.947	0.316		0.053
naive Bayesian	0.901	0.049	0.088	0.099

These results can be compared to others reported in the literature for a number of different classifiers on the same dataset (Table 3). In most cases the same 10-fold crossvalidation was used, with the exception of Lim et al. [5], who, instead of reporting predictive accuracy, reported on the rate of misclassification for the algorithms they tested. Other statistics are also given where provided by the original paper. While data to perform statistical tests are not available, the difference in the predictive accuracy suggests that the immune-based algorithm outperforms all of these algorithms.

4.2   Classification of HTML documents

Having validated our immune based classifier, we then applied it to a web document classification task. Figure 3 shows the predictive accuracy of the immune-based and NBC learning algorithms trained on the Syskill & Webert data, and using a number of different training set sizes(as explained in Sect. 3.5) for each of the four data sets.

The graphs show that both classifiers have a somewhat lower predictive accuracy on this classification problem compared with that of the voting problem of the previous section. This is not suprising as text classification problems are generally considered to be relatively hard classification problems, and in an informal survey of the literature on text classification we found the best classifiers to be achieving predictive accuracies of around 0.70 whatever the text classification problem. In this context, the performance of both classifiers is more than reasonable on the four problems. The results presented for the NBC are also similar to those of the NBC-based system of Pazzani et al. [8], offering an indication that the performance levels are due to the nature of the problem and not a result of implementation problems. This said, there are marked differences in the predictive accuracies of the NBC and immune-based classifiers on the four problems, with the immune-based algorithm at first sight appearing to generally perform better on all data sets, except the bands set, than the NBC. As before, significance tests were per-

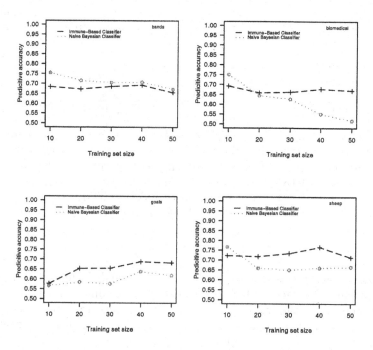

Fig. 3. Classifier performance (document classification). Predictive accuracy is plotted against training set size for each of the four datasets used.

formed on the results, confirming that in general, the immune-based classifier produces better classifiers than those produced by the NBC algorithm. Summaries of these results, including standard deviations and confidence intervals are given below in Table 4.

Also interesting to note is the relatively constant performance of the immune-based classifiers over a range of training set sizes. Increasing the size of the training set seemed to produced little increase in classifier performance for this algorithm, while for the NBC there was a much greater fluctuation in classifier performance on an increase in training set size. This constancy of performance is particularly useful on problems such as this one because classifiers which perform well on a small training set size would be advantageous as users would be able to rate less pages but still obtain accurate predictions.

5   Discussion

While the AIS classifier outperforms all the learning algorithms it has been compared against in many cases, there are marked differences in the levels

Table 4. Performance of immune-based and NBC classifiers on the four data sets, for training set sizes from 10 to 50 documents. Table entries are in the form: mean predictive accuracy (standard deviation). **Bold** figures mark where the performance of the NBC (or immune-based classifier) is significantly better (Student T-test)

size	classifier	Bands	Biomedical	Goats	Sheep
10	Immune	0.698(0.056)	0.685(0.080)	**0.605(0.065)**	0.734(0.070)
	NBC	**0.750(0.000)**	**0.754(0.002)**	0.562(0.055)	0.738(0.134)
20	Immune	0.690(0.047)	**0.706(0.047)**	**0.650(0.083)**	**0.725(0.062)**
	NBC	0.707(0.067)	0.661(0.095)	0.590(0.061)	0.627(0.142)
30	Immune	0.686(0.074)	**0.686(0.056)**	**0.664(0.054)**	**0.747(0.049)**
	NBC	0.696(0.084)	0.597(0.137)	0.601(0.066)	0.662(0.097)
40	Immune	0.690(0.081)	**0.689(0.048)**	**0.679(0.087)**	0.751(0.067)
	NBC	0.689(0.056)	0.556(0.155)	0.624(0.066)	0.724(0.104)
50	Immune	0.694(0.116)	**0.679(0.070)**	**0.717(0.095)**	**0.743(0.106)**
	NBC	0.683(0.105)	0.516(0.137)	0.649(0.073)	0.629(0.153)

of predictive accuracies it achieves on the voting and document classification problems. One factor which contributes to these differences is the variation in the size of the training sets available for each problem. In terms of available data, the voting problem represents a fairly data-rich classification problem, with 435 samples available, whereas the number of available samples for the document classification problem, between 61 and 131, makes it a relatively data-sparse problem. This leads to less samples being available for classifier training, and so it is expected that performance will generally be lower on this task. A second factor contributing to the difference in performance across the two problems is the feature extraction algorithm. In this paper we use a statistical feature extraction algorithm which extracts features based on their expected information gain and, while necessary to provide a principled comparison with the NBC-based system of Pazzani et al. [8], such a method is far from ideal in the context of HTML document classification. The coarse-grained document representation produced by our feature extractor introduces a fair amount of noise into the system, making classification generally that much harder than in the voting problem. Using a more fine-grained feature extractor, such as that described by Cohen [3], which also exploits meta-features of HTML documents such as hyperlinks and tags, would potentially increase classifier performance on the document classification problem and close the gap in performance difference.

In practical terms, choice of a classifier not only depends upon the predictive accuracies it is able to achieve, but also on the amount of time taken in training the classifier. No matter how good the results achieved, users would often be unwilling to wait for more than a few seconds for these results, and definitely not, for example, the several days or even months taken by some of the algorithms reported by Lim et al. [5]. While our immune based classifier is obviously more computationally expensive than the NBC, for all problems examined here training times were measured in seconds. Thus as well as achieving better predictive accuracies than that of many other classifiers, our immune-based classifier is able to do so in a time which allows it to be applied to real-world problems.

In summary, we have produced a novel, working system built on an immune-based learning algorithm, which is able to perform better than currently available learning algorithms.

5.1   Future work

Future work could include dynamic classification tasks. Take for example a system in which users have a collection of documents from which the concepts of 'related' and 'unrelated' are learnt. Over time, users may add or remove documents to and from this collection. Instead of relearning the concepts from scratch each time a document is added, as is necessary with NBC, it would be interesting to see if the AIS concept learner was able to produce accurate hypotheses starting from the previously learned concept, and so potentially offer savings in the amount of training time necessary.

Another possibility is to change the form and general properties of the antibodies produced by the AIS classifier; for example the encoding and matching algorithm. The possibility of using a more fine-grained feature extraction algorithm has already been mentioned.

In terms of the mechanisms at work in the human immune system those of the artificial immune system described in this paper are, to say the least, simplistic, and present a very crude analogy to their biological counterparts. Nevertheless, even from such humble an analogy it has been shown that a powerful concept learning system can be created. Perhaps with increased fidelity to its biological counterpart, for example through processes akin to affinity maturation and clonal selection, yet further increases in performance can be gained.

6   Acknowledgements

We thank the anonymous reviewer for useful comments. Computing facilities were provided by the HP Labs Bristol BICAS Research Group. We'd like to thank members of that group for valuable comments on the manuscript.

The BICAS website is at: http://www.hpl.hp.com/research/bicas

References

1. R. Baeza-Yates and B. Ribeiro-Neto. *Modern Information Retrieval.* ACM Press, New York, 1999.

2. C. L. Blake and C. J. Merz. UCI Repository of Machine Learning Databases. Department of Information and Computer Sciences, University of California, Irvine, CA. *http://www.ics.uci.edu/~mlearn/MLRepository.html*.

3. W. W. Cohen. Automatically extracting features for concept learning from the web. In P. Langley, editor, *Proc. of the Seventeenth Int. Conf. on Machine Learning (ICML-2000)*, pages 159–166. Morgan-Kaufmann, San Francisco, CA, 2000.

4. D. Dasgupta and F. A. González. Evolving complex fuzzy classifier rules using a linear genetic representation. In L. Spector, D. Whitley, D. Goldberg, E. Cantu-Paz, I. Parmee, and H. Beyer, editors, *Proc. of the Int. Conf. on Genetic and Evolutionary Computation (GECCO-2001)*, pages 299–305. Morgan-Kaufmann, San Francisco, CA, 2001.

5. T. Lim, W. Loh, and Y. Shih. A comparison of prediction accuracy, complexity, and training time of thirty-three old and new classification algorithms. *Machine Learning*, 40(3):203–228, 2000.

6. M. Pazzani. Syskill and Webert web page ratings. UCI Repository of Machine Learning Databases, Department of Information and Computer Sciences, University of California, Irvine, CA. *http://www.ics.uci.edu/~mlearn/MLRepository.html*.

7. M. Pazzani and D. Billsus. Learning and revising user profiles: the identification of interesting web sites. *Machine Learning*, 27:313–331, 1997.

8. M. Pazzani, J. Muramatsu, and D. Billsus. Syskill and Webert: identifying interesting websites. In W. J. Clancey and D. Weld, editors, *Proc. of the Thirteenth Amer. Nat. Conf. on Artificial Intelligence (AAAI-96)*, volume 1, pages 54–61. AAAI Press, Portland, OR, 1996.

9. M. A. Potter and K. A. De Jong. Cooperative coevolution: an architecture for evolving coadapted subcomponents. *Evolutionary Computation*, 8(1):1–29, 2000.

10. M. A. Potter and K. A. De Jong. The coevolution of antibodies for concept learning. In A. E. Eiben, T. Bäck, M. Schoenauer, and H. Schwefel, editors, *Proc. of the Fifth Int. Conf. on Parallel Problem Solving from Nature (PPSN-98)*, pages 530–539. Springer-Verlag, Amsterdam, 1998.

11. J. Twycross. An immune system approach to document classification. Master's thesis, COGS, University of Sussex, U.K., 2002.

12. Andrew Watkins and Lois Boggess. A New Classifier Based on Resource Limited Artificial Immune Systems. In *Proceedings of Congress on Evolutionary Computation, Part of the 2002 IEEE World Congress on Computational Intelligence held in Honolulu, HI, USA, May 12-17, 2002*, pages 1546–1551. IEEE, May 2002.

Part II

Recommenders and Text Classifiers

Entish: Agent Communication Language for Service Integration

Stanislaw Ambroszkiewicz[1][2] *

[1] Institute of Computer Science, Polish Academy of Sciences,
  al. Ordona 21, PL-01-237 Warsaw,
[2] Institute of Informatics, University of Podlasie,
  al. Sienkiewicza 51, PL-08-110 Siedlce, Poland
  sambrosz@ipipan.waw.pl; www.ipipan.waw.pl/mas/

Abstract. A new technology for service description and composition in open and distributed environment is proposed. The technology consists of description language (called Entish) and composition protocol called entish 1.0. They are based on software agent paradigm. The description language is the contents language of the messages that are exchanged (between agents and services) according to the composition protocol. The syntax of the language as well as the message format are expressed in XML. The language and the protocol are merely specifications. To prove that the technology does work, the prototype implementation is provided. It is still under testing. However, it is available for use and evaluation via web interfaces starting with the website www.ipipan.waw.pl/mas/. The specifications were created on the basis of the requirements produced by the Service Description and Composition Working Group (www.ipipan.waw.pl/mas/sdc-wg) of Agentcities.NET project. Related work was done by WSDL + BPEL4WS + (WS-Coordination) + (WS-Transactions), and DAML-S. Our technology is based on different principles. The language Entish is fully declarative contrary to BPEL4WS and DAML-S. A task (expressed in Entish) describes the desired static situation to be realized by the composition protocol.

1   What are web services?

Perhaps the most popular definition can be found in IBM's tutorial [8]:

Web services are self-contained, self - describing, modular applications that can be published, located, and invoked across the Web. Web services perform functions that can be anything from simple requests to complicated business processes ... Once a Web service is deployed, other applications (and other Web services) can discover and invoke the deployed service.

In order to realize this vision simple and ubiquitous protocols are needed. From service providers' point of view, if they can setup a web site they could join global community. From a client's point of view, if you can click, you could access services.

* The work was supported partially by KBN project No. 7 T11C 040 20

Web services are supposed to realize the Service-Oriented Architecture (SOA) in a global networked environment. SOA provides a standard programming model that allows software components, residing on any network, to be published, discovered, and invoked by each other as services. There are essentially three components of SOA: Service Provider, Service Requester (or Client), and Service Registry. The provider hosts the service and controls access to it, and is responsible for publishing a description of its service to a service registry. The requester (client) is a software component in search of a component to invoke in order to realize a task. The service registry is a central repository that facilitates service discovery by the requesters.

The following stack of protocols: SOAP, WSDL, UDDI, and BPEL4WS is positioned to become Web services standards for invocation, description, discovery, and composition. SOAP (Simple Object Access Protocol) is a standard for applications to exchange XML - formatted messages over HTTP. WSDL (Web Service Description Language) describes the interface, protocol bindings and the deployment details of the service. UDDI (Universal Description, Discovery and Integration) provides a registry of businesses and web services. A UDDI service description consists of physical attributes such as name and address augmented by a collection of tModels, which describe additional features such as, for example, reference to WSDL document describing the service interface, and the classification of the service according to some taxonomies.

BPEL4WS (BPEL for short) is a process modeling language designed to enable a service composer to aggregate Web services into an execution. There are abstract and executable processes. Abstract processes are useful for describing business protocols, while executable processes may be compiled into invokeable services. Aggregated services are modeled as directed graphs where the nodes are services and the edges represent a dependency link from one service to another. Canonical programmatic constructs like SWITCH, WHILE and PICK allow to direct an execution's path through the graph. BPEL was released along with two others specs: WS-Coordination and WS- Transaction. WS-Coordination describes how services can make use of predefined coordination contexts to subscribe to a particular role in a collaborative activity. WS-Transaction provides a framework for incorporating transactional semantics into coordinated activities. WS-Transaction uses WS-Coordination to extend BPEL to provide a context for transactional agreements between services.

There is also a DARPA project DAML-S. It is a DAML+OIL ontology for describing Web Services. It aims to make Web services computer-interpretable, i.e., described with sufficient information to enable automated Web service discovery, invocation, composition and execution monitoring. The DAML-S Ontology comprises ServiceProfile, ServiceModel, and Service-Grounding. ServiceProfile is like the yellow page entry for a service. It relates and builds upon the type of content in UDDI, describing properties of a service necessary for automatic discovery, such as what the services offers, and its inputs, outputs, and its side-effects (preconditions and effects). ServiceModel describes the service's process model, i.e., the control flow and data-flow involved in using the service. It relates to BPEL and is designed to enable automated composition and execution of services. ServiceGrounding connects the process model description to communication- level protocols and message descriptions in WSDL. The main limitation of DAML+OIL is its lack of a definition of well formed formulae and an associated theorem prover. So that there is a problem with expressing (as formulas) preconditions, effects, and output - input constrains.

Coming back to the SOA architecture. What is service provider and what is service requester (client) in the technologies described above? The question about service provider is in fact the question about the service architecture. Application that is connected as a service can be an object with public methods exposed as operations in a WSDL interface. Usually, the operations are accessible via SOAP. WSDL interface can be generated from application code. It is not possible to specify (in WSDL) the abstract functions implemented by the operations, i.e., what the operations do. Service provider itself must classify these operations according to the taxonomies available in UDDI.

Client to a service is a user (via its GUI) or another application that invokes the service. The task the client performs is not explicitly stated and is hard encoded in the client implementation. BPEL extends the client capability to compose service invocations into a workflow. A client realized in BPEL can itself be exposed as a service, i.e., has its own WSDL interface. Client's task is specified in an imperative way, that is, by specifying (in BPEL) *How to realize?* There is no way to specify the task in a declarative way, i.e., to specify formally *What to realize?* Actually, there is no need to do so if the imperative approach has been chosen.

The question is whether the technologies presented above are simple, ubiquitous, and widely deployed? Note, that there is a criticism of WSDL and UDDI, e.g., [3,4] and [7]. There is also an attempt to revise the notion of service architecture, see [6], by adding next protocol, called CS-WS (conversation support), on the top of BPEL.

Perhaps, it is time to revise the concept of service and client architecture as well as to take into account the declarative approach. The technology we propose realizes the declarative approach.

2   Our approach

We follow the idea of layered view of service architecture introduced in [5,6]. Our service architecture comprises the following three layers: conversation layer, functionality layer, and database management layer. The database management layer is the same as in [5]. However, the next two layers have different meaning. The functionality layer has exactly two interrelated components: raw application, and so called filter associated with the raw application. Raw application implements a single operation, i.e., given input resources, it produces the output resource according to the operation specification. Note, that operation has exactly one output, although it may have several inputs. The associated filter works as follows. Given constrains on the output resource, it produces the constrains on the input resources. That is, given a specification of the desired output, the filter replies with properties that must be satisfied by the input in order to produce the desired output by the raw application. It is clear that these constrains must be expressed in one common language. The conversation layer implements a conversation protocol to arrange raw application invocation, as well as input / output resource passing to / from the raw application. The conversation protocol specifies the order for message exchange. Message contents is expressed in the common language.

Since our service architecture is different than the one of WSDL and UDDI, we must revise the concept of service description language as well as the concept of service registry. It is natural that service description language should describe the types of service input / output resources as well as attributes of these types to express constrains. Note, that the language is supposed merely to *describe* resource types in terms of theirs attributes, not to construct data types as it is done in WSDL. It is also natural to describe *what service does* in the language, i.e., the abstract function implemented by the operation the service performs. Note, that this is a job of UDDI. We include this job in our description language.

Since service has additional functionality performed by filter (i.e., service may be asked if it can produce output resources satisfying some properties), the description language should be augmented with a possibility to formulate such questions as well as answers. Moreover, the clients' tasks should be expressed in the language.

We also want to describe some static properties of service composition such as intentions, and commitments; this corresponds to the functionality of WS-Coordination.

The final requirement is that the language must be open and of distributed use. It means that names for new resource types, their attributes, and names for new functions can be introduced to the language by any user, and these names are unique (e.g., URIs [11]). This completes the requirements for the description language called Entish. Since our technology is supposed to realize the declarative approach, we need a universal protocol for realizing the tasks specified in our description language. The realization is to be done by service discovering, arranging composition and coordination. The protocol we propose is called entish 1.0.

To prove that the requirements for the service description language and composition protocol can be satisfied we provide the prototype implementation.

3   Walk-through example

The working example presented below constitutes an intuitive introduction
to the description language and the composition protocol. The services described in the example are implemented and are ready for testing via the
www interfaces.

A client was going to book a flight from Warsaw to Geneva; the departure
was scheduled on Nov. 31, 2002. It wanted to arrange its task by Nov. 15,
2002. With the help of TaskManager (TM for short), the client expressed the
task in a formal language; suppose that it was the following formula:
phi =
"invoice for ticket (flight from Warsaw to Geneva,
departure is Nov. 31, 2002) is delivered to TM by
Nov. 15, 2002"

Then, the task formula (i.e., *phi*) was delegated to a software agent, say
agent0. The task became the goal of the agent0. The agent0 set the task
formula as its first intention, and was looking for a service that could realize
it. First of all, the agent0 sent the query:
"agent0's intention is phi"
to a service registry called infoService in our framework. Suppose that infoService replied that there was a travel agent called FirstClass that could
realize agent0's intention. Then, the agent sent again the formula *"agent0's*
intention is phi" however, this time to the FirstClass. Suppose that First-
Class replied with the following commitment:
"FirstClass commits to realize phi,
if (order is delivered to FirstClass by Nov. 15, 2002 and
the order specifies the flight (i.e., from Warsaw to
Geneva, departure Nov. 31,2002)
and
one of the following additional specification of the
order is satisfied:
(airline is Lufthansa and the price is 300 euro)
or
(airline is Swissair and the price is 330 euro)
or
(airline is LOT and the price is 280 euro))"

Let *psi* denote, the formula after *"if"* inside (...) parentheses. The formula
psi is the precondition of the commitment. Once the agent0 received the info
about the commitment, the agent0 considered the intention *phi* as arranged
to be realized by FirstClass, and then the agent0 put the formula *psi* as its
current intention, and looked for a service that could realize it. Let us notice
that the order specified in the formula *psi* could be created only by the client
via its TM, that is, the client had to decide which airline (price) should be
chosen, and the complete order was supposed to include details of a credit
card of the client. Hence, the agent0 sent the following message to TM:

"agent0's intention is psi"
Suppose that TM replied to the agent:
"TM commits to realize psi, if true "
The agent0 considered the intention *psi* as arranged to be realized by TM. Since the precondition of the TM commitment was the formula *"true"*, a workflow for realizing agent0's task was already constructed. Once TM created the order and sent it to FirstClass, the FirstClass would produce the invoice and send it to TM. It was supposed (in the protocol) that once a service realized a commitment, it sent the confirmation to the agent0. Once the agent0 received all confirmation, it got to know that the workflow was executed successfully. In order to complete this distributed transaction, the agent sent synchronously the final confirmation to the all services engaged in the workflow. This completes the example.

4   The prototype implementation

Before we go into details of the language and the protocol, let us present how the prototype works. The system resulting from the prototype implementation is fully functional. It allows providers to join their application as services as well as to formulate tasks by clients, and delegate the tasks to the system for realization.

The system is available for testing and evaluation via three www interfaces starting with http://www.ipipan.waw.pl/mas/ . The first interface called *EntishDictionary* serves for introducing names for new data types, their attributes, and new functions to the language Entish. The second interface called *serviceAPI* is for joining applications (that implement the new functions) as networked services to our system. The third interface called TaskManager is devoted for a client to specify its task in Entish, and provide initial resources for the task realization. Hence, from the outside, i.e., from service providers and clients point of view, the system consists of the three interfaces: EntishDictionary, TaskManager, serviceAPI. What is inside, that is, the system engine is transparent for the providers and clients. The engine implements the language, and the protocol that realizes the clients' tasks by discovery of appropriate services, their composition and invocation. The www interfaces are user friendly so that to use the system almost no knowledge on XML, Entish syntax, and entish 1.0 is needed.

EntishDictionary (ED for short) serves also as an ordinary dictionary, i.e., for looking at the existing ontologies as well as for explanation of names used in the language. Ontology is meant as a collection of names of resource types, their attributes, and functions defined on these types. It is supposed that they come from one application domain. Any user can introduce its own ontology. There is no conflict with names, because short names introduced by users are automatically extended to long names that are URIs [11]. ED has also additional functionalities. It allows a service provider to create Entish formula that describes operation type performed by its application. The formula along with some additional information about the host on which the application is running is sent automatically to the system (actually to serviceServer) for registration and publication.

The second functionality of ED is that it allows a user to create task; usually it is a composition of abstract functions. The task is sent automatically to TaskManager for realization. It is important to note, that task is merely an abstract description of what is to be realized, so that it does not indicate what services could realize it. Note, that EntishDictionary realizes some functionality of UDDI concerning service classification. However, service registration and publication is done inside the system.

Generally there is no restriction on the type of resources that can be defined in the dictionary. However for the purpose of system demonstration we assume that resource type has flat XML format, i.e., it has several elements that are of type `xsd:string` .

TaskManager is a GUI that, given a task from a user, creates and manages appropriate interfaces for delivering (by the user) initial resources needed for the task realization. Then, the task is sent to the system (actually to *agentServer*) for realization. If the system is ready to realize the task (it means that a workflow consisting of appropriate services has been already constructed), the TaskManager asks the user to provide the initial resources according to the constrains returned by the system. More sophisticated tasks can be generated directly from a repository of typical tasks provided by the TaskManager.

The interface serviceAPI provides Java classes and explanation for creating service (according to our architecture) by a service provider.

4.1   Middleware

What was described intuitively in the examples is realized by the engine of the prototype system. Since the engine is transparent for clients and service providers, it may be regarded as a *middleware*. The www interfaces were created only to facilitate the use of the middleware. The middleware is an implementation of the language Entish and the protocol entish 1.0. The language and the protocol constitute the technology we propose.

Service provider creates a service according to serviceAPI, and runs it on its host. Then, the provider registers its service to the middleware via EntishDictionary. Client creates its task via TaskManager, and the task is sent to the middleware. If the task could be realized, the middleware returns constrains on the initial resources needed for task realization. The constrains are displayed to client by TaskManager. The client creates (via GUI) initial resources according to the constrains, sends it to the middleware and waits for the final resource. What is going on inside the system is transparent to clients and service providers. From their point of view only result is important, that is, services are to be used, and tasks are to be realized.

The middleware consists of three basic elements: agentServer, service-Server, and infoService that exchange messages according to the protocol entish 1.0. The contents language of the messages is Entish, so that these three components speak one common language. The middleware can be distributed, i.e., the components may be scattered on different hosts; several instances of the same component may be running at the same time; they may be implemented independently.

The first component, i.e., agentServer represents clients in the middleware. Once a task is delivered to the agentServer from TaskManager, a process dedicated to the task realization is created. The process is called agent, and has its own state for expressing its goal, intentions, and knowledge.

The second component called serviceServer represents service providers in the middleware. Once a service is registered to the serviceServer, a dedicated process is created that represents this service in the middleware for arranging invocation, transactions, etc. The process is equipped with its own state for expressing commitments and knowledge about the current status of conversation with agents and other services.

The third component, i.e., infoService is a global database for storing info about services, that is , what is the operation type of a service, and what is its name and the communication address.

The protocol entish 1.0 is divided into the following phases:

1. serviceServer publishes a service at infoService;
2. agent discovers services at infoService;
3. services are composed into a workflow by an agent;
4. workflow is executed;
5. transactional semantics is realized.

5   The language Entish

The language is supposed to describe the resources (electronic data, documents), their types, attributes, locations, and processing by abstract functions. Thus, it serves for expressing tasks by clients. The language also describes services (implementing the abstract functions) and agents (for composing services) in terms of goals, intentions, and commitments. They are essential for arranging services into workflow in order to realize the clients' tasks. Hence, the language also describes workflow formation and execution process, however, the language is fully declarative, i.e., no explicit actions are used.

There are three kinds of objects the language describes: resources, services, agents. Since the language is supposed to be open and of distributed use, all names must be URIs [11]. So that we must provide namespaces for resources, services, and agents. Resources are supposed to be passed from one service to another. Hence, their names are URLs, and their transport is realized by the GET method of HTTP protocol.

Services and agents communicate and speak the language Entish, i.e., they exchange messages with contents expressed in Entish. The message format is defined at high level of abstraction, i.e., at the level of logical communication between agents and services, so that the message transport is not essential and is not specified in the language. However, we assume that the message transport identifier is an integral part of the names of agents and services. Since SOAP is a standard for exchanging XML - formatted messages, we introduce the following name format:

```
soap://host_name:8080#agent1
soap://host_name:8080#service1
```

for agents and services.

Agent (or service) is a processes running on agentServer (or serviceServer) and equipped with its own state. The format of the state is common for agent and service. It is defined as XML schema in the document state.xsd, see the documentation of the enTish project.

This concludes the short introduction to project enTish, and corresponding technology for service description and composition. The specifications of the proposed technology are in version 1.0, and were designed to be extremely simple for the purpose of understanding the idea behind them as well as for the prototype implementations. So that, several important features, not essential for the basic functionality, were skipped and left to be completed in the next version. These are for example security, and service quality. However, transactional semantics is included in the specifications.

6   Acknowledgments

I am grateful to Steven Willmott and N.C. Narendra for their cooperation within the framework of Service Description and Composition Working Group of Agentcities.NET project. My thanks to Prof. Andrzej Barczak for his encouragement and support from Institute of Informatics of the University of Podlasie. I am also grateful the members of enTish team (Dariusz Czerski, Pawel Jurski, Dariusz Mikulowski, Michal Rudnicki and Leszek Rozwadowski) for their assistance. The work was done within the framework of KBN project No. 7 T11C 040 20.

References

1. Ambroszkiewicz S. Entish: a simple language for Web Service Description and Composition, In (eds.) W. Cellary and A. Iyengar. Internet Technologies, Applications and Societal Impact. Kluwer Academic Publishers. pp. 289- 306, 2002.
2. Service Description and Composition Working Group of Agentcities.NET, www.ipipan.waw.pl/mas/sdc- wg/

3. M. Gudgin and T. Ewald. All we want for Christmas is a WSDL Working Group. www.xml.com/pub/a/2001/12/19/wsdlwg.html

4. M. Gudgin and T. Ewald. The IDL That Isn't. www.xml.com/pub/a/2002/01/16/endpoints.html

5. F. Leymann and D. Roller.Workflow-based applications. IBM Systems Journal, Volume 36, Number 1, 1997 Application Development http://researchweb.watson.ibm.com/journal /sj/361/leymann.html

6. Santhosh Kumaran and Prabir Nandi. Conversational Support for Web Services: The next stage of Web services abstraction. http://www- 106.ibm.com/developerworks/webservices/library/ws- conver/?dwzone=webservices

7. Tarak Modi. WSIL: Do we need another Web Services Specification?Explaining the difference between UDDI. http://www.webservicesarchitect.com/content/ articles/modi01.asp

8. IBM's tutorial http://www- 4.ibm.com/software/solutions/webservices

9. KQML - Knowledge Query and Manipulation Language, http://www.cs.umbc.edu/kqml/

10. FIPA ACL http://www.fipa.org

11. Naming and Addressing: URIs, URLs, ... http://www.w3.org/Addressing/

Automated Classification of Web Documents into a Hierarchy of Categories

Michelangelo Ceci, Floriana Esposito, Michele Lapi, and
Donato Malerba

Dipartimento di Informatica, Universitá degli Studi, via Orabona 4,
70126 Bari, Italy
{ceci, esposito, lapi, malerba}@di.uniba.it

Abstract. In this paper, the problem of classifying a HTML documents into a hierarchy of categories is investigated in the context of cooperative information repository, named WebClassII. The hierarchy of categories is involved in all aspects of automated document classification, namely feature extraction, learning, and classification of a new document. Innovative aspects of this work are: a) an experimental study on actual Web documents which can be associated to any node in the hierarchy; b) the feature selection process; c) the automated selection of thresholds for the score returned by a classifier; d) the comparison of three different techniques (flat, hierarchical with proper training sets, hierarchical with hierarchical training sets); e) the definition of new measures for the evaluation of system performances. Results show that the use of hierarchical training sets improves the hierarchical techniques.

Keyword: web content mining, hierarchical document classification.

1 Introduction

In cooperative information repositories the manual indexing of documents is effective only when all information providers have a thorough comprehension of the underlying shared ontology. However, an experimental study on manual indexing for Boolean information retrieval systems has shown that the degree of overlap in the keywords selected by two similarly trained people to represent the same document is not higher than 30% on average [3]. Therefore, to facilitate document sharing, it is important to develop tools that assist users in the process of document classification.

WebClass [8] is a client-server application that has been designed to support the search activity of a geographically distributed group of people with common interests. It works as an intermediary when users browse the Web through the system and categorize documents by means of one of the classification techniques available. Automated classification of Web pages is performed on the basis of their textual content and may require a preliminary training phase in which document classifiers are built for each document category on the basis of a set of training examples.

A simplifying assumption made in the design of WebClass is that document categories are not hierarchically related. This permits the system to build either a unique classifier for all categories or a classifier for each category independently of the others (*flat classification*). However, in many practical situations categories are organized hierarchically which is essential when the number of categories is quite high, since it supports a thematic search by browsing topics of interests.

In this work we present an upgrade of some techniques implemented in WebClass to the case of Web documents organized in a *hierarchy of categories*, that is, a tree structure whose nodes and leaves are document categories. The upgrading of the techniques involved all aspects of automated document classification, namely:

- the definition of a document representation language (*feature extraction*),
- the construction of document classifiers (*learning*), and
- the *classification* of a new document according to the hierarchy of categories.

The paper is organized as follows. In the next section, we introduce some issues related to upgrading the document classification techniques. In Section 3 we describe a new feature selection process for document classification, while in Section 4 we present the naïve Bayes and the centroid-based classification techniques as well as the automated threshold determination algorithm. Section 5 is devoted to the explanation of the document classification process. All these techniques have been implemented in a new version of the WebClass system, named WebClassII. Finally, some experimental results are reported and commented in Section 6.

2   Hierarchical document classification

In flat classification, a unique set of features is extracted from the training documents belonging to several distinct categories [8]. The uniqueness of the feature set permits the application of several statistical and machine learning algorithms defined for multi-class problems. However, this approach is impractical when the number of categories is high. In the case of hierarchically related categories it would be difficult to select a proper set of features, since documents concerning general topics are well represented by general terms like "mathematics", while documents concerning specific topics (e.g., trigonometry) are better represented by specific terms like "cosine". By taking into account the hierarchy of categories, it is possible to define several representations (sets of features) for a document. Each of them is useful for the classification of a document at one level of the hierarchy. In this way, general and specific terms are not forced to coexist in the same feature set.

As for the learning process, it is possible to consider the hierarchy of categories in the definition of the training sets. In particular, training sets can be specialized for each internal node of the hierarchy by considering only documents of the subhierarchy rooted in the internal node (*hierarchical training set*). This is an alternative to using all documents for each learning problem like in flat classification.

In the classification process, considering the problem hierarchically reduces the number of decisions that each classifier has to take and increases its accuracy. Indeed, the problem is partitioned into smaller subproblems, each of which can be effectively and efficiently managed on the contrary, in flat classification with r categories the classifier has to choose one of r. This can be difficult in the case of large values of r and may lead to inaccurate decisions.

Some of these aspects have been considered in several works [7], [9], [10], [4], [5]. Our work differs from previous studies in several respects. First, documents can be associated to both internal and leaf nodes of the hierarchy. Surprisingly, this aspect is considered only in [10]. However, differently from Mladenic's work, we consider actual Web documents referenced in the Yahoo! ontology, and not only the items which briefly describe them in the Yahoo! Web directories (see Fig. 1). This is the situation that we expect to have in cooperative web repositories indexed by a hierarchy of categories.

A second difference is in the feature selection process for each internal category. It is based on an upgrade of the technique implemented and tested in WebClass, named TF-PF^2-ICF. Indeed, a comparison with other two well-known feature selection measures showed better results in the case of flat classification [8].

The third innovative contribution is in the development of a technique for the automated selection of thresholds both for posterior probabilities (in the case of naïve Bayes classifiers) and for similarity measures (in the case of centroid-based classifiers). The thresholds are used to determine whether a document has to be passed down to the one of the child categories during the top-down classification process.

The fourth difference is the comparison of system performances on two types of training sets definable for a given category: i) a *proper* training set, which includes documents of the category (positive examples) and documents of the sibling categories (negative examples), and ii) a *hierarchical training set*, which includes documents of the subtree rooted in the category (positive examples) and documents of the sibling subtrees (negative examples).

Finally, we define new measures for the evaluation of the system performances so to capture some aspects related to the "semantic" closeness of the predicted category from the actual one.

3   The feature selection process

In WebClassII each document is represented as a numerical feature vector, where each feature corresponds to the occurrence of a particular word in the document. In this representation, no ordering of words or any structure of text is used. The feature set representation is unique for each category. It is automatically determined by means of a set of positive and negative training examples. All training documents are initially tokenized, and the set of tokens (words) is filtered in order to remove HTML tags, punctuation marks, numbers and tokens of less than three characters. Standard text pre-processing methods are used to:

1. remove words with high frequency, or *stopwords*, such as articles,
2. determine equivalent stems (*stemming*) by means of Porter's algorithm for English texts [11].

Only relevant tokens are considered in the feature set. Many techniques have been proposed in the literature on information retrieval for the identification of relevant words to be used as index terms of documents [12]. However, they are not appropriate for the task of document classification, where words are selected for discrimination purposes and not indexing. For two-class problems, Mladenic [10] compared scoring measures based on the *Odds ratio* and those based on *information gain*, leading her to favor the former. For multi-class problems, as in the case of WebClassII, Malerba et al. [8] developed a feature selection procedure based on an extension of the well-known TF-IDF measure. Here we briefly present its extension to the case of hierarchical training sets.

Let c be a category and c' one of its children in the hierarchy of categories, that is, $c' \in SubCategories(c)$. Let d be a training document from c', w a feature extracted from d (after the tokenizing, filtering and stemming steps) and $TF_d(w)$ the relative frequency of w in d. Then, the following statistics can be computed:

- the maximum value of $TF_d(w)$ on training documents d of category c',

$$TF_{c'}(w) = \max_{d \in Training(c')} TF_d(w) \tag{1}$$

- the percentage of documents in c' where w occurs (*page frequency*)

$$PF_{c'}(w) = \frac{occ_{c'}(w)}{|Training(c')|} \tag{2}$$

where $Training(c')$ is the training set for category c' (see section 6).

The union of feature sets extracted from Web pages of c' defines an "empirical" *category dictionary* used by documents on the topics specified by that category. By sorting the dictionary with respect to $TF_{c'}(w)$, words occurring frequently only in one long HTML page might be favored. By sorting each category dictionary according to the product $TF_{c'}(w) \cdot PF_{c'}(w)^2$, the effect of this phenomenon is kept under control.[1] Moreover, common words used in

---

[1] The plain $PF_{c'}(w)$ factor was also used, but was found to reduce performance slightly. For small sets of training documents, the term $PF_{c'}(w)$ might not be small enough to reduce the effect of very frequent words in single documents.

documents of c' will appear in the first entries of the corresponding category dictionary. Some of these words are actually specific to that category, while others are simply common English words (e.g., "information", "unique", and "people") and should be considered as *quasi-stopwords*. In order to move quasi-stopwords down in the sorted dictionary, the value $TF_{c'}(w) \cdot PF_{c'}(w)^2$ is multiplied by a factor $ICF_c(w) = 1/CF_c(w)$, where $CF_c(w)$ (*category frequency*) is the number of categories $c'' \in SubCategories(c)$ in which the word w occurs. In this way, the relevant features that discriminate documents of c' from documents of its sibling categories can be found in the first entries of the category dictionary for c'. If n_{dict} is the maximum number of features selected for a category dictionary, then $Dict_{c'} = [(w_1, v_1), (w_2, v_2), \dots, (w_k, v_k)]$ such that $\forall i \in [1 \cdots k]$ w_i is a feature extracted from some document d in c', $v_i = TF(w_i) \cdot PF_{c'}^2(w_i) \cdot 1/CF_c(w_i)$ and $k \leq n_{dict}$ (with k=n_{dict} when at least n_{dict} features can be extracted from training documents of category c'). The feature set ($FeatSet_c$) associated to a category c is the union of $Dict_{c'}$ for all its subcategories c'. It contains features that appear frequently in many documents of one of the subcategories but seldom occur in documents of the other subcategories (*orthogonality of category features*). In other words, selected features decrease the intra-category dissimilarity and increase the inter-category dissimilarity. Therefore, they are useful to classify a document (temporarily) assigned to c as belonging to a subcategory of c itself. It is noteworthy that this approach returns a set of quite general features for upper level categories, and a set of specific features for lower level categories.

Once the feature set has been determined, training documents can be represented as numerical feature vectors whose values are frequencies.

4   The learning process

WebClassII has two ways of assigning a Web document d to a category c_i:

1. By computing the similarity between d and the *centroid* of c_i.
2. By estimating the posterior probability for c_i given d (*naïve Bayes*).

Therefore, a training phase is necessary either to compute the centroids of the categories or to estimate the posterior probability distributions. Let us suppose that d has been temporarily assigned to a category c. We intend to classify d into one of the subcategories of c. According to the Bayesian theory, the optimal classification of d assigns d to the category $c_i \in SubCategories(c)$ maximizing the posterior probability $P_c(c_i|d)$. Under the assumption that each word in d occurs independently of other words, as well as independently of the text length, it is possible to estimate the posterior probability as follows (adapted from [6]):

$$P_c(c_i|d) = \frac{P_c(c_i) \cdot \prod_{w \in FeatSet_c} P_c(w|c_i)^{TF(w,d)}}{\sum_{c' \in SubCategories(c)} P_c(c') \cdot \prod_{w \in FeatSet_c} P_c(w|c')^{TF(w,d)}} \quad (3)$$

where the prior probability $P_c(c_i)$ is estimated as follows:

$$P_c(c_i) = \frac{|Training(c_i)|}{\sum_{c' \in SubCategories(c)} |Training(c')|} = \frac{|Training(c_i)|}{|Training(c)|} \quad (4)$$

and the likelihood $P_c(w|c_i)$ is estimated according to Laplace's law of succession:

$$P_c(w|c_i) = \frac{1 + PF(w, c_i)}{|FeatSet_c| + \sum_{w' \in FeatSet_c} PF(w', c_i)} \quad (5)$$

In (3)-(5), $TF(w, d)$ and $PF(w, c)$ denote the absolute frequency of w in d and the absolute frequency of w in documents of category c, respectively.

The *centroid* of a category is defined as a feature vector whose components are computed by averaging on the corresponding feature values of all training documents of the category. In order to classify a document d, the centroid most similar to the description of d is sought. Similarity is measured by means of *cosine correlation*, which computes the angle spanned by two feature vectors (d and the centroid):

$$P_c(w, c_i) = \frac{\sum_{d \in Training(c_i)} TF_d(w)}{|Training(c_i)|} \quad (6)$$

$$Sim_c(c_i, d) = \frac{\sum_{w \in FeatSet_c} P_c(w, c_i) \cdot TF_d(w)}{\sqrt{\sum_{w \in FeatSet_c} P_c(w, c_i)^2 \cdot \sum_{w \in FeatSet_c} TF_d(w)^2}} \quad (7)$$

The cosine correlation is particularly meaningful when features define orthogonal directions and vectors are highly dimensional. Both conditions are satisfied in WebClassII, though orthogonality refers to the group of features extracted from each category dictionary rather than to the single features.

The above formulation of both the naïve Bayes and the centroid-based classifiers assigns a document d to the most probable or the most similar class, independently of the absolute value of the posterior probability or similarity measure. However, we should expect that WebClassII users try to classify documents not related to any category in the hierarchy. In this case we expect a "reject" of the document. By assuming that documents to be rejected have either a low posterior probability for all categories or a low similarity to the centroids of all categories, the problem can be formulated as defining a threshold for the value taken by a naïve Bayes or centroid-based classifier [2].

5   The classification process

The classification of a new document is performed by means of a general graph search of the hierarchy of categories. The system starts from the root and selects the nodes to be expanded such that the score returned by the classifier is higher than the threshold determined by the system. At the end of the process, all explored categories are considered for the final selection. The winner is the explored category with the highest score. It is noteworthy that the application of a classifier is always preceded by a change in the document representation according to the set of selected features. This corresponds to abstracting the same document at increasing levels of specificity.

6   Experimental results

In this section we study the performance of WebClassII on a set of Web documents. The data source used in this experimental study is Yahoo! ontology. We extracted 1026 actual Web documents referenced at the top two levels of the Web directory http://dir.yahoo.com/Science. There are 7 categories at the first level and 28 categories at the second level (see Fig. 1). A document assigned to the root of the hierarchy is considered "rejected" since its content is not related to any of the 35 subcategories. All the results are averages of five cross-validation folds. The size of the feature set ranges from 5 to 50 features per category. Features are extracted using hierarchical training sets. Collected statistics concern centroid-based and naïve Bayes classifiers trained according to one of the following three techniques:

1. *flat*, that is, by considering all subcategories together and neglecting their relations;
2. *hierarchical with proper training sets*, that is, by assigning only documents of category c to *Training(c)*;
3. *hierarchical with hierarchical training sets*, that is, by assigning documents of either category c or one of its subcategories to *Training(c)*.

To evaluate both flat and hierarchical classification techniques, we begin by considering the macro-weighted-average of both precision and recall measures [13]. They are computed as the normalized weighted sum of the corresponding microaverage measure. Weights are the percentages of documents per category, while the normalization factor is the number of categories.

Results reported in Fig. 2 (a) and (b) show that the flat technique always outperforms the hierarchical technique with respect to precision, while the naïve Bayes with hierarchical training sets has the best recall for increasing feature set size. Moreover, the centroid-based classifiers almost always have a lower accuracy and recall than the corresponding naïve Bayes classifiers, independently of the adopted technique (flat or hierarchical) and of the feature set size. Another interesting point is that the use of hierarchical training sets improves both performance measures independently of the method and the feature set size (see lines with suffix _SubCat).

In fig. 2 (c) the percentage of "rejected" documents is reported. All training and test documents belong to a category of the hierarchy; therefore, it

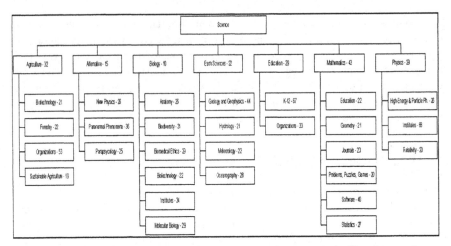

Fig. 1. A segment of Yahoo! ontology considered in this work. Numbers refer to the number of training documents for each category

would be desirable to have very low percentages of rejected documents, as in the case of naïve Bayes method with hierarchical training sets ($< 30\%$).

Intuitively, misclassifications of a document into categories similar to the correct one should be preferred to misclassifications into totally unrelated categories. Therefore, other three evaluation measures can be defined:

1. the *generalization error*, which computes the percentage of documents in c misclassified into a supercategory c';
2. the *specialization error*, which computes the percentage of documents in c misclassified into a subcategory c';
3. the *misclassification error*, which computes the percentage of documents in c misclassified into a category c' not related to c in the hierarchy.

The macro-weighted-average of the misclassification error is reported in Fig. 2 (d). Bayesian methods are the most prone to misclassification errors. However, in the case of hierarchical training sets, the increase of misclassification rate with respect to the centroid-based approach is about 7%, while the decrease of the rejection rate is above 23% with 50 features per category.

The graphs in fig. 2 (e) and (f) show the generalization and specialization errors. The flat approaches have the lowest generalization error, since they simply ignore the relations among categories. The centroid-based method with hierarchical training sets tend to overgeneralize. On the contrary, all methods have a low specialization error rate.

Finally, it is noteworthy that the centroid-based method is more computational efficient of the naïve Bayes only in the flat approach, while they have almost equal learning time in the hierarchical approaches. Also, the learning time of hierarchical techniques is lower than the flat technique, independently of the learning method.

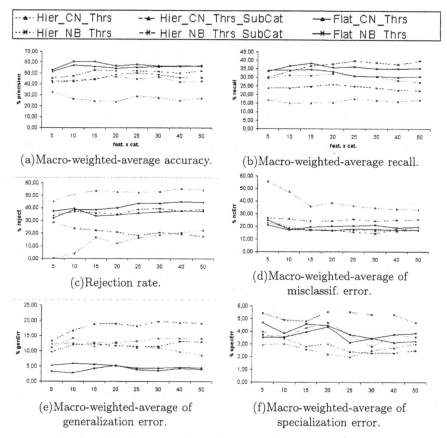

(a) Macro-weighted-average accuracy.

(b) Macro-weighted-average recall.

(c) Rejection rate.

(d) Macro-weighted-average of misclassif. error.

(e) Macro-weighted-average of generalization error.

(f) Macro-weighted-average of specialization error.

Fig. 2. Results achieved

7   Conclusion

In this paper, the problem of automatically classifying HTML documents into a hierarchy of categories has been investigated in the context of a client-server application developed to support Web document sharing in a group of users with common interests. We studied the use of the hierarchy of categories both in the feature extraction and in the construction of the classifiers and in the classification process. As to feature extraction, a novel technique for the selection of relevant features from training pages has been presented. For the learning step, two classifiers have been considered and a thresholding algorithm has been proposed in the case of a reject class. For the classification, a generalized graph search technique has been considered. Experiments have been performed on a set of Web documents indexed in the Yahoo! ontology. Three techniques have been compared: i) flat classification; ii) hierarchical classification with proper training sets and iii) hierarchical classification with hierarchical training sets. Results show that for hierarchical techniques it is better to use hierarchical training sets. Furthermore, the naïve Bayes classifier with hierarchical training sets has high overall recall and efficient learning process. Our results are comparable to those reported in [9], which is the only other work where Web documents indexed by Yahoo! ontology are actually used. However, in our study we have obtained an overall recall of 39.58% using only 875 features, while McCallum et al.'s method required about 13,000 features.

As future work, we intend to investigate a multistrategy approach, where different criteria and techniques are used at different levels of the ontology.

References

1. Almuallim H., Akiba Y., and Kaneda S.(1996) An efficient algorithm for finding optimal gain-ratio multiple-split tests on hierarchical attributes in decision tree learning. Proc. of the Nat. Conf. on Artificial Intelligence (AAAI'96), 703-708

2. Ceci M., Malerba D. (2003) Web-pages Classification into a Hierarchy of Categories, in Proceedings of the BCS-IRSG 25th European Conference on Information Retrieval Research (ECIR '03)

3. C. Cleverdon (1984) Optimizing convenient online access to bibliographic databases. Information Services and Use, **4**, 37-47

4. D'Alessio S., Murray K., Schiaffino R., and Kershenbau A.(2000) The effect of using hierarchical classifiers in text categorization, Proc. of the 6th Int. Conf. on "Recherche d'Information Assiste par Ordinateur" (RIAO), 302-313

5. Dumais S. and Chen H.(2000) Hierarchical classification of Web document. Proc. of the 23rd ACM Int. Conf. on Research and Development in Information Retrieval (SIGIR'00), 256-263

6. Joachims T.(1997) A probabilistic analysis of the Rocchio algorithm with TFIDF for text categorization. Proc. of the 14th Int. Conf. on Machine Learning, 143-151

7. Koller D. and Sahami M.(1997) Hierarchically classifying documents using very few words. Proc. of the 14th Int. Conf. on Machine Learning ICML'97, 170-178

8. Malerba D., Esposito F., and Ceci M.(2002) Mining HTML Pages to Support Document Sharing in a Cooperative System. In R. Unland, A. Chaudri, D. Chabane and W. Lindner (Eds.) XML-Based Data Management and Multimedia Engineering - EDBT 2002 Workshops, Lecture Notes in Computer Science, 2490, 420-434.

9. McCallum A., Rosenfeld R., Mitchell T.M., Ng A.Y.(1998) Improving text classification by shrinkage in a hierarchy of classes. Proc. of the 15th Int. Conf. on Machine Learning (ICML'98), 359-367

10. Mladenic D.(1998) Machine learning on non-homogeneus, distribuited text data, PhD Thesis, University of Ljubjana

11. Porter M. F.(1980) An algorithm for suffix stripping. Program, **14(3)**, 130-137

12. Salton G.(1989) Automatic text processing: The transformation, analysis, and retrieval of information by computer. Reading, MA: Addison-Wesley

13. Sebastiani F. (2002) Machine Learning in Automated Text Categorization. ACM Computing Surveys **34**, 1-47

Extraction of User Profiles by Discovering Preferences through Machine Learning

Marco Degemmis, Pasquale Lops, Giovanni Semeraro, and Fabio Abbattista

Dipartimento di Informatica, Università degli Studi, Via E. Orabona, 4 - 70126
Bari - Italy
{degemmis, lops, semeraro, fabio}@di.uniba.it

Abstract. The recent evolution of e-commerce has emphasized the need for services to be suitable to the needs of individual users: as a consequence, *personalization* has become an important strategy to improve access to relevant products.
This work presents a personalization process based on a text categorization method,
which exploits the textual descriptions of the products in online catalogues, in order
to discriminate between interesting and uninteresting items for the customer. Experimental results encourage the integration of the method in the personalization
component we developed in the COGITO project[1].

1 Introduction

In the recent years, the Internet has experienced a rapid shift from information and entertainment to electronic commerce. Enterprises are developing
new business portals and providing huge amounts of product information,
which in many cases is heterogeneous, not structured and needs to be dealt
in a personalised way. The main challenge is to support Web users in order to facilitate searching in extremely large Web repositories, such as online
product catalogues. The complexity of the problem could be lowered by the
automatic construction of machine processable profiles that can be exploited
to deliver personalized content to the user [8], fitting his or her personal
choices. This is called *user modelling* process.

User modelling [12] simply means ascertaining a few bits of information
about each user, processing that information quickly and providing the results to applications, all without intruding upon the user's consciousness. The
final result is the construction of a user model or a user profile. Roughly, a
user profile is a structured representation of the user's needs which can be
exploited by a retrieval system in order to autonomously pursue the goals
posed by the user. In short, the user model should be able to recognize the
user, know why the user did something, and guess what he or she wants to
do next.

---

[1] The COGITO project is funded by the European Commission under contract
IST-1999-13347.

For this purpose, we adopt machine learning techniques that extract permanent features of a given user from interaction data and construct a corresponding user profile.

2   Related work

During recent years several systems have been designed to offer personalized services and to deliver user-tailored Web content. In this context, various learning approaches have been applied to discover user preferences and to construct user profiles. A text categorization method is adopted by Mooney and Roy [6] in their LIBRA system. It makes content-based book recommending by applying automated text categorization methods to the product descriptions in Amazon.com, using a naïve Bayes text classifier.

Lee et al. [4] present a system able to provide personalized recommendations for some types of product that a consumer may purchase frequently, such as books, CDs, DVD films. The system uses an evolutionary method to model the preferences of a customer for DVD film recommendations. The examples used to train the system are the feature vectors (the keywords associated with a film) of the products more recently collected during the browsing history of a user on a movie site. This learning strategy focuses on the capacity to adapt to the customer's changes of interests.

Data mining methods are used by the 1:1Pro system [2] in order to construct individual profiles. One part of the profile contains facts about a customer, and the other part is made of rules describing the customer's behavior. The behavioral part of the profile is derived from transactional data, representing the purchasing and browsing activities of each user. Using rules to describe customer behavior has the advantage of being an intuitive way to represent the customer's needs. Moreover, the rules generated from a huge number of transactions tend to be statistically reliable. For these reasons, the learning mechanism of our profiling system exploits transactional data to discover rules describing the preferences of a user. Since this method does not use any strategy that categorizes the products into the classes of customer-likes and customer-dislikes, in this paper we try to combine the rule-based approach with a text categorization method applied to semi-structured text, the product description as in [6], in order to build more detailed user profiles. According to this idea, we propose a two-step process for generating profiles: in the first step, the system learns coarse-grained profiles in which the preferences are the product categories the user is interested into. In the second step, profiles are refined by adding a probabilistic model of each preferred product category, induced from the descriptions of the products the user likes in these categories. The final outcome of the process is a more specific fine-grained profile able to discriminate between interesting and uninteresting products for the user.

3   Discovering user preferences from transactional data

The Profile Extractor (Figure 1) is a highly reusable personalization module which employs supervised learning techniques to dynamically discover user preferences from data collected in log files during past interactions between users and the Web site [11]. In the COGITO project, the system was

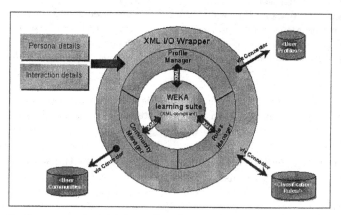

Fig. 1. Profile Extractor Architecture. In addition to the profiling sub-modules, the figure shows the *Community Manager* that groups usage sessions to infer usage patterns exploited for understanding trends useful to further market studies

tested on the German virtual bookshop of the Bertelsmann Online company (www.bol.de): the user's preferences automatically inferred by the system are the ten main book categories the BOL product database is subdivided into, ranked according to the degree of interest computed by the learning system (Figure 2). The Profile Extractor automatically assigns each customer to these predefined classes.

The input to the Profile Extractor is represented by an XML file containing all the interactions and personal details of users (Users History). The *XML I/O Wrapper* performs the extraction (from the Users History) of data required to set up the examples used to train the learning component (number of searches and purchases for each product category, number of connections, etc.). This information is arranged into a set of unclassified instances, where each instance represents a single customer. The instances are pre-classified by a domain expert (each customer is associated with a subset of the BOL book categories) and processed by the *Rules Manager*, which induces a classification rule set for each book category. The learning scheme is based on the PART algorithm [3] and is implemented by the WEKA learning suite [13]. After the training phase, the 10 rule sets are used by the *Profile Manager* to predict whether a user is interested in each book category. All these classifications, together with the interaction details, are gathered to form the

Profile for User: 117

CONNECTIONS_NUM	23
SEARCH_NUMBelletristik	3
SEARCH_FREQBelletristik	0.2
PURCHASE_NUMBelletristik	23
PURCHASE_FREQBelletristik	0.35
SEARCH_NUMComputer_und_Internet	1
SEARCH_FREQComputer_und_Internet	0.2
PURCHASE_NUMComputer_und_Internet	13
PURCHASE_FREQComputer_und_Internet	0.24
SEARCH_NUMKinderbucher	0
SE·'?CH FREOK'nd·rbuchar	n

Belletristik	yes	0.9902	no	0.0098
Computer_und_Internet	yes	1.0	no	0.0
Kinderbucher	yes	0.0	no	1.0
Kultur_und_Geschichte	yes	0.7902	no	0.2098
Nachschlagewerke	yes	0.0	no	1.0
Reise	yes	0.0038	no	0.9962
Sachbuch_und_Ratgeber	yes	0.6702	no	0.3298
Schule_und_Bildung	yes	0.0	no	1.0
Wirtschaft_und_Recht	yes	0.0	no	1.0
Wissenschaft_und_Technik	yes	0.0	no	1.0

Fig. 2. A COGITO coarse-grained profile is composed by two main frames: the frame of user data (personal+interaction data) and the frame of user's interests

user profile. In the COGITO context, it is used to fit the search in the BOL product database to the user's interests [1].

4 Refinement of profiles by learning from textual annotations

The coarse-grained profiles inferred by the COGITO system contain the book categories preferred by a user. Our intention was to enhance the profiles by taking into account the user's preferences in each category, in order to achieve more precise book recommendations. Thus, we adopt an algorithm for learning to classify the textual descriptions of the books, based on the naïve Bayes classifier [5,10]. Our prototype, called ITem Recommender (ITR), is able to classify books belonging to a specific category as interesting or uninteresting to a particular user: for example, the system could learn the target concept "book descriptions the user finds interesting in the category *Computer and Internet*".

In our learning problem, each instance is represented by a set of *slots*. Each slot is a textual field corresponding to a specific feature of a book. The slots used by ITR are: *title*, *authors* and *textual annotation*. A simple pattern-matcher, the *Item Extractor* (Figure 3), analyzes the book descriptions and extracts the words, the *tokens* to fill each slot (it also eliminates stopwords and applies stemming). The text in each slot is a collection of words (bag of word, *BOW*) processed taking into account their occurrences in the original text. Thus, each instance is represented as a vector of three BOWs, one for each slot. Moreover, each instance is labelled with a discrete 1 to 10 rating provided by a user, according to his degree of interest.

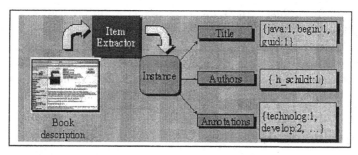

Fig. 3. The BOW extraction process

According to the Bayesian approach to classify natural language text documents, given a set of classes $C = \{c_1, c_2, \ldots, c_{|C|}\}$, the probability of a class c_j, given a document d is calculated as follows:

$$P(c_j|d) = \frac{P(c_j)}{P(d)} P(d|c_j)$$

In our problem, we have only 2 classes: c_+ represents the positive class (user-likes, ratings from 6 to 10) and c_- the negative one (user-dislikes, ratings from 1 to 5). Since instances are represented as a vector of documents, one for each BOW, the category probabilities of an instance d_i are computed using:

$$P(c_j|d_i) = \frac{P(c_j)}{P(d_i)} \prod_{m=1}^{|S|} \prod_{k=1}^{|b_{im}|} P(t_k|c_j, s_m)^{n_{kim}} \tag{1}$$

where $S = \{s_1, s_2, \ldots, s_{|S|}\}$ is the set of slots, b_{im} is the BOW in the slot s_m of the instance d_i, n_{kim} is the number of occurrences of the token t_k in b_{im}.

To calculate (1), we have to estimate the probability terms $P(c_j)$ and $P(t_k|c_j, s_m)$, from the training data, where each instance is weighted according to the user rating r:

$$w_+^i = \frac{r-1}{9}; \qquad w_-^i = 1 - w_+^i \tag{2}$$

The weights in (2) are used for estimating the two probability terms from the training set TR, according to the following equations:

$$\hat{P}(c_j) = \frac{\sum_{i=1}^{|TR|} w_j^i}{|TR|} \qquad \hat{P}(t_k|c_j, s_m) = \frac{\sum_{i=1}^{|TR|} w_j^i n_{kim}}{\sum_{i=1}^{|TR|} w_j^i |b_{im}|} \tag{3}$$

In (3), n_{kim} is the number of occurrences of the term t_k in the slot s_m of the ith instance and the denominator denotes the total weighted length of the slot s_m in the class c_j.

This approach allows the refinement of the profiles by including words most indicative of user preferences for each preferred book category the system was trained on. An example of profile obtained by rating books about "Computer and Internet" is given in Figure 4. The features are ranked according to a measure (strength) that indicates the discriminatory power of a word in classifying a book.

5   Experimental sessions

Two different experiments were performed: the former consisted in observing the accuracy of the ITR system, the latter was conducted to evaluate the combination of the COGITO profiles with the ITR ones.

User ID: 117

Category: Computer &
            Internet

Class Priors: P(YES)=
                  0.6169947952025346

Slot: title   P(NO)=
                   0.3830052047974654

Feature	Strength
gam	2.4155710451915797
directx	2.2707400973131238
enterpris	1.6517008889069003
edit	1.5504909869753871
gem	1.4822827369488536
...	
th	-2.303970881190259
sympos	-2.303970881190259
iee	-2.4741920310501264

Fig. 4. An example of ITR user profile

For both experiments, five book categories at *uk.bol.com* were selected. For each category, a set of book descriptions was obtained by analyzing Web pages using the Item Extractor (Figure 3).

Table 1 describes the extracted information. For each category we considered:

- *Book descriptions* - number of books extracted from the Web site belonging to the specific category;
- *Books with annotation* - number of books with a textual annotation (slot annotation not empty);
- *Avg. annotation length* - average length (in words) of the annotations;

Table 1. Dataset information

Category	Book descriptions	Books with annotation	Avg. annotation length	User
Computer & internet	5414	4190 (77%)	42.39	User1
Fiction & literature	6099	3378 (55%)	35.54	User2
Travel	3179	1541 (48%)	28.29	User3
Business	5527	3668 (66%)	42.04	User4
SF, horror & fantasy	667	484 (72%)	22.33	User5
Total	**20886**	**13261**		

- *User* - the COGITO user that rated books in the category.

The system has been trained for each category by a specific COGITO user that rated approximately 90 books (chosen among those contained in the dataset). In this way, a training set of roughly 450 classified instances is obtained.

5.1 Measuring the accuracy of ITR

The accuracy of ITR system was measured using a 10-fold cross-validation and several metrics were used in the testing phase. In the evaluation phase, the concept of *relevant book* is central: a book d_i in a specific category is considered *relevant by the system* if $P(c_+|d_i)$, calculated as in equation (1), is greater than 0.5. A book d_i in a specific category is considered *relevant by a user* if w_+^i, calculated as in equation (2), is greater than 0.5. Classification effectiveness is measured in terms of the classical Information Retrieval (IR) notions of *precision* (Pr) and *recall* (Re), adapted to the case of text categorization [9].

Let TP (true positive) be the number of relevant test documents correctly classified, that is documents both the system and the user deemed relevant. Then, recall and precision are computed as follows:

$$Re = \frac{TP}{number\ of\ documents\ the\ user\ deemed\ relevant}$$

$$Pr = \frac{TP}{number\ of\ documents\ the\ system\ deemed\ relevant}$$

Also used is F-measure, which is a combination of precision and recall:

$$F = \frac{2 \times Re \times Pr}{Pr + Re}$$

In addition to these classical measures, we adopted also:

- *Normalized Distance-based Performance Measure (NDPM)* - the distance between the ranking imposed by the user ratings and the ranking predicted by the system, as defined by Yao [14]. Values range from 0 (agreement) to 1 (disagreement);
- *Spearman's Rank Correlation (Rs)* - a statistic measure used to discover the strength of the association between two sets of data. In our experiments we will look at the strength of the link between the the ranking imposed by the user and the ranking obtained using the profiles. Given a list of elements sorted according two different measures, Spearman's Rank Correlation compares the two sortings and computes a correlation coefficient that tells us how *close* the two measures are. Say we have n elements sorted on two measures. The following formula computes Rs, where d is the difference between the ranks given to each element by the two measures.

$$Rs = 1 - \frac{6 \times \sum d^2}{n^3 - n}$$

Coefficient ranges from -1 (completely reversed rankings)and +1 (exactly same rankings). A correlation coefficient of 0.3 to 0.6 is considered as moderate and above 0.6 is considered strong;
- *Error (E)* - it is calculated as an average of the absolute difference between the user ratings and those predicted by the system.

All results of the experiment are reported in Table 2.

Table 2. 10-fold cross validation results

Category	Pr	Re	F	NDPM	Rs	E
Computer & internet	0.8500	0.5476	0.6660	0.3241	0.5499	0.3498
Fiction & literature	0.5971	0.7033	0.6459	0.4458	0.0676	0.3489
Travel	0.8100	0.8900	0.8481	0.3322	0.4683	0.2885
Business	0.7364	0.6800	0.7070	0.3741	0.3466	0.3576
SF, horror & fantasy	0.4695	0.7833	0.5871	0.3583	0.3970	0.4105
Avg.	0.6926	0.7209	0.6909	0.3670	0.3659	0.3611

Values of Pr, Re and F provide evidence that the system produces accurate recommendations. NDPM is fairly consistent, while looking at Rs we observe that there is at least a moderate correlation for each category.

5.2   Coarse-grained profiles vs. fine-grained profiles

In the second experiment, each user has been requested to submit 3 different queries to ITR. Then, a feedback is given to the system by rating the 20 top ranked books in each result set. The experiment has been modelled on the basis of two different scenarios. In the first scenario, books are ranked according to the COGITO profile, whereas in the second scenario the ranking is performed using the COGITO profile integrated with the ITR one. For both scenarios feedback evaluation results are given in Table 3.

For pairwise comparison of methods, the non-parametric Wilcoxon signed rank test is used [7], since the number of independent trials is relatively low and does not justify the application of a parametric test, such as the t-test. In this experiment, the test is adopted in order to evaluate the difference between the effectiveness of the different profiles by means of the metrics pointed out in Table 3, requiring a significance level $p < 0.05$.

Table 3. Results of the comparison between the COGITO and the ITR profiles

User	Query	Pr		NDPM		Rs	
		COGITO	ITR	COGITO	ITR	COGITO	ITR
1	Java	0.50	0.90	0.594	0.423	-0.288	0.300
1	Graphics	0.30	0.70	0.465	0.328	0.156	0.490
1	Security	0.80	0.75	0.636	0.410	-0.412	0.278
2	Realism	0.35	0.50	0.421	0.400	0.258	0.329
2	romanticism	0.60	0.55	0.505	0.636	-0.053	-0.362
2	Science fiction	0.65	0.55	0.468	0.476	0.042	0.109
3	Islands	0.65	0.90	0.600	0.536	-0.288	-0.136
3	Guides	0.40	0.60	0.539	0.694	-0.130	-0.581
3	restaurants	0.30	0.35	0.505	0.415	0.037	0.338
4	Business manager	0.35	0.60	0.513	0.494	-0.074	0.018
4	enterprise solution	0.20	0.30	0.365	0.292	0.405	0.595
4	investment	0.50	0.70	0.547	0.605	-0.118	-0.312
5	s_king	0.30	0.60	0.589	0.197	-0.261	0.806
5	Space	0.10	0.40	0.447	0.184	0.178	0.839
5	King	0.70	1.00	0.550	0.326	-0.154	0.517
	Avg.	0.45	0.63	0.516	0.428	-0.047	0.215
	W=	130		-74		72	

On the basis of the values of the W statistic calculated above, we can deduce that there is a consistent statistically significant difference in performance among the two different profiles.

6   Conclusions

In this paper we evaluated a simple approach, based on the naïve Bayes machine learning method, to build user profiles for a content-based book recommender system. We presented a prototype system, called Item Recommender, able to refine user profiles by adding a list of words to each book category preferred by a specific user. Our goal was to integrate the prototype in an already existing personalization system, the Profile Extractor, which employs machine learning techniques to infer the book categories preferred by a user, and stores them in a user profile. Experiments confirm that the use of the integrated profiles has a significant positive effect on the quality of the recommendations made by the system.

References

1. Abbattista F., Degemmis M. et al. (2002) Improving the Usability of an E-commerce Web Site through Personalization. In Recommendation and Personalization in Ecommerce, Ricci F. and Smyth B. (Eds.) Proceedings of the Workshop on Recommendation and Personalization in Ecommerce, 2nd International Conference on Adaptive Hypermedia and Adaptive Web Based Systems, RPeC'02, Malaga, Spain, 20–29
2. Adomavicius G., Tuzhilin A. (2001) Using Data Mining Methods to Build Customer Profiles. IEEE Computer $34(2)$, 74–82
3. Frank E., Witten I.H. (1998) Generating accurate rule sets without global optimization. Proceedings of the 15th International Conference on Machine Learning, Morgan Kaufmann, 144–151
4. Lee W.-P., Liu C.-H. et al. (2002) Intelligent Agent-based Systems for Personalized Recommendations in Internet Commerce. Expert Systems with Applications $22(4)$, 275–284
5. Mitchell T. (1997) Machine Learning. McGraw-Hill, New York
6. Mooney R. J., Roy L. (2000) Content-Based Book Recommending Using Learning for Text Categorization. Proceedings of the 5th ACM Conference on Digital Libraries, San Antonio, USA, 195–204
7. Orkin M., Drogin R. (1990) Vital Statistics. McGraw-Hill, New York
8. Pazzani M., Billsus D. (1997) Learning and Revising User Profiles: The Identification of Interesting Web Sites. Machine Learning $27(3)$, 313–331
9. Salton G., McGill M.J. (1983) Introduction to Modern Information Retrieval. McGraw-Hill, New York
10. Sebastiani F. (2002) Machine Learning in Automated Text Categorization. ACM Computing Surveys $34(1)$, 1–47
11. Semeraro, G., Abbattista, F. et al. (2002) Agents, Personalization, and Intelligent Applications. In R. Corchuelo, A. Ruiz-Cortés, R. Wrembel (Eds.), Technologies Supporting Business Solutions, Part IV: Data analysis and Knowledge Discovery, Chapter 7. Nova Science Publishers (to appear)
12. Webb G. I., Pazzani M. et al. (2001) Machine Learning for User Modeling. User Modeling and User-Adapted Interaction 11, 19–29
13. Witten I. H., Frank E. (1999) Data Mining: Practical Machine Learning Tools and Techniques with Java Implementations. Morgan Kaufmann Publishers, San Francisco, USA
14. Yao Y. Y. (1995) Measuring Retrieval Effectiveness Based on User Preference of Documents. Journal of the American Society for Information Science $46(2)$, 133–145

Safer Decisions Against A Dynamic Opponent

Wojciech Jamroga

Parlevink Group, University of Twente, Netherlands
Institute of Mathematics, University of Gdansk, Poland

Abstract. A hierarchy of beliefs for an agent has been proposed in [3]. The aim of this paper is to investigate the performance of such 'multi-model' agents with some experiments in the simplest possible case. The experiments consist of the agent's interactions with simulated agents, acting as customers of an imaginary Internet banking service.

Keywords: autonomous agents, user modeling, machine learning, decision making, beliefs, game theory.

1 Introduction

A software agent may clearly benefit from having an up-to-date model of her environment of activity. The model may, for example, include actual users' profiles or some assumptions being accepted by default. However, the agent doesn't have to stick to one model only, she can possess a set of complementary beliefs, both learned and assumed. The agent may then switch to the most appropriate one when making her decisions, or even combine several models under certain assumptions [3]. The aim of this paper is to investigate the performance of such 'multi-model' agents in the simplest possible case, i.e. in the case of an agent using exactly two alternative models of reality. The output of such a hybrid agent can be then compared with the performance of both 'single-model' agents alone to see if (and when) a software agent can really benefit from using a more complicated belief structure and decision making scheme.

The controversy between normative models (like non-cooperative equilibria from Game Theory) and adaptive models (obtained through some kind of learning) has been another inspiration for this paper. The adaptive solutions are more useful when the domain is cooperative or neutral; they also allow the agent to exploit deficiencies of her adversaries. The 'best defense' assumptions are still tempting, though, in a situation when the agent risks real money. Even one opponent who plays his optimal strategy persistently can be dangerous then. This paper presents another attempt to integrate both approaches. The main model used by the agent in the experiments is a profile of the user; the other model is based on the maxmin equilibrium.

1.1 The Game and the Agent

The experiments were inspired by the following scenario: a software agent is designed to interact with users on behalf of an Internet banking service; she can make an offer to a user, and the user's response determines her output at this step of interaction. In the actual experiments the agent has had 3 possible offers at hand: the 'risky offer', the 'normal offer' and the 'safe offer', and the customer could respond with: 'accept honestly', 'cheat' or 'skip'. The complete table of payoffs for the game is given below. The 'risky offer', for example, can prove very profitable when accepted honestly by the user, but the agent will lose much if the customer decides to cheat; as the user skips an offer, the bank still gains some profit from the advertisements etc.

	accept	cheat	skip
risky offer	30	-100	0.5
normal offer	10	-30	0.5
safe offer	0.5	0	0.5

The banking agent is a 1-level agent, i.e. an agent that models other agents as stochastic (0-level) agents. The user is simulated as a random 0-level agent – in other words, his behavior can be described with a random probabilistic policy. The agent estimates the user's policy p with a relative frequency distribution $\hat{p}$, counting the user's responses. At the same time the agent computes a confidence value C for the profile acquired so far. The default model is defined in the Game Theory fashion: the user is assumed an enemy who always cheats.

Fig. 1. The simplest hierarchy: two models of reality

The motivation behind the confidence C is the following: if a numerical evaluation can be computed for every decision with respect to a particular model (the expected payoff, for instance), then the agent's decision may be based on a linear combination of the evaluations, with the confidence providing weights. If the agent trusts the user's profile in, say, 70% – the final evaluation may depend on the profile in 70%, and the remaining 30% can be derived from the default model: $eval(a) = C\ eval_{profile}(a) + (1 - C)\ eval_{default}(a)$. In consequence, the decision is based on both models at the same time, although in different proportions – weighting the partial evaluations with the confidence the agent has in them [3].

2   Confidence

In order to provide the agent with a way of computing her 'self-confidence', two measures are combined: the log-loss based confidence [4] and the datasize-related measure proposed by Wang [9].

Let $\hat{p}_i(\cdot|a)$ be the model of the user at his ith response to a, and b_i^* – his actual response at that step. To detect changes in the pattern of the user's behavior, a confidence measure based on the log-loss function [7] can be used. The one-step loss is defined as $l_i = -\log_2 \hat{p}_i(b_i^*|a)$, and the confidence is based on the average deviation of the actual loss from the expected optimal loss in n steps, re-scaled with respect to the minimal and maximal possible deviation value [4]. To implement a simple forgetting scheme, temporal decay with a decay rate of $\lambda \in [0,1]$ is introduced to make the recent loss values matter more than the old ones [5,6,4]:

$$C_{log}^{\lambda} = 2^{-\left|\frac{\Delta_n^{\lambda}}{\Delta_n^{max,\lambda} - \Delta_n^{min,\lambda}}\right|}$$

$$\text{where:} \quad \Delta_n^{\lambda} = M_{\lambda}\left(-\log_2 \hat{p}_i(b_i^*|a) + \sum_b \hat{p}_i(b|a)\log_2 \hat{p}_i(b|a)\right)_{i=1..n}$$

$$\Delta_n^{max,\lambda} - \Delta_n^{min,\lambda} = \max_{(b_1^*..b_n^*)}\{-M_{\lambda}(\log_2 \hat{p}_i(b_i^*|a))\} - \min_{(b_1^*..b_n^*)}\{-M_{\lambda}(\log_2 \hat{p}_i(b_i^*|a))\}$$

$$= M_{\lambda}\left(\log_2 \frac{\max_b \hat{p}_i(b|a)}{\min_b \hat{p}_i(b|a)}\right)_{i=1..n}$$

$$M_{\lambda}(X_{i=1,...,n}) = \frac{\sum_{i=1}^n \lambda^{n-i} X_i}{\sum_{i=1}^n \lambda^{n-i}}$$

It is easy to prove that if $\hat{p}_i(\cdot|a)$ are frequency distributions and $\hat{p}_1(\cdot|a)$ is uniform, then for every $n > 1$: C_{log}^{λ} is defined and $0.5 \leq C_{log}^{\lambda} \leq 1$. C_{log}^{λ} helps to detect changes in the user's policy, but it's unreliable when the number observations is small. This disadvantage can be tackled, though, with a very simple (but efficient) measure: $C_{Wang} = n/(n+1)$ [9,2]. Now, the agent is confident in her knowledge if she has enough data and she detects no irregularities in the user's behavior: $\mathbf{C} = \min(\mathbf{C_{log}^{\lambda}}, \mathbf{C_{Wang}})$. The decay rate λ was set to 0.9 throughout the experiments.

3   The Simulations

To investigate performance of the hybrid agent, several series of experiments were run. The agent played with various kinds of simulated 'users', i.e. processes displaying different dynamics and randomness. Those included:

- static (or stationary) 0-level user with a random policy,
- 'linear' user: a dynamic 0-level agent with the initial and the final preferences p_0, p_{100} generated at random, and the rest evolving in a linear way: $p_i = p_0 + (p_{100} - p_0)/100$,
- 'stepping' user: same as the 'linear' one except that the preferences change after every 30 steps: $p_i(b) = p_0(b) + (i \text{ div } 30)(p_{100}(b) - p_0(b))/3$,
- 'cheater': a user that chooses the action 'cheat' with probability of 1.0,
- 'malicious': an adversary 0-level user with a stationary random policy for the first 30 rounds, then switching to the 'cheater' policy.

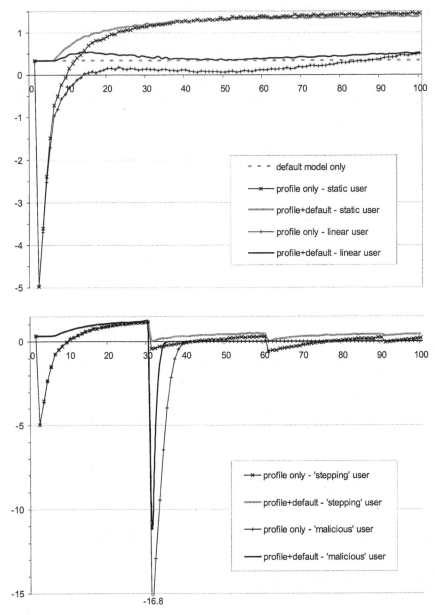

Fig. 2. Hybrid agents vs. single-model agents: the average payoffs for single-minded users

1000000 independent random interactions (a sequence of 100 rounds each) have been simulated for every particular setting; the average results are presented in the following subsections.

3.1 Playing Against Single-Minded Users

The user has been assumed rather simple-minded in the first series of experiments, in order to get rid of the exploration/exploitation tradeoff. Thus, it has been assumed that the user's response doesn't depend on the actual offer being made: $p(cheat)$, $p(accept)$ and $p(skip)$ are the same regardless of the offer (if he's dishonest, he cheats for a small reward as well as a big one, for instance). In consequence, no specific exploration strategy is necessary – every action the agent can choose will reveal exactly the same about the user's policy – so the agent can just maximize $eval(a)$ when making her decisions. The results for various types of users are presented on figure 2. The hybrid agent is almost never worse than the agent using only the user's profile (even for the static user), and in the most risky moments she plays much safer than the latter. At the same time, she has the potential to play positively better than the 'default model' agent. Some simulations were also run for a modified version of the banking game (representing a situation in which the agent's decisions involve less risk) with similar results (see figure 3).

The 'single-mindedness' assumption looks like a rough simplification. On the other hand, the preferences of a particular user (with respect to different offers) are hardly uncorrelated in the real world. For most human agents the situation seems to be somewhere between both extremes: if the user tends to cheat, he may cheat in many cases (although not all by any means); if the user is generally honest, he'll rather not cheat (although the temptation can be too strong if the reward for cheating is very high). Therefore the assumption that the user has the same policy for all the agent's offers may be also seen as the simplest way of collaborative modeling [10]. Section 3.3 gives some more rationale for this kind of assumption, while in the next section users with multi-dimensional (uncorrelated) policies are studied to complete the picture.

3.2 Experiments for Users with More Complex Policies

In this section users are simulated with no restriction on the relation between their conditional policies $p(\cdot|safe)$, $p(\cdot|normal)$ and $p(\cdot|risky)$. Boltzmann exploration strategy is used to deal with the exploration-exploitation problem: the agent chooses action a with probability $P(a) = e^{eval(a)/T} / \sum_{a'} e^{eval(a')/T}$ [1]. As the possible rewards span a relatively large interval (we are using the first payoff table again), the initial temperature parameter is relatively high: $T_0 = 100$, and the decay factor is 0.8. Thus $T_i = T_0 * (0.8)^i$. The results on figure 4 show that the double-model agent has some problems with efficient exploration – in consequence, she plays *too* safe against a stationary user. On the other hand, she is much better protected from sudden changes in the user's behavior. Moreover, the double-model agent plays much better against a 'cheater': she loses 86.9 less than the profile-based agent in the first 15 steps (after that both agents fare almost the same).

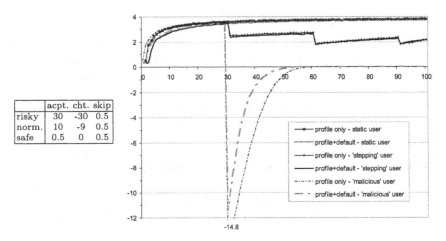

	acpt.	cht.	skip
risky	30	-30	0.5
norm.	10	-9	0.5
safe	0.5	0	0.5

Fig. 3. Results for the modified game

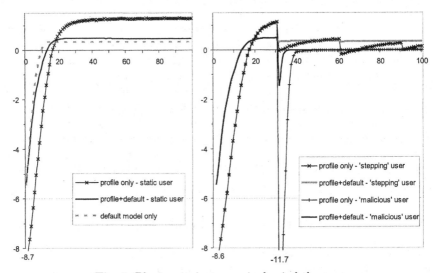

Fig. 4. Playing against non-singleminded users

3.3 Matrix Games with No Pure Equilibrium

Let us go back to section 3.1 and to the assumption that the user's response doesn't depend on the actual action from the agent. Note that the assumption makes perfect sense when the user simply cannot know the agent's action in advance. This is the case, for instance, when the negotiation process is longer and consists of multiple steps, or when some hidden policy of the bank is concerned (instead of particular 'offers'). The game is a matrix game then, and the 'default' strategy pair (safe offer,cheat) is the maxmin equilibrium [8]. The games from section 3.1 are somewhat special since they have their equilibria within the set of pure strategies (i.e. single decisions of the agents). For most matrix games that's not the case. However, every game has its maxmin equilibrium within the set of *mixed* strategies (i.e. probabilistic policies). The set is infinite, but in principle only a finite subset really matters: if the agent can guess the opponent's current (mixed) strategy approximately, then there is a pure strategy with the best expected payoff; otherwise, the agent should choose her maxmin. An example of such game is presented below. The agent's maxmin strategy for this game is $S_D = [0.4, 0.4, 0.2]$. If any of the players plays his/her maxmin, the expected output of the game is 0.

	b1	b2	b3
a1	-1	2	0
a2	0	-2	5
a3	2	0	-10

Note that in the case of mixed strategies, the strategies can be combined directly instead of combining the evaluations. Thus in the experiments the hybrid agent has been choosing the strategy $S = C\, S_{profile} + (1 - C)\, S_D$ where $S_{profile}$ is the strategy with the best estimated payoff. A different way of decision making calls for a modified confidence measure: the confidence is now $C' = C$ for $C \leq 0.4$, and $C' = \max(0.4, 3C - 1.9)$ otherwise. The results (figure 5) reveal that the hybrid agent is again too cautious when the user is random and stationary. However, the bottom line in the game is drawn by a user who can guess the agent's current strategy S somehow (it must a 2-level agent rather than 0-level, since the banking agent is a 1-level one). The 'malicious' user here is defined this way: he uses a random policy for the first 30 steps, and after that starts choosing the most dangerous action (the one with the minimal payoff), 'guessing' the agent's strategy in advance. Playing against a user who chooses the most dangerous action all the time, the hybrid agent was 93.6 better off than the profile-based agent after the first 50 steps.

4   Conclusions

The experiments presented in this paper suggest that a software agent can combine machine learning with Game Theory solutions to display more profitable (or at least safer) performance in many cases. The confidence measure used here is not perfect, and it shows in the results of the simulations. Further experiments should include also agents using more sophisticated learning methods.

References

1. B. Banerjee, R.Mukherjee, and S. Sen. Learning mutual trust. In *Working Notes of AGENTS-00 Workshop on Deception, Fraud and Trust in Agent Societies*, pages 9–14, 2000.

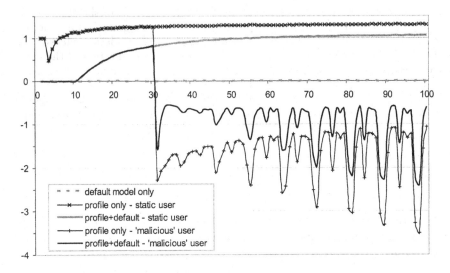

Fig. 5. Results for the matrix game: static user, 'malicious' user

2. W. Jamroga. Datasize-based confidence measure for a learning agent. In *Benelearn 2002*, 2002.
3. W. Jamroga. Multiple models of reality and how to use them. In H. Blockeel and M. Denecker, editors, *BNAIC 2002*, pages 155–162, 2002.
4. W. Jamroga. A confidence measure for learning probabilistic knowledge in a dynamic environment. In *Proceedings of International Conference on Computational Intelligence for Modelling Control and Automation*, 2003.
5. I. Koychev. Gradual forgetting for adaptation to concept drift. In *Proceedings of ECAI 2000 Workshop "Current Issues in Spatio-Temporal Reasoning"*, pages 101–106, 2000.
6. S. Kumar. Confidence based dual reinforcement q-routing: an on-line adaptive network routing algorithm. Master's thesis, Department of Computer Sciences, The University of Texas at Austin, 1998. Tech. Report AI98-267.
7. N. Merhav and M. Feder. Universal prediction. *IEEE Trans. Inform. Theory*, IT-44(6):2124–2147, 1998.
8. J. von Neumann and O. Morgenstern. *Theory of Games and Economic Behaviour*. Princeton University Press: Princeton, NJ, 1944.
9. P. Wang. Confidence as higher-order uncertainty. In *Proceedings of the Second International Symposium on Imprecise Probabilities and Their Applications*, pages 352–361, 2001.
10. I. Zukerman and D.W. Albrecht. Predictive statistical models for user modeling. *User Modeling and User-Adapted Interaction*, 11:5–18, 2001.

Bayesian Nets for Recommender Systems

Mieczysław A. Kłopotek and Sławomir T. Wierzchoń

Institute of Computer Science, Polish Academy of Sciences, Warsaw, Poland

Abstract. The paper introduces a new concept of a collaborative filtering system, based on application of Bayesian Networks for joint classification of documents and clustering of clients.

Keywords: Bayesian networks, recommender systems

1 Introduction

An intelligent agent, called recommender [23], is an assistant to the user in search for useful information exploiting machine learning techniques and data mining technologies.

There exist two principal approaches to the construction of a recommender systems:

1. content-based approach to document filtering [7,18]
2. collaborative approach, called also social learning, collaborative or social filtering [12]

The first approach rooted in the tradition of information search is applicable if the system deals with text only. The system seeks information similar to that preferred by the user. If a user is interested in some knowledge areas (represented by documents described by some keywords, phrases), then the recommender looks for documents with similar content to that articulated. The basic problem here is to capture all essential contents of a document in various areas. Even with restriction to text documents most representations can capture only some aspects of document content which results in weak system quality.

The second approach relies on exploiting reactions of other users to the same document (e.g. a course, educational path, a film, etc.). The system looks for users with similar interests, capabilities etc. and recommends them necessary information or items. The best known example of such a recommender is probably the "people who bought this <book> also bought " feature of the internet company Amazon, [1].

This approach allows for posing questions like "show me information I've never seen but it turned interesting to people like me". Personalized information is provided in an iterative process where information is presented and their ranking is being asked for, just allowing to determine the user profile. This profile is used to locate other users with similar interests in order to distinguish groups with similar interests.

This approach is characterized by two features: First of all the document relevance is determined in the context of the group and not of a single user. Second, evaluation of the document is subjective. Hence one can handle complex and heterogeneous evaluation schemas.

A restricted number of users compared to the number of documents causes a restricted coverage of any evaluation questionnaire. New documents require collection of information from different users.

From the algorithmic point of view important and effective tools supporting social filtering processes are clustering algorithms (e.g. classical k-NN algorithm), Bayesian classifiers and Bayesian networks.

Bayesian networks [8,25] are more promising with respect to achieved results both compared to Bayesian classifiers and also to more traditional vector methods. The reason is that these networks take into account relationships among variables influencing the decision process, while the Bayesian classifiers assume their independence, and on the other hand vector classifiers have to assume a common metrics for all variables while this is not required for Bayesian networks.

2   Basic elements of a social filtering system

The task of social filtering is to predict the utility of an item to a particular user (called also active user) based on a database of user votes from a sample of population of other users, [4]. Typically, the items to be recommended are treated as "black boxes": user's recommendations are based purely on the votes of his neighbors and not the content of the item. The preferences of a user, usually a set of votes on an item, constitute a user profile, and these profiles are compared to build a neighborhood. The key decisions to be made are concerned with data encoding (applied to represent user's profile) and similarity measure (needed to compare profiles). We can distinguish between memory-based algorithms which operate over the entire user database to make predictions and model-based social filtering where the user database is exploited to learn a model which is then used for prediction. Examples of this last approach are cluster models and Bayesian networks. More detailed exposition of different approaches to social filtering can be found in [10].

Generally speaking, two elements are needed in the process of social filtering (assume for clarity that we are interesting in recommending a document to a user):

1. clustering of users/documents in groups of similar ones
2. classifying documents into distinguished categories

In each case the documents have to be described by attributes which may refer to contents (e.g. presence/absence of a keyword) or to an evaluation by the user group (e.g. percentage of users that find the document useful, percentage of users that obtained some capabilities when studying the document etc.). In a similar way we can describe a user in terms of documents that he found interesting/non-interesting.

2.1   Clustering of documents and users

A naive approach to clustering is as follows. Let Z be an unobserved variable (say, a class variable) taking on a relatively small number of discrete values (states), z_k $(1 \leq k \leq K)$. Assuming that the probability of votes are conditionally independent given membership in Z, the probability model relating joint probability of class and votes $v_1, , v_n$ takes the "standard" Bayes formulation

$$P(Z = z_k, v_1, ..., v_n) = P(Z = z_k) \times \prod_{i=1,...,n} P(v_i | Z = z_k).$$

Since we never observe the class variable in the database of users, we must apply methods that can learn parameters for modes with hidden variables, e.g. the Expectation-Maximization algorithm, or EM, [21].

A more complicated model offers the so-called Probabilistic Latent Semantic Analysis (PLSA) [14] where a factored multinomial model is build based on the assumption of an underlying document generation process. The starting point is the term-document matrix N of word counts, i.e., N_{ij} denotes how often a term (single word or phrase) t_i occurs in document d_j. PLSA performs a probabilistic decomposition which is closely related to the non-negative matrix decomposition. Each factor is identified with a state z_k $(1 \leq k \leq K)$ of a latent variable (corresponding somehow to a topic, group in the document collection) with associated relative frequency estimates $P(t_i | z_k)$ for each term in the corpus. A document dj is then represented as a convex combination of factors with mixing weights $P(z_k | d_j)$, i.e., the predictive probabilities for terms in a particular document are constrained to be of the functional form $P(ti|dj) = \sum_k P(ti|zk)P(zk|dj)$, with non-negative probabilities and two sets of normalization constraints $\sum_i P(t_i | z_k) = 1$ for all k and $\sum_k P(z_k | d_j) = 1$ for all j.

Both the factors and the document-specific mixing weights are learned by maximizing the likelihood of the observed term frequencies. More formally, PLSA aims at maximizing $L = \sum_{i,j} N_{ij} \times \log \sum_k P(t_i | z_k) P(z_k | d_j)$. Since factors zk can be interpreted as states of a latent mixing variable associated with each observation (i.e., word occurrence), the EM algorithm can be applied to find a local maximum of L.

Empirically, different factors usually capture distinct "topics" of a document collection; by clustering documents according to their dominant factors, useful topic-specific document clusters often emerge (using the Gaussian factors of LSA, this approach is known as "spectral clustering").

PLSA differs from standard mixture model approach in that it assumes that a single document d_j may belong to different states z_k to different extents. Hence one can think of clustering induced by PLSA as a non-disjoint one.

By analogy, we can treat users as a kind of "documents" that contain "terms" being the items to be recommended. Each user group may be identified with a state z_k $(1 \leq k \leq K)$ of a latent variable (corresponding somehow to a user group) with associated relative frequency estimates $P(d_i|z_k)$ for each document in the corpus (the probability that a user from the group finds the document interesting). A user u_j is then represented as a convex combination of factors with mixing weights $P(z_k|u_j)$, i.e., the predictive probabilities for document interestingness for a particular user are constrained to be of the functional form $P(d_i|u_j) = \sum_k P(d_i|z_k)P(z_k|u_j)$, with non-negative probabilities and two sets of normalization constraints $\sum_i P(d_i|z_k) = 1$ for all k and $\sum_k P(z_k|u_j) = 1$ for all j.

Both the factors and the user-specific mixing weights are learned by maximizing the likelihood of the observed term frequencies. More formally, PLSA aims at maximizing $L = \sum_{i,j} N_{ij} \log \sum_k P(d_i|z_k)P(z_k|u_j)$, with N_{ij} meaning if the jth user is interested in the ith document. Since factors z_k can be interpreted as states of a latent mixing variable associated with each observation, the EM algorithm can be applied again to find a local maximum of the above function L.

Notice that not only an attribute like interestingness, but also training effectiveness, perceived clarity, terms occurring in all documents of interest for the user, links occurring in documents of interest for the user (in analogy to PHITS of [7]), or even document group interestingness for a group of users etc. may be the basis for a similar clustering approach. In fact we may combine them to a single L function of the form $L = \sum_{i,j} N'_{ij} \times \log \sum_k P'(d_i|z_k)P(z_k|u_j) + \sum_{i,j} N''_{ij} \times \log \sum_k P''(d_i|z_k)P(z_k|u_j) + \sum_{i,j} N'''_{ij} \times \log \sum_k P'''(d_i|z_k)P(z_k|u_j) + ...$, where the primes refer to different evaluation aspects. The EM algorithm will be applied here in an analogous way as done in [6].

2.2   Classifying new documents into the categories

For a long time already Bayesian networks have been applied for free text classification [19] and other classification tasks. The basic strategy of classification consists in creation of Bayesian networks for each class of known documents and in calculating the probability of membership to each class of a new document (perhaps weighed by a priori class membership). The class with highest a posteriori probability is considered as the proper class for the new document.

Notice that, given we are knowing the degree of interestingness of a group of documents for a group of users, we can easily derive the degree of interestingness of the document for the group of users from the document class membership.

However, a major problem with Bayesian networks is that the number of attributes (terms) to be considered is beyond learning capabilities of any Bayesian network learning algorithm (for a review of the algorithms consult

[13]). Therefore new algorithms have been devised specially for applications with huge number of attributes [15,17]. Basic concepts of Bayesian networks and a new algorithm are briefly introduced in Section 3 below.

2.3   The social network hypothesis

In a number of studies personal opinions play important role in evaluation of broadly understood documents. The expression of the opinion (vote) is in the simplest case either positive or negative, and usually a "no opinion" possibility is also permitted. Failures of public opinion pools of prediction of votings especially in cases when more than 50

We are currently investigating a different independence hypothesis exploring the "no opinion" phenomenon. We assume that the decision of "no opinion" results from a clash of two types of factors: on the one hand the matter under consideration, on the other hand the social pressure (opinions of neighbors, living conditions etc.) which imply a positive or negative mood. Now the actual vote is positive if both the "mood" and the matter considered are both positive, the actual vote is negative if both the "mood" and the matter considered are both negative. Otherwise there is a clash leading to a "no opinion" vote.

For simplicity we assume that both the matter and the mood factors are independent ($P(q_i, r_j) = P(q_i) \cdot P(r_j)$). But we allow for interdependence between the matters q_i as well as between the respondees r_j. In particular both sets of interdependencies may be described e.g. by a Bayesian network. In case of users we will call it a "social network". The notion of a social network would introduce a kind of fuzzyfied clustering of respondees. Each respondee is a center of its own cluster to which other belong to the smaller degree the less we can predict their "mood" knowing the mood of the cluster center, that is the further they are in the social network.

In spite of the above-mentioned assumption of independence between the matter and the mood factors, the actual variables representing the content of response of the respondee r_j to the matter q_i are dependent because of the mixing of the two factors in the response. Hence, given the table R with entries R_{ij} equal to 1, 0, or -1 representing positive, no opinion and negative votes responder r_j to the matter q_i cannot be used directly to create Bayesian networks of matters and responders, because column/row relative frequencies do not approximate probabilities of corresponding r_j and q_i variables. Separating the impacts of both sets of variables is simple, however, in this case. Out of the table Rij we can calculate the "expected relative number of positively mooded responders" EPR_{pos} and "expected relative number of positively answerable questions" EPQ_{pos}, the same for negative ones, from the quadratic equation system:

$$EPR_{pos} \cdot EPQ_{pos} = rel - card(R_{ij} = 1)$$
$$(1 - EPR_{pos}) \cdot (1 - EPQ_{pos}) = rel - card(R_{ij} = -1)$$

(rel - number of n-tuples in the relational table). To approximate relative frequencies of r_j variables, one needs to divide the relative frequencies from the table R_{ij} by EPQ_{pos}, if positive frequencies are calculated, or by (1-EPQ_{pos}) if negative ones. Then the methods of Bayesian network learning described below are applicable.

We still need to elaborate a more complex model of interaction between subject matter networks and social networks taking into account possible missing independence.

3   Bayesian belief networks

A Bayesian network is a directed acyclic graph with nodes representing variables (events), where nodes are annotated with conditional probability distribution of the node given its parents in the network [13]. The product of node annotations over the whole network represents the joint probability distribution in all variables. Bayesian belief networks were primarily intended to model a typical causal situation: in a causal chain (network) subsequent events occur with some probability given one or more preceding ones. One could learn the model of dependencies among the events given a sufficiently large database describing identical sets of events governed by the same model, with one set of events being independent of the other.

The most important issue in modeling via Bayesian networks is the way they are constructed. One can distinguish situation where both the structure (the node interconnections) and conditional probability distributions are provided by the expert. The other extreme is where both are unknown to the expert. Situations in-between are usually the case (experts may have excellent knowledge about some dependencies among variables, whereas other may be partially unclear). The real challenge is when the structure at least partially is unknown, though estimating conditional probabilities in presence of hidden variables may also be a hard task. For instance, in [4] the next strategy was used: It was assumed that each node in the network corresponds to an item in the domain and the states of each node correspond to the possible vote values for each item; also a state "no vote" corresponding to the lack of data was introduced. Then an algorithm for learning Bayesian networks from training data was applied. Each item in the resulting network has a set of parent items that are the best predictors of its votes.

The clue behind the scenes, when classification is our goal, is that we need to learn and predict at the same time. Hence not all Bayesian network learning methods known from literature are of comparable value. Some learning methods, like SGS are strongly oriented towards detecting intrinsic causal relationships among variables. However, reasoning with structures discovered in this way may be of prohibitive complexity. On the other hand, algorithms like Chow/Liu or Markov-model learning algorithms learn structures suitable for efficient reasoning. However, though they possibly either oversimplify the relationships or overlook too many conditional dependencies, their time and space complexity may turn too high. For applications with natural texts 40,000 to 100,000 words is not unusual.

The ETC algorithm [15,17], developed recently in our research group, overcomes these difficulties by working with a special construct called edge-tree, representing a tree-like Bayesian network as a tree of edge removals. When building the Bayesian network, a new node is not compared to all nodes of the network included so far, but rather to those in one of the subtrees which reduces the comparison time to logarithmic one.

Experimental implementations of CL (Chow/Liu) and ETC were tested on identical artificial data sets generated from tree-like Bayesian networks with binary variables. Networks with 100 up to 2,000 nodes were considered. Conditional probabilities of success on success and failure on failure of the variables were varied from 0.6 to 0.9. Branching factors of the underlying trees were chosen in the range from 2 to 8. Sample sizes ranged from the number of variables to the tenfold of the number of variables. The sequence of variable inclusions was randomized. The ETC algorithm exhibited consistently overwhelming advantage in execution time advantage over CL algorithm. In separate runs (for ETC only, because the runtime of CL exceeded 1 hour for 2,000 variables on a PC Pentium) it has been verified that the construction time of twenty 45,000 node Bayesian networks lasts 30 minutes or less on a conventional PC Pentium (twenty networks needed for classification into 20 classes).

In preliminary experiments classification of documents into about 20 classes is performed with 70% accuracy.

4   Envisaged application

The recommender as described above is intended to be foundation of an e-Advisor system planning educational paths based on information about knowledge acquired by a student, the educational goals and the interests and capabilities of a student.

We assume that student knowledge may be described by a vector representing degree of knowledge acquired in competence areas of interest. Such a vector may be obtained objectively as a result of properly designed tests or subjectively as an evaluation of an interviewer. A similar vector may be established to describe interests and capabilities of a student. Also educational goals are described by vectors of degrees of competences targeted at. This vector is predefined by goals of course participant.

Student descriptions will be used to cluster students either by creating new profiles (based on some degree of similarity, e.g. cosine, overlap between the profiles) or to fit them into predefined profiles (like "well prepared student", "poorly prepared student" etc.). Clustering approaches from section 2.1. or respectively classifying methods from section 2.2 are to be applied. As an alternative, application of fuzzy clustering methods [3,20] is considered. These methods can be used to identify preliminary categorization of the set of students before and during the training process.

In a similar way one may define the required profiles after the first, second etc. period of training. Based on similarity measures we can identify degree of achieving required capabilities and recommend additional training staff. Initially, the recommender parameters would be set manually, and later, after collecting sufficient number of cases, the recommender will be trained on data.

Specifically, a Bayesian network (section 3) for each type of course available for educational path construction would join vectors describing the initial and final state of knowledge (profiles or sub-profiles) trained by data of people that completed the course already. For a new person, we would be capable of calculating probability of attaining required capabilities. Given a set of alternative courses we can find one that would be most successful given the initial knowledge and/or skills of the candidate. As such Bayesian networks can be combined to a cascade, in an analogous way we can evaluate the formation of a longer educational path.

5   Concluding remarks

This paper presented a Bayesian network approach to collaborative filtering. Bayesian network technology is proposed especially for classification task. The user and document clustering is primarily to be performed by PLSA related techniques. However, if one inspects the EM algorithm as applied in the literature, he/she can notice unrealistic assumption made about e.g. independence of terms given the user/document groups. Here also the tree-like Bayesian networks may play an important role for improving reflection of real-world data onto the collaborative filtering models. Improvements in performance are still needed to handle realistic applications, however.

References

1. Amazon.com Recommendations: http://www.amazon.com
2. Balabanovic, M., Shoham, Y.: Content-based, Collaborative Recommendation. In: Communications of ACM, 40(3), p. 66-72, March 1997
3. Bezdek J. C.: Pattern Recognition with Fuzzy Objective Function Algorithms. Plenum Press, NY 1981
4. Breese, J.S., Heckerman, D., and Kadie, C.: Empirical analysis of predictive algorithms for collaborative filtering. Proceedings of the 14th Conference on Uncertainty in AI, 1998, pp. 43-52
5. Cadez, I., Gaffney, S., Smyth, P.: A general probabilistic framework for clustering individuals. Tech. Rep. 00-09, Dept. of Information and Computer Science, Univ. of California, Irvine
6. Cohn, D. and Hofmann, T.: The missing link - a probabilistic model of document content and hypertext connectivity, in T. K. Leen, T. G. Dietterich and

V. Tresp (eds), Advances in Neural Information Processing Systems, Vol. 10, 2001. http://citeseer.nj.nec.com/cohn01missing.html

7. Cohn D., Chang H.: Learning to probabilistically identify authoritative documents. In Proceedings of the 17th International Conference on Machine Learning, 2000.

8. Cowell, R.G., David, A.P., Lauritzen, S.L., Spegelhalter, D.J. Probabilistic Networks and Expert Systems. 1999 Springer-Verlag, NY Inc.

9. Cooper G.F., Herskovits E.: A Bayesian method for the induction of probabilistic networks from data, Machine Learning, 9(1992), 309-347

10. Fisher, D., Hildrum, K., Hong, J., Newman, M., and Vuduc, R.: SWAMI: a framework for collaborative filtering algorithm development and evaluation, http://guir.berkeley.edu/projects/swami/swami-paper/

11. Gokhale A.: Improvements to Collaborative Filtering Algorithms, http://citeseer.nj.nec.com/gokhale99improvements.html

12. Goldberg D., Oki B., Nichols D., Terry D.B.: Using Collaborative Filtering to Weave an Information Tapestry. Communications of the ACM, December 1992, Vol 35, No 12, pp. 61-70.

13. Heckerman D. A tutorial on learning with Bayesian networks. In: M.I. Jordan, ed., Learning in Graphical Models, 1999 The MIT Press, London, England, pp. 301-354

14. Hofmann T. (1999):. Probabilistic latent semantic analysis. In Proceedings of the 15th Conference on Uncertainty in AI, pp. 289-296.

15. Kłopotek M.A.: A New Bayesian Tree Learning Method with Reduced Time and Space Complexity. Fundamenta Informaticae, 49(no 4)2002, IOS Press, pp. 349-367.

16. Kłopotek M.A., Wierzchoń S.T.: Empirical Models for the Dempster-Shafer Theory. in: Srivastava, R.P., Mock, T.J., (Eds.). Belief Functions in Business Decisions. Series: Studies in Fuzziness and Soft Computing. VOL. 88 Springer-Verlag. March 2002. ISBN 3-7908-1451-2, pp. 62-112

17. Kłopotek M.A.: Minig Bayesian Networks Structure for Large Sets of Variables. in M.S.Hacid, Z.W.Ras, D.A. Zighed, Y. Kodratoff (eds): Foundations of Intelligent Systems, Lecture Notes in Artificial Intelligence 2366, Springer-Verlag, pp.114-

18. Maes P., Agents That Reduce Work and Information Overload, Communications of the ACM 37, No. 7, 3040 (July 1994)

19. Meila M.: An accelerated Chow and Liu algorithm: fitting tree distribu-tions to high-dimensional sparse data. http://citeseer.nj.nec.com/363584.html

20. Nasrouin O., Krishnapuram R.: A robust estimator based on density and scale optimization, and its application to clusterin. IEEE Conf. on Fuzzy System, New Orleans, Sept. 1996, 1031-1035

21. Nigham, K., McCallum, A.K., Thun, S., and Mitchell, T.: Text classification from labeled and unlabeled documents using EM algorithm. Machine Learning 16(1999)1-34.

22. Resnick P. et al. GroupLens: An Open Architecture for Collaborative Filtering of Netnews, Internal Research Report, MIT Center for Coordination Science, March 1994.

23. Resnik P., Varian H. R.: Introduction (to the special section on recommender systems). Communications of the ACM, 40(3):56-59, 1997.

24. Resnick P. "Filtering Information on the Internet". Scientific American, March 1997. An overview of non-collaborative automated filtering system (PICS). On-line copy: http://www.sciam.com/0397issue/0397resnick.html
25. Wierzchoń, S.T., Kłopotek, M.A. Evidential Reasoning. An Interpretative Investigation. Wydawnictwo Akademii Podlaskiej, Siedlce 2002, PL ISSN0860-2719, 304 pp.

Implementing Adaptive User Interface for Web Applications *

Tadeusz Morzy, Marek Wojciechowski, Maciej Zakrzewicz, Piotr Dachtera, and Piotr Jurga

Poznan University of Technology, ul. Piotrowo 3a, Poznan, Poland

Abstract. Adaptive web sites automatically improve their organization and presentation to satisfy needs of individual web users. The paper describes our experiences gained during designing and implementing an adaptive extension to a web server - AdAgent. AdAgent is based on the adaptation model, where lists of recommended links are dynamically generated for each browsing user, and embedded in web pages. AdAgent consists of two components: the off-line module using web access logs to discover knowledge about users' behavior, and the on-line module extending the web server functionality, responsible for dynamic personalization of web pages.

1   Introduction

In the last few years, adaptive web sites have focused more and more attention from data mining researchers [2,5–7]. Adaptive web sites dynamically improve their structure and presentation in order to satisfy needs of individual web users. Various techniques are used to recognize individual user's expectations. In some applications users are *explicitly* questioned about their preferences, and then the preferences are used in the future to select the best delivery format. A different idea is to employ data mining techniques to *implicitly* gather preferences of web users. The data source for data mining is usually a web log file, which stores information about all visits to the web site made by users. This technique is the most promising one since it does not require users to fill out any additional web questionnaires.

Designing and implementing adaptive web sites pose many new research problems. Web log files must be transformed into a logically readable form, where all complete web access paths of users are identified. Then, the web access paths must be cleaned in order to remove unwanted noise [4]. The most typical (frequent) paths must be extracted and clustered into user categories. New web users must be monitored and their current access paths must be compared to the most typical ones from the past. Finally, the structure of web pages must be dynamically changed in order to follow user's behavior predictions.

* This work was partially supported by the grant no. 4T11C01923 from the State Committee for Scientific Research (KBN), Poland.

In this paper we describe our experiences on adaptive web sites implementation. We have designed a functional model of an adaptive web interface and implemented a generic web server extension called AdAgent that handles web site adaptivity. Web designers can use the AdAgent functions by means of a new tag library which allows them to define presentation features of automatic recommendations. AdAgent offers various types of recommendation lists and gives the web designer full control over types and locations of the lists. Moreover, AdAgent does not require any changes to the web server, and it can cooperate with popular servers like Apache.

1.1   Related Work

Using web access log mining for web site adaptivity is an area of active research. [7] described the problem of analyzing past user access patterns to discover common user access behavior. User access logs were examined to discover clusters of users that exhibit similar information needs. The clusters were used for better understanding of how users visit the web site, what lead to an improved organization of the web documents for navigational convenience. The authors suggestion was to extend a web server in order to dynamically generate recommended links. In [5,6] the problem of index page synthesis was addressed. An index page is a web page consisting of links to a set of pages that cover particular topics. The described PageGather algorithm was used to automatically generate index pages by means of web access log information analysis. [2] proposed to use frequent itemset clustering for automated personalization of web site contents. The authors chose ARHP algorithm (Association Rule Hypergraph Partitioning) for generating overlapping usage clusters, which were used to automatically customize user sessions.

2   Architecture of AdAgent Environment

AdAgent was designed to allow web designers to make their sites adaptive by providing automatically generated lists of recommended links to be embedded in web pages. The lists are generated based on the browsing (navigation) history of the current user and on the knowledge extracted from access histories of previous users. In our approach, the web designer is given a choice of several recommendation list types, and is responsible for placing them in appropriate places within the web pages. The users interact with the adaptive service normally and do not have to perform any specific actions (like filling out a questionnaire, etc.) to get a personalized view of the web service. A very important feature of AdAgent is that it does not require any changes to the web server, popular web servers can be used together with AdAgent.

Architecture of the AdAgent system is presented in Fig. 1. Typically for adaptive web site solutions, AdAgent consists of two components. The *offline module* analyzes the web server log, and is responsible for discovering knowledge about users' behavior in the form of clusters of access paths (see Sect. 3 for details). The *on-line module* tracks users' requests and uses knowledge provided by the off-line module for dynamic generation of recommended links (see Sect. 4 for details). The off-line module is run periodically (e.g., once a week), and after completion it notifies the on-line module that its knowledge buffer has to be refreshed as more recent information is available. Both AdAgent components are implemented in Java: the off-line module as a standalone application, the on-line module as a servlet.

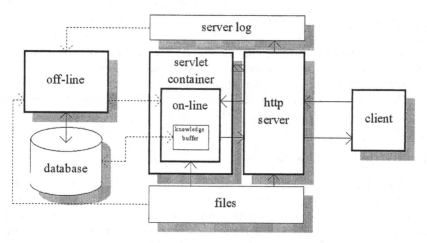

Fig. 1. AdAgent architecture

The off-line module does not cooperate with the web server directly, it just reads and analyses the log and writes discovered knowledge to the database (any database server can be used - in our implementation MS Access was used). The on-line component extends the functionality of the web server - it tracks user requests and adds recommendation lists to web pages served. The web server has to be able to cooperate with some servlet container and has to support the ECLF log format (in our implementation Apache 1.3.13 with ApacheJServ 1.1.2 was used). HTML files requested by users from the web server are processed by the on-line AdAgent module. Positions where dynamic recommendation lists should be embedded are indicated by special tags. To handle those extra tags the Freemarker package (version 2.0) is used by the on-line module.

3 Off-line Segmentation of Web Access Paths

The off-line component is responsible for segmentation of web access paths registered in the web server's log. The actual segmentation is performed using a clustering algorithm. Before the clustering algorithm can be applied, the log has to be cleaned and access paths have to be extracted from it. The three general phases of the segmentation process are described below.

3.1 Log Cleaning

Preprocessing is required for clustering of web access sequences as in case of any advanced analyses of web logs. The log contains entries corresponding to requests, ordered according to request time. Each entry contains a set of fields describing one request. AdAgent uses the following fields from ECLF format: IP address of the client, request time, request type (e.g., GET, POST), requested URL, referrer's URL, browser type, and HTTP status code. Some of the fields are used to filter out information that is not relevant, some are then used to extract access sequences. AdAgent offers an interface to specify which log entries should be filtered out. In the default configuration, only entries corresponding to successful accesses to HTML documents using the GET method are considered for further analysis.

3.2 Access Path Extraction

The first step in access path extraction is identifying blocks of requests coming from the same IP address and the same browser type. Each block contains one or more access paths but each path is guaranteed to be contained in exactly one block. In the second step, blocks are analyzed independently. Access paths are extracted from blocks based on referrer's URL (used to find "previous" pages). The problem with this approach is possibility of "path crossing" where determining which of the matching paths should be extended is not possible (AdAgent does not rely on any additional session tracking mechanisms). To avoid the risk of generating false paths, AdAgent cuts the crossing paths before the crossing point, and starts a new path from there. The result is the set of reliable access paths, some of which may in fact be fragments of actual ones.

3.3 Access Path Clustering

For access path clustering a slightly modified version of the POPC-J algorithm ([3]) is applied. POPC-J clusters sequences based on co-occurring frequent sequences (sequential patterns), and can be decomposed into three phases: Pattern Discovery Phase, Initialization Phase and Merge Phase. Pattern Discovery Phase can be regarded as a preprocessing step, and consists in finding all frequent subsequences (for some specified frequency threshold, e.g., by means of AprioriAll algorithm [1]). In the Initialization Phase, for each frequent pattern a group of web access sequences containing the pattern is built (sequences that do not support any frequent pattern are ignored).

Each pattern, together with the list of sequences supporting it, constitutes a cluster. In the Merge Phase, the similarity matrix for all possible pairs of clusters is built and maintained. The similarity of two clusters is expressed as a Jaccard coefficient of cluster contents (each cluster is a set of sequences). The Jaccard coefficient is one of the well-known set similarity measures:

$$f(C_1, C_2) = \frac{|C_1 \cap C_2|}{|C_1 \cup C_2|} \qquad (1)$$

Clusters are iteratively merged (according to the agglomerative hierarchical clustering scheme), in each iteration the two most similar clusters are merged to form a new larger cluster. The result is a hierarchy of possibly overlapping clusters. Each cluster is described by a set of patterns that are typical for cluster members.

The original POPC-J algorithm builds a complete cluster hierarchy or can stop when a desired number of clusters is reached. AdAgent modifies the stop condition by requiring a given number of "useful" clusters, i.e., clusters containing more than a given number of sequences.

4   On-line Recommendations for Navigation

The discovered segments of typical web access paths are used to dynamically recommend popular navigation paths to new users. The basic idea of dynamic recommendations is to follow the new user's web access path and to match the path against descriptions of the discovered segments. The segment, to which the user's web access path is the most similar is then chosen, and the navigation styles it contains become new recommendations for further navigation.

4.1   Web Access Path Classification

The current web access path of the user is compared to all the frequent web access paths from all segments and the corresponding similarities are evaluated. The segment, for which the sum of the similarities is maximal, is chosen as the source for the recommendations.

The similarity $sim(C, S)$ between the current web access path $C = <c_1\ c_2\ ...\ c_n>$ and a discovered frequent path $S = <s_1\ s_2\ ...\ s_m>$ is defined as follows:

if there exist integers $i_1 < i_2 < ... < i_m$,
such that $s_1 = c_{i_1}$, $s_2 = c_{i_2}$, ... , $s_m = c_{i_m}$, then
    $sim(C, S) = f(i_1) + f(i_2) + ... + f(i_m),$
else
    $sim(C, S) = 0.$

The function $f(x)$ is a *relevance function*, used to decrease the significance of user actions that occurred relatively long before. Example implementations of the $f(x)$ function include $f(x) = x/n$, $f(x) = n - x$, $f(x) = x$, etc. To ignore the aging of current web access paths, we may choose $f(x) = 1$.

Example. Assuming $f(x) = x/n$, the similarity between the current web access path $C = < N\ A\ D\ F\ E\ P\ C\ G >$ and a discovered frequent path $S = < A\ F\ C >$ is the following:

$$sim(C, S) = f(2) + f(4) + f(7) = 2/8 + 4/8 + 7/8 = 1.6$$

4.2   Selecting the Best Recommendations

Recommendations are generated from frequent paths describing the best matching segment. The ordering of the paths depends on two factors: their support values and their mean distances. The two factors multiplied form the ordering key. The paths having the highest values of the key are used as recommendations.

4.3   Nesting Recommendation Directives Inside HTML Pages

In order to offer the generated recommendations to users, a web designer uses a set of additional HTML tags. The tags define where on the HTML page the recommendations will be displayed and what kind of generated recommendations will be used. We have defined four types of recommendations: (1) recommended global links, (2) recommended personal links, (3) recommended frequent links, and (4) recommended hit links. *Global links* represent general trends discovered among all users who have visited a given web page. All users receive identical recommendations. *Personal links* represent trends discovered among those users who followed a web access path similar to the path of the current user. Thus, different groups of users may receive different recommendations. *Frequent links* are the logical sum of global links and personal links. Hit links include the most popular web pages on the whole web site. They do not depend on the user's web access path. An example of an HTML page containing the recommendation tags is given below.

```
<body>
...
<h2>Our recommendations</h2>
<ul>
 <list freqlinks("5") as my_recmnd>
  <li><a href="${my_recmnd.href}">${my_recmnd.name}</a>
 </list>
</ul>
...
</body>
```

The above HTML page displays a list of five recommended frequent links. The page will be processed by the AdAgent and the required dynamic contents will be included in the resulting page to be sent to the user. The user will receive a document similar to the following one.

```
<body>
...
<h2>Our recommendations</h2>
<ul>
 <li><a href="cameras.html">Our new video products</a>
 <li><a href="/company/about.html">HFC Profile</a>
 <li><a href="/faq/index.html">Frequently Asked Questions</a>
 <li><a href="dvd_sales.html">DVDs on sale!!!</a>
 <li><a href="cnd30.html">Canon D30 Specifications</a>
</ul>
...
</body>
```

5   Conclusions

We have presented the AdAgent environment for creating adaptive web sites. AdAgent is based on the adaptation model, where lists of recommended links are dynamically generated for each browsing user, and embedded in web pages. AdAgent's general architecture is rather typical for adaptive web site solutions: the system consists of two components: the off-line module discovering knowledge about users' behavior, and the on-line module extending the web server, responsible for dynamic personalization. However, there are several characteristics of AdAgent distinguishing it from previous proposals, which we believe make it an attractive solution. Firstly, our system offers various types of recommendation lists and gives the web designer full control of which kinds of link lists and where will be embedded. Secondly, the clustering algorithm used by AdAgent's off-line module takes into consideration sequential dependencies between requests. Finally, AdAgent does not require any changes to the web server, and can cooperate with popular servers like Apache.

References

1. Agrawal, R., Srikant, R. (1995) Mining Sequential Patterns. Proceedings of the Eleventh International Conference on Data Engineering, Taipei, Taiwan, 3–14
2. Mobasher, B., Cooley, R., Srivastava, J. (1999) Creating Adaptive Web Sites Through Usage-Based Clustering of URLs. Proceedings of the 1999 IEEE Knowledge and Data Engineering Exchange Workshop, Chicago, Illinois, 19–25

3. Morzy, T., Wojciechowski, M., Zakrzewicz, M. (2001) Scalable Hierarchical Clustering Method for Sequences of Categorical Values. Proceedings of the Fifth Pacific-Asia Conference on Knowledge Discovery and Data Mining, Hong Kong, China, 282–293
4. Pitkow, J. (1997) In search of reliable usage data on the www. Proceedings of the Sixth International World Wide Web Conference, Santa Clara, CA, USA, Computer Networks and ISDN Systems **29**, 1343–1355
5. Perkowitz, M., Etzioni, O. (1997) Adaptive web sites: an AI challenge. Proceedings of the Fifteenth International Joint Conference on Artificial Intelligence, Nagoya, Japan, 16–23
6. Perkowitz, M., Etzioni, O. (1999) Towards Adaptive Web Sites: Conceptual Framework and Case Study. Proceedings of the Eighth International World Wide Web Conference, Toronto, Canada, Computer Networks **31**, 1245–1258
7. Yan, T. W., Jacobsen, M., Garcia-Molina, H., Dayal, U. (1996) From User Access Patterns to Dynamic Hypertext Linking. Proceedings of the Fifth International World Wide Web Conference, Paris, France, Computer Networks **28**, 1007-1014

Web-based Intelligent Tutoring System with Strategy Selection Using Consensus Methods

Janusz Sobecki, Krzysztof Morel, and Tomasz Bednarczuk

Wroclaw University of Technology, Department of Information Systems
Wybrzeze S. Wyspianskiego 27, 50-370 Wroclaw, Poland.
sobecki@pwr.wroc.pl

Abstract. In this paper we present web-based implementation of the Intelligent Tutoring System. In the system the strategy selection is made by application of consensus methods. In the system the tutoring strategy is adapted for the particular student basing on the successful strategies of other similar students. Domain knowledge of the system contains traffic regulations in Poland. This area was chosen to attract potentially many users.

Key words: Web-based systems, Intelligent Tutoring System, Consensus method

1 Introduction

Web-based DL (Distant Learning) environments are attracting increasing number of very differentiated users. The great tempo of the whole civilization development necessitates continuous education for those people who wish to keep pace with it. The DL is a method of conducting of a didactic process which according to [5] has the following properties: pupils and teachers are separated, they use modern media to communicate, they communicate bi-directionally using modern media, the process is controlled by some educational institution.

The TS (Tutoring Strategy) could be defined as a kind of combination of means and methods used in the whole didactic process to increase its effectiveness. In most of CAL (Computer Assisted Learning) or DL environments only single TS is applied. Some specialists, i.e. teachers or tutors, prepare this strategy for so called typical student — the user of the CAL or DL system. The main disadvantage of such systems is keeping the same strategy despite differences in students' progresses. To overcome this disadvantage of the traditional environments ITS (Intelligent Tutoring System) were developed [3]. Their goal was to develop such CBT (Computer Based Tutoring) system that could offer every individual optimal conditions for his or her education by generating the content and controlling the knowledge transfer. ITS is build from three basic modules [6]: the domain, the student and the teacher modules. The domain module is responsible for delivering knowledge about the domain of the course that could be represented i.e. in forms of facts and rules. The module is responsible for both delivering the knowledge to the student as well as testing his or her progress. The students module records data about the of students knowledge acquisition progress. Finally, the teacher module contains knowledge about different TS's.

2   Tutoring strategies

The differences in TS's are caused by differences in students' learning styles, which in turn are consequences of differences in their cognitive styles. Finding out the students' learning styles preferences is quite a difficult task. In the literature we can find at least several approaches to this problem. In the paper [1] for each learning style perspective one of the two contradictory values (given in brackets) has been assigned: perception (sensory or intuitive), input (visual or auditory), organization (inductive or deductive), processing (active or reflective) and understanding (sequential or global). For each learning perspective corresponding teaching style has been assigned accordingly: content (concrete or abstract), presentation (visual or verbal), organization (inductive or deductive), student participation (active or passive) and perspective (sequential or global).

Taking into account all the above mentioned TS perspectives with assigned values, several different TS's could be developed. In our application we distinguished the following TS's: text, graphic, animated and active. All the TS's have the same content, but they use different means of presentation and students participation. They are supposed to offer their students the same knowledge in the domain of the Polish traffic regulation.

Text strategy is based on the Traffic Code regulations presented in the text form. This strategy was prepared for those users who could easily interpret the content of the legal regulations, i.e. people with the legal education. Other people that could benefit this strategy are those who have some practical knowledge in the field and need only some precise information in the matter or to be used in the repetitions. In those cases where the student has some knowledge in the field that were acquired from other sources this strategy could be used to present also the new concepts.

In the graphic strategy, the main means of presentation is a static picture taken from the 3D scene that illustrates the presented regulation. Together with the image the text describing the situation shown in the picture is presented. This strategy was prepared for those users which prefer learning using the graphical examples, for example, they could. have visual input preference, instead of verbal input, but could have some problems with too complicated interface. In these cases users will pay more attention to the interface itself than to the presented content.

In the animated strategy the 3D static graphics with the 2D animations are used. The animation enables the student to imagine how the real traffic problem should be solved. Additionally the strategy offers the possibility to control the animation with the standard video control panel, having also additional ability to rotate the views. The animation illustrates precisely the correct behavior of each vehicle and pedestrian from the scene.

The active strategy requires from a student answering the multiple choice test questions with three answers. This strategy was designed to resemble the theoretical test that students have to pass making their driving licences in Poland, because this official test is being given using PC's. In this strategy, students acquire their knowledge by try and error method, however in the case of bad answer the student receives the detailed information explaining why his answer is incorrect and which is correct. Not to omit important information in case of giving the correct answer the student is also offered some information on his good decisions. To avoid the problem of discouraging students in case of giving too many incorrect answers the strategy was equipped with the hint mechanism.

3   Consensus based tutoring strategy selection

In the consensus methods we are seeking for such solution that is the most representative for some subset of the whole set of elements [8]. In the papers [2,3,7] the adaptive algorithm and the consensus based tutoring strategy selection method were presented. It was assumed that in the CAL environment the population of students are tutored one course that is divided into lessons. The TS selection is made after finishing each lesson by taking into account the test results accomplished by the student. The TS is prepared for each new lesson or repetition (obligatory in case of failing the test). The history of the whole system usage, i.e. TS for each student as well as his/her test scores are stored and used in the consensus based tutoring strategy selection. The strategy is called a sequence belonging to Cartesian product $(K \cup T)^n$ for n being a natural number, where K is a set of knowledge piece presentations and T is a set of test questions. Let also Str be the set of strategies. Let S be a given student who is characterized by 3 parameters: $Begin\_Grade_S$, St_S: a sequence of N strategies and End_Grade_S. One should find a $N + 1$ strategy from Str that is optimal for student S.

Procedure: Determining the $(N + 1)$-th strategy
Step 1: Create set $STUD_S$ from set $STUD$, which consists of such tuples $stud = (Begin\_Grade, St, End\_Grade)$, where St is restricted to N strategies for which distance $d_1(stud_S, stud) \leq \varepsilon$ (where ε is some threshold)
Step 2: On the basis of set $STUD_S$ create set $STUD'_S$ where
$\overline{STUD}'_S = \{(N + 1)$-th strategy of sequence St where $St \in STUD_S\}$
Step 3: Determine such strategy St^* from Str as the consensus for set $\overline{STUD}'_S$, such that

$$\sum_{s' \in STUD_S} d_2(St^*, s') = \min_{s \in Str} \sum_{s' \in STUD_S} d_2(s, s')$$

Step 4: Take St^* as the $(N+1)$-th strategy for the student S.

4   Implementation of the traffic regulation course

In our implementation several modifications have been introduced to the strategy selection procedure presented above. First, the set $STUD_S$ is determined on the basis resemblance of the explicit preferences given by the students. To consider only the successful strategies we can take into account only those students, whose tests results are above the desired threshold. Here, the distance function d_1 equals Euclidean distance among several preference values.

The implemented course has been divided into 7 lessons. Each lesson covers several topics. Each topic could be presented using different scenes, which could belong to one of the three strategy types. To preserve the consistency of each lesson it was necessary to enumerate all possible scenes combinations for each lesson. We stored all the scene combinations in the database and we denote them by the set Str. As the same topic could be presented using different number of scenes, null scenes were entered to obtain equal topic lengths. To reduce the computation time of the application we also computed all the distances between all the combinations of scenes for all the lessons in advance and stored them in the database. We used the distance function:

$$d_2\left(x,y\right) = \sum_{i=1}^{n}\left(x_i \circ y_i\right),$$

where: $a \circ b = 0$, if $a = b$ and $a \circ b = 1$, if $a \neq b$; n is the number of scenes in the longest combination; x, y are scenes combinations; x_i and y_i are scene identifiers. Having defined the distance between combinations, we can define the distance function for the consensus profile [8]. In our case the profile is denoted by Str^* and contains the combinations from Str that were studied by students from the set $STUD_S$:

$$d_3\left(x, Str^*\right) = \sum_{i=1}^{n} d_2\left(x, y_i\right),$$

where: x is the element of the set Str; n is the number of elements in the profile Str^*; y_i is i-th element of the profile Str^*. So the consensus function in the space (Str, d_2) is defined as follows:

$$c(Str^*) = \left\{x \in Str : d_3\left(x, Str^*\right) = \min_{y \in Str} d_3\left(y, Str^*\right)\right\},$$

In consequence the strategy we are looking for belongs to the set $c(Str^*)$.

To test the knowledge acquired by the students, after each lesson students have to pass the multiple choice test with randomly selected questions. Each question from the test is associated with the particular topic. Then, if the student makes too many errors or fails to answer more than one question associated with the same topic, the whole topic should be repeated. The strategy selection is repeated each time the student starts a new lesson or has to repeat its fragment. In case of repeating the whole lesson from the *Str* space the recent strategy is removed and the whole process is repeated.

After completing all the seven lessons the student is tested with 18 questions test generated randomly from the set of about 110 questions. After completing the test student is able to see his or her results.

5 Summary

The application was implemented in the 3-layer architecture. The presentation layer was implemented in Macromedia Flash, the application layer was implemented in PHP and database layer in MS-SQL server. This type of architecture is the most flexible and enables easier modification. Nowadays the method should be verified with a larger group of users. So far we were able to test the application with about 20 users but we considered mainly the usefulness of the strategies and the utility of the application itself. In the near future we have to evaluate the efficiency of the TS adaptive selection consensus method by comparing final test results.

References

1. Felder R. M., Silverman L. K.: *Learning and Teaching Styles in Engineering Education.* Engineering Education (1988) 78 (7) 674–681.
2. Kukla E., Nguyen N.T., Sobecki J., *Towards the model of tutoring strategy selection based on consensus procedure in hypermedia Web systems,* Proc. of 15th Int. Conf. on System Science, Wroclaw (2001) 3, 404–411.
3. Kukla E., Nguyen N. T., Sobecki J.: *The consensus based tutoring strategy selection in CAL systems,* World Transactions on Engineering and Technology Education, Melbourne (2001) 1, 44–49.
4. Lennon J. A.: *Hypermedia Systems and Applications — World Wide Web and Beyond.* Berlin: Springer-Verlag (1997).
5. Nielsen J.: *Hypertext And Hypermedia,* Boston: Academic Press Professional (1993).
6. Nwana H. S.: *Intelligent tutoring systems: an overview.* Artificial Intelligence Rev. (1990) 3, 251–277.
7. Sobecki J.: *An algorithm for tutoring strategy selection in CAL systems.* Technical Reports of Dept. of Information Systems, SPR nr 30, Wrocław (2000).
8. Sobecki J., Nguyen N. T., *Using Consensus Methods to User Classification in Interactive Systems.* Advances in Soft Computing. Physica Verlag (2002) 331–339.

Discovering Company Descriptions on the Web by Multiway Analysis

Vojtěch Svátek[1], Petr Berka[2], Martin Kavalec[1], Jiří Kosek[2], and Vladimír Vávra[1]

[1] Department of Information and Knowledge Engineering,
University of Economics, Prague, W. Churchill Sq. 4, 130 67 Praha 3, Czech Republic
`svatek@vse.cz, kavalec@vse.cz, vavra@vse.cz`
[2] Laboratory for Intelligent Systems, Prague, Czech Republic
`berka@vse.cz, jirka@kosek.cz`

Abstract. We investigate the possibility of web information discovery and extraction by means of a modular architecture analysing separately the multiple forms of information presentation, such as free text, structured text, URLs and hyperlinks, by independent knowledge-based modules. First experiments in discovering a relatively easy target, general company descriptions, suggests that web information can be efficiently retrieved in this way. Thanks to the separation of data types, individual knowledge bases can be much simpler than those used in information extraction over unified representations.

1 Introduction

Until recently, performance improvements of web search engines have mostly been connected with enhancement and accumulation of hardware resources. This was partly due to the fact that the engines represented documents, in accordance with traditional *information retrieval* (IR), by keyword indices, regardless their internal structure and connectivity. The first serious alternative to this 'brute force' approach was the introduction of a 'link—counting' algorithm into Google and related efforts in hub/authority analysis, including term weighing according to the position in an HTML tag or interlinked pages. Still, the essence of these approaches was the retrieval of single, whole documents, according to some generic, numerical relevance measure on uniform document representation.

On the other hand, the discipline of *information extraction* (IE), which aims at retrieving textual information *below* the document level, relies on task–specific, symbolic extraction patterns[1]. In principle, IE can improve both the *accuracy* and *completeness* of web IR. Instead of whole pages, only the fragments containing desired information can be returned to the user. These may not be all parts of text containing the *query term*, and even

---

[1] We consider pattern–based extraction from loosely structured or unstructured pages, not wrapper–based extraction from database–like pages [10,11].

may not contain it at all (nor its thesaurus synonyms). For example, when seeking the *address* of a particular company, we are not interested in phrases such as *"we address the whole range of problems related to..."* but specific semi–structured text in a certain page of the website. The address could be recognised by specific phrases occurring in its neighbourhood ("Where can you find us?") or merely by its internal structure (particular combinations of lower/uppercase text, numerals and special symbols).

A clear disadvantage is however the high computational cost of IE, which (in contrast to optimised search of keyword indices in IR) has to parse the intricate structure of source documents. Approaches that take into account the complexity of structures such as HTML trees and link connectivity tend to use powerful representational languages based on first–order logic [8]; this however entails the use of complex (and thus slow) reasoners. Therefore, IE was mostly assumed to be applied offline, with the extracted text being stored in a database. Conversely, if IE were to be adopted by *online* search engines, without dramatic performance degradation, extraction patterns should be simple and/or used only selectively. In fact, simpler (and thus more comprehensible) patterns are preferable even if there is no time pressure.

The principle we suggest is the decomposition of web IE tasks according to the types of data (structures) to be analysed; we denote it as *multiway* analysis. The configuration of extraction system for a particular task will consist of several (generic) analytical modules equipped with (task–specific) knowledge bases, exchanging information by means of messages conforming to a shared ontology. In this paper, we explore a very simple scenario of multiway analysis of company web pages targeting their general descriptions (company profiles). In section 2 we outline the principles and report on the current state of our *Rainbow* system for multiway web analysis. In section 3 we present some empirical results for extraction of company descriptions. In section 4 we compare our approach with the state of the art. Finally, in section 5 we summarise the contents and outline the future work.

2   The *Rainbow* Architecture

The central idea of the *Rainbow*[2] system (Reusable Architecture for INtelligent Brokering Of Web information access) is the separation of different web analysis tasks according to the *syntactical type of data* involved. In this way, the natural complementarity and/or redundancy of information inferable from different types of data co–existing on the web should give way to robust and reusable applications. The prospective suite of metadata–acquisition components will consist of specialised modules for

---

[2] For more information see http://rainbow.vse.cz. Beyond the acronym, the name is motivated by the idea that the individual modules for analysis of web data should synergistically 'shed light' on the web content, in a similar way as the different colours of the rainbow join together to form the visible light.

- analysis of the (free–text–sentence) linguistic structure
- analysis of the HTML mark–up structure
- analysis of URL addressing
- analysis of explicit metadata (in META tags and RDF)
- analysis of the link topology
- analysis of images

plus additional modules handling source data as well as interaction with users and external services.

The syntactic (data–type–oriented) labour division enables:

1. to reuse *low–level analysis routines*, which do not significantly differ across the tasks and application domains
2. to exploit the possibly deep–and–narrow *specialisation of application developers* (e. g. in linguistics, graphs, or image processing)
3. to perform *data mining* (for important abstraction patterns) on simple, *homogeneous representations* of data, and thus more efficiently[3].

The implementation of the architecture is currently limited to a quite primitive running prototype. The functionality of modules is realised by means of *web services* a priori defined by the WSDL [4] interface description. The model of communication is limited to synchronous requests and answers represented by means of the SOAP [3] protocol.

The current version of the *linguistic* module is able to extract interesting *sentences*[4]. The criterion for selecting sentences is the presence of one of a family of 'indicators', which have been collected, for the company description problem addressed in this paper, by offline data mining in a web directory (Open Directory [1]). The key idea of this mining is the fact that indicators are often syntactically linked, in the text, to the 'heading' terms of directory pages pointing to the respective company pages. This enables to bypass manual labelling of cases for extraction pattern learning. For more details see [9].

Other analytical modules have either been implemented within *Rainbow* in a trivial manner (selection of useful types of META tags), exist as a stand–alone program (URL analysis, see [14], and HTML mark–up analysis), as a dedicated, domain–specific application (image analysis) or merely as design models plus informal knowledge bases (link topology analysis).

---

[3] An obvious risk related to the last point is that of missing some aspects of information perceivable only in combination of multiple types of data. We however hope that a large proportion of such combinations can be captured at the level of inter–module communication (ontology concepts), without mixing the low–level representations.

[4] The version of linguistic module capable of term–level extraction and distinguishing specific information classes is forthcoming.

In addition to the analysis modules, a *source data* component has been developed, which is responsible for the acquisition (incl. the removal of duplicate pages), cleaning, storage and provision of source data. For the acquisition of data, a usual web spider has been created. The cleaning amounts to a conversion of common HTML to well–formed XHTML. The data are then stored in a relational database and sent to the other modules on request.

Finally, a (testbed) *navigation interface* has been built, in the form of plug–in panel in the open–source Mozilla browser. Every time a new page is opened in the browser, it is downloaded and pre–processed (provided not yet in the database), the analysis modules are invoked, and their results are displayed. The three sections of the panel currently correspond to the sentences provided by linguistic analysis, to the content of selected META tags, and to the listing of 'similar' pages provided by Google (as an additional, external web service).

The process of building a knowledge base for a particular analysis module, within a particular *Rainbow*–based application, should rely on the analysis of problem domain (usually including interaction with domain experts and/or data mining) by the responsible developer—knowledge engineer specialised in the particular type of data. The result should ultimately have the form of knowledge base connecting high–level semantic concepts (such as categories of web pages, images, or textual 'messages') to data–type–specific features (such as sets of 'synonymous' URL strings, free–text terms, or skeletal HTML structures). In order for the whole collection of knowledge bases to be consistent, the developers should interact with each other 'on behalf of their modules' and unify their views of the domain; this unification will be materialised in a *formal ontology*. Currently, a top–level ontology of web objects (with 30 classes and 12 slots) has been developed in the OIL language [7], which enables automatic consistency testing. The top–level ontology has one task–specific extension for pornography recognition (with 12 classes); the extension for the OOPS domain (focal in this paper, see next section) is also being elaborated. For more details, see [15].

3   Discovery of Company Descriptions

As starting point for testing our multiway approach, we selected a rather simple task: discovery of general company descriptions (profiles) from websites of 'organisations offering products or services' (we therefore use the acronym OOPS). The descriptions delimit the areas of expertise of the company and the generic types of products and services. They can be presented as free–text paragraphs, as HTML structures such as lists or tables, or as content of META tags such as *keywords* or *description*[5]. The profile usually occurs either immediately at the main page of the site, or at a page directly referenced

---

[5] The former are lists of terms while the latter are either lists or free–text paragraphs.

Table 1. Presence of 'profile' information about the company

Information present in	Main page	Follow–up (profile) page	*Only* follow–up page	Overall	Overall (% of sample)
META tags	28	10	4	32	89
Free text	11	15	12	23	64
Free text, discovered (FTD)	8	8	7	15	42
FTD while not in META	2	5	5	3	8
HTML–structured text	3	3	3	6	17

by this page. The URL of such profile–page is very likely to be 'indicative'. This favourises 'navigational' access: rather than analysing exhaustively the whole website, the link can be followed from the main page.

In our experiment, we randomly chose 50 websites, whose main page was referenced by the 'Business' part of Open Directory [1]. Of these, 36 sites could be considered as directly belonging to 'OOPS'; other were multi–purpose portals or information pages. We exploited the *Rainbow* system in its current state, i.e. the source data module to acquire and pre–process the pages, the linguistic and metadata analysis to extract information, and the navigational assistant to view the results. The simple task–specific knowledge base for linguistic analysis contained only the 10 most frequent 'indicators' such as 'offer', 'provide' or 'specialise in'. Since the URL analyser (originally developed for the VSEved meta–search tool [2]) was not yet operationally connected to *Rainbow*, we simulated its behaviour by observing the links on the page and following them manually. As 'knowledge base' for URL–based detection of profile page, we set up[6] a collection of four significant strings, namely: *about, company, overview, profile.*

Table 1 lists the numbers of websites (among the 36 in the pre–selected sample) that contained the target information in the respective forms. For free text, we distinguish the cases where the presence of company profile was verified only *visually*, those where it has been detected by *Rainbow* (denoted as 'FTD'), and, most specifically, those where the target terms from free text (successfully detected by *Rainbow*) were not contained in META tags. Note that, in this particular row, the fourth column is not equal to the sum of first and third, since cases when information in free text and META tags matched *accross* the two pages also had to be ignored.

We can see that profile information is often contained in META tags as well as in free text, but quite rarely in structured HTML form. It might seem

---

[6] We arrived at this collection after having processed the first 20 pages and have not changed it afterwards. Admittedly, this is not a sound cross–validation method-ology, it could be however tolerated given the triviality of the task.

Table 2. Link to the profile page

Recognisable by	Cases	Cases (%)
Both anchor text (or ALT) and URL	15	62
Anchor text, not URL	4	17
URL, not anchor text	2	9
Neither anchor text nor URL	3	12
Total (page exists)	24	100

that the added value of linguistic analysis is low compared to META tags (which are available more easily). Note however that in most cases, META tags contain an unsorted mixture of keywords including both generic and specific terms related to the company, products as well as customers. Parsing the free–text sentences can help distinguish among semantically different categories of important terms.

Table 2 shows the availability and accessibility of a specialised 'profile' page. We can see that only 3 of the 24 pages did not have the link denoted by one of the *four* terms from our set. Analysis of URL (as specific data structure) is more-or-less redundant here since the same information can mostly be obtained from the *anchor* text of the link, or from the ALT text if there is an image instead of textual anchor. Never mind, our hypothesis of 'informative' URLs has been confirmed.

4 Related Work

As mentioned earlier, information extraction has already been applied on loosely–structured web pages [8,13], however, without clear separation of different types of data. The representation is typically mixed, which increases the expressive power but decreases tractability and maintainability. For example, an usual extraction pattern for our task might look, in Horn logic, like

```
profile\_of(S,O) IF
    main_page(P1) AND has_owner(P1,O) AND has_url(P1,U) AND
    AND refers(U,P2) AND contains(P2,S) AND sentence(S) AND
    AND contains(S,T) AND term(T) AND semantic_of(offering,T).
```

In contrast, we can decompose it to several parts processed by different modules (with optimised internal representation) that only exchange instances of ontology concepts such as 'profile–page' or 'profile–sentence'. Dealing with explicit ontology concepts leads to 'folded' and thus much simpler representation. Clearly, this approach disables 'global' learning of extraction patterns; we rather rely on careful knowledge modelling assisted with targeted data mining.

The particular task of company profile extraction has been addressed by Krötzch [12]. Their approach is similar to ours in the attention paid to multiple modes of information presentation on the web, namely HTML structures and phrasal patterns. They concentrated on a specific domain, casting technology; compared to our domain–neutral approach, their technique is thus more precise and comprehensive but not directly reusable for other domains. Ester [6] also account for different classes of company pages: they apply probabilistic techniques (Naive Bayes and Markov chains) to classify whole company websites based on classes of individual pages. There is however no extraction of information below the page level.

5 Conclusions and Future Work

We have described an architecture for multiway analysis of websites, with modules specialised in different types of data, as well as an experiment targeted at general company information (profiles). Since the experiment was a small–scale one, the concrete results should not be generalised as 'typical company website patterns'. Rather, the goal was to witness the potential benefit of multiway analysis. In this sense, the variability of 'indicators' (be it phrasal terms or URL patterns) at company websites seems to be reasonably low to be covered by simple knowledge bases. We however do not believe that it can be covered by a few *hardwired* heuristics: knowledge engineering as well as data mining effort is required to obtain satisfactory performance.

In the nearest future, we would like to proceed to more complex problems, in particular those requiring thorough analysis of *HTML structures*; a new module dealing with this sort of tree–structured data is already being built. For complex tasks, however, the modules cannot be as tightly coupled as in the current implementation. Instead, we propose the adoption of *problem–solving methods* (PSM) in order to design (navigational) web IE applications 'on the fly' with minimal effort. The PSM technology has recently been used to construct semi–automatically component–based applications for traditional reasoning tasks such as diagnosis or planning [5]. We however assume that it can be used for IE as well, thanks to the knowledge–intensive nature of this task. Finally, we plan to couple one of the future versions of Rainbow with a powerful *fulltext database* system (developed by our partner university), which could help focus the actual IE to the most promising parts of a website.

Acknowledgements

The research has been partially supported by the grant no. 201/03/1318 (Intelligent analysis of web content and structure) of the Grant Agency of the Czech Republic.

References

1. Open Directory (2002) online at http://www.opendir.org
2. Berka, P., Sochorová, M., Svátek, V., Šrámek, D. (1999) The VSEved System for Intelligent WWW Metasearch. In: (Rudas I. J., Madarasz L., eds.:) INES'99 – IEEE Intl. Conf. on Intelligent Engineering Systems, Stara Lesna 1999, 317-321.
3. Box, D. et al. (2000) Simple Object Access Protocol (SOAP) 1.1 – W3C Note. W3C, 2000. http://www.w3.org/TR/SOAP/
4. Christensen, E. et al. (2001) Web Services Description Language (WSDL) 1.1, W3C Note, 2001. http://www.w3.org/TR/2001/NOTE-wsdl-20010315
5. Crubézy, M., Lu, W., Motta, E., Musen, M. (2001) The Internet Reasoning Service: Delivering Configurable Problem–Solving Components to Web Users. Workshop on Interactive Tools for Knowledge Capture, First International Conference on Knowledge Capture (K-CAP 2001), Victoria, Canada. 2001.
6. Ester, M., Kriegel, H.–P., Schubert, M. (2002) Web Site Mining: a new way to spot Competitors, Customers and Suppliers in the World Wide Web. In: Proc. 8th Int. Conf. on Knowledge Discovery and Data Mining (KDD 2002), Edmonton, Alberta, Canada, 2002.
7. Fensel, D. et al. (2000) OIL in a nutshell. In: Proc. EKAW2000, Juan–les–Pins. Springer Verlag, 2000.
8. Freitag, D. (1998) Information Extraction From HTML: Application of a General Learning Approach. In: Proceedings of the 15th National Conference on Artificial Intelligence (AAAI–98).
9. Kavalec, M., Svátek, V., Strossa, P. (2001) Web Directories as Training Data for Automated Metadata Extraction. In: (G.Stumme, A.Hotho, B.Berendt, eds.) Semantic Web Mining, Workshop at ECML/PKDD–2001, Freiburg 2001, 39-44.
10. Knoblock, C. et al. (1998) Modeling Web Sources for Information Integration. In: Proceedings of the 15th National Conference on Artificial Intelligence (AAAI–98).
11. Kushmerick, N., Weld, D. S., Doorenbos, R. (1997) Wrapper Induction for Information Extraction. In: Intl. Joint Conference on Artificial Intelligence (IJ-CAI), 1997.
12. Krötzch, S., Rösner, D. (2002) Ontology based Extraction of Company Profiles. In: Workshop DBFusion, Karlsruhe 2002.
13. Soderland, S.: (1999) Learning Information Extraction Rules for Semi–Structured and Free Text. Machine Learning, Vol. 34, 1999, 233–272.
14. Svátek, V., Berka, P. (2000) URL as starting point for WWW document categorisation. In: (Mariani J., Harman D.:) RIAO'2000 – Content–Based Multimedia Information Access, CID, Paris, 2000, 1693–1702.
15. Svátek, V., Kosek, J., Vacura, M. (2002) Ontology Engineering for Multiway Acquisition of Web Metadata. TR LISp 2002–1, Laboratory for Intelligent Systems, University of Economics, Prague, 2002.

Part III

Natural Language Processing for Search Engines and Other Web Applications

Passage Extraction in Geographical Documents

F. Bilhaut, T. Charnois, P. Enjalbert and Y. Mathet

GREYC, University of Caen, Campus II, 14032 Caen, France
{fbilhaut, charnois, enjalbert, mathet}@info.unicaen.fr

Abstract. This paper presents a project whose aim is to retrieve information in geographical documents. It relies on the generic structure of geographical information which relates some phenomena (for example of sociological or economic nature) with localisations in space and time. The system includes semantic analysers of spatial and temporal expressions, a term extractor (for phenomena), and a discourse analysis module linking the three components altogether, mostly relying on Charolles' discourse universes model. Documents are processed off-line and the results are stored thanks to an XML markup, ready for queries combining the three components of geographical information. Ranked lists of dynamically-bounded passages are returned as answers.

1 Introduction: Querying Geographical Document

This paper is concerned with Infomation Retrieval (IR) from geographical documents, i.e. documents with a major geographic component. They constitue an important source of geographical information and are massively produced and consumed by academics as well as state organisations, marketing services of private companies and so on. The GeoSem[1] project starts from the observation of a strong, characteristic, structure of graphical information which relates some *phenomena* (possibly *quantified*, either in a numeric or qualitative manner) with a *spatial* and, often, *temporal* localisation. Fig. 1 gives and example extracted from our favourite corpus [Hérin, 1994], relative to the educational system in France (note that we are mainly interested in human geography, where the phenomena under consideration are of social or economic nature). As a consequence, a natural way to query documents will be through a 3-component topic based on the "phenomena", space and time aspects, and the three should co-operate to elaborate the answer. A second general remark concerns the form of geographical documents: they are often long or very long (books, long reports or articles) and typically include both a textual and graphical component (maps or various statistical tables). A

---

[1] Semantic Analysis of Geographical Documents (GeoSem) : collaboration between GREYC, ESO (Caen), ERSS (Toulouse), EPFL (Lausanne), supported by the CNRS program "Société de l'Information".

consequence is that answers to IR queries should consist of *passages* rather than whole documents. Co-operation between text and illustration should be taken into consideration as well, though this aspect will not be presented in this paper (see [Malandain, 2001]).

De 1965 à 1985, le nombre de lycéens a augmenté de 70%, mais selon des rythmes et avec des intensités différents selon les académies et les départements. Faible **dans le Sud-Ouest et le Massif Central,** modérée **en Bretagne et à Paris,** l'augmentation a été considérable **dans le Centre-Ouest, et en Alsace.** [...] Intervient aussi **l'allongement des scolarités,** qui a été plus marqué dans les départements où, **au milieu des années 1960, la poursuite des études après l'école primaire** était loin d'étre la règle.

*From 1965 to 1985, the number of high-school students has increased by 70%, but at different rythms and intensities depending on academies and departments. Lower in **South-West and Massif Central,** moderate in **Brittany and Paris,** the rise has been considerable in **Mid-West and Alsace.** [...] Also occurs **the schooling duration increase** which was more important in departments where, **in the middle of the 60's, study continuation after primary school** was far from being systematic.*

Fig. 1. Excerpt from [Hérin, 1994]

The typical task will then be to answer queries bearing on the three components of geographical information such as show in Fig. 2. The answer should consist of a ranked list of passages in a collection of documents. From the technical viewpoint, as indicated by the name of the project, we lay stress on a deep linguistic analysis and especially on semantic aspects, including the discourse level. Documents have to be processed off-line and the results stored thanks to an XML semantic markup, exploited in subsequent queries. Hence, to situate the project among current research, we see that the goals are those of document retrieval, but at an intra-document level, selecting passages [Callan, 1994]. But the methods are rather (though not exclusively) those of information extraction in the sense of MUC's [Pazienza, 1997] and we are quite close to answer extraction in the sense of [Molla, 2000].

The paper is organised as follows. Section 2 describes the semantic analysers. Section 3 shows how phenomena are linked to localisations in space and time. Section 4 presents the application of these semantic analyses to passage retrieval. We conclude on a brief discussion of the results and intended future work.

Retard scolaire dans l'Ouest de la France depuis les années 1950 - *Educational difficulties in West France since the 50's*

Politiques de sécurité maritime dans la Manche - *Navigational security policies in the Channel*

Fig. 2. Typical queries on geographical documents

2 Semantic analyses

It concerns the three aspects : space, time and phenomena but we insist here on the most characteristic of geographical information, spatial analysis. Fig. 1 provides some typical examples of spatial expressions (noun or prepositional phrases) found in geographical documents and Table 1 gives a rough approximation of the informational structure which they convey. The general idea is that they refer to a set of "places" in some "zone". The QUANT part is a quantification expressed by a determiner. The TYPE column describes which kind of "places" is considered and can include an administrative characterisation (town, district, region, ...) and a further qualification, either sociological (rural, urbanised, more or less densely populated, ...) or physical (near seaboard, mountainous, ...). Finally the ZONE part refers to a "named entity", like in MUC tradition, but which is here of geographical nature; and the "position" gives the situation of the "places" under consideration w.r.t. this entity.

QUANT	:	TYPE		:	ZONE	
		: *administrative* :	*qualification*	:	*position*	: *named geo. entity*
(1a) Le quart des	:			:		:
(1b) Tous les		: départements :		:	du nord de	: la France
(1c) Quelques	:			:		:
(1d) Quinze	:	:		:		:
(2) Quelques	:	villes	: maritimes	:		: de la Normandie
(3) Les		: départements :	les plus ruraux	: situés au sud de :		la Loire

(1a/b/c/d): The quarter of / All / Some / Fifteen / districts of north of france
(2) Some seabord towns of Normandy
(3) The most rural districts situated from south of Loire

Table 1. Structure of spatial expressions

The analyser is quite classical, using local, semantic, grammars. We assume a tokenisation and a morphological analysis of the text: presently we use Tree-Tagger [Schmid, 1994] which delivers the lemma and part-of-speech (POS) categorisation. This is turned into a form acceptable by Prolog (list of terms) and a definite clause grammar (DCG) performs altogether syntactic and semantic analyses. Prolog proves to be an interesting choice here since it allows complex semantic computations to be integrated in the grammar, and unification on feature structures thanks to GULP [Covington, 1994]. The semantics of extracted phrases (represented as feature structures) are examplified in Fig. 3. Example (1.b) stipulates an exhaustive determination selecting all entities of the given TYPE (departments) located in ZONE. This zone matches with the northern half inside the named geographic entity (France). In (2) the determination (induced by "quelques / *some*") is relative, i.e. only a part of the elements given by the type is to be considered. Here, TYPE stipulates that we only keep from ZONE (Northern Normandy) the "towns" which are "seabord". Note that the actual model of spatial semantics is in fact significantly more complex, allowing notably recursivity ("les villes maritimes des départements ruraux du nord de la France / *the seaboard towns of rural districts in north of France*"), geometrically defined zones ("le triangle Avignon-Aix-Marseille / *the Avignon-Aix-Marseille triangle*") and different kinds of enumerations ("dans les départements de Bretagne et de Normandie / *in the departments of Brittany and Normandy*").

Tous les départements du nord de la France / Quelques villes maritimes de la Normandie

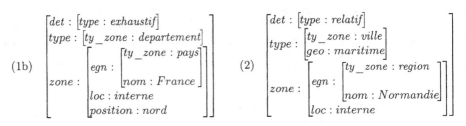

Fig. 3. Spatial expressions accompanied by their semantic representation

The grammar contains 160 rules, 200 entries in an internal lexicon and one hundred thousand terms in an external lexicon of geographical names. It was designed by observation of [Hérin, 1994], and a qualitative evaluation on several other texts seems to indicate that we captured correctly the general structure of spatial expressions. However a more precise, quantitative, evaluation on a wide and diversified corpus is still an open question, left for further work as discussed in Section 5. How the computed semantic structures allows spatial querying of the document is an important question, discussed in Section 4. Temporal analysis is quite similar and qualitatively easier ; again we have to analyse phrases such as reproduced in Fig. 4.

Finally, we also proceed to the extraction of expressions that are supposed to denote phenomena. Adopting a terminological point of view, we assume

De 1985 à 1994 (*From 1985 to 1994*) / Au début des années 90 (*In the begining of the 90's*)

$$\left[periode : \begin{bmatrix} debut : [annee : 1985] \\ fin : [annee : 1994] \end{bmatrix}\right] \qquad \left[annees : \begin{bmatrix} type : debut \\ annee : 1990 \end{bmatrix}\right]$$

Fig. 4. Temporal expressions accompanied by their semantic representation

that phenomena are denoted by noun phrases, and we extract them using a DCG that uses no lexicon, but operates after a preliminary POS tagging. In the example of Fig. 1, it would extract phrases like *l'allongement des scolarités* or *la poursuite des études après l'école primaire*. Note that we also compute semantic values for some phenomena using a domain-dependant lexical semantic base, but this point is beyond the scope of this paper and will not be described here.

3 Linking phenomena to spatio-temporal localisations

The previous section explained how three classes of expressions can be extracted and semantically represented. This part will consider these objects in the discourse flow in order to study their interactions and produce a composite indexation at the passage level. Our method is based on some assumptions that apply specifically to geographical documents, and that led to a representation of themes as 3-component items involving phenomena, space and time. Unlike many thematic analysis methods, this one is not focused on text segmentation, but will instead assume that the discourse is firstly divided into independant segments, either using structural demarcations or any higher-level segmentation approach.

In the proposed method, several models are combined into a three-phase process. The first phase involves extraction of phenomena and spatio-temporal entities as described before, in order to identify potentially relevant parts of the discourse. The following ones are set out to evaluate thematic relevance of extracted entities (represented by numeric values called *thematic weights*), and to link relevant items together. This process, which is described below, will be illustrated using the passage of Fig. 1, where phenomena and spatio-temporal localisations are highlighted.

The second phase aims to obtain a first evaluation of the thematic relevance of each phenomenon, using a distributional approach similar to segmentation methods described in [Ferret, 1997]. The first step selects representative forms, and computes the distributional significance of each of them relatively to each segment using the $tf \cdot idf$ coefficient, which evaluates the importance of a given form considering its occurrence count in the segment relatively to its occurrence count in the rest of the document. Relying on the fact that the distributional significance of a form is representative of its thematic relevance, the distributional weight is used as a first approximation of thematic weight of phenomena.

The purpose of the third phase is to link phenomena with their spatio-temporal localisations, as well as refining thematic weights of phenomena. To achieve this goal, we use two different and supplementary approaches, combining syntactic dependancies and Charolles' discourse universes theory [Charolles, 1997]. In his theory, Charolles describes how some particular forms can introduce frames that constrain the interpretation of a text segment. In the example of Fig. 1, the phrase *De 1965 à 1985* introduces a temporal frame that will determine the interpretation of the rest of the sentence, and probably of a larger following text span. Charolles defines many frame and introducer types, but temporal and spatial frames obviously take on partic-ular importance in our case. Note should be taken that the work presented here takes place in a wider study of the thematic structure of discourse, but we focus here on Charolles' frames model which is particularly well suited for the addressed task, and for which we developped and implemented an automatic analysis method.

The critical point regarding discourse frames is to locate their bound-aries, in order to link phenomena that are contained in a frame to its intro-ducer. Since temporal and spatial introducers can be identified quite easily (mostly using positional criteria), the difficult point is the identification of fi-nal bounds of frames, which is much more subjective. This is achieved thanks to a combination of various linguistic clues, including enunciative criteria (like verb tenses cohesion, see [Le Draoulec, 2001]) and semantic computa-tions (for instance, any spatio-temporal expression that is encountered in a frame is semantically compared to this frame's introducer, in order to test semantic cohesion).

Even if Charolles' frame model is very important in this method, it does not match some spatio-temporal localisations that do not behave as frame introducers, but may still locate punctually one or several phenomena. At the end of the passage of Fig. 1, we could for example link the temporal phrase *au milieu des années 1960* to the phenomena *la poursuite des études après l'école primaire*. To handle these cases, our method also integrates a syntactic approach. Since spatio-temporal phrases usually act as temporal or locative complements, and phrases denoting phenomena as subject or object comple-ment, these dependancies do not directly match syntactic links, but involve *transitivity* along these links. Since an exhaustive syntactic analysis would be superfluous in our case, these dependancies are established relying on a *waiting/resolution* paradigm: as soon as a phenomenon or a localisation is encountered, it is either stacked, waiting for a complementary item, or linked to an already waiting one. To avoid irrelevant links, this method involves intra-phrastic scope management techniques, mostly based on inter-clausal marks (like relative pronouns).

As it was mentioned before, the process of linking phenomena to localisations also provides a way to refine their thematic weight. This mechanism relies on the fact that in geographical discourse, thematically relevant phenomena are likely to be localised spatially and/or temporally. Our method handles this aspect thanks to the attribution of thematic weights to spatio-temporal entities, that are propagated to phenomena they are linked to.

4 Document querying for passage retrieval

The process we described here produces, for each discourse passage, a composite index composed by relevant phenomena along with their spatio-temporal localisations and thematic weight, as shown in Table 2. IR systems can make use of these results as an index to obtain *passages* that are relevant to a 3-component query such as shown in Fig. 2. Our prototype operates on plain text queries and proceeds to passage retrieval using semantic structures matching when available, or falling back to lemma matching in last resort (only for phenomena, since exploiting surface forms is irrelevant for spatio-temporal expressions).

It should be noted that since phenomena and spatio-temporal expressions are precisely marked in documents as well as discourse frames and syntactic links, the search engine is able to dynamically select relevant passages, and to *dynamically* establish the bounds of passages for each particular query instead of using predefined ones (like paragraphs).

l'école primaire	jusqu'aux années 1950	9.0
le certificat d'études	de 1940 à 1950	3.9
le collège	maintenant	3.0
les instituteurs	*not localised*	2.0
la loi d'orientation	de 1989	2.0

Table 2. Composite indexation of a text passage

The search engine matches temporal expressions quite simply, turning their abstract semantic forms into time intervals, but the matching of spatial localisations is much more difficult. However, we will outline the way our system, linked with a geographical information system (GIS)[2], would be able to compute ranked relevant passages thanks to the structures previously obtained. Even if some expressions would need a strong spatial model (cf.*the Avignon-Aix-Marseille triangle*), we believe that queries into a GIS are enough in general case to get relevant results. Assuming that a text contains the sentences (1a) to (1d) of Table 1 (page 123), let us see how the system could compute the answer to the following requests:

(r_1) Which passages address Calvados district?
(r_2) Which passages address Tarn district?

As these requests address districts, we have to focus only on semantics structures having "district" type. For each structure, it must be known to what extent the zone (here, district) is relevant regarding the request. First, we have to know whether the district is included or not in the zone corresponding with the semantic representation of "north of France", common to examples (1a) to (1d). Therefore, the GIS is requested whether Calvados (for r_1), or Tarn (for r_2) is located in the northern half of France. For r_2, GIS answers *no*. Hence, no passage is selected. In opposite, as GIS answers *yes* for r_1 we have to investigate deeper in order to compute a relevance weight for examples (1a) to (1d). Finally, we obtain the following ranked list for r_1: (1b), (1d), (1a), (1c) and an empty list for r_2.

(1a) The semantics of "the quarter" give (no GIS needed) a weight equal to 25%.
(1b) The semantics of "all" give a weight equal to 100%.
(1c) The semantics of "some" indicate that few entities are concerned. In this case, we stipulate a number of 5 entities (that's a heuristic). This leads us a weight equals to $\frac{5}{n}$, n being the number of districts included in the zone. A request to the GIS gives $n = 52$. Hence, the weight is $\frac{5}{52} = 9.6\%$.
(1d) In the same way, we obtain here $\frac{15}{52} = 29\%$.

These principles are partial and prospective. In particular, a relevance computation method based on a possible change of granularity (towns, districts, regions, etc.) between request and answer would be interesting. For instance, we can obviously say that "north-west of France" deserves a lower rank than "the *districts* of the north-west of France" as an answer to request r_1 since Calvados is a district.

---

[2] Database management system that relies on geographical models, allowing spatial localisation of entities such as towns or districts.

5   Conclusion and future work

The general purpose of the GeoSem Project is to design a system for information retrieval in geographical documents and we have presented in this paper a first realisation going in that direction. It includes several analysers and combines linguistic methods (with focus on the semantic and the discourse level) with more standard statistical IR ones. The whole process takes about 30' on our favourite corpus [Hérin, 1994], a 250 pages book. The system is already able to treat queries with both a temporal and phenomenon component, with a ranked list of passages as answer. A generic workshop, LinguaStream, was designed for the project, which allows cumulative annotations of a document according to different analysers.

Several directions of research will be followed in the near future. For one, we must extend the query system to include the spatial component as explained in Section 4. Though processing time is not so crucial for off-line analysis, we also want to improve the system's efficiency: working on the grammars and their implementation techniques (such as bottom-up parsing and compilation of feature structures as described in [Covington, 1994]), we hope to gain a factor 2 or 3. Other, possibly more efficient, parsing methods could also be considered if necessary, provided a good integration in the LinguaStream platform is preserved. Another important aspect concerns evaluation of the semantic analysers, esp. the spatial one. We have to compare the semantic structures computed by the system with expected ones and hence to define a relevant and robust measure of adequation between complex feature structures. Finally we want to run the system on a wider corpus in order to evaluate robustness of the analysers and the cost of transfer to different kinds of geographical documents.

References

[Callan, 1994] James P. Callan, 1994, *Passage-Level Evidence in Document Retrieval*, Proc. 7th Ann. Int. ACM SIGIR Conference on Research and Development in Information Retrieval, Dublin, Ireland.

[Charolles, 1997] Michel Charolles, 1997, *L'encadrement du dicours - Univers, champs, domaines et espace*, Cahier de recherche linguistique, 6.

[Covington, 1994] Michael A. Covington, 1994, *GULP 3.1: An Extention of Prolog for Unification-Based Grammar*, Research Report, AI.

[Ferret, 1997] Olivier Ferret, Brigitte Grau, Nicolas Masson, 1997, *Utilisation d'un réseau de cooccurrences lexicales pour améliorer une analyse thématique fondée sur la distribution des mots*, Actes 1ères Journées du Chapitre Français de l'ISKO, Lille, France.

[Hérin, 1994] Robert Hérin, Rémi Rouault, 1994, *Atlas de la France Scolaire de la Maternelle au Lycée*, RECLUS - La Documentation Française, Dynamiques du Territoire, 14.

[Le Draoulec, 2001] Anne Le Draoulec, Marie-Paule Péry-Woodley, 2001 *Corpus-based identification of temporal organisation in discourse*, Proceedings of the Corpus Linguistics 2001 Conference, Lancaster, pp. 159-166.

[Malandain, 2001] Nicolas Malandain, Mauro Gaio, Jacques Madelaine, 2001, *Improving Retrieval Effectieveness By Automatically Creating Some Multiscaled Links Between Text and Pictures*, Proceedings of SPIE, Document Recognition and Retrieval VIII, vol. 4307, pp. 88-89, San Jose, California, USA.

[Molla, 2000] Diego Molla, Rolf Schwitter, Michael Hess, Rachel Fournier, 2000, *Extrans, an Answer Extraction System*, Traitement Automatique des Langues, Hermes Science Publication, 41-2, 495-522.

[Pazienza, 1997] Maria Teresa Pazienza (Ed.), 1997, *Information Extraction*, Springer Verlag.

[Schmid, 1994] Helmut Schmid, 1994, *Probabilistic Part-of-Speech Tagging Using Decision Trees*, Intl. Conference on New Methods in Language Processing. Manchester, UK.

Morphological Categorization using Attribute Value Trees and XML

Stefan Diaconescu

SOFTWIN Str. Fabrica de Glucoza Nr. 5, Sect.2, 72246 Bucharest, ROMANIA
sdiaconescu@softwin.ro

Abstract. Abstract. In order to handle information into the natural language processing systems the morphological categorization of parts of speech POS must be represented in two kinds of forms: external - most human readable form and internal - most computer readable form. This document presents a General Model that contains an external form to represent morphological categorization based on attribute value trees AVT and two internal forms based on XML: one for general morphological information and one for particular information attached to a POS. Finally it is presented a LIR model that can efficiently be used for strong inflected languages to generate inflected forms of POS, to spell and annotate POS in a text and to full indexing a text.

1 Introduction

From a certain point of view we can consider that there are two directions in which the NLP (Natural Language Processing) is developed: one stress on the lexicon and the other one stress on the syntax. If we try to put too much linguistic information in the lexicon, we can arrive in the situation where for each POS (Part of Speech) we must consider not only information about its intrinsic value but also a lot of information about its relation with the others POS. If we try to put too much linguistic information in the syntax description, we can find ourselves in the situation to consider all the POS of a language with all their inflected forms as terminals. In both cases the rule numbers (or the description volume) we need becomes too great and difficult to manage. We consider that there is a middle way, where the syntactic description is detailed only to the level of pseudo terminals that have attached a set of categories with their values. The morphological description will contain all the rules that are observed by morphological categories and will associate these categories and their values to the information from the lexicon. In this paper we will describe a General Model that contains a mode to structure morphological information. This model (section 2) and AVT (Attribute Value Trees) [1] and XML [3]. The three steps of the model are described in the sections 3,4,5. The morphological information from the General Model can then be exploited in a particular model. In the section 6 we will describe such a model named LIR (Lexicon, Inflection, Roots).

2   The General Model

We will present here a language processing model, the General Model that can be used to process and store linguistic information. The main components of the General Model are (see the figure 1):

- *The Morphological Configurator Creating (1)*. It has as input the general morphological knowledge (usually taken from the classical grammars) and has as output an AVT morphological configurator (2).
- *The conversion from the AVT format to XML format (3)*. It has as inputs the AVT morphological configurator and the morphological configurator DTD. It has as output the morphological configurator in XML format (4).
- *The morphological knowledge acquisition (5)*. It has as inputs the XML morphological configurator (4), the particular morphological knowledge (POS from the lexicon and rules and exceptions taken from the grammars) and the Inflected POS DTD. It has as output the detailed morphological knowledge in XML format (6).
- *The particular model generation (7)*. Using the detailed XML morphological knowledge (6) different particular models can be generated (8).

3   AVT Morphological Configurator

A Morphological Configurator is a formal description of the morphological structure of a language. It is based on AVT (Attribute Value Tree). We have not enough space here to give all the syntax of the AVT representation. We will indicate only few hints about this representation.

The Morphological Configurator contains an AVT that has associated to each node attributes (category) and attribute values. Each attribute has associated one or many values. Each value can have zero, one or many attributes.

Each attribute has associated some features like:

- inflection: indicates if the category is inflected or not;
- treatment: indicates the name of a procedure that must make the treatment of the category in a Natural Language Processing System (NLP).

Each value has also associated some features like:

- lemma: indicates if the category value will be associated to the POS lemma;
- lexicon: indicates if the category value is associated to a POS that will be an entry in the lexicon (if lexicon = entry) a supplementary entry in the lexicon (if lexicon = supplement) or not an entry in the lexicon (if lexicon = no);

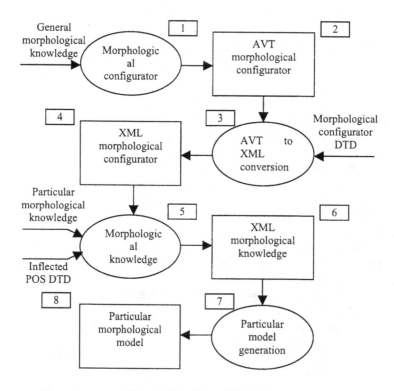

Fig. 1. The General Model

- indicative: indicates if the category value is associated or not to a POS that will be used by an automatic treatment in a NLP to generate the other inflected forms of this POS;
- treatment: indicates the name of a procedure that must make the treatment of the category value in a NLP.

An attribute can have associated a label (the definition) that can be used later in a place where it will be substituted by its definition. An attribute list associated to an attribute value can have also a label that can be used later in a place where it will be substituted with its definition. Using this mechanism (similar with the HPSG reentrancy) the AVT description is much more compact. For example, the AVT for Romanian language morphological configurator is about four times smaller.

The Morphological Configurator contains also a label dictionary section that gives in different languages the categories, category values, some abbreviations and comments. Therefore the AVT description can be prepared to be used in different exploitation systems implemented in different languages.

4 XML Morphological Configurator

The AVT morphological configurator can be converted to an XML format compatible with a specific DTD. We present here this DTD named Configurator.

```
<?xml version="1.0" encoding = "UTF-8"?>
<!ELEMENT configurator   (attribute, labelDictionary*)>
<!ELEMENT attribute      (name, value*)>
<!ELEMENT name           (NMTOKEN)>
<!ELEMENT abbreviation   (NMTOKEN)>
<!ELEMENT value          (name, attribute*)>
<!ELEMENT labelDictionary (label)+>
<!ELEMENT label          (name, translation, abbreviation)>
<!ELEMENT translation    (#PCDATA)>
<!ELEMENT abbreviation   (#PCDATA)>
<!ATTLIST configurator
          sourceLanguage NMTOKEN #REQUIRED>
<!ATTLIST attribute
          inflection     (yes | no)              #REQUIRED
          treatment      CDATA                   #IMPLIED
          comment        CDATA                   #IMPLIED>
<!ATTLIST value
          lemma          (yes | no)              #REQUIRED
          lexicon (entry | supplement | no)      #REQUIRED
          hint           (yes | no)              #REQUIRED
          comment        CDATA                   #IMPLIED>
<!ATTLIST labelDictionary
          usingLanguage  NMTOKEN                 #REQUIRED>
```

The tag significance can be easily deduced from the previous section.

5 XML Morphological Knowledge

Based on the Inflected POS DTD and XML morphological configurator a linguist can generate the XML morphological knowledge using the particular morphological knowledge (specific to a certain language). This XML morphological knowledge is compatible with a specific DTD. We will present now such an inflected POS DTD named Wordlist that describes the structure of the information about each word from lexicon.

```
<!ELEMENT wordlist       (word+, labelDictionary*)>
<!ELEMENT word           (form, (attribute, value)+)>
<!ELEMENT form           (#PCDATA)>
```

```
<!ELEMENT attribute      (NMTOKEN)>
<!ELEMENT value          (NMTOKEN)>
<!ELEMENT labelDictionary (label)+>
<!ELEMENT label (name, translation, abbreviation)>
<!ELEMENT name           (NMTOKEN)>
<!ELEMENT translation    (#PCDATA)>
<!ELEMENT abbreviation   (#PCDATA)>
<!ATTLIST form
          lexicon (entry | supplement | no) #REQUIRED>
<!ATTLIST wordlist
          source_language NMTOKEN #REQUIRED>
<!ATTLIST attribute
          inflection      (yes | no)              #REQUIRED
          treatment       CDATA                   #IMPLIED
          comment         CDATA                   #IMPLIED>
<!ATTLIST value
          lemma           (yes | no)              #REQUIRED
          lexicon (entry | supplement | no)       #REQUIRED
          hint            (yes | no)              #REQUIRED
          comment         CDATA                   #IMPLIED>
<!ATTLIST labelDictionary
          usingLanguage NMTOKEN                   #REQUIRED>
```

The filling of the XML file with the particular morphological knowledge for each POS is not at all an easy task. It can be realized with a special application that works partially automatic (for the regular rules) and partially manual (for exceptions).

6   LIR Model

Starting from the information that we have in the XML morphological knowledge we can generate different natural language processing models that we will use in different linguistically purposes. We will present here very shortly such a model that we will name LIR because it is based on three main structures: the Lexicon, the Inflection rule collection and the Root collection.

a) The Lexicon contains usually the word lemmas, but also other supplementary forms. We will consider that the lemma is the form of the word on which applying some transformations we will obtain all the other inflected forms of the word. Each lemma will have associated all the corresponding attributes - value pairs. From the previous section results how the lexicon entries can be deduced.

b) The Inflection Rule Collection is a set of rules that explain how each inflected form of the word can be obtained from its lemma. Thus the Inflection Rule Collection will contain a rule for each lemma. Many lemmas can have associated the same rule. Each rule will contain an elementary rule for each inflected form. Each elementary rule contains a set of transformations that applied to the lemma give the inflected form. There are many possibilities to describe these elementary rules like the two level model [9]. We will present here an XML DTD that can be used to describe the structure of the Inflection Rule Collection.

```
<?xml version="1.0" encoding = "UTF-8"?>
<!ELEMENT inflectionRules (rule+, labelDictionary*)>
<!ELEMENT rule (inflection+)>
<!ELEMENT inflection ((attribute, value)+,
                      ((insert | delete | replace | add)+)?)>
<!ELEMENT attribute      (NMTOKEN)>
<!ELEMENT value          (NMTOKEN)>
<!ELEMENT insert         (#PCDATA)>
<!ELEMENT delete         (EMPTY)>
<!ELEMENT replace        (#PCDATA)>
<!ELEMENT add            (what)>
<!ELEMENT what           (word)>
<!ELEMENT word           (#PCDATA)>
<!ELEMENT labelDictionary (label)+>
<!ELEMENT label (name, translation, abbreviation)>
<!ELEMENT name           (NMTOKEN)>
<!ELEMENT translation    (#PCDATA)>
<!ELEMENT abbreviation   (#PCDATA)>
<!ATTLIST inflection_rules
          source_language NMTOKEN           #REQUIRED>
<!ATTLIST insert
          from             (begin | end)    #REQUIRED
          position         CDATA            #REQUIRED>
<!ATTLIST delete
          from             (begin | end)    #REQUIRED
          to               (begin | end)    #IMPLIED
          position         CDATA            #REQUIRED
          length           CDATA            #REQUIRED>
<!ATTLIST replace
          from             (begin | end)    #REQUIRED
          to               (begin | end)    #IMPLIED
          position         CDATA            #REQUIRED
          length           CDATA            #REQUIRED>
<!ATTLIST add
          where            (before | after) #REQUIRED>
<!ATTLIST word
          exclusive        (yes | no)       #REQUIRED>
<!ATTLIST label_dictionary
```

```
using_language   NMTOKEN              #REQUIRED>
```

The transformations described by this DTD consist of:

- The insertion (insert) of some characters after a number (position) of lemma's characters numbered from (from) the beginning (begin) or from the end (end) of the lemma.
- The deletion (delete) of some characters in the lemma. The first position (from) is indicated by the character number (position) numbered starting from the beginning (begin) or the end (end) of the lemma. The second position (to) is indicated also by the number of characters (length) numbered starting from the beginning (begin) or the end (end) of the lemma. If the attribute to is missing, then the attribute length indicates the length of the string to be deleted.
- The replacement (replace) of some characters (the attributes from, to, position, length have the same meaning as in case of delete.)
- The adding (add) of some new words (word). These words can be added (where) before the lemma (before) of after the lemma (after). The exclusive attribute indicates if the corresponding word can be followed (yes) by a non specified sequence or not (no).

c) The Root Collection. By root we understand here a character sequence common for many inflected forms of a word. A root can be defined in many ways. We consider generally that the common root of set of words is the longest common sequence of characters between the words of the set. A word can have many roots if the inflected forms are very different. For example, for the French verb aller we have for indicative present the inflected forms: (je) vais, (tu) vas, (il) va, (nous) allons, (vous) allez, (ils, elles) vont. If we consider a minimum length of a root to one character, then there will be the roots: v- (for the inflected forms vais, vas, va, vont) and all- (for the inflected form allez). If we consider a minimum length of a root to two characters, then there will be the roots: va- (for the inflected forms vais, vas, va), and all- (for the inflected form allez) and vont (for the inflected form vont). The Root collection contains all the roots considered for all the inflected forms that are in the lexicon. For each root there is a pointer to the corresponding word in the lexicon.

In a speller the root collection can be used as follows. We consider different possible roots for the word that we want to check. Each generated root is searched in the root collection. For each found root in the root collection we search the corresponding word in the lexicon. For each different form found in the lexicon we take the inflected rules form the Inflection Form Collection and we generate with these rules the inflected forms that can be generated. If the checked word is found among these generated inflected forms, then the word is a correct form.

In an annotation process of a text we proceed in the same way but we can keep also from the Inflection Form Collection the corresponding AVT trees associated to the inflected forms that match the checked word. We will obtain one or more interpretations of the checked word.

7   Conclusions

The General Model and the LIR Model we presented here are based on three basic data structures: AVT morphological configurator, Morphological Configurator XML DTD and Inflected POS XML DTD. These data structures are conceived to be used in a computer natural Language processing system. In such a system it the linguistic information belongs to a language and the exploitation user interface belongs possibly to another language. We presented also an Inflection Rule DTD that can be used to store the inflection rules and also the corresponding AVT associated with each inflected form. The AVT morphological configurator formalism was introduced in a more general language GRAALAN (Grammar Abstract Language) that are designated to describe the linguistic knowledge used in machine translation systems. A complete AVT morphological configurator for Romanian language (that is a high inflected language) was realized. The Romanian AVT morphological Configurator (version 3) contains more than 800 category nodes and more that 1400 category value nodes.

References

1. Diaconescu, Stefan (2002) Natural Language Understanding using Generative Dependency Grammar, in Proceedings of the twenty-second Annual International Conference of the British Computer Society's Specialist Group on Artificial Intelligence (SGES), (ES2002), Liverpool, UK
2. EAGLES (1996) EAGLES Recommendation on Subcategorisation, EAGLES DOCUMENT EAG-CLWG-SYNLEX, Version of 31st August 1996
3. Extensible Markup Language (XML) 1.0 (Second Edition), W3C Recommendation, 6 October 2000.
4. IANA-LANGCODES (Internet Assigned Numbers Authority) Registry of Language Tags, ed. Keld Simonsen et al. (See http://www.isi.edu/in-notes/iana/assignements/language).
5. IETF RFC 1766 IETF (Internet Engineering Task Force). RFC 1766: Tags for the Identification of Languages, ed. H. Alvestrand. 1995. (See http://www.ietf.org/rfc/rfc1766.txt).
6. ISO 639 (International Organization for Standardization). ISO 639:1988 (E). Code for the representation of names of languages. [Geneva]: International Organization for Standardization, 1988
7. ISO 3166 (International Organization for Standardization). ISO 3166-1:1997 (E). Codes for the representation of names of countries and their subdivisions – Part 1: Country codes [Geneva]: International Organization for Standardization, 1997
8. Koskenniemi, K. (1983) Two-level morphology: A general computational model for word-form recognition and production, Publication No. 11, Department of General Linguistics, University of Helsinki.

Towards a Framework Design of a Retrieval Document System Based on Rhetorical Structure Theory and Cue Phrases

Kamel Haouam[1,2], Ameur Touir[2], and Farhi Marir[1]

[1] Knowledge Management Research Group,School of Informatics & Multimedia Technology, London Metropolitan University (UK), f.marir@londonmet.ac.uk
[2] Computer Science Dept. College of Computer and Information Sciences, King Saud University (SA), {haouam,touir}@ccis.ksu.edu.sa

Abstract. The amount of information available on the Internet is currently growing at an incredible rate. However, the lack of efficient indexing is still a major barrier to effective information retrieval on the web. This paper presents the design of a technique for content-based indexing and retrieval of relevant documents from a large collection of documents such as the Internet. The technique aims at improving the quality of retrieval by capturing the semantics of the documents. It introduces a thematic relationship between parts of text using a linguistics theory called Rhetorical Structure Theory (RST) based on cue phrases to determine the set of rhetorical relations. Once these structures are determined, they can be saved into a database. We can then query that collection using not only keywords, as traditional Information retrieval systems, but also rhetorical relations. The indexing and retrieval technique described in this paper is under development and initial results on a small number of documents have been very successful.

1   Introduction and Previous Work

Day by day, the Internet is becoming more accessible, computers are becoming faster, and memory is becoming cheaper. As a consequence, even more documents are placed on the web. The Internet is currently growing at the rate of 300% per annum and if it maintains its high development growth rate then retrieval of relevant information will become more of a crucial issue than what it is today.

A lot of research has gone into developing retrieval systems on the Web [1],[2],[3]. Despite all that, using current indexing techniques, it has been reliably estimated that on average only 30% of the returned documents are relevant to the user's need, and that 70% of all relevant documents in the collection are never returned [4]. These results are far from ideal considering the user is still presented with thousands of documents pertaining to a keyword query in milliseconds. Existing indexing techniques, mainly used by search engines, are keyword-based. In other words, each document is represented by a set of meaningful terms (also called *descriptors*, *index terms* or *keywords*) that are believed to express its content. These keywords are assigned some weights depending on factors such as their frequency of occurrence (i.e. using Boolean vector based, or probabilistic methods [5], [6], [7]). The major drawback to keyword-based retrieval methods is that they only use a small amount of the information associated with a document as the basis for relevance decisions. As a consequence, irrelevant information that uses a certain word in a different context might be retrieved or information where different words about the desired content are used might be missed. To achieve better performance, more semantic information about the documents needs to be captured. Some attempts at improving the traditional techniques using Natural Language Processing [8], logic [9] and document clustering [10] have offered some improvements.

The aim of the work presented in this paper is to develop a technique that analyses the document for content based indexing and retrieval using a computational and linguistic technique called Rhetorical Structure Theory (RST) [11]. On the basis of the cue phrases, the rhetorical relations between units of text are identified. Once these structures are identified, they can be saved into a database. We can then query that collection using not only keywords, as traditional information retrieval systems (IRs), but also rhetorical relations. The technique will focus on capturing the content of the documents for accurate indexing and retrieval resulting in an enhanced recall. The paper is composed of four sections. The first section is devoted to the introduction and previous work, the second section is devoted to the explanation of the RST using cue phrases and the third section will be devoted to the work which will followed by the conclusion and future work.

2   Rhetorical Structure Theory (RST)

2.1   Background Information

Efficient document structuring goes back as far as Aristotle [13], who recognised that in coherent documents, parts of text can be related in a number of ways. A number of researchers have pursued this idea and developed theories to relate sentences. Amongst these theories, the theory developed by Mann & Thompson, called Rhetorical Structure Theory (RST) has a number of interesting characteristics [11]. It postulates the existence of about twenty-five (25) relations and is based on the view that these relations can be used in a top down recursive manner to relate parts and sub parts of text. RST determines relationships between sentences and through these relationships the term semantics can be captured. In Table 1 we give some of the relationships used in RST. Also, these relations can be identified by cue words in the text. This top down nature means that the documents can be decomposed into sub-units containing coherent sub-parts with their own rhetorical structure, and therefore opens up the possibility of extracting only relevant information from documents. RST is a descriptive theory of a major aspect of organisation of natural text. It is a linguistically useful method for describing texts and characterising their structure. It explains a range of possibilities of structure by comparing various sorts of "building blocks" which can be observed in documents. Using RST, two spans of text (virtually always adjacent, but exceptions can be found) are related such that one of them has a specific role relative to the other. A paradigm case is a claim followed by an evidence for the claim. The claims span a *nucleus* and the evidence spans a *satellite*. The order of spans is not constrained, but there are more likely and less likely orders for all of the relations.

Table 1. Some common relationships between spans

Relation Name	Nucleus	Satellite
Contrast	One alternative	The other alternative
Elaboration	Basic information	Additional information
Background	Text whose understanding is being facilitated	Text for facilitating understanding
Preparation	Text to be presented	Text which prepares the reader to expect and interpret the text to be presented.
Antithesis	Ideas favoured by the author	Ideas disfavoured by the author
Circumstance	Text expressing the events or ideas occurring in the interpretative context	An interpretative context of situation or time
Condition	Action or situation resulting from the occurrence of the conditioning situation	Conditioning situation

Four (4) parameters/constraints are used in describing RST relationships, which are listed as follows:

(i) Constraints on the nucleus
(ii) Constraints on the satellite
(iii) Constraints on the combination of nucleus and satellite
(iv) The effect

Text coherence in RST is assumed to arise due to a set of constraints and an overall effect that are associated with each relation. The constraints operate on the nucleus (N), the satellite (S), and the combination of nucleus and satellite (N+S). A Volitional Cause relation defined in [11] can be seen as a constraint data model that operates on nucleus, satellite, and their combination. Fig. 1 shows how this interaction is defined.

All the rhetorical relations can be assembled into rhetorical structures trees (RS-trees) organized into the five schemas as shown in Fig. 2.

Relation name: Volitional Cause
Constraints on N: presents a volitional action or else a situation that could have arisen from a volitional action
Constraints on S: none
Constraints on the N+S combination: S presents a situation that could have caused the agent of the volitional action in N to perform that action; without the presentation of S, R might not regard the action as motivated or know the particular motivation; N is more central to W's purposes in putting forth the N-S combination than is S.
The effect: R recognizes the situation presented in S as a cause for the volitional action presented in N **Locus of the effect**: N and S

Fig. 1. Definition of the Volitional Cause relation

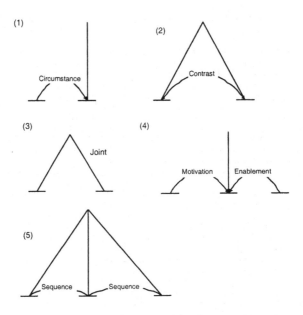

Fig. 2. RST five (5) schemas

Four criteria determine the well formedness of an RST tree [11]:

Completeness: a single tree covers the entire text.
Connectedness: each text span in the text, with the exception of the text span, which covers the entire text, is a node in the tree.
Uniqueness: text spans have a single parent.
Adjacency: only adjacent text spans can be grouped together to form larger text spans.

2.2   Cue Phrases

Cue phrases are words and phrases that connect two or more spans, and add structure to the discourse in text. Example of cue phrases are: "first", "and", "now", "accordingly", "actually", "also", "although", "basically", "because", "but", "essentially", "except", "finally", "first", "further", "generally", "however", "in conclusion", "neither ... nor", "either ... or", "as well as", "rather than", etc. Marcu created a set of more than 450 cue phrases [18]. Also Simon H Corston-Oliver describes a set of linguistic cues that can be identified in a text as evidence of discourse relations [19]. Mann and Thompson recognize that rhetorical relations are often signaled by cue words and phrases, but emphasize that rhetorical relations can still be discerned even in the absence of such cues [11]. That connective can be used in order to determine rhetorical relations that hold between elementary units and between large spans of text. Using only knowledge of cue phrases, an algorithm may be able to hypothesize the rhetorical relations. The relation *Contrast* (see Table 1) can be hypothesized on the basis of the occurrence of the cue word "but", "however", etc. Also, "sometimes", signals an *Elaboration* relation. Table 2 presents a sample set of the cue phrases used in our system to identify rhetorical relations.

Table 2. Set of cues phrases

Contrast	Whereas, but, however
Elaboration	Also, sometimes, usually, for-example
Circumstance	After, before, while
Condition	If, unless, as long as
Cause	Because, since
Concession	Although, without, even-though
Sequence	Until, before, and, later, then
Purpose	In order to, so, that

3   The RSTIndex System

As mentioned earlier, the conventional indexing techniques lack in keyword semantics. In these techniques, both documents and queries are represented by keywords. For retrieval, a similarity computation is performed between the two sets of keywords (of a document and query) and if they are sufficiently similar then the document is retrieved. Since these techniques are keywords-based, therefore, they also retrieve irrelevant documents. Also, these techniques lack of *semantic relationships* between different parts of texts.

We propose a technique that enhance the keyword-based retrieval techniques by capturing the relationships between units (or parts) of texts wherever keywords in text. These relationships are established by using RST. RST has previously been used on text generation [12], but we are using it in our proposed technique for the text indexing. RST provides an analysis of a coherent and carefully written text, and this analysis provides a motivated account of why each element of the text has been included by the author. RST gives an account of textual coherence that is independent of the lexical and grammatical forms of the text. Further investigation on Natural Language Understanding techniques can be found in [15]. These techniques aim at resolving ambiguity and determining the theme of the text in exploring the roles of certain keywords in a text. This could also add a major refinement to the number of documents retrieved resulting in an enhanced retrieval precision.

With the growing number of the documents available to users of the Web and the advance in the Internet technology, more robust and reliable document retrieval systems are needed. There is a growing need to understand the content of a document, compare it with the meanings of the query and select it only if it is relevant. The proposed approach takes into account the semantics, context and the structure of documents instead of considering only keywords (or index terms) for the indexing. This approach creates indices of a document in three phases, i.e., *segmentation, parsing* and *retrieval.* The first phase determines boundaries of elementary unit using cue phrase, the second phase determines the rhetorical relations among the elementary units; and third phase captures the content of the documents for accurate indexing and retrieval result. The working of these three phases is described in details in the next three sections. We have borrowed two documents examples (see Fig. 3) from [18] and these example documents are used as running examples in this paper.

Document 1
John likes sweets. Most of all, John likes ice cream and chocolate. In contrast, Mary likes money and fruits. Especially bananas and strawberries.
Document 2
Because Mary is such a generous man, whenever he is asked for money, he will give whatever he has, for example: he deserves the "Citizen of the year" award.

Fig. 3. Example documents

3.1   Segmentation

Marcu provides a first-order formalization of RST trees, along with an algorithm for constructing all the RST trees compatible with a set of hypothesized rhetorical relations for a text [16]. Main function of the segmentation phase is to identify cue words and phrases that are compatible with various rhetorical relations (see Table 1). This identification of cue words and phrases is performed by a shallow analysis, essentially pattern matching based on regular-expressions. For example, Document 1 (in Fig. 3) can be segmented into four units, as shown below. The elementary units are delimited by square brackets [18].

[John likes sweets.^{A1}][Most of all, John likes ice cream and chocolate.^{B1}] [*In contrast*, Mary likes money and fruits.^{C1}] [*Especially* bananas and strawberries. ^{D1}]

Whereas Document 2 can be broken down into five units, as shown below.

[Because Mary is such a generous man.^{A2}][- whenever he is asked for money,^{B2}] [he will give whatever he has, for example^{C2}][- he deserves the "Citizen of the year" award.^{D2}]

3.2 Parsing

The second phase, after determining the rhetorical relations, performs a computational discourse parser that is used to get the analyses of large text fragments. On the basis of the cue phrases, the principle of the parser consists in positing rhetorical relations between the identified clauses. These rhetorical relations are then used to assemble RST representations [17]. Rhetorical Relations (RR) extracted from the text are expressed as a triplet: *(relation name, span 1, span 2)*.

In Document 1 example, the parser hypothesizes rhetorical relations between the obtained units. *In contrast* signalled a *Contrast* relation. It holds between unit A1 and C1. (for simplicity, we omit the exclusive disjunctive hypothesis presented by Marcu [18]), we obtain the set given below.

RR(CONTRAST,A1,C1)
RR(ELABORATION,B1,A1)
RR(ELABORATION,D1, C1)

For Document 2, the following result is obtained using the principle of the non-contiguous spans of the rhetorical relations as shown in schema 5 of Fig. 2:

RR(EVIDENCE, A2, D2)
RR(CIRCUMSTANCE,B2,C2)
RR(EXAMPLE, C2, A2)

3.3 Retrieval

The previous two sections 3.1 and 3.2 enable us to determine the rhetorical structure of the text. Once these rhetorical relations are built, and saved, they can be easily queried. This allows the user to pose queries using rhetorical relations techniques in addition to the traditional keywords-based retrieval principle. Rhetorical Relation's Query (RRQ) is derived from a query in the same manner that the corpus text is processed. Consider the query:A3 = "is there anything that contradicts Mary?" An RRQ is then derived: RR(Contrast,A3,?). Using the relationship "Relations" (Fig 4), the aim consists of replacing "?" with the right spans that lets the RRQ true.

3.4   The database

The out of the parsing phase is stored in a relational database. Fig. 4 shows the Entity-Relationship Diagram (ERD) of the DB schema.

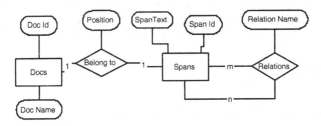

Fig. 4. The ERD of the DB and its schema

Table 3 and Table 4 show the output of the segmentation phase and the Parsing phase, respectively.

Table 3. Spans Text resulting from the segmentation

Span Id	Span Text	Doc Id	Pos.
A1	John likes sweets	Doc. 1	1
B1	Most of all, John likes ice cream and chocolate	Doc. 1	20
C1	In contrast, Mary likes money and fruits.	Doc. 1	69
D1	Especially bananas and strawberries.	Doc. 1	111
A2	Because Mary is such a generous man	Doc. 2	1
B2	Whenever he is asked for money	Doc. 2	38
C2	he will give whatever he has, for example	Doc. 2	69
D2	he deserves the "Citizen of the year" award	Doc. 2	111

3.5   Query Data Model

In our model, a query can be atomic or complex. An atomic query is of the form of an RRQ. A complex query is a sequence and composition of atomic

Table 4. Rhetorical relations resulting from the parsing

Relation Name	Id1	Id2
CONTRAST	A1	C1
ELABORATION	B1	A1
ELABORATION	D1	C1
EVIDENCE	A2	D2
CIRCUMSTANCE	B2	C2
EXAMPLE	C2	A2

queries connected with Boolean operators. An RRQ can be derived either from :

- a keyword–based query: The RRQ will be of the form RR(Any, Span, ?). Suppose that we want to find a document that contains the word " money". The RRQ will be of the form: RR(Any, "money",?). Document 2 will be returned as a result of the query.
- or rhetorical–based query: The RRQ will be of the form of RR(Relation name, Span, ?). If we apply this principle for the query A3 of section 3.3, the RRQ will be of the form: RR(*contrast*, A3,?) The first search of Mary in spans table gives us C1 and A2 as a result; so the keyword Mary exists in two different documents: Document 1 and document 2. Tuning the search, in accessing the table Relations (using a Join operation), allows us to reject A2 and thus Document 2. Consequently, the set of irrelevant document is reduced.

4   Conclusion and Future work

In this paper, we have presented a computational and linguistic technique for content-based indexing of documents in a large collection of text documents. We introduced an indexing method, which is a text description theory called rhetorical structure theory based on cue phrases. RST was created in the late eighties for text generation primarily, we proposed to employ it for text indexing. Applying RST showed potential to improve the quality of current information retrieval systems. A system that indexes relations is able to obtain higher precision than a system that utilizes keywords. The segmentation, the extraction of the rhetorical relations and the storing in the database could involve some pre-processing time during indexing. This time might be saved during the retrieval process. It is an undergoing project as it is at its conceptual level and so far small experiments have been conducted. Future work includes the development of some experiments based on different document topics and sizes. A comparative study on bigger collections will follow.

The study will compare the results of the agent to other retrieval systems including the ones used by search engines.

References

1. Pollitt, A. S, Information Storage and Retrieval Systems. Ellis Horwood Ltd., Chichester, The UK, 1989.
2. Salton Gerard, Automatic Text Processing, Addison-Wesley, USA, 1989.
3. Frants, Valery.I, Shapiro, Jacob and Voiskunskii, Vladimir G, Automated Information Retrieval: theory and Methods, Academic Press, California, 1997.
4. Sparck-Jones, Karen & Willet, Peter, Readings in Information Retrieval. Morgan Kauffman, California, USA, 1997.
5. Liu, G.Z. Semantic vector space model: implementation and evaluation. Journal of the American Society for Information Science, 48(5), 395-417, 1997.
6. Korfhage, R. Information Storage and Retrieval. John Wiley and Sons, London, 1997.
7. Losee, R.M. Comparing boolean and probabilistic information retrieval systems across queries and disciplines, Journal of the American Society for Information Science, 48(2), 143-156, 1997.
8. Smeaton, A.F. Progress in the application of natural language processing to information retrieval, The Computer Journal, 35, 268-278, 1992.
9. Lalmas, Mounia and Bruza, Peter, D. The use of logic in information retrieval modelling. The Knowledge Engineering Review, 13(3), 263-295, 1998.
10. Hagen, Eric, An Information Retrieval System for performing Hierarchical document Clustering, Thesis, Dartmouth college, 1997
11. Mann, W.C., & Thompson, S.A. Rhetorical Structure Theory: Towards a functional theory of text organization. Text , 8 (3). 243-28, 1988.
12. Rosener,D and Stede, M. Customizing RST for the Automatic Production of Technical Manuals. Lecture Notes in AI, 587, 199-214, 1992.
13. Aristotle. The Rhetoric, in W. Rhys Roberts (translator), The Rhetoric and Poetics of Aristotle, Random House, New York, 1954.
14. www.sil.org/linguistics/rst/index.htm
15. Vadera, Sand Meziane, F. From English to Formal Specifications. The Computer Journal, 37(9), 1994
16. Marcu, D. 1996 . Building up Rhetorical Structure Trees. In Proceedings of the Thirteenth National Conference on Artificial Intelligence, vol.2. 1069-1074, Portland, Oregon, August 1996
17. Marcu, D. 1997 a. The rhetorical parsing of natural language texts. In Proceedings of the 35th Annual Meeting of the Association for Computational Linguistics (ACL-97), 96-103.
18. Marcu, D. 2000 The theory and practice of discourse parsing and summarization. In Proceedings of the 35th Annual Meeting of the Association for Computational Linguistics (ACL-97), 96-103.
19. Simon H Corston-Oliver. 1998. "Computing representations of the structure of written discourse". In Technical report MSR-TR-98-15, Microsoft research, Microsoft corporation, One Microsoft way, Redmond, WA 98052. April 3, 1998

Adaptive Translation between User's Vocabulary and Internet Queries

Agnieszka Indyka–Piasecka[1] and Maciej Piasecki[2]

[1] Wrocław University of Technology, Information Science Department
[2] Wrocław University of Technology, Computer Science Department,
   Wybrzeże Wyspia nskiego 27, Wrocław, Poland

Abstract. The paper starts with a short overview on areas of application for user profiles. Subsequently a method to represent user profile in the field of document retrieval by using query terms and weighted terms of retrieved documents is defined. The method is based on evaluating the pertinence of retrieved documents by the user. The created sub–profiles are used to express the translation between terminology used by user and terminology accepted in some field of knowledge. Finally, the preliminary experiments are discussed.

1 Introduction

In today's Internet reality, the common facts are: increasingly growing number of documents in Internet, high frequency of their modifications and, as consequence, the difficulty for users in finding important and valuable information. These problems caused that much attention is paid to helping user in finding important information in nowadays Internet information retrieval systems. Individual characteristic and user's needs are taken under consideration, what leads to system personalization. System personalization is usually achieved by introducing *user model* into the information system. User model might include information about user's preferences and interests, attitudes and goals [3,10], knowledge and beliefs [6], personal characteristics [7], or history of user's interaction with system [13]. User model is also called *user profile* in domain of information retrieval. Profile represents user information needs, such as interests and preferences.

In literature, few types of application of the user profile in the process of information retrieval can be distinguished, below.

- Profile can be used for ranking documents received from the information retrieval system. Such ranking is usually created due to degree of the similarity between the query and a document [8].
- In system of information filtering, profile can be used as a query in the process of information filtering. Such profile represents user information need, relatively stable in time [1].
- There are propositions in the literature to use user profile for query expansion, base on explicit and implicit information obtained from the user [5,16].

The main issue, in the domain of user profile for information retrieval, is the representation of user information needs and interests. Usually user interests are represented as a set of keywords or n-dimentional vector of keywords, where every keyword's weight or position in vector represents importance of keyword in representing user interests [8,11]. The approaches with more sophisticated structures for representing knowledge about user's preferences are also applied:

- stereotypes — the set of characteristics of prototype user of some class of users, sharing the same interests [4], or
- semantic net, which discriminates subject of user interests with underlining the main topic of interests [2].

The approaches to determine user profile can be also divided into few groups. The first group includes the methods where user's interests are stayed explicitly by the user in specially prepared forms or during answering standard questions [4,8], or in the example piece of text, written by the user [12]. The second group can be these approaches, where user profile is based on the analysis of terms frequency in user queries directed to information retrieval system [8]. There is an assumption for these methods that the interest of the user, represented by a term, is higher as the term is more frequent in the user query. Analysis of the queries with the use of genetic algorithms [14] or semantic nets [2] are the extension to this approach. The third group of approaches includes methods, where the user evaluates documents retrieved by the system. From documents assessed as interesting (pertinent) by the user, additional index terms, describing user interests, are added to the user profile [4,8].

Most of the research in domain of user modelling for information retrieval considers only user information needs stayed explicitly by the user using the information retrieval system. The user difficulty in precise expressing the real information need is frequently neglected. In other words, the fact is ignored that the user usually does not know, which words he should use to formulate his interests to receive valuable documents from information retrieval system. We claim that user can express user's preferences by valuation of relevance of retrieved documents.

The purpose of this paper is to present the user profile, which represents the translation between the terminology used by the user and the terminology accepted in some field of knowledge. This translation is supposed to describe the meaning of words used by the user in context fixed by pertinent documents.

2   User profile

Information retrieval system is defined in this paper by four elements: set of documents D, user profiles P, set of queries Q and set of terms in the dictionary T. There is retrieval function $\omega : Q \rightarrow 2^D$, for each $q \in Q\,\omega(q) \subset D$. Retrieval function returns the set of documents, which is the answer for the query q. The set T contains terms from documents, which have been indexed in the retrieval system (for WWW — in search engine). Set T is called dictionary, where terms are ordered by index $i = 1, \ldots, n$. User profile is an object p from set P, where P is set of all possible user's profiles. Profile p is described by function π, which maps: user's query q, set of retrieved documents and previous user profile, into a new user profile. Thus, the profile is the following structure determined by function $\pi : \pi(q, D_{qPer}, p_{n-1}) = p_n$. Function π is responsible for profile modifications. The function π is valid for arguments (q, D_{qPer}, p_{n-1}) and returns profile p_n, where q is the question, p_{n-1} and p_n are profiles before and after modification respectively, $D_{qPer} = \omega(q)$ is the set of the pertinent documents among the documents retrieved for query q, $D_{qPer} \subseteq D$. For user profile we define also the set of *user subprofiles* $SubP$ (see below).

User profile is created on the base of information received from the user after *user verification of documents* retrieved by the system. During verification user points out these documents which he considers pertinent for him.

User query pattern s_j we call a Boolean statement, the same as user query $q : s_j = r_{j_1} \wedge r_{j_2} \wedge \ldots \wedge r_{j_n}$, where r_{i_j} is a term $t_{i_j} \in T$ or negated term $(\neg t_{i_j})^1$, and for $j = 1, 2, \ldots, n$ $i_j < i_l$ if $j < l$. User query pattern s_j indicates subprofile and is connected with only one subprofile.

User subprofile $sp \in SubP$ we call n-dimensional vector of weight of terms from pertinent documents: $sp = (w_{j,1}^{(k_i)}, w_{j,2}^{(k_i)}, w_{j,3}^{(k_i)}, \ldots, w_{j,n}^{(k_i)})$, where $SubP$ is the set of subprofiles.

User profile $p \in P$ we define as the following structure:

$$p = (s_1\ s_2\ \ldots\ s_l) \begin{pmatrix} (w_{1,1}^{(k_1)}, w_{1,2}^{(k_1)}, w_{1,3}^{(k_1)}, \ldots, w_{1,n}^{(k_1)}) \\ (w_{2,1}^{(k_2)}, w_{2,2}^{(k_2)}, w_{2,3}^{(k_2)}, \ldots, w_{2,n}^{(k_2)}) \\ \cdots\cdots\cdots\cdots\cdots\cdots\cdots \\ (w_{l,1}^{(k_m)}, w_{l,2}^{(k_m)}, w_{l,3}^{(k_m)}, \ldots, w_{l,n}^{(k_m)}) \end{pmatrix}$$

where: $n \in N$ — number of terms in dictionary T, $w_{j,i}^{(k)}$ — weight of significant term t_i^z in user profile (the weight is calculated according to the frequency of term t_i^z in pertinent documents retrieved by the system in k-th retrieval and the frequency of this term in all collection documents, called *cue validity*) and the number of modifications made so far for the user profile ($k - 1$ modifications) are also respected, s_j — user query pattern, $(w_{j,1}^{(k_i)}, w_{j,2}^{(k_i)}, w_{j,3}^{(k_i)}, \ldots, w_{j,n}^{(k_1)})$ — user subprofile (user query pattern indicates one user subprofile univocally).

---

[1] Further instead of designation t_{i_j} we will use symbol t_i. Double indexes were used for underlining the order in the set T.

Position of weight $w_{j,i}^{(k)}$ in subprofile (its co–ordinate in vector of subprofile) indicates the significant term $t_i^z \in T$. There is an order introduced for the set T. Terms from dictionary T are the indexing terms, that index documents retrieved for the query q_i and those terms belong to those pertinent documents.

Weight of term t_i in profile is calculated according to the following formula, proposed in [9]:

$$w_{j,i}^{(k)} = \frac{1}{n+1}(n\,w_{j,i}^{(k-1)} + z_i) \tag{1}$$

where: n — number of retrieval of documents made so far for this subprofile ($n = k - 1$), $w_{j,i}^{(k)}$ — weight of significant term t_i^z in profile after k-th modification of subprofile, which is indicated by the pattern s_j (i.e. after k-th document retrieval with use of this subprofile), z_i — weight of significant term t_i^z in k-th selection of these terms.

3   Modification of user profile

User profile expresses the translation between terminology used by user and terminology accepted in some field of knowledge. This translation describes the meaning of word used by user in context fixed by pertinent documents and it is described by assigning to the user's query pattern s_j a subprofile ('translation') created during the process of selection of significant terms t_i^z from pertinent documents. We assume following designations: q_i — i-th user query, $D_q^{(i)}$ — set of documents retrieved for user query q_i, $D_q^{(i)} \subseteq D$, $D_{qPer}^{(i)}$ — set of documents pointed by the user as pertinent documents among documents retrieved for user query q_i, $D_{qPer}^{(i)} \subseteq D_q^{(i)}$.

As it was described above, user profile p_n is the representation of user query q_i, set of pertinent documents $D_{qPer}^{(i)}$ and previous (former) user profile p_{n-1}. After every retrieval and documents verification made by user, the profile is modified. The modification is performed according to following procedure: $p_1 = \pi(q_1, D_{qPer}^{(1)}, p_0)$, $p_n = \pi(q_n, D_{qPer}^{(n)}, p_{n-1})$, where p_0 — initial profile, in which weights of all significant terms t_i^z are set to value 0, p_1 — profile created after the first time query was asked and the analysis of pertinent documents was made, p_n — profile after n-th time the same query was asked and the analysis of pertinent documents was made.

Traditionally user profile is represented by one n — dimension vector of terms describing user interests. User interests change, and so should the profile. Usually changes of profile are achieved by modifications of weights of terms in vector. After appearance of queries from various domains, modifications made for this profile can lead to unpredictable state of the profile. By unpredictable state we mean disproportional increase in vector representing the profile the weights of some terms, what could not be connected with increase of user interests in domain represented by these terms. The weights of terms can grow, because of high frequency of these terms in whole collection of documents, regardless of domain of actual retrieval.

Representation of user profile as one vector could also cause ambiguity during the use of this profile for query modification. At certain moment user query refers only to one domain of user's interests. To use user profile mentioned above for current query modification we need a mechanism of choosing from vector of terms representing various user' interests only these terms that are connected with domain of current query. To obtain this information, usually knowledge about relationship between terms from query and profile, and between terms in profile is needed. In literature, this information is obtained from co–occurrence matrix created for collection of documents [15] or from semantic net [11]. One of disadvantages of the presented approaches is that the two structures, namely user profile and structure representing term dependencies, should be maintain and manage for each user. The other is that creating the structure representing term relationships is difficult for so diverging and frequently changing environment as Internet.

There are no such problems for user profile p created in this paper. After each retrieval, only weighs of terms from user subprofile identified by pattern s_j (identical as user's query) are modified, not weighs of all terms from user profile. Similarly, when profile is used to modify user's query, direct translation between current user query q_i and significant terms from domain connected with the query is used. In user profile p, existing mapping between one user query pattern s_j and one subprofile represents this translation.

In information retrieval system user profile is created during a period of time — during sequence of retrievals. There could appear a problem how many subprofiles should be kept in user profile. We have decided that only subprofiles that are frequently used for query modifications should not be deleted. If subprofile is frequently used, it is important for representing user's interests.

Modification of user subprofile sp is made always when from the set of pertinent documents pointed out from retrieved documents by the user the significant terms t_i^z are determined. In the appropriate subprofile modifications of weights are made only for these terms. Modification of user subprofile equals to actualisation of weight $w_{j,i}^{(k)}$ of term t_i^z in subprofile identified by user query pattern s_j. Weight of term t_i^z is calculated according to formula (1). After each process of retrieval modification takes place in one subprofile for all significant terms t_i^z obtain during k-th selection of these significant terms from pertinent documents retrieved for query q_j, which was asked k-th time. If the modification took place for significant terms t_i^z in whole user profile, it would cause disfigurement of importance of significant term for single question.

4 Application of user profile

User profile contains terms selected from pertinent documents. These terms are good discriminators distinguishing pertinent documents among other documents of collection and these terms represent whole set of pertinent documents.

Application of user profile p is performed during each retrieval for user query q. One of the main problems is selection of significant terms t_i^z for query modification. Not all significant terms in subprofile will be appropriate to modify next user's query (in this paper, understood as "to replace"), because the query becomes to long.

If user asked new query q_j to the retrieval system, new pattern s_j and subprofile identified by this pattern are added to the profile. Subprofile is determined after analysis of pertinent documents. If user asks the next query and this query is the same as the previous query q_j the given query is modified basing on user profile. Modified query is asked to information retrieval system, retrieved documents are verified by user and subprofile in user profile is brought up to date. After each next use of the same query as query q_j, subprofile identified by pattern s_j better represents user's interests described at beginning by the query q_j. Each next retrieval, with use of subprofile identified by pattern s_j, leads to query narrowing, decrease in the number of retrieved documents, increase in the number of pertinent documents.

User profile can be used for query modification if pattern s_j existing in profile is identical to current query q_i or *similar* to current query q_i. For example for queries: $q_a = t_1 \wedge t_2 \wedge t_3 \wedge t_4$, $q_b = t_1 \wedge \neg t_2$, patterns: $s_1 = t_1 \wedge t_2 \wedge t_3 \wedge t_4$, $s_2 = t_1 \wedge \neg t_2$ are identical to queries q_a, q_b, respectively, and patterns: $s_2 = t_2 \wedge t_4$, $s_4 = t_1 \wedge t_2$, $s_5 = t_1 \wedge t_3$, $s_6 = t_2$ are similar to query q_a.

If pattern s_j is identical to current user query q_i, current user query q_i is replaced by r_1 best significant terms t_i^z from subprofile identified by pattern s_j. If in user profile there are few patterns that are similar to current user query q_i, all significant terms t_i^z from all subprofiles identified by these patterns are taking under consideration. The weights of all significant terms t_i^z from subprofiles identified by similar patterns are summing. The n-dimensional vector of $R = (r_1, r_2, \ldots, r_n)$ is created. Coordinates of vector R are ordered by weight, not by order of dictionary T. The ranking of all these significant terms is made and r_2 best significant terms, which weights are over $\tau_{profile}$ threshold, replace current user query q_i. Parameter r_1, r_2 and other parameters for choose best significant terms have been set experimentally.

Second situation with replacement of query in case of similar patterns has a name *retrieval hypothesis*. We formulate here a hypothesis that if user query patterns are similar to query, the significant terms t_i^z from subprofiles identified by these patterns have the same sense as terms used by user in current query and thereby could be appropriate terms to replace current user query.

5   Experiments

User profile was implemented as part of Web search engine. The user profile is used as the mechanism for personalisation of retrieval process. Personalisation is performed by the query modifications, which appear during information retrieval. Modification of user query takes place as a result of the analysis of user interaction with search engine (i.e. documents verification). During interaction with the user, system automatically asks the modified query to the search engine and presents the answer.

The experiments are forked into two directions. In the first, preliminary case, the aim is to prove the usefulness of proposed profile in a test environment, where the simulation of retrieval process is arranged. In the second case — to verify the usefulness during retrievals evaluated by users.

The first case should show that for any field of knowledge the profile converges. It means that starting from any random query, the proposed analyse of set of pertinent (for the user) documents, the selection method of important terms, and the methods of profile creation and query modification will lead to the set of pertinent documents. In the test environment the sets of pertinent documents and the set of random queries were established. For every random query one experiment was made. The query was asked to the search engine. The retrieval process was run. If in the answer there were pertinent documents, the random query was modified — the significant terms (from pertinent documents that were found) replaced the random query. The modified query was automatically asked to the search engine and next pertinent documents were found. Each stage of the described cyclic process is called *iteration*. The iterations were repeated until all pertinent documents from the set were found.

The main problem in experiment was to determine the values for parameters described in the preceding chapter. These parameters decide whether or not all pertinent documents will be found. Initially for every random query, the values of parameters were constant during whole experiment. But in some cases it was observed that there is no improvement of retrieval for some random queries, in the same iteration number. We noticed that the parameters should change according to several factors. There are for instance: number of pertinent documents in the answer (one, two or more), number of iteration (the number of repetition for the given pertinent document in earlier retrievals), length of the pertinent document (number of terms). The heuristics H were proposed to solve this problem.

The percent of pertinent documents retrieved at every iteration is the measure of improvement of proposed method. The experiments without and with heuristics were considered. The results are presented in Table 1.

Table 1. Average percent (%) of retrieved pertinent documents in the subsequent iterations.

Iteration	1	2	3	4	...	11	...	50
without H	15,63	25,53	49,26	44,40	...	52,73	...	52,73
with H	27,76	50,00	100,00	100,00	...	100,00	...	100,00

6 Discussion

User profile presented in this paper is a new approach to representation of user's interests and preferences. By introducing structure of user query patterns and user subprofiles the translation between terminology used by the user and terminology accepted in some field of knowledge is described. The preliminary experiments are encouraging. User of WWW search engine receives support during query formulation, even in the cost of 'hidden iterations' of searching process. The query is modified in such a way that, for more cases of retrievals, in next retrievals user will receive set of retrieved documents which is smaller and consists of better documents. In order to verify finally the usefulness of presented user profile, more in–depth experiments need to be done and, especially, the experiments from the second case in which real users will take part in retrieving and assessing the documents.

References

1. Ambrosini L., Cirillo V., Micarelli A. (1997) A Hybrid Architecture for User-Adapted Information Filtering on the World Wide Web. Proc. of the 6th International Conference on User Modelling UM'97, Sardinia, Springer Wien New York.
2. Asnicar F., Tasso C. (1997) ifWeb: a Prototype of User Model-Based Intelligent Agent for Document Filtering and Navigation in the World Wide Web. Proc.of the Workshop "Adaptive Systems and User Modeling on the World Wide Web" 6th International Conference on User Modelling, Sardinia.
3. Billsus D., Pazzani M. (1999) A Hybrid User Model for News Story Classification. Proc. of the 7th International Conference on User Modeling, UM'99, Banff, Canada, pg. 99–108.
4. Benaki E., Karkaletsis A., Spyropoulos D. (1997) User Modeling in WWW: the UMIE Prototype. Proc. of the Workshop "Adaptive Systems and User Modeling on the World Wide Web, 6th International Conference on User Modelling UM'97, Sardinia.
5. Bhatia S.J. (1992) Selection of Search Terms Based on User Profile, Communication of the ACM.
6. Bull S. (1997) See Yourself Write: A Simple Student Model to Make Students Think, Proc. of the 6th International Conference on User Modeling, UM'97, Sardinia, pg. 315–326, Springer Wien New York.
7. Collins J.A., Greer J.E., Kumar V.S., McCalla G.I., Meagher P., Tkatch R. (1997): Inspectable User Models for Just–In Time Workplace Training, Proc.

of the 6th International Conference on User Modeling, UM'97, Sardinia, pg. 327–338, Springer Wien New York.

8. Daniłowicz Cz. (1994) Modelling of user preferences and needs in Boolean retrieval systems. Information Processing and Management, vol. **30**, no. **3**.

9. Daniłowicz Cz. (1998) Reprezentacja preferencji użytkownika końcowego w modelach informacyjnych agentów. I Krajowa Konferencja: Multimedialne i Sieciowe Systemy Informacyjne, Wrocław.

10. Daniłowicz Cz. (2000) Możliwości i problemy wyszukiwania informacji w otwartym systemie WWW, Technical Report no. 27, Wrocław University of Technology.

11. Davies N. J., Weeks R., Revett M. C. (1997) Information Agents for World Wide Web. In H. S. Nwana, N. Azarni (Eds.) *Software Agents and Soft Computing*, Springer.

12. Jeapes, B. (1996) Neural Intelligent Agents. Online & CDROM Review, vol. **20** no. **5**.

13. Maglio P.P., Barrett R. (1997) How to Build Modeling Agents to Support Web Searchers. Proc. of the 6th International Conference on User Modeling, UM'97, Sardinia, pg. 5–16, Springer Wien New York.

14. Moukas A., Zachatia G. (1997) Evolving a Multi–agent Information Filtering Solution in Amalthaea. Proc. of the Conference on Agents, Agents'97, ACM Press.

15. Qiu Y. (1996) Automatic Query Expansion Based on a Similarity Thesaurus. PhD. Thesis, Swiss Federal University of Technology, Zurich, Swiss.

16. Seo Y.W., Zhang B.T. (2000) A Reinforcement Learning Agent for Personalised Information Filtering. Proceedings of the 2000 International Conference on the Intelligent User Interfaces, New Orleans, LA USA, ACM Press, January 9–12, 2000, str. 248–251.

Aspect Assignment in a Knowledge-based English-Polish Machine Translation System

Anna Kupść

[1] Polish Academy of Sciences, Institute of Computer Science, 21, Ordona, 01-237 Warszawa, Poland

[2] Carnegie Mellon University, Language Technologies Institute, 5000, Forbes Ave., Newell-Simon Hall, 4527, Pittsburgh, PA 15213-3891, USA

Abstract. The paper presents a prototype of a knowledge-based English-Polish machine translation (MT) system, focusing on issues related to aspect assignment. We propose a set of heuristic rules based on interlingua (IR) representation provided by the system. Our method obtains 78%–88% accuracy of aspect assignment on test data, which in the absence of a true semantic component, is a very encouraging result.

1   Introduction

This paper presents a prototype of a knowledge-based English-Polish machine translation (MT) system, focusing on issues related to aspect assignment. Although there is no agreement among linguists as to its precise definition, aspect is pretty much a semantic category which strongly affects overall text understanding. Aspect is a result of a complex interplay of verb semantics, tense, mood and pragmatic context but in English it is not overtly marked on a verb. On the other hand, in Polish, aspect is overtly manifested and incorporated into verb morphology. This difference between the two languages makes English-Polish translation particularly difficult as it requires contextual and semantic analysis of the English input in order to derive an aspect value for the Polish output.

The MT system presented in this paper takes advantage of a knowledge-based interlingua (IR) representation in order to assign aspect in Polish translation. In particular, we define a set of heuristic rules based on this representation. Aspect assignment in our system does not refer to verb semantics. Lexical semantic databases for Polish, such as the one reported in [3], are not publicly available, whereas developing such a database of one's own is time-consuming. Of course, lack of such a semantic component does not result in the optimal system performance. However, our account offers a reasonable trade-off between development of an expensive and time-consuming semantic-based approach and the current behaviour of publicly available commercial MT systems where aspect assignment is highly neglected.

2 System description

The English-Polish MT project presented in this paper is an extension of the existing multilingual KANTOO system (a reimplementation of the KANT system, cf. [7], [6]) developed at Carnegie Mellon University. KANTOO is a knowledge-based, domain-specific MT system (in the English-Polish MT project, the domain is restricted to printer manuals) and it uses Interlingua (IR) as a semantic representation, see [5]. The system takes as an input a text written in constrained English (controlled language), which limits vocabulary and grammar of sentences accepted by the system, cf. [4]. Example (1) presents a sample English input along with the IR representation and its Polish translation provided by the system.

(1)   The printer prints pages.

```
(*A-PRINT
  (agent
    (*O-PRINTER
      (number singular)
      (reference definite)))
  (argument-class agent+theme)
  (mood declarative)
  (punctuation period)
  (tense present)
  (theme
    (*O-PAGE
      (number plural)
      (reference no-reference))))
```

    Drukarka drukuje strony.

The Polish generation module, which we are most interested in here, consists of four components: a mapper, a unification grammar (a type of context-free grammar), a morphological generator and a post-processing module. Mapping rules transform the IR semantic representation of the source text into a syntactic structure corresponding to the output in the target language. The structure is a functional structure or FS in the LFG (Lexical Functional Grammar) formalism, cf. [1]. Generation grammar rules convert this FS into a list of lexical tokens (FS frames), which are then fed to the morphology module responsible for generating appropriate inflected forms. Finally, a set of post-processing rules is applied to produce the resulting surface form of translation by cleaning up spacing, capitalization, inserting punctuation, etc. In order to develop the current system, a small corpus of about 280 English sentences from a printer manual has been examined. This corpus served as a baseline to develop heuristic rules presented in the paper.

3   Aspect Assignment

As mentioned above, aspectual marking is incorporated into verb morphology in Polish. Polish verbs may have two aspect forms: imperfective, e.g., *drukuje* 'prints', or perfective, e.g., *wydrukuje* 'will print$_{3.sg}$ (out)'. Aspect is independent of tense or mood as it is also present on infinitives: *drukować* 'to print$_{imperf}$' and *wydrukować* 'to print$_{perf}$ (out)', or on gerunds: *drukowanie* 'printing$_{imperf}$' and *wydrukowanie* 'printing$_{perf}$ (out)'.

Since English does not have morphological aspect, we consider lexical concepts corresponding to English verbs ambiguous, i.e., they can be translated by either a perfective or an imperfective verb, see (2).

```
(2)   *A-PRINT = ([?verb] drukowa/c-v)
      drukowa/c-v = (*OR* ((morph verb-imperf) (root drukuje))
                          ((morph verb-perf) (root wydrukuje))),
```

The appropriate aspect value is chosen according to several ordered context-sensitive heuristic rules specified in the mapper module. This means that the aspect value, introduced into FS, is based on the interlingua specification. Aspect assignment rules proposed in the system are discussed below.

3.1   Declarative Mood

For finite verbs in declarative mood, aspect assignment is primarily based on tense specification. In the system, all continuous forms, marked as (progressive +) in IR, are translated as imperfective. Next, forms of perfective tenses, i.e., (perfective +), are translated as perfective. Then, verbs in simple past or future simple tenses (marked as (tense past) or (tense future) in IR) are translated as perfective. Similar assignment rules have been independently proposed in [2].

Additionally, we assume that certain types of subordinate conjunctions, e.g., 'while', 'once', 'before', etc., impose aspect requirements on a verb in the subordinate clause. The following assignments have been proposed:

- 'while': imperfective
  (3)   You can send an electronic fax **while** the printer makes copies.
        Można   wysłać   elektroniczny   faks,   podczas gdy   drukarka
        can   send$_{inf}$ electronic   fax   while   printer
        robi   kopie.
        makes$_{imperf}$ copies
- 'once', 'after', 'before', 'in order to', 'until': perfective; additionally clauses introduced by conjunction 'until' have to be negated in Polish
  (4)   Jobs also queue and wait **until** another job finishes.
         Zadania także ustawiają się   w kolejce i   czekają, dopóki
         jobs   also   stand   REFL   in queue and   wait   until
         inne   zadanie nie skończy   się.
         another job   not finishes$_{perf}$ REFL

- 'by'+gerund: imperfective; such clauses are translated into Polish by a contemporary adverbial participle derived only from imperfective verbs, see [9]

(5)   Close the document **by selecting** <content> Close </content> from the <content> File </content> menu.

  Zamknij   dokument   wybierając   <content>   Zamknij
  close   document   selecting$_{imperf}$ <content>   Close
  </content> z   menu <content> Plik </content>.
  </content> from menu <content> File </content>

If none of the above cases hold, we assume that aspect of present tense verbs is imperfective. This assignment is valid also for gerunds as they are represented in IR as present tense verbs with an additional feature (nominal +).

This specification, however, is not unproblematic. Let us consider the two sentences below:

(6)   If it prints properly, then adjust the contrast **before** copying.
  Jeśli drukuje się   właściwie, dostosuj kontrast przed kopiowaniem.
  if   prints   REFL properly   adjust   contrast before copying$_{imperf}$

(7)   Allow alcohol to dry completely **before** closing the printer and plugging in the power cord.
  Przed zamknięciem drukarki i   włączeniem   przewodu pozwól
  before closing$_{perf}$ printer   and plugging$_{perf}$ cord   allow
  alkoholowi całkowicie wyschnąć.
  alcohol   completely dry

In (6) and (7), the preposition 'before' (not the subordinate conjunction discussed above) has a gerund complement. However, gerunds in the two sentences differ in aspect value: 'copying' in (6) appears in imperfective, whereas perfective is used in (7). This difference is triggered by distinct semantics of the two phrases. In (6), 'copying' indicates a continuous action, whereas 'closing' and 'plugging in' in (7) are integrated short actions. As discussed at length in [2], whenever an event / situation is a distinct, integrated whole perfective aspect is used; otherwise aspect is imperfective. The current system is not able to capture the semantic difference between (6) and (7) and imperfective is assigned in both cases.

3.2   Imperative Mood

We assume that in imperative mood, aspect depends on negation. This observation is confirmed by a brief analysis of Polish technical documentation. We have examined 3 short instructions for changing ink cartridge, available from http://www.balta.pl/katalog/file/ (SC005FC3.jpg (Doc1), SH649FC3(51625).jpg (Doc2), SH625FC3.jpg (Doc3)). The documents contained, respectively, 32, 37 and 28 imperative verbs. The summary of obtained results is presented in Fig. 1.

ASPECT	NEGATED				NON-NEGATED			
	Doc1	Doc2	Doc3	**All**	Doc1	Doc2	Doc3	**All**
imperfective	6.25%	5.4%	3.57%	**5.15%**	9.38%	21.62%	14.29%	15.46%
perfective	3.12%	0%	0%	1.03%	81.25%	72.97%	82.14%	**78.35%**

Fig. 1. Statistics for imperative forms in a sample of Polish technical documentation

As these results indicate, in technical documentation, negated imperatives more often appear with imperfective forms (5.15%), whereas perfective aspect prevails with non-negated imperatives (78.35%).

Heuristic rules used in the system conform with the above statistics: we translate non-negated imperatives as perfective, (8a), and negated imperatives as imperfective verbs, (8b).

(8) a. Print a test page.
         Wydrukuj stronę próbną.
         print$_{perf}$ page test
     b. Do not move the lever after the scanner has begun sending the page.
         Nie przesuwaj dźwigni, gdy skaner zaczął wysyłanie strony.
         not move$_{imperf}$ lever   when scanner started sending   page

The above rules, void of verb semantics, (9), and pragmatic context, (10), are not perfect:

(9) Keep in mind that this technology is not perfect.
     Pamiętaj,   że ta technika   nie jest doskonała.
     remember$_{imperf}$ that this technology not is   perfect
(10) (When you are using the OCR software, follow these guidelines in order to obtain the best possible results:)
     Process only pages that have crisp, clear text.
     Przetwarzaj tylko strony, które mają czytelny, wyraźny tekst.
     process$_{imperf}$ only pages which have crisp   clear   text

Both (9) and (10) require imperfective rather than perfective aspect as would be assigned by our rule. In (9), imperfective is correctly obtained by the system because 'keep in mind' is considered a fixed phrase and, hence, exceptionally translated by an imperfective verb. In (10), however, no such assumption can be made. In this case, aspect is triggered by the context in the previous sentence: since guidelines usually refer to general repetitive actions, imperfective rather than perfective is used in (10).

These examples further indicate that semantic information (meaning of 'keep in mind' and 'guideline') is necessary in order to correctly resolve aspect in Polish translations. Since such information is not available in KANTOO, the system cannot assign the correct imperfective form in (10) and the perfective form *przetwórz* is used instead.

3.3 Infinitives

Infinitives have no mood or tense specified and we need separate rules to resolve aspect of these forms. In general, English infinitives appear as either complements of other verbs, e.g., modals, or they head infinitive clauses introduced by a conjunction such as 'in order to'. We assume that in the former case, aspect of the infinitive depends on the governing verb while in the latter — on the subordinate conjunction.

Only some verbs such as 'start', 'finish' or 'continue' provide a clear aspect specification for their infinitival complement and require an imperfective verb in Polish. Other verbs can occur with both forms and the choice of aspect on their complements depends on the situation / event semantics, see [2]. Since such a deep semantic representation is unavailable in the IR, we propose several tentative heuristic rules based on behaviour of infinitives in the English corpus.

For the conjunction 'in order to', we assume that it requires a perfective infinitive argument (see § 3.1). Aspect assignment for complements of modals is more complex.

Modal verbs are represented in IR by a set of semantic features such as `ability`, `possibility`, `tentativity`, `necessity`, `obligation`, see [5]. The following aspect assignment has been adopted for infinitive complements of modals:

- 'can': (`ability +`) or (`possibility +`) → perfective
- 'cannot': (`ability +`) (`negation +`) → imperfective
- 'cannot': (`possibility +`) (`negation +`) → perfective
- 'could': (`possibility +`) (`tentativity medium`) → perfective;
- 'may': (`possibility +`) (`tentativity low`) → imperfective;
- 'must': (`obligation medium`) → perfective;
- 'must': (`necessity +`) → imperfective;
- 'should': (`expectation +`) → perfective;

Again, this method is not flawless as the choice of aspect strongly depends on semantic context. Let us consider the following example:

(11) (This product is specifically designed to allow you to do many tasks simultaneously.)
   For example, you can print a document while you send a fax.
   a. Na przykład, można drukować   dokument podczas wysyłania
      on example   can   print$_{imperf}$ document while   sending
      faksu.
      fax
   b. Na przykład, można wydrukować dokument podczas wysyłania
      on example   can   print$_{perf}$ document while   sending
      faksu.
      fax

If taken out of context, both (11a) and (11b) can be translations of the English sentence: the former is focused more on simultaneity of the two tasks whereas the latter indicates that one task is completed while the other is performed. If we take into account the previous context, however, the imperfective form in (11a) is preferred by native speakers. This contextual information is unavailable to the system and, hence, (11b) is generated instead.

4   Results

This section presents quantitative results obtained by our current system. Accuracy of aspect assignment in the system has been compared with human translations of the corresponding English texts. Results obtained on the training corpus (280 sentences) are presented in Fig. 2.

	TRAINING		HELD-OUT	
	#	%	#	%
correct	430	88.1%	53	88.3%
incorrect	58	11.9%	7	11.7%

Fig. 2. Accuracy on the training and held-out data from the same manual

The heuristic aspect assignment rules have been developed in order to accommodate data in the training corpus. Therefore, results obtained on this corpus are likely to be biased by the data. In order to obtain a more objective verification of the proposed rules, we tested the system performance on two sets of unseen data. The first held-out set contained 24 sentences from the same manual. The results obtained on this corpus are very similar to those achieved for the training corpus, see Fig. 2.

Following a suggestion of one of the reviewers, we also checked performance of the system on a different manual. The aim of this second test was to limit a potential bias introduced by the training manual. Since no other printer manual written in the controlled language was accessible, we decided to take any related text available on-line and re-write it to the controlled language using the tool described in [8]. Time constraints for preparing the final version of the paper disallowed any substantial extension of the lexicon and/or syntactic coverage of the Polish module. Hence, we focused our attention on short texts, such as instructions or leaflets available on-line. The text which served for the present task was taken from http://www.praxis.pl/html/main_instrukcje.html, instruction SC005FC3 for changing ink cartridge. The text contained 14 English sentences with their translations in 7 languages, including Polish. However, we did not rely on those translations as most of them were quite indirect or incomplete and therefore difficult to compare with an automatic translation.

Most of the sentences in the instruction (11 out of 14; 78.57%) had to be re-written to comply with the syntax and domain representation accepted by KANTOO. After pre-processing, 12 sentences successfully obtained IR representation and were transferred to the Polish generation module. Performance of aspect assignment rules on this set is presented in Fig. 3. Obviously, the

	#	%
correct	18	78.26%
incorrect	5	21.74%

Fig. 3. Accuracy on the additional held-out corpus

results of this test are lower (about 10%) than those obtained on the text taken from the manual the system has been trained on. The results in Fig. 3, however, have to be considered as very preliminary. First, the test sample was too small (23 verbs) to draw any definite conclusions. Second, the test was made without any adjustments of the Polish grammar which would be necessary to ensure full translations. Taking into account these imperfections, the present results are quite encouraging.

5   Conclusions

In this paper we discussed aspect assignment in a prototype of an English-Polish MT system. The system is designed to translate domain-specific technical texts written in a controlled language. KANTOO uses a relatively rich interlingua representation, which allows us to quite successfully deal with the complex issue of aspect assignment. High accuracy of the proposed method (78%–88%) is a very encouraging result as for an approach which does not rely on semantics.

The success of the system has to be partially credited to the fact that KANTOO is a domain-specific system. The controlled language also simplifies the task by restricting the scope of phenomena accepted by KANTOO. It would be interesting to explore how our heuristic rules perform in a different domain or in a general purpose (open-domain, unconstrained language) MT system. The preliminary results of testing the system on an initially unconstrained text, Fig. 3, give hope that at least some of the rules could be carried over to other systems. Also, in the absence of a strong semantic component, data-driven statistic rules offer an attractive alternative to text analysis. Hence, if the current rules turn out not to be directly applicable, statistic techniques can be applied to improve performance (e.g., aspect assignment on infinitives) and to provide further generalisations. This is left as a topic for further study.

References

1. Bresnan, J., editor (1982). *The Mental Representation of Grammatical Relations.* MIT Press Series on Cognitive Theory and Mental Representation. The MIT Press, Cambridge, MA.
2. Gawrońska, B. (1993). *An MT Oriented Model of Aspect and Article Semantics.* Lund University Press.
3. Gawrońska, B. (2001). PolVerbNet: an experimental database for Polish verbs. In *Proceedings of EAMT Summit VIII, 2001.*
4. Kamprath, C., Adolphson, E., Mitamura, T., and Nyberg, E. (1998). Controlled language for multilingual document production: Experience with Caterpillar technical English. In *Proceedings of the Second International Workshop on Controlled Language Applications (CLAW '98).* Available from http://www.lti.cs.cmu.edu/Research/Kant/claw98ck.pdf.
5. Leavitt, J. R., Lonsdale, D. W., and Franz, A. M. (1994). A reasoned interlingua for knowledge-based machine translation. In *Proceedings of CSCSI-94.* Avalibale from: http://www.lti.cs.cmu.edu/Research/Kant/.
6. Mitamura, T. and Nyberg, E. (1992). The KANT system: Fast, accurate, high-quality translation in practical domains. In *Proceedings of COLING-92.*
7. Mitamura, T., Nyberg, E., and Carbonell, J. (1991). An efficient interlingua translation system for multi-lingual document production. In *Proceedings of the Third Machine Translation Summit.*
8. Mitamura, T., Nyberg, E., Baker, K., Svoboda, D., Torrejon, E., and Duggan, M. (2001). The KANTOO MT system: Controlled language checker and knowledge maintenance tool. In *Proceedings of NAACL 2001 (Demonstration).*
9. Saloni, Z. and Świdziński, M. (1985). *Składnia Współczesnego Języka Polskiego.* Państwowe Wydawnictwo Naukowe, Warszawa, 2nd edition.

An Approach to Rapid Development of Machine Translation System for Internet

Marek Łabuzek[1] and Maciej Piasecki[1]

Wrocław University of Technology, Computer Science Department,
Wybrzeże Wyspiańskiego 27, Wrocław, Poland

Abstract. The paper presents methods applied in the construction of a bi–directional Polish-English machine translation system. Since the system was created as a wide-scale commercial product, such aspects as speed of translation, ungrammatical texts and time and cost of developing the system are of great importance. An attempt of solving these problems by use of Machine Learning techniques in parsing and tagging is discussed.

1 Introduction

The inspiration for the work presented here has come from the problems encountered during implementation of a wide–scale commercial machine translation (MT) system, mainly designed for fully automatic rough translation of Web pages. The system has been finally developed by *Techland* company, and its main 'engine' has been used for the first time in a product called *Internet Translator*. Because the domain of translation is practically unlimited, it has been impossible to apply typical MT methods, based on corpus analysis, detailed grammar construction (sometimes semantic representation, as well) and 'deep' parsing. The paper presents the results of investigations into methods and techniques facilitating the relatively fast and economical (regarding resource consumption) development of the intended system. The paper is focused mainly on one, better developed, direction of translation, namely English to Polish.

The directions of the investigations have been determined by the following assumptions. According to the intended use in the dynamic area of Internet, the data sets of the system must be exhaustive and easy to extend. Because the commercial system was developed, it was impossible to use most publicly available tools e.g. Charniak parser [2]. Because of the assumption of limited resources, especially time, the construction of the shallow parser has been based on application of *Machine Learning* methods in as large extent as possible. Here, the starting point was the work of Hermjakob [4] on parser and lexical transfer both based on inductive learning (details below), which gives first result with very few learning examples and seemed to be suitable for incremental construction of a parser. However, he assumed availability of a dictionary of rich, hand coded, semantic information, which is unrealistic for large scale *MT* systems. Because the shallow parsing often produces a partial description of the syntactic structure, very often including mistakes, the transfer process has been distributed among many local 'subroutines'.

The constructed system has the typical architecture of the *MT* system based on transfer with the following subsequent stages of processing (described in the following sections):

1. *Text segmentation* based on grammar (*Finite States Automata*).
2. *Morphological analysis.*
3. *Part of Speech Tagging* (PST).
4. *Parsing* based in large extent on inductive learning of 'parsing strategy'.
5. *Transfer:* implemented in the form of a set of rules.
6. *Target syntactic structure synthesis.*
7. *Word form generation.*

2   Text Segmentation

The text is first segmented into blocks, which are words, punctuation marks, numbers and others. Each kind of block is described by a regular expression, which is compiled into a finite state automaton. The text is given as input to all automatons and the one which accepts the longest part of the text is chosen. It also returns the position of the last accepted character and searching for a next block starts with the successor of this character. For each block, the information about its format (font size and type, colour) is stored. It is used to preserve the layout of the text. The blocks most difficult to describe are abbreviations with periods and various symbols. The former have to be listed one by one for each language (to be properly segmented and translated). The latter are further divided into subcategories (e.g. Internet addresses, dates, Roman digits) and for each there is one or more regular expressions describing it. After obtaining each block, it is checked whether it terminates the sentence. It is determined by one hand-crafted finite state automaton. It deals with balanced delimiters (parenthesis, quotes). The very hard problem is period not terminating a sentence (e.g. after ordinal numbers, like in "1." or in abbreviation at the end of a sentence, where the period plays two roles: being part of an abbreviation and terminating a sentence). A set of rules which check one word before and after a period was created.

3   Morphological Analysis

The morphological analysis unit uses a monolingual dictionary (MDic) storing all known words together with morphosyntactic information. It will be called a compressed layer. To present the information in the lexicon to other parts of the system in a consistent way, a second universal layer is created on demand for given words. Unknown words are treated simply as nouns. In the case of

Polish it would be possible to try to guess some morphological information for unknown words but it has not been implemented yet.

The compressed layer consists of three parts: inflection, lexical information and derivations. The inflection part of the MDic is implemented as a transducer (based on simplified Daciuk's [3] algorithm and dedicated techniques of binary compression of files) translating between a word form and a pair: an identifier of a lexeme and a code of inflection form. The lexical information part is a simple table of compressed values of lexical aspects. The derivation part of the MDic stores links between lexemes.

The universal layer describes words in a hierarchical way. One syntactic element, describing one word, consists of a word form and a set of syntactical alternatives. A syntactic alternative consists of an identifier of lexeme, a code of basic syntactic category and a set of morphological alternatives. And a morphological alternative is a set of pairs: attribute and value, where attribute can be either inflectional or lexical. An example of the representation is presented below (from the Polish morphological dictionary):

```
Surface: "chodzenie"
(    Syntactic Category: ODS-NOUN
     Semantic Class: 14060
     ( PERSON: F-THIRD-P, CASE: F-NOM, NUMBER:
            F-SING, GENDER: F-NEUT, NEG: F-NEG-N )
     ( PERSON: F-THIRD-P, CASE: F-ACC, NUMBER: F-SING,
            GENDER: F-NEUT, NEG: F-NEG-N )
     ( PERSON: F-THIRD-P, CASE: F-VOC, NUMBER: F-SING,
            GENDER: F-NEUT, NEG: F-NEG-N ) )
```

A set of *basic syntactic categories* (BSC) was proposed. The set is extended in comparison to a typical list of Polish *parts of speech* and the categories from the set are also a part of the grammar of the parser. All syntactic categories are organised into hierarchy by explicitly defined *subsumption* relations, e.g.:

```
NOUN: ODS-NOUN, PN, PRON
PRON: PER-PRON, PRON-NPER, PRON-ZPR,PRON-ZWR,PRON-NEG,PRON-DEM.
```

In the example NOUN has three subcategories: deverbal noun, proper noun and pronoun. PRON, in turn, has six subcategories: personal pronoun, indefinite pronoun, interrogative pronoun and others. The subsumption relations determine the set of morpho-syntactic attributes assigned to the categories. Some categories of the higher levels are motivated by the correspondence between Polish and English grammar or contitute semantic subcategories. The subsumption relation is used in machine learning algorithms.

4   Part of Speech Tagging

The initial version of the system did not have a tagger — it was implicitly included in the parser. However, problems with the quality of the parser, made it necessary to look for improvements by the introduction of a tagger. This late decision have made a lot of problems with adaptation of the purchased *Penn Tree Bank* (*PTB*), which had been chosen as the base for the construction of the tagger. The system of syntactic categories of the parser, being close to the one proposed in *XTAG* [8], had been defined before purchasing *PTB*. It was necessary to convert syntactic categories of *PTB* to the assumed standard. The problem was so serious that a sophisticated expert system had to be constructed to perform the task, which it could not completely do. The patterns of subcategorisation of multiword verbs and phrasal verbs concerning the tagging of words as prepositions and adverbs are not explicitly given in *PTB*. Moreover, they seem to be very unstable across the corpus.

PTB includes also a lot of mistakes of different types strongly influencing the final quality of the tagger. There are quite a big number of ambiguities left in *PTB* tags, at least two different historical versions of the system of taggs can be met (e.g. word "to" earlier tagged as TO, now having two different tags) and, finally, many simple mistakes (e.g. pronouns tagged as determiners and vice versa). Besides construction of the expert system, in order to correct mistakes, a lot of manual disambiguations and corrections had to be done (up to 1.5% of all words of *PTB*) resulting in, at most, a half of *PTB* being usable.

The operation of the constructed English tagger is combination of the purely statistical, *Hidden Markov Model* based solution in initial phase and Brill's tagger associated with hand coded rules in the main phase. The overall accuracy of the tagger is almost 97%. The achieved result meets the current good standards but the tagger still makes about one mistake in each sentence, which is a serious problem for the parser.

The construction of the Polish tagger appeared to be a much more difficult task, mainly because of the lack of big, annotated corpus of Polish. Some activities have been undertaken in order to build the Polish corpus. Annotation of the corpus gave also a good opportunity to eliminate mistakes from the monolingual dictionary together with the introduction of many new lexemes of 'internet jargon' (what is positive according to the typical area of application of the system). Nevertheless, the present size of the corpus of about 65.000 tagged and corrected words appeared to be too small for the construction of the tagger. The first estimation gave the size of the full tag set to be about 1600 different tags. The initial experiments with a statistical tagger (similar to the English one) resulted in about 86% accuracy counted in a standard way (in relation to all words), but the percentage of mistakes among the ambiguous words was very high. It seems that the better solution would be introduction of more hand-crafted, disambiguating rules.

5   Parsing

Parsing is divided into the three subsequent phases: preprocessing (identification of some 'constant' phrases), main parsing (based on inductive learning) and 'the correction' (during which an attempt to correct some obvious mistakes made by the main parser is undertaken).

During preprocessing, a sequence of analysed blocks is checked to contain simple phrases (mostly idioms), words and phrases from a *user dictionary*, words for which user changed translations and syntactic alternatives with small probability (e.g. "take" as noun, "father" as verb). A user dictionary stores words and phrases together with their syntactical category and translations which were entered by a user of the system. This makes possible to properly translate words and phrases not known by the system, change translation of known words or enforce their syntactical category. The simple phrases are changed to one block with attributes properly set, and information about their translation is remembered. For words from a user dictionary and words with a changed translation, the proper syntactic alternative is chosen and information about a translation is also remembered. Finally, morphological alternatives with small probability are simply removed.

The main parser is based on the architecture proposed in [4]. It follows the general shift-reduce scheme but uses the hierarchical structure of *decision trees* instead of a control table. The decision trees are constructed by the application of a version of C4.5 algorithm of inductive learning. Additionally Hermjakob introduced heuristically optimising techniques for construction of decision trees. There is no one big tree but a structure of trees. He introduced *decision structure* built up hierarchically of decision trees and *decision structure lists*. In the list each decision structure is responsible for parse actions belonging to some similarity class e.g. shift actions. The similarity classes are defined manually by predicates which map parse actions to true or false. In the decision structure, if the action compatible with the predicate can be undertaken, then it is done. And in opposite case, the next decision structure in the list is activated.

The learning set includes pairs consisting of a vector of values of *features* and a parsing *action*, which was performed. The feature values in each case describe partially the state of the parser (the stack and the input list) in which the given action was performed. The elements of the input list can be ambiguous in respect of their syntactic category, all other can be ambiguous only in respect of morphological attributes.

In comparison with [4], the set of actions has been reduced to only four main types. Two of them are 'standard': *shift* and *reduce*. However restrictions put on reduce are weak. It can be applied to many arguments, not necessarily located on the top of the stack (even changing their order). The 'non–standard' *add into* action is similar to reduce, but it can insert one node into any place in the structure of the other. The action of creating gaps produces an 'empty copy' of some node (i.e. a co–indexed clone of it) e.g.:

1. S I-EN-HAVE
2. R (-3 -1) TO VP AS MOD PRED AT -2
3. A (-2 -1) TO (NP -1 BEFORE -2) AS CONJ COMPL

The action 1 shifts specific lexeme to the stack (it must be one of alternatives). The action 2 reduces elements of the stack on positions 3 and 1 (counting from the top) to VP node, assigning to these elements appropriate roles. The result is put into position 2 of the stack. The action 3 adds two elements (2 and 1) into the existing tree. It is identified as the first NP-tree below the position 2 of the stack.

The parser cannot backtrack but has a limited possibility of sending elements back from the stack to the input list by means of the special shift-out action.

The features express such rich syntactic information as:

- values of morphological attributes such as number, gender, verb form etc. (some of them ambiguous in most of parse nodes),
- details of the structure of nodes e.g. presence of some branch described by the syntactic (and semantic) role or values of attributes of some role filler,
- possible agreement in values of attributes between some nodes,
- *matching*: between subcategorisation pattern of some node and a possible argument (the feature returns syntactic role or category of the filler).

The last type is based on detailed subcategorisation dictionary (SCD) and, in the case of ambiguity, a ranking of patterns is heuristically calculated. Some features from the feature vector for English parser are:

1. `synt of -3 at verb`
2. `np-vp-match of 1 with 2`
3. `syntrole of vp -1 of -2`
4. `morphp of f-ger of mod of -1`

The feature 1 gets the syntactical category of an element of the stack at position 3 from the range of immediate subcategories of the verb category. The feature 2 checks whether elements on positions 1 and 2 on input list (counting from the left) match syntactically as subject and verb predicate of a sentence. The feature 3 checks whether an element on position 2 on the stack matches the first not filled argument of some subcategorisation pattern of the first VP node on the stack. The feature 4 returns true or false depending on whether the value of the appropriate attribute of the subnode with role `mod` of the node on the position 1 equals to `f-ger`.

The features expressing the proximity of matching together with the features expressing semantic matching, based on relatively rich semantic information (semantic class of the lexem and semantic role according to the matching of subcategorisation), decided about the reported in [4] good quality of parsing of sentences from *Wall Street Journal* corpus. However, the creation of the detailed, hand-coded semantic dictionary for unlimited domain of Web pages is impossible. The semantic information used in the parser is reduced only to some classes based on *WordNet* categories of location, time etc. The only sophisticated features are features based on syntactic SCD. The size of SCD for unlimited domain must be relatively very large[1]. Unfortunately, increasing number of ambiguities is correlated with decreasing quality of heuristic matching. Entries in SCD are tree structures with distinguished leaf (signed PRED) node. The identifier of a lexeme kept in the PRED node is used as an index for retrieval. Besides the structure, each tree describes several requested elements like *subtrees, syntactic roles*, requested values of morphological attributes (e.g. _G annotating the case) and *specific lexemes* (extending the key for retrieval). The English subcategorisation patterns has been based on *XTAG* solutions where the structure and the description of the trees have been little simplified (according to the implicit grammar assumed in the system). The Polish subcategorisation dictionary, based on [6], codes additional information concerning the optional elements and groups of optional elements (where at least one element of the group must be realised) and sequences with fixed order e.g.:

SNT{SUBJ NP_N} {PRED VP{strzec PRED VERB} {się MOD PART}
{OBJ NP_G}}
SNT{SUBJ NP_N} {PRED VP{dawać PRED VERB}{IOBJ NP_D} {OBJ NP_A}}}

The decision structure is built on the base of prepared examples. However, the initial expectations that there would be no necessity to create a detailed grammar of the source language, appeared to be false. The practice of teaching showed with the increasing number of examples[2] the probability of inconsistency is rapidly increasing. Each inconsistency in decisions made during the teaching causes formation of a 'strange rule' e.g. the parsing action of an object to VP can be activated by the presence of an adverb in some remote position on the stack. Obviously, such strange associations result from the large number of features (in case of English, 135 with the tagger, above 256 without the tagger) in comparison to the number of examples. In case of inconsistency in a learning dataset the resulting 'strange rules' cannot be eliminated by the machine learning algorithm itself. From the point of view of learning algorithm it has to deal with to different learning cases.

But negative proportion between examples and features can be hardly changed. The preparation of one example of the parsing of the sentence containing about 30 words takes more than 0.5 hour by an experienced person. Anyway, the number of features can not be further reduced. Some features are important only for some specific constructions. But in case of inconsistency, the parser tries to solve the problem using any part of the vector of features. In that way, the teacher must constantly remember what is the 'range of view' of the parser i.e. the window of ±3 elements of the input list and the stack. The feature selection is very difficult and probably the best way is by empirical reduction of them. The initial fast improvement in syntax covering

---

[1] The present state is more than 18 000 entries in English SCD.

[2] More than 2000 different English sentences, where even the parsing of the sentences consisting of several words can include more than 30 actions

of the parser, slows down quickly. Thus, monitoring the constructions being presented in the parser and preserving consistency makes it necessary to develop some kind of the grammar.

For this purpose a special tool was created that performs all parsing actions stored in a learning set and checks whether they conform to a set of constraints. It helps to find all possible situations of inconsistencies.

According to [4], the parser never backtracks. It always tries to find the solution in a single run and it always constructs only one solution - the best one, at least in theory. But such idealistic assumption appeared to be very hard to maintain in the case of an unlimited domain, having only very simplified semantic information. As the result, the parser trying to find the best analysis, makes mistakes and stops in any case of unknown combinations of feature values. Very often the stops are the result of a wrong parsing decision, regarding some ambiguous construction, undertaken somewhere earlier.

A mechanism of 'pushing forward' by a special shift action has been introduced. The positive is that after 'pushing' some parts of the sentence, the words following the problematic construction can be analysed properly. Moreover, the parser is extremely fast (several sentences per second on PC). It consumes much less time than other parts of the MT system!

In the first view, the quality of the parser is far from being satisfactory. In the test assessed by the human operator, only 147 on 325 test sentences (typical 'textbook' examples) were parsed without errors. But, from the other point of view, a lot of remaining sentences had large, well parsed fragments and included errors that have not influenced the overall result of the translation. The latter was possible because of the existence of special *reconstruction procedures* included in the rules of transfer. Each rule of transfer is associated with some type of the source language constructs.

Original, Hermjakob's version of the parser implicitly includes a POS tagger in its decision structure: nodes on the input list are ambiguous according to their categories and the shift action has to choose between the alternatives. The quality of this implicit tagger is comparable to 'stand alone' taggers ($\tilde{9}5\%$) but it needs a lot of additional learning features. The application of the tagger (described earlier) before parser allowed for significant reduction of the features ($256 \rightarrow 135$), which resulted in identification of many inconsistencies and elimination of many 'strange rules'. Anyway, the relatively good quality of the tagger has brought a little improvement in the parser: 97% of accuracy still means that there is almost one mistake in each sentence!

Because the final state of the parser's stack contains very often not a single tree of complete analysis, it was necessary to introduce a special, simplified, correcting parsing. It is based on some kind of expert system with powerful rules which try to recognise some more obvious mistakes and join all partial structures into one tree. The last operation facilitates transfer in assigning the proper case to arguments of the verb.

6   Transfer

The main goals of the transfer are to transform a tree created by a parser to a tree, which has rough target language structure, and to translate words and phrases constituting the tree. It is implemented as recursive functions assigned to the particular syntactic categories. The functions start from the root of the tree and continue their work, mainly in depth-first style, towards lower and lower parts of the tree. In this phase some typical parser errors can also be corrected.

The bilingual dictionary used by transfer is compiled form texts files, containing word and phrases annotated with part of speech and other information necessary for identifying lexemes and forms of words. The file is compiled to binary files: phrase bases, which store phrases in the form of parse trees and a proper bilingual dictionary, which stores pairs of identifiers of words and phrases. Currently, we have 125,000 translations in the English–Polish dictionary (not counting virtual entries for derivations, which are translated through verbs).

A lot of entries for both English and Polish monolingual dictionary have been taken from texts from Internet. These texts were first extracted from HTML, then converted to lists of words. The words in the list is then checked for the presence in dictionary at a current state. Next a special tool facilitates manual marking of the new words so that they provide necessary information for the generation of binary data. The acquired entries in monolingual dictionary were then given translations and added to bilingual dictionaries.

Translation of a lexeme only on the basis of a bilingual dictionary is very often ambiguous or simply wrong. The results can be significantly improved when we use the subcategorisation context during translation. A bilingual subcategorisation dictionary suits these needs. Moreover, during the transfer we should not only translate a lexeme on the basis of its subcategorisation but also we should transform, during the transfer, the whole structure described by the tree from the dictionary.

The bilingual subcategorisation dictionary associates pairs of the trees and delivers additional information controlling the transfer e.g. the information defining the corresponding pairs of arguments, marking arguments to be deleted or controlling the process of transformation of the source argument into the target in case when their categories differ significantly. A special tool with a graphical interface has been created in order to facilitate the process of definition of the entries in the bilingual subcategorisation dictionary. Presently, the subcategorisation dictionary contains mainly the verb trees (probably the most important between all other types for the parsing). But also, there are some trees describing adverbial constructions, some compound adverbs, prepositions and conjunctions. An important part of the dictionary, constantly growing, is formed by trees describing *multiword idioms*.

7 Target Language Generation

In this phase a tree produced by the transfer is modified to fully conform to target language syntax. The modifications are of three kinds: adding function words like particles (Polish reflexive "się" or English "do"–support), correcting the order of words (e.g. in questions) and setting morphological attributes. The last task is the most complicated: it must not only take into account values set by transfer but also all agreements which especially in Polish a very plentiful and complicated. The values set by transfer are usually propagated to the proper child nodes of a given tree while in case of agreements, morphological values of a main word are firstly raised to the root of the phrase and then propagated to proper child nodes. On the base of values of attributes set in leaves of the whole tree, target word forms are generated.

8 Further Development

The system is constantly developed. Various works are being conducted. Some of them concern dictionaries and the transfer rules are also improved. New solutions for parsing are being sought, too. One of the biggest problems is to find the proper balance between the usage of Machine Learning techniques (MLT) and hand coding. Here important are such aspects like resources both human and linguistic necessary to develop the parser, the quality and speed of parsing and the easiness of improving it. It seems that for a wide–scale translation system MLT are indispensable but they should be carefully designed and supported with significant amount of hand-crafted knowledge.

References

1. Bień J.S. (1991) Koncepcja słownikowej informacji morfologicznej i jej komputerowej weryfikacji. Wyd. UW, Warszawa.
2. Charniak E. (2000) A Maximum-Entropy-Inspired Parser. Proceedings of NAACL-2000.
3. Daciuk J. (1998) Incremental Construction of Finite-State Automata and Transducers, and their Use in the Natural Language Processing. PhD thesis, Technical University of Gdańsk.
4. Hermjakob U. (1997) Learning Parse and Translation Decisions From Examples With Rich Context PhD dissertation. University of Texas, Austin.
5. (1999) Nowy słownik poprawnej polszczyzny ed. Markowski A., PWN, Warszawa.
6. (1984) Słownik syntaktyczno–generatywny czasowników polskich. ed. Polański K., Instytut Języka Polskiego PAN, Kraków.
7. Saloni Z., Świdziński M. (1998) Składnia współczesnego języka polskiego. PWN, Warszawa.
8. (1999) A Lexicalized Tree Adjoining Grammar for English. The XTag Group of Institute for Research in Cognitive Science, University of Pennsylvania. www.cis.upenn.edu/~xtag

Hierarchical Clustering of Text Corpora Using Suffix Trees

Irmina Masłowska and Roman Słowiński

Institute of Computing Science, Poznań University of Technology
Piotrowo 3A, 60-965 Poznań, Poland

Abstract. We present a novel method for hierarchical clustering of text corpora, which proves especially suitable for online clustering. *Information overload* – the current phenomenon in electronic document repositories and the Internet in particular – constitutes an unceasing challenge for researchers. Clustering has been proposed as a comprehensive information access method. We describe a system, which automatically builds a navigable hierarchy of meaningful document groups. We claim that our system addresses two chief needs of the Web users: the need for efficient access to the up-to-date information on every available topic and the need for an organized and meaningful presentation of the desired information.

1 Introduction

Information overload is a salient feature of the present-day Internet – a resource, which over the decades has become one of the largest and most diverse sources of information. The dynamic, untamed growth of this resource has long ago lead to a point where it is unthinkable to handle the available information manually.

Various services for supporting the information retrieval process have emerged. They could be very roughly categorized as alterations of two opposite approaches: manually crafted catalogues and fully automatic search engines. The pros and cons of the two approaches become apparent to anyone who had a chance to use them extensively.

The main advantage of manually crafted catalogues, such as e.g. Yahoo (www.yahoo.com), is that an army of truly intelligent agents – humans – organize documents into a hierarchy, grouping adequate ones together under a meaningful description. Paradoxically, the same aspect may sometimes prove to be a disadvantage as human decision maker is always subjective, thus one person's adequate document assignment can seem inadequate for another. The chief disadvantage of those catalogues, however, is that they are chronically obsolete – not being able to keep up with the speed at which new documents, and new topics appear.

On the contrary, the mechanisms behind automatic search engines, like e.g. Google (www.google.com), enable them to stay relatively up-to-date, thus making them the top choice for most Internet users. Their strength, however, can also turn into a weakness because of the overwhelming number of the returned documents. Presenting such results in an automatically created ranked list, without providing the user with appropriate navigation tools can turn the process of finding adequate information into a tedious job.

Our aim is to propose a method that marries the benefits of the two opposite approaches, offering a robust clustering technique which produces a navigable hierarchy of document groups delivered with meaningful, yet concise (one phrase) descriptions. At the same time, it shall be fully automatic and independent of any predefined categories, thus being able to keep the pace with the document resources growth.

2   State of the art in the field

The little popularity of traditional clustering in today's search support can be explained with the fact that for years it had been extensively investigated (see [22] for a review) for the purpose of *cluster search*, i.e. improving document search by pre-clustering the entire corpus according to the *cluster hypothesis*, which states that similar documents tend to be relevant to the same queries [15]. It seems rather obvious that such methods could not prove effective when the corpus in mind is the whole Internet.

Our study follows a different approach, where clustering is considered a comprehensive method for accessing information [4] and is applied as a post-retrieval browsing technique [1,6,8,10,23].

The probably still most recognized algorithm for hierarchical document clustering is the Agglomerative Hierarchical Clustering (AHC) which uses a similarity function to iteratively consider all pairs of clusters build so far, and merge the two most similar clusters into a single one, which thus becomes a node in the *dendrogram* (the tree-like hierarchy). Typically, the processed documents – constituting the bottom level of the hierarchy – are represented as sparse TF-IDF (*term frequency inverse document frequency*) vectors which unfortunately ignore the dependencies existing between words in the documents [11]. When comparing two documents the *cosine measure* usually serves as the similarity function [17]. When comparing two clusters, a *single-linkage* method is preferred (which defines the similarity as the maximum similarity between any two individuals, each taken from one of the considered clusters), mainly because of its relatively low complexity $O(n^2)$, n being the number of documents [20].

The commonly recognized shortcomings of the AHC procedure, especially when applied to texts, are: its sensitivity to a stopping criterion (usually being the desired number of clusters [12]), its greedy manner of choosing the clusters to be merged, and its unsuitability for clustering data of considerably big size [2]. Another limitation is the partitioning character of the method (i.e. one document cannot be placed in more than one branch of the dendrogram – a postulate described thoroughly in the next section.

Scatter/Gather [4,6] is a more recent, AHC-based clustering method tailored specially for dealing with large text corpora. Its authors, however, managed to address only the time-complexity aspect of the original AHC algorithm. Due to operating on small random samples of the documents Scatter/Gather achieves the complexity $O(kn)$, k being the desired number of clusters. One can agree with the authors claiming that "in an interactive session it is vital for the clustering algorithm to run as quickly as possible, even at the expense of some accuracy" [4].

An interesting probabilistic framework for the unsupervised hierarchical clustering of large-scale sparse high-dimensional data collections is proposed in [19]. Unfortunately, its authors failed to provide it with the means for giving readable descriptions of the created clusters. Moreover, as it is typical for probabilistic approaches, they use the *bag-of-words* representation of the processed documents, which naturally looses important information about the original texts.

Several other, quite recently introduced clustering methods [3,9,13,14] are designed for building a flat partition of documents, and it is not straightforward to adapt them for hierarchical clustering without deteriorating their running times.

Yet another interesting clustering algorithm for text documents called Suffix Tree Clustering (STC) has been introduced in [23] and adapted for the Polish language in [21]. STC not only fulfills the postulate of forming readably described clusters, but also scales well to text corpora of considerable size due to its excellent time complexity $O(n)$. Its disadvantage is that it is not suitable for hierarchical clustering.

In the next section of this paper we describe aims and scope of our work, then we propose an adequate algorithm which uses the basic mechanisms of STC for building a hierarchy of documents clusters, finally we present an implementation of the proposed methodology and give our conclusions on its performance on Web documents.

3   Aims and scope

The shortcomings of the philosophy behind traditional clustering algorithms (developed in the field of data mining) when applied to text documents can be identified in the very definition of clustering.

Definition 1

Clustering is a process of unsupervised partitioning of a set of objects into meaningful groups called clusters, where the similarity of objects belonging to the same cluster should be significantly stronger than the similarity of objects belonging to different clusters.

The assumption in the above definition is that an object cannot belong to more than one cluster at a time. This assumption proves inadequate when faced with a diversity of text documents, especially ones coming from such an unsystematized source as the Web. It seems rather natural that a document can be classified as belonging to more than one thematic category, as e.g. a text about French cuisine could be placed as well in a "Cooking" category as in a "French culture" category (see [5]).

Although the multi-membership is allowed in *fuzzy clustering*, we preferred to avoid the problematic quantification of membership degree in multiple categories, thus proposing an algorithm specially dedicated to processing text documents, aimed at producing a navigable hierarchy of meaningful groups of text documents. By *meaningful* we understand groups labeled with a short (one word or a phrase) description corresponding to the documents topic. Such a method applied to results of a search engine should be able to support the user, who could at a glance eliminate groups of uninteresting documents and focus on the most promising branch of the hierarchy, perhaps drilling down to narrow the search even more.

4   Proposed methodology

We adapt the STC algorithm introduced in [23] by Zamir and Etzioni who construct a *suffix tree* for the considered corpus of documents. By using a suffix tree they are able to uncover groups of documents, which share a common phrase (an ordered sequence of words). It must be stressed that such data structure can be constructed in $O(n)$ [18]. We find the concept of grouping documents based on their common content very appealing and adopt this step of STC in our method.

We believe that sharing the same phrase (excluding phrases that are too commonly used to be topic-specific) is a strong similarity criterion for text documents. As it is stated in [7]:

"We have every right to believe that the vocabulary used by an individual expresses de facto her/his preferences towards certain real-life aspects. Therefore, if we are to measure the similarity of text documents with the intersection of the words used in them, we have a great chance that the documents will in fact be grouped by topic".

Having identified all *base-clusters*, i.e. groups of documents which share a phrase, the authors of the STC algorithm assign a score to each of the groups. The score is a function of the number of documents the base-cluster contains, and the number of words (excluding stop-words) that make up its phrase. Only some arbitrarily chosen number k of top scoring base clusters is then processed further. Due to this simplification the processing time can be kept as low as it is required for online applications.

As it can be easily noticed, some base-clusters may overlap (as for the documents contained) or even be identical with one another. This is due to the fact that documents may share more than one phrase. The STC algorithm performs further merging of overlapping clusters based on a similarity measure calculated for each pair of base-clusters. The similarity measure

$Sim(B_r, B_s)$ for two base-clusters B_r and B_s, with sizes $\mid B_r \mid$ and $\mid B_s \mid$ respectively, is defined as follows [23]:

Definition 2

$\mid B_r \cap B_s \mid / \mid B_r \mid > \alpha \wedge \mid B_r \cap B_s \mid / \mid B_s \mid > \alpha \Leftrightarrow Sim(B_r, B_s) = 1$, otherwise $Sim(B_r, B_s) = 0$, where α is an arbitrarily chosen merge threshold.

Base-clusters are merged to form final clusters using the equivalent of a single-link clustering algorithm where a predetermined minimal similarity (equal to 1) serves as the halting criterion. Naturally, by using different values for the merge threshold α one can expect to obtain few distinct clusters or a lot of similar ones. Thus, although it is claimed in [23] that the performance of STC is not very sensitive to this threshold, empirical evaluation performed in [21] has shown that its sensitivity is highly related to the number of base-clusters.

The Hierarchical Suffix Tree Clustering algorithm (called hereafter HSTC) proposed by the authors is based on two observations:

- The similarity measure defined in Def. 2 yields a similarity relation between pairs of base-clusters, which in the original STC algorithm is treated as an equivalence relation. However, it does not fulfill all the requirements for an equivalence relation, namely it is not transitive.
- A set of documents grouped in one base-clusters can be a subset of another base-cluster, just as a more topic-specific group of documents can be a subgroup of a more general group.

The second observation adds a special merit to our approach. It implies that by identifying the base-clusters in the considered corpus we potentially could concurrently identify the nodes of the target dendrogram. The only thing left to do is to arrange the nodes appropriately in a hierarchic structure. This is done by applying the HSTC algorithm, which builds an oriented graph whose vertices represent identified base-clusters and arcs represent existence of an inclusion relation between pairs of clusters.

Before proceeding to the algorithm we formulate two useful definitions.

Definition 3

Kernel of an oriented graph $G = (V, A)$, composed of a set of vertices V and a set of arcs A, is a set K of vertices $K \subseteq V$, which fulfills the following two stability conditions [16]:

- *Internal stability* – no two vertices belonging to the kernel of the graph are connected by an arc:
  $\forall v_i, v_j \in K \Rightarrow a_{ij} \notin A$;
- *External stability* – for every vertex from outside the kernel there exists at least one arc from a kernel vertex to the considered vertex:
  $\forall v_l \in V \setminus K \Rightarrow \exists a_{kl} \in A: v_k \in K$

Definition 4

C_i *includes* C_j (denoted by $C_i \to C_j$) $\Leftrightarrow | C_i \cap C_j| / |C_j| >= \alpha$,
C_i, C_j are clusters of documents, and $\alpha \in (0.5; 1]$ is an inclusion threshold.

Setting the inclusion threshold to a sample value of 0.9 implies that at least 90% of a cluster's documents should be contained in another cluster to decide that the former is included in the latter. Obviously such defined inclusion is a fuzzy concept unless the inclusion threshold is set to 1.

The outline of the HSTC algorithm is as follows:

- input: a set of text documents after standard preprocessing (stemming, identifying stop-words),
- output: a hierarchy of groups containing somehow related documents, each group identified by a (set of) phrase(s) characterizing the shared content of the documents.

1. Using the STC technique create set B containing base-clusters of documents, each identified by a phrase.
2. Merge identical (as for the documents contained) base clusters, creating set C=$\{C_1, C_2,\ldots,C_h\}$ of clusters, each identified by one or more phrases.
3. Create an oriented inclusion graph $G=(V, A)$ where:
   V is a set of vertices corresponding one-to-one to clusters created in step 2; $V=\{v_1, v_2,\ldots,v_h\}$,
   A is a set of arcs, each representing the existence of inclusion relation (denoted $\to$) between two clusters, defined as follows:
   $\forall a_{ij} \in A \Leftrightarrow v_i, v_j \in V \wedge C_i \to C_j$.
4. Identify cycles in graph G and eliminate them progressively; replace each cycle O with a single vertex and preserve the arcs connecting the vertices of O with other vertices of G, thus creating graph $G'=(V', A')$ of merged clusters.
5. Identify the kernel of the acyclic graph G' .
6. Build a hierarchy of clusters putting the kernel clusters at the highest hierarchy level, and adding subclusters accordingly to the arcs of G' depth-in, until an end cluster (terminal vertex) is met. The fact that graph G' is acyclic guaranties that for every branch of the dendrogram its construction process ends with a cluster that does not include any more subclusters.

Choosing only kernel clusters for the top of the hierarchy ensures that:

- no top-hierarchy cluster is a direct subgroup of another top-hierarchy cluster (due to the internal stability of the kernel);
- every non-kernel cluster is placed as a subcluster of at least one top-hierarchy cluster (due to the external stability of the kernel).

Step 4 of the HSTC algorithm is necessary because only an acyclic graph has exactly one kernel; otherwise there may exist no kernel or more than one kernel. Unfortunately, the process of replacing all the clusters joined in a cycle with a single cluster carries potential risk of merging many clusters, which are not directly mutually connected with one another. Avoiding such situation should be subject for further research. However, the empirical evaluation reported in the next section had shown that, usually, the cycles in the inclusion graph constitute a fully connected graph (a clique).

As it can be noticed, our approach takes the 'several topics per document' postulate [5] one step further: not only can one document belong to more than one thematic group, but a whole group of documents can also prove to be an adequate subgroup for several more general groups residing in different branches of the dendrogram.

5   Results of an experiment

We have deployed the HSTC algorithm for processing the results of a popular search engine returning a typical ranked list of snippets. The method has been embedded as a module within a java-based Carrot system [21] that implements the original STC algorithm to process English and Polish document repositories.

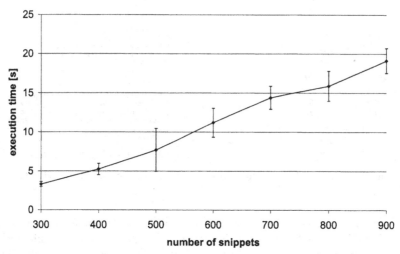

Fig. 1. HSTC execution time averaged over 10 snippet collections as a function of the collection size (the inclusion threshold $\alpha=0.8$, the maximum number of processed base clusters k=400)

To date we have managed to perform experiments testing the execution time of HSTC while clustering snippet collections of various sizes (300 to 900

snippets). The average snippet length was 32 words. The obtained results
are presented in Fig. 1. The experiment has shown that HSTC is fit for
online applications, since the average processing time for the sets of 900
snippets (and that is the maximum some search engines deliver) was 19.12 s
with standard deviation of 1.59 s on Windows 2000 machine with Celeron
366MHz, 128MB RAM.

We also found out that a reasonable value for the inclusion threshold
should be $\alpha \in [0.7; 1]$, meaning that at least 70% of a cluster's documents
should be contained in another cluster, to decide that the former is included
in the latter.

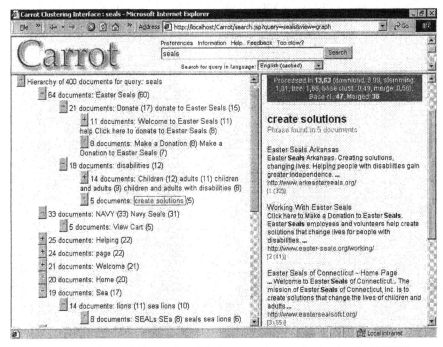

Fig. 2. A navigable hierachy of document groups obtained using HSTC

Figure 2 presents the outcome of the HSTC algorithm for an example
query formulated as "seals". The left panel presents the obtained hierarchy
with some nodes expanded to show the bottom levels of the hierarchy. The
right panel presents documents in the selected node (here: Easter Seals →
disabilities → create solutions). We can easily recognize at least three distinguished most general topic groups: (i) Easter Seals organizations helping
people with disabilities, (ii) Navy SEa, Air and Land forces, and (iii) sea
wildlife. The possibly useless "page", "Welcome", and "Home" groups have
been formed because we did not ignore Web-specific vocabulary by adding it

to the stoplist. Our rationale was that queries related to Web subjects should not be limited.

The example shows a typical hierarchy with a few major branches, each extending up to four or five levels, and several minor ones (hidden in the example figure because of space constrains), which correspond to groups of outliers. In the described case those were groups labeled with phrases such as e.g.: "book", "oil seals", and "mechanical".

The considered example reveals also the positive aspects of applying stemming at the preprocessing stage, since documents containing phrase "creating solutions" are grouped together with ones containing phrase "create solutions".

The obtained results also seem to justify the claim made in [23] that STC is snippet-tolerant. Naturally the quality of snippets affects the quality of the final hierarchy.

An extensive empirical evaluation of the system is being designed, in which we want to pay particular attention to estimating the system usefulness to users. The usefulness could be measured with the time saved on accessing the desired information in the hierarchy (as opposed to browsing the ranked list) and the users' subjective satisfaction.

Naturally, we also need to evaluate the system using the standard Information Retrieval metrics like precision and recall [17], although we believe that the additional aspect of phrases describing document groups should not be overlooked. Comparative tests with available commercial tools of concealed technology are also planned.

Acknowledgments

We wish to express our gratitude to Dawid Weiss for making his implementation of STC available as open source, thus providing us with an excellent platform for implementing HSTC. We also acknowledge the financial support from the State Committee for Scientific Research, KBN research grant no. 8T11F 00619.

References

1. Allen R.B., Obry P., Littman M. (1993) An interface for navigating clustered document sets returned by queries. Proceedings of the ACM Conference on Organizational Computing Systems, 166-171
2. Boley D., Gini M. et al. (1999) Partitioning-based clustering for web document categorization. Decision Support Systems 27 (3), 329-341
3. Choudhary B., Bhattacharyya P. (2002) Text Clustering Using Semantics. World Wide Web Conference (http://www2002.org/CDROM/poster/79.pdf) Hawai
4. Cutting D.R, Karger D.R. et al. (1992) Scatter/Gather: A Cluster-based Approach to Browsing Large Document. Proceedings of the 15th Int. ACM SIGIR Conference on Research and Development in Information Retrieval, 318-329

5. Hearst M.A. (1998) The use of categories and clusters in information access interfaces. T. Strzalkowski (ed.), Natural Language Information Retrieval. Kluwer Academic Publishers

6. Hearst M.A., Pedersen J.O. (1996) Reexamining the Cluster Hypothesis: Scatter/Gather on Retrieval Results. Proceedings of the 19th Int. ACM SIGIR Conference on Research and Development in Information Retrieval, 85-92

7. Kłopotek M.A. (2001) Inteligent Internet Search Engines. Akademicka Oficyna Wydawnicza EXIT, Warszawa (in Polish)

8. Leouski A.V., Croft W.B. (1996) An evaluation of techniques for clustering search results. Technical Report IR-76, Department of Computer Science, University of Massachusetts, Amherst

9. Liu X., Gong Y. et al. (2002) Document clustering with cluster refinement and model selection capabilities. Proceedings of the 25th Int. ACM SIGIR Conference on Research and Development in Information Retrieval, 191-198

10. Maarek Y.S., Fagin R. et al. (2000) Ephemeral document clustering for Web applications. IBM Research Report RJ 10186, Haifa

11. Masłowska I., Weiss D. (2000) JUICER – a data mining approach to information extraction from the WWW, Foundations of Computing and Decision Sciences **25** (2), 67-87

12. Milligan G.W., Cooper M.C. (1985) An examination of procedures for detecting the number of clusters in a data set. Psychometrika **50**, 159-79

13. Neto J.L., Santos A.D. et al. (2000) Document clustering and text summarization. Proceedings of the 4th Int. Conference on Practical Applications of Knowledge Discovery and Data Mining, 41-55

14. Pantel P., Lin D. (2002) Document clustering with committees. Proceedings of the 25th Int. ACM SIGIR Conference on Research and Development in Information Retrieval, 199-206

15. van Rijsbergen C.J. (1979) Information Retrieval, Butterworths, London

16. Roy B. (1969) Algèbre moderne et théorie des graphes orientées vers les sciences économiques et sociales, Dunod

17. Salton G. (1989) Automatic Text Processing. Addison-Wesley

18. Ukkonen E. (1995) On-line construction of suffix trees, Algorithmica **14**, 249-260

19. Vinokourov A., Girolami M. (2000) A probabilistic hierarchical clustering method for organising collections of text documents. Technical Report 5, University of Paisley, Department of Computing and Information Systems, Paisley

20. Voorhees E.M. (1986) Implementing agglomerative hierarchical clustering algorithms for use in document retrieval. Information Processing and Management **22**, 465-76

21. Weiss D. (2001) A Clustering Interface for Web Search Results in Polish and English. Master Thesis, Poznań University of Technology (http://www.cs.put.poznan.pl/dweiss/site/publications/download/dweiss-master-thesis.pdf)

22. Willett P. (1988) Recent trends in hierarchical document clustering: A critical review. Information Processing & Management **24** (5), 577-597

23. Zamir O., Etzioni O. (1998) Web Document Clustering: A Feasibility Demonstration. Proceedings of the 21st Int. ACM SIGIR Conference on Research and Development in Information Retrieval, 46-54

Semantic Indexing for Intelligent Browsing of Distributed Data

M. Ouziri, C. Verdier, and A. Flory

Laboratory for Information Systems Engineering
INSA of Lyon
69621 Villeurbanne Cedex, France
{mouziri, cverdier, flory}@lisi.insa-lyon.fr

Abstract. We present in this paper a semantic indexing technics based on description logics. Data to be indexed are semantically organized as a Topic Map. The index is constructed according to data organization, user profile, and data distribution (association rules). This way leads to obtain a more efficient index and represents more semantics. The index is well adapted to jointly query and navigate in the topic map. DL allows to represent semantics and performs powerful reasoning. The index structure is based on subsumption relationships (for intra-concept indexing) and roles (for inter-concepts indexing).

1 Introduction

Intelligent information systems represent a combination of intelligent computer-human interfaces and intelligent data integration, management, indexing and querying. The design of such systems is complex. It requires to jointly use databases and knowledge representation techniques [5].

Database management systems (DBMS) are made to manage efficiently a large amount of data. They are based on a conceptual model to organize data. Knowledge representation techniques focus on data management according to a particular context (i.e. knowledge). So, intelligent reasoning are developed on knowledge.

In this paper, we propose a semantic indexing technique for intelligent computer-human interface based on Topic Map [10] to access distributed data. Querying heterogeneous and distributed data involves the integration, organization and presentation of multi-sources data. The data integration step is out of the scope of this paper. The integrated data are organized as an extended Topic Map and are presented in an adaptive multi-facet interface [12]. To improve the system efficiency, an index is used to organize data into partitions, in order to minimize the query search spaces.

In traditional databases, indexes are used to improve the queries. The next generation of indexes, called semantic indexes, are not performed on the data level but they cluster the data according to their conceptual models. So, an index is a concept-address pair instead of value-address pair.

In our approach, we create an index structure over data organized following a semantics represented with a Topic Map. Traditional index is not adapted to this framework. That is, data is generally queried according to a specific domain knowledge. Although, traditional index can be used at lower level as a secondary index. Our semantic index is a tree of concepts. The structure and the concepts of the index are defined according to the Topic Map-based data organization, the user profile and association rules. Therefore, the index represents more semantics and its use-ratio is increased. The index is represented in a description logic-based knowledge base. It allows to perform conceptual reasoning when we construct and maintain the index and when we use the index for the query evaluation.

2   Data organisation

As a semantic network, a Topic Map is an artificial intelligence technique for knowledge organization [8]. A TopicMap is a paradigm used to formalize and organize human knowledge in order to support easier creation and retrieval in computer processing. It is a mechanism used for representing and optimizing the information. It organizes the information space into two layers: the abstract domain which is composed of topics and the concrete domain containing the occurrences linked to the topics. A topic can be anything that is a subject and it is the formal representation of any subject in a computer system. It can be used to group together repository objects which relate to a single abstract subject [11]. Each repository object may be defined as an occurrence (in the concrete domain) to the topic (in the abstract domain). A role may be assigned to occurrences defining relationship with the parent topic. In a Topic Map, topics can be related together by associations expressing therefore a given semantic. In an association, each topic role may be defined. Other characteristics assigned to topics, associations and occurrences will not be presented, see [10] for more details.

In our system, the distributed data are organized around a global topic map. This topic map is obtained by integration of the distributed data. Topics represent the concepts coming from the integration of data sources schemas. Occurrences represent the concepts instances extracted from the data sources. In order to have more intelligence, we extend Topic Map with the following functionalities :

- The Topic Map is exactly cardinalized. This means that exact cardinalities of topics associations are computed and registered on the Topic Map associations. We think that it gives an essential information when the user navigates onto the Topic Map.
- The Topic Map is a confused representation of conceptual schemas (abstract domain) and instances (concrete domain). That is, the topic map represents an integrated schema structured in accordance with the instances. For example, in the global schema, if the concept Patient is asso-

ciated with the concept Doctor with the association Exam, and there is
no patients examined by doctors in the data sources, thus this association
will not be represented in the topic map. In the same way, if there is not
no occurrence for a concept, it is deleted from the topic map.

• This last extension leads to a dynamically adaptive topic map. When a
  user query is evaluated, the topic map is adjusted to query result. If the
  query "select Students with age $\leq$ 18" produces the result set {John},
  then the topic map is centered on the student John. Thus, the occurrences
  of the topic Car will only represent the cars bought by John. If no car is
  bought by John, the topic Car is omitted from the topic map.

So, we call this extension of the Topic Map: the Extended Topic Map.

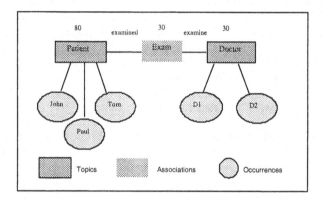

Fig. 1. A topic map

This Extended Topic Map assigns intelligence to the system interface. To
improve interface efficiency, we compute effectively the cardinalities and the
new form of the topic map each time it is necessary. It means that the extended Topic Map is adapted, step by step, according to the user navigation.
To optimize the system, we use an intelligent index which divides the information space according to the user profile, the data distribution (datamining),
and the structure of the topic map (the data organization).

3   Index construction and use

We propose to construct an index according to the data organization, the
user practice (represented by the user profile), and the data distribution. The
information space to be indexed is organized into a Topic Map. The instances
are assigned to topics. This organization consists of a previous partitioning
of the information space (figure 1). The role of the index is to optimize query
evaluation and cardinalities. In semantic index, an index element is a concept

description which is a widened notion of key value used in traditional index. For each indexing concept, we assign some partitions of the information space.

To represent a semantic index, we choose the description logic (DL) formalism [4]. It provides powerful reasoning at the terminological level (onto descriptions) like testing the satisfiability of a description, computing subsumption relationship, etc. The DL organizes a knowledge base (KB) into two parts: the intentional part (TBox) and the extensional part (ABox). The extensional part represents the concrete objects, called individuals. The intentional part is an abstraction of the extensional one. It contains the concepts definitions (descriptions) of the individuals and the definitions of the roles (a role is a binary relationship between two concepts).

The description logic is well adapted to the problems involving the process of the data at their conceptual level. Indeed, it provides powerful reasoning facilities on KB conceptual part (TBox) rather than reasoning on individuals (ABox). The significant ones in data management are [3]:

- Consistency/Coherence of a description
- Subsumption between two descriptions
- Equivalence between two descriptions
- Disjunction between two descriptions

There are other useful reasoning such as [2] *lcs* (*Least Common Subsumer* between two or more descriptions), *msc* (*Most Specific Concept* of an individual), *glb* (*Greatest Lower Bound*), etc.

The suitability of description logic-based semantic index is illustrated through diverse applications. It is used for video indexing [6], for higher level description of video data (sequence, shot, etc.) and for describing and classifying multimedia objects [7], [9]. In [14] semantic indexing is used for information retrieval. A concepts (describing document structure or paragraph) hierarchy covering a domain is used to references documents (or fragments of documents) where the concepts are referred to. A connected approach is presented in [13]. In this approach, the index is constructed using the subsumption and disjoint relationships between indexing concepts. In our approach, we index the concepts and the roles. The roles indexing is useful especially for queries implying multiple concepts and for navigation purpose.

3.1   Index construction

In our approach, we carry out the index on two levels constructed like "Russian doll patterns", namely: the intra-concept indexing and the inter-concepts indexing. Firstly, in the intra-indexing level, the topic map concepts are indexed separately. This indexing is performed using subsumption relationship. The result of this indexing level consists of several indexing trees. Secondly, these indexing trees are connected by performing the inter-concepts indexing. Indeed, the indexing concepts of different trees are linked using indexing roles. That is, the intra-concept indexing is performed by indexing concepts and the inter-concepts indexing is performed by indexing roles. If a higher level of granularity is considered, we can view the index as a "hyper" tree in which a "hyper" node is represented by an intra-concept indexing tree and a "hyper" edge is represented by all the indexed roles linking any pairs of concepts belonging to the intra-concept trees (figure 2).

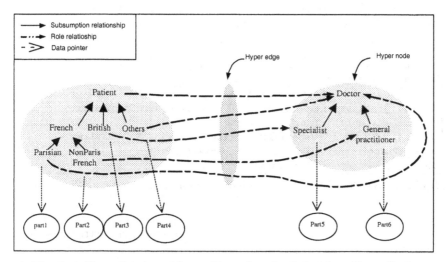

Fig. 2. A "hyper" index with two "hyper" nodes linked by a "hyper" edge

We use description logics to represent the index. The index knowledge base is constructed using the concrete domains of attributes and roles, as Integers and real numbers, to define concepts and roles. However, in the basic description logics, based on KL-ONE, the terminological knowledge has to be defined on an abstract logical level. So, we use the $\mathcal{ALC}(\mathcal{D})$ [1] description logic which integrates concrete domains to concept and assertion languages and provides decidable reasoning. The syntax, semantic and properties of the terminological constructors of $\mathcal{ALC}(\mathcal{D})$ are detailed in [1] and will not be presented in this paper.

So an index is a description logic-based knowledge base (table 1). An index node is a concept description in the knowledge base. The roles are explicitly expressed in terms of descriptions and are defined in the same way as concepts (TBox = concepts descriptions + roles descriptions). The subsumption relationships are automatically generated by DL reasoning. Then, the "hyper" nodes are calculated as connected nodes while we consider only nodes subsumption links (while we are unaware of roles links). An "hyper" edge which links an "hyper" node $hn1$ to an "hyper" node $hn2$ represents the set of roles defined in the nodes of $hn1$ and having domains one of the nodes descriptions of $hn2$. The data partitions are referenced from leaves. The physical storage management is not studied in this paper.

With this index, two types of cardinalities are maintained: the indexing concept cardinality and the role (inter-concepts association) cardinality. The former computes the number of instances for each term description (which corresponds to the tree node) and the later computes the number of instances for each roles. Inverse roles cardinalities are included on roles cardinalities. These cardinalities are viewed in the topic map and allow the user to direct his navigation without submitting any query.

```
Patient ⊑ ∀ n°.Number ⊓ ∀ nationality.String ⊓∀ examinedBy.Doctor ⊓···
French ⊑ Patient ⊓ nationality = "France"⊓
          ∀ examinedBy.GeneralPractioner
British ⊑ Patient ⊓ nationality = "UK" ⊓ ∀examinedBy.Specialist
Others ⊑ Patient ⊓ ¬French ⊓ ¬British
Doctor ⊑ ∀n°.Number ⊓ ∀speciality.String ⊓ ∀examinedBy⁻¹.Patient ···
GeneralPractitioner ⊑ Doctor ⊓ = 0 speciality
Specialist ⊑ Doctor ⊓ ≥ 1 speciality
```

Table 1. A DL-based Knowledge base representing the semantic index of figure 2

The indexing concepts French, British represent the restrictions (sub-concepts) of the concept Patient over the attribute nationality. The restriction attribute and values are determined in a semi-automatic way using the user profile, the volume of the data and its distribution as explained below. Afterwards, we detail the construction of the index. Indeed, the index construction requires to define the indexing concepts, the intra-concept trees and the indexing roles. This index construction is semantics-oriented and is performed as follows:

1. User profile: in the user profile we keep the information about the interest centers of the user such as the most asked queries, the user preferences, etc. That type of information is essential in the index construction. Indeed, to select indexing concepts and criteria are greatly affected by the user queries. For example, if the user queries frequently the concept patient and asks conditions about patients nationalities, then, the patient objects will be indexed according to their nationalities. The index descriptions are represented in a DL-based knowledge base as follows:
Patient ⊑ ∀n°.Number ⊓ ∀Nationality.String ⊓ ∀examinedBy.Doctor
French ⊑ Patient ⊓ Nationality = "France"
British ⊑ Patient ⊓ Nationality = "UK"
The DL reasoning constructs a descriptions hierarchy using the subsumption reasoning. This hierarchy represents the intra-Patient index.
2. Volume of the data: for the concepts which concern a huge volume of objects, the system indexes the concepts in order to reduce the queries

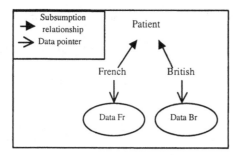

Fig. 3. Intra-concept index tree of Patient

search space and compute more quickly the associations cardinalities. The data are partioned by introducing subsumed concepts of concept indexing. For this, we must define the best attribute criteria for subsumption. For the concept Patient, we can construct an intra-concept index using the role age. The index on Patient is then:

OldPatient $\sqsubseteq$ Patient $\sqcap$ age $\geq$ 65

YoungPatient $\sqsubseteq$ Patient $\sqcap$ age $<$ 65

3. Data distribution: we use association rules extracted with datamining techniques to construct the index. The data distribution is given as rules and their validity (percentage of individuals which verify the rule) extracted from the concrete data using datamining techniques. An association rule has the form:

$if(a_1$ and a_2 and ... and $a_n)$ Then $(b_1$ and b_2 and ... and $b_m)$ where $a_i, b_j \in not(x), x = y, x < y, ...$

For example the following rules can be extracted from a medical database:

$if(Patient.nationality = "France")$ then $Doctor.specialty = \emptyset$ at 95% of precision. This rule means that 95% of French patients are examined by general practitioner.

$if(Patient.name = "Tom"$ and $Doctor.speciality = "Cardio")$ then $Examen.storedIn = "ECG"$ at 100% of precision. This rule means that all cardiologic examinations of Tom are ECG.

We use these association rules to perform an inter-concepts indexing. Therefore, the inter-concepts indexing involves the intra-concept indexing. That is, we must construct intra-concept hierarchies before linking sub-concepts (**British** with **Specialist** for example). The inter-concepts index is based on roles.

An association rule is not valid for all the individuals. But it gives a correlation between individuals. However, the index must be precise at 100%. Thus, the association rules (which are not precise at 100%) are refined by adding more attributes constraints until having 100% precision. In the above example, the rule precision is 95%. This rule is refined by excluding the 5% of patients who don't verify the rule. If these individuals are those who are living in Paris, the rule is modified into the following one:

$if(Patient.nationality = "France"$ and $Patient.address = "Paris")$
then $Doctor.specialty = \emptyset.$

This rule becomes valid at 100% withh the definition of a new indexing concept as:

NonParisFrench $\sqsubseteq$ French $\sqcap$ address $\neq$ "Paris"

Another indexing concept corresponding to the result part of the rule is defined as:

GeneralPractionner $\sqsubseteq$ Doctor $\sqcap$ = 0 speciality

Description Logics are very flexible to express the semantics. The condition Doctor.specialty $= \emptyset$, which represents non-specialist doctors is effectively expressed in DL. The two intra-indexes are then linked using roles (role based indexing) as follows:

NonParisFrench $\sqsubseteq$ French $\sqcap$ address $\neq$ "Paris"$\sqcap$
examinedBy.GeneralPractionner

which means that all non Parisian French patients are examined by general practitioners.

3.2 Use of the Index

The extended Topic Map based interface is adaptive according to user queries. It allows the user to navigate and query the data simultaneously. The index is used to optimize the computing of:

- Concepts and associations instances, and
- Concepts and associations cardinalities.

The user can simply navigates without consulting the data. Then, it is not necessary to compute the exact instances each time the user navigates or specifies a query. But, to compute (or to estime) the links and concepts cardinalities is essential for each user action.

The query evaluation using the index is based on *glb* (*Greatest Lower Bound*) reasoning. The *glb* of a concept represents the most generic concepts which it subsumes. The semantic representation of a query is classified in the index hierarchy using the subsumption relationship and the *glb* reasoning, and then its roles are adapted consequently. Depending on the query position in the index hierarchy, its evaluation is firstly processed as follows:

- The query Q subsumes the indexing concepts $IC_1, IC_2, ... IC_n$. The query extension contains at least the union of indexing concepts extensions. That is, $Ext(IC_1) \cup Ext(IC_2) \cup ... \cup Ext(IC_n) \subseteq Ext(Q)$

If $IC_1 \sqcap IC_2 \sqcap ... \sqcap IC_n \sqsubseteq \perp$ then the cardinality of Q is given by : $card(Q) \geq \sum_{i=1}^{n}(card(IC_i))$

- The query is subsumed by the indexing concepts $IC_1, IC_2, ...IC_n$. Then the query extension is included in the intersection of the indexing concepts extensions. That is, $Ext(Q) \subseteq Ext(IC_1) \cap Ext(IC_2) \cap ... \cap Ext(IC_n)$, and

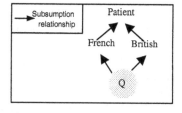

$card(Q) \leq Min_{i=1}^n(card(IC_i))$
If $IC_1 \sqcap IC_2 \sqcap ... \sqcap IC_n \sqsubseteq \perp$, then $Ext(Q) = \emptyset$

- The query is subsumed by an indexing concept IC and has disjoint relationship with indexing concepts $IC_1, IC_2, ...IC_n$. The query extension is included in the IC indexing concept extension while removing the instances of $IC_1, IC_2, ...and IC_n$.

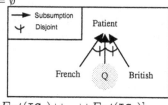

That is, $Ext(Q) \subseteq Ext(IC) - [Ext(IC_1) \cup Ext(IC_2) \cup ... \cup Ext(IC_n)]$
If $IC_1 \sqcap IC_2 \sqcap ... \sqcap IC_n \sqsubseteq \perp$, then the cardinality of Q is :
$card(Q) \leq card(IC) - \sum_{i=1}^n(card(IC_i))$

- The combination of the previous rules optimizes query evaluation and restricts the query cardinality interval.

Secondly, the inter-concepts indexing process adapts the roles implied in the query and estimates the cardinalities according to the result of the query classification (performed in the intra-concept indexing). Indeed, since a query has been subsumed by an indexing concept, the domain of each role of the query is reduced to the domain of the same role of the subsuming indexing concept. For example, we use the previous index (table 1 and figure figure 2) to process the query: $Q \sqsubseteq$ Patient $\sqcap$ nationality $=$ "France" $\sqcap$ age ≥ 65
The query Q is subsumed by the (*glb* of Q) concept French. Therefore, the role examined of the query Q must be subsumed by the same role, examined, of the indexing concept French.

To verify the role subsumption: examinedBy$_Q$ $\sqsubseteq$ examinedBy$_{French}$, the domain of examinedBy$_Q$ becomes GeneralPractitioner. The role cardinality is computed as presented previously for the query cardinality considering only roles subsumption relationships.

4   Conclusion

An index divides data set into smaller data spaces. Therefore, it reduces the queries search spaces. We have presented in this paper a description logic-based semantic indexing for our extended Topic Map-based interface. This approach allows to use more complex indexing elements (complex concept description) based on data semantics. The data are interrogated jointly using queries and navigation in the extended Topic Map. For this, we perform two levels indexing: intra-concept indexing and inter-concepts indexing. These two indexing levels contribute to optimize data queries and navigation.

The semantic index is constructed taking into account the user profile, the data distribution and the data volume. It increases the efficiency of the index.

The use of DL to formalize the index allows to perform intelligent reasoning by evaluating queries and facilitating index updating. Then, as a perspective of this work, we propose to update the index progressively, to be adapted to the user requirements and the data updates. Frequently queries can be added to the index according to their semantics. This update facility will be performed automatically using the DL. We will also consider the problems involved by the decidability of the DL reasoning, particularly, those related to the concrete domains.

References

1. Baader F., Hanschke, P. (1991) A Scheme for Integrating Concrete Domains into Concept Languages. International Joint Conferences on Artificial Intelligence, 452–457
2. Baader, F., Ksters, R. (1998) Computing the least common subsumer and the most specific concept in the presence of cyclic ALN-concept descriptions. O. Herzog and A. Gunter, editors, Proceedings of the 22nd KI-98, **1504**, 129–140
3. Borgida, A. Description Logics in data management. IEEE Trans. on Knowledge and Data Engineering, **7(5)**, 671–682
4. Brachman, R., Schmolze, J. (1985). An overview of the KL-ONE knowledge representation system. Cognitive Science, **9**, 171–216
5. Bresciani, P. (1996) Some research trends in KR and DB. Proceedings of the 3rd Workshop KRDB'96, CEUR Workshop Proceedings, **4**
6. Carrive, J., Pacher, F., Ronfard, R. (1998) Using Description Logics for Indexing Audiovisual Documents. Proceedings of The International Workshop on Description Logics (DL'98), 116–120
7. Fan, J., Zhu, X., Hacid, M-S., Elmagarmid, A. K. (2001) Model-Based Video Classification Toward Multi-level Representation, Indexing and Accessing. Multimedia Tools and Applications, **17(1)**, 97–120
8. Fresse, E. (2000) Using Topic Maps for the representation, management and discovery of knowledge. XML Europe 2000, Palais des congres Paris, 12–16
9. Goble C. A., Haul, C., Bechhofer, S. (1999) Describing and classifying multimedia using description logic GRAIL. Department of Computer Science, Storage and Retrieval for Image and Video Databases (SPIE)
10. ISO/IEC 13250. (1999) Topic Maps. ISO/IEC FCD
11. Kal, A. (2000) Topic Maps for repositories. XML Europe 2000, Palais des congres Paris, 12–16
12. Ouziri, M., Verdier, C., Flory, A. (2002) Utilisation des TopicMaps pour l'interrogation et la visualisation du dossier médical distribué. Document Virtuel Personnalisable (DVP'02)
13. Schmiedel, A. (1994) Semantic Indexing Based on Description Logics. Proceedings of the 1st Worshop on Knowledge Representation meets DataBases (KRDB'94)
14. Todirascu, A., de Beuvron, F., Rousselot, F. (1999) Using Description Logics for Document Indexing. Proceedings of the 14th IAR'99 Workshop, 109–116

An Improved Algorithm on Word Sense Disambiguation

Gabriela Şerban[1] and Doina Tătar[2]

University "Babeş-Bolyai", Department of Computer Science
1, M. Kogalniceanu Street, Cluj-Napoca, Romania

Abstract. The task of disambiguation is to determine which of the senses of an ambiguous word is invoked in a particular use of the word [4]. Starting from the algorithm of Yarowsky [6,5,9,10] and the Naive Bayes Classifier (NBC) algorithm, in this paper we propose an original two-steps algorithm which combines their elements. This algorithm preserves the advantage of principles of Yarowsky (*one sense per discourse and one sense per collocation*) with the known high performance of NBC algorithms. We design an Intelligent Agent, who learns (based on the algorithm mentioned above) to find the correct sense for an ambiguous word in some given contexts.

Keywords: Word sense disambiguation, corpus, agents, learning.

1 Introduction

The word sense disambiguation (WSD) is probably one of the most important open problem and it has now already a long "history" in computational linguistics [3,2]. WSD problem has direct applications in some fields of text understanding as *information retrieval, text summarization, machine translation*.

The problem that arises in natural language is that many words (called polysemic), have several meanings or senses. These senses depend on what context they occur. The task of disambiguation is to determine which of the senses of an ambiguous word is invoked in a particular use of the word [4]. WSD is necessary whenever a system's actions depend on the meaning of the text being processed.

Our algorithm relies dictionary based disambiguation (which we will present in the following section) and can be considered as intermediary between supervised and unsupervised disambiguation [4,7].

2 A Bootstrapping Algorithm Based on the Principles: One Sense Per Discourse and One Sense Per Collocation (BA)

The BA algorithm begins by identifying a small number of training contexts. This could be accomplished by hand tagging with senses the contexts of w

for which the sense of w is clear because some *seed collocations* [6,9,10] occur in these contexts.

The tagging is made on the base of dictionaries or by using the known on-line dictionary of senses WordNet [14]. This initial set of annotated contexts is used for learning a *naive bayesian classifier*. This NBC will help in annotating new contexts. By repeating the process, the annotated part of corpus grows. We will stop when the remaining unannotated corpus is empty or any new context can't be annotated.

The notational conventions are as above:

- w is the polysemic word
- $S = \{s_1, s_2, \cdots, s_K\}$ are possible senses for w, as in a dictionary, or as obtained with WordNet.
- $C = \{c_1, c_2, \cdots c_I\}$ are contexts (windows) for w, as obtained for w with an on-line corpus tool (for example Cobuild [13]). Each c_i is of the form:

$$c_i = w_1, w_2, \cdots, w_t, w, w_{t+1}, \cdots, w_z$$

where $w_1, w_2, \cdots, w_t, w_{t+1}, \cdots, w_z$ are words from the set $v_1, \cdots, v_J$ and t and z (usually $z = 2t$) are selected by user.

Let us consider that the words $V = \{v^1, \cdots, v^l\} \subset \{v_1, \cdots, v_J\}$, where l is small (for example 2) are *surely* associated with the senses for w, such that the occurrence of v^i in the context of w determines the choice of a sense s^i for w (one sense per collocation). Here $\{s^1, \cdots, s^l\}$ is a subset of $\{s_1, \cdots, s_K\}$.

We mention that the set of words (V) used in the BA algorithm as contextual features for the disambiguation is very important; the disambiguation results are improved as the set V grows. This represents the first important characteristic of the BA algorithm.

For example, for the word *plant*, the occurrence in the same context of the word *life* has a sense (lets say A), while the occurrence in the same context of the word *manufacturing* has another sense (lets say B). These rules can be done generally as a decision list:

if v^i occurs in a context c of w then *the sense of c is s^i*, $s^i \in S$ (1)

So, from the set of contexts obtained as query results with (for example) Cobuild, some contexts can be solved.

For our algorithm, we define a relation $\delta \subset W \times \mathcal{P}(W)$, where W is the set of all words and $\mathcal{P}(W)$ is the power set of W. If $w \in W$ is a word and $c \in \mathcal{P}(W)$ we say that $(w, c) \in \delta$ if $w \in c$ or, else, if there is a word $w1 \in c$ so that the words w and $w1$ have the same grammatical root (particularly c is a context).

So, a corresponding decision list has the following form:

if $(v, c) \in \delta$ and v has the sense s_i
$\qquad$ **then** *the sense of the context c is s_i*   (2)

The decision list (2) improved with the relation δ represents the second important characteristic of the BA algorithm.

Algorithm

$C_{res} = \Phi$, determine the set $V = \{v^1, \cdots, v^l\}$

For *each context c in C apply the rules:*

  if $(v^i, c) \in \delta, \Rightarrow$ *sense* s^i, $i = 1, \cdots, l, C_{res} = C_{res} \cup \{c\}$

EndFor

$C_{rest} = C \backslash C_{res}$

While $C_{rest} \neq \Phi$ **do** :

  Determine a set V^* *of words with a maximum frequency in* C_{res}

  $Define\, V = V \cup V^* = \bigcup_{j=1}^{K} V_{s_j}$,

    where V_{s_j} *is the set of words associated with the sense* s_j

    (some V_{s_j} *can be* Φ)

    (If $v \in V^*$, *the context c solved with the sense* s_j *and*

    $(v,c) \in \delta$, *then* $v \in V_{s_j}$, *according with the principle*

    "one sense per discourse")

  For *each* $c_i \in C_{rest}$ *apply the BNC algorithm* :

$$s_i^* = argmax_s P(s \mid c_i) = argmax_s \frac{P(c_i|s) \times P(s)}{P(c_i)} = \tag{3}$$
$$= argmax_s P(c_i \mid s) \times P(s)$$

  EndFor

  $C_{res}^* = \{c_i \mid P(s_i^* \mid c_i) > N, N\, fixed\}$

  $C_{res} = C_{res}^* \cup C_{res}$

  $C_{rest} = C_{rest} \backslash C_{res}$

EndWhile

In this algorithm:

$P(c_i \mid s) = P(w_1 \mid s) \cdots P(w_t \mid s) P(w_{t+1} \mid s) \cdots P(w_z \mid s)$

and $P(w_i \mid s_j) = \begin{cases} 1 & if (w_i, V_{s_j}) \in \delta \\ \frac{nr.occ.w_i}{nr.\,total\,of\,words} & else \end{cases}$

3   The Agent for Words Disambiguation

The application is written in JDK 1.4 and implements the behavior of an Intelligent Agent, whose purpose is to find the correct sense for a given word (the target word) in some given contexts using the algorithm described in the previous section. In fact it is a kind of semi-supervised learning; the agent starts with an initial knowledge (the senses of the target word and a set of words using as contextual features for the disambiguation) and learns to disambiguate the word in the given contexts.

The environment of this agent consists in some information which the agent reads from an input text file "in.txt":

- the target word(w);

- the possible senses for w;
- the contexts for w;
- the words used as contextual features for w's sense disambiguation.

On the basis of its environment, using the algorithm described in the previous section, the agent learns to find the correct sense of the target word in the given contexts.

The main class of the application is the class **Agent**, which implements the agent behavior and the learning algorithm.

We notice that all the representations of data structures used in implementation are linked, which means that there are no limitations for the structures' length (number of contexts, number of words in a context).

4   Experiments

4.1   First Experiment

Our aim is to solve some contexts in which the word *band* appears, from a set of contexts obtained as query results with Cobuild [13]. Using the application we accomplish the training of the agent in the following environment (given in the text file "band.txt").

The file "band.txt":

- the target word
band
- the senses of the target word (as in Wordnet)
set music ring strip
- the words used as contextual features for Q's disambiguation and the indexes of the corresponding sense of the target word
song 2 sing 2 paper 4 dance 2 club 2 release 2 jazz 2 member 1 jewel 3 sound 2 rock 2
- the contexts of the target word

1. going to happen, we're not that kinda *band*. [p] I don't write 15 songs in a
2. fiber, more dust, a broken rubber *band*, a paper clip, a penny, more dust, a
3. Hickman conducts an all-woman's *band* and choir, the next she sings
4. olde worlde part of town where the *band* are staying. All hideous new Europe
5. studio-dusty shrouds. Finally, the *band* that had me dancing 'til
6. to reunite musicians of a famous soul *band* who have not played for 30 years.

7. fan club show. It's a rowdy night-the *band* first played here in '87 with the

8. and 'You Love Us'-are the best the *band* have released so far, claim Jeff and

9. the more esoteric brands of big-*band* jazz in favour of a lively

10. The Commitments I've never been in a *band*, know nothing about it. [p] Adams:

11. Maker, the signatures of each *band* member, lovingly inscribed in non-

12. Cert 15 [p] 15 (18) RESERVOIR DOGS: A *band* of foiled jewel robbers reassemble

13. it right up here. (Fade down) FX *BAND* SIX: MUSIC. 'STYLE Fade up. Fade

14. to none - a recent reviewer said: The *band* sounds learner than before, the

15. a new idea they cannot, like a young *band*, simply book the games equivalent of

16. Radiation put together a rockabilly *band*, The Tearjerkers, while Panter

17. Before Us is by The Albion Dance *Band*-with the emphasis on 'dance'. Live,

18. Tonight the *band* plays at the Hotel,

19. present a famous soul *band*, with more than 10 albums,

20. Maker, the signatures of each *band* member, lovingly inscribed in non-

21. W. Axl know that the letters in his *band*-name spell 'Nor See Gnus'? Probably

22. to a junior officer, 'no more loyal *band* of brothers than the Grand Staff of

23. the North Sea) and Britain's growing *band* of ethical fund managers are

24. government of Trinidad. [tref] *BAND* ONE: Tony Blair, Shadow Employment

25. That's tough, but that's life. A *band* out of time. [p] DAVE JENNINGS [p]

After the agent reads the information from the environment, he applies the disambiguation algorithm for the given contexts. The result is shown in Table 1. Each context is followed by the sense for the target word found by the agent after the disambiguation, and by the correct sense of the target word for the given context.

From Table 1, we observe than the Agent learns to find the correct sense of the word *band* in the contexts 5, 12, 13, 16, 18, 19, 20, 21, 23, 24, 25. The sense of the word in the other contexts is deduced from the set of words used

Table 1. The result of disambiguation

Context 1 — *music*	— music
Context 2 — *strip*	— strip
Context 3 — *music*	— music
Context 4 — *music*	— set
Context 5 — *music*	— music
Context 6 — *music*	— music
Context 7 — *music*	— music
Context 8 — *music*	— music
Context 9 — *music*	— music
Context 10 — *music*	— set
Context 11 — *set*	— set
Context 12 — *ring*	— ring
Context 13 — *music*	— music
Context 14 — *music*	— music
Context 15 — *ring*	— set
Context 16 — *music*	— music
Context 17 — *music*	— music
Context 18 — *music*	— music
Context 19 — *music*	— music
Context 20 — *set*	— set
Context 21 — *set*	— set
Context 22 — *ring*	— set
Context 23 — *set*	— set
Context 24 — *set*	— set
Context 25 — *set*	— set

as contextual features for the disambiguation. For example, from the context 7, the agent learns to associate the word *play* with the sense *music*, and from context 6 the agent learns to associate the word *soul* with the sense *music*. The accuracy of the BA algorithm in the proposed experiment is **84%**. We note that the *accuracy* of the disambiguation algorithm is calculated with the following formula

$$A = \frac{number\,of\,correctly\,solved\,contexts}{number\,of\,contexts} \tag{4}$$

In fact, in our case, the accuracy is the learning rate of the algorithm.

The experiment at Hearst (1991) shows that to achieve a high precision in word sense tagging, the initial set must be large (20-30 occurrences for each sense).

We have to mention that, in our experiment, we associated a single occurrence for each sense (for an easier evaluation of the algorithm).

Experimental Comparison with the NBC Algorithm. In the case of the algorithm described in section 2 (BA - Bootstrapping Algorithm), the relation δ described in equation (2) is very important. In order to illustrate the efficiency of the BA algorithm (with an without δ), we ran at the same time the NBC algorithm for the experiment proposed in subsection 3.2. The obtained results (including the algorithms' accuracies) are shown in Table 2. The correct results after the disambiguation are bold. We note that "BA without δ" is the BA algorithm (Section 2), in which a decision list has the form described in Equation 1.

Table 2. Comparative experimental results between the disambiguation algorithms BA and NBC

	BA with δ	BA without δ	NBC algorithm
The result	Context 1 **music**	Context 1 *ring*	Context 1 *strip*
of contexts'	Context 2 **strip**	Context 2 **strip**	Context 2 **strip**
disambiguation	Context 3 **music**	Context 3 *ring*	Context 3 *strip*
	Context 4 *music*	Context 4 *music*	Context 4 *strip*
	Context 5 **music**	Context 5 *ring*	Context 5 *strip*
	Context 6 **music**	Context 6 *ring*	Context 6 *strip*
	Context 7 **music**	Context 7 **music**	Context 7 **music**
	Context 8 **music**	Context 8 *set*	Context 8 *strip*
	Context 9 **music**	Context 9 **music**	Context 9 **music**
	Context 10 *music*	Context 10 *ring*	Context 10 *strip*
	Context 11 **set**	Context 11 **set**	Context 11 *strip*
	Context 12 **ring**	Context 12 **ring**	Context 12 **ring**
	Context 13 **music**	Context 13 **music**	Context 13 *strip*
	Context 14 **music**	Context 14 **music**	Context 14 *strip*
	Context 15 *ring*	Context 15 *ring*	Context 15 *strip*
	Context 16 **music**	Context 16 *ring*	Context 16 *strip*
	Context 17 **music**	Context 17 **music**	Context 17 **music**
	Context 18 **music**	Context 18 **music**	Context 18 *strip*
	Context 19 **music**	Context 19 *ring*	Context 19 *strip*
	Context 20 **set**	Context 20 **set**	Context 20 *strip*
	Context 21 **set**	Context 21 *ring*	Context 21 *strip*
	Context 22 *ring*	Context 22 *ring*	Context 20 *strip*
	Context 23 **set**	Context 23 *music*	Context 23 *strip*
	Context 24 **set**	Context 24 *ring*	Context 24 *strip*
	Context 25 **set**	Context 25 **set**	Context 25 *strip*
Accuracy	84%	44%	20%

From Table 2 results very clearly the efficiency of the BA algorithm, improved with the relation δ. The comparative experimental results are shown in Figure 1. In Figure 1, we give, for each algorithm, a graphical representation of accuracy/context. More exactly, for a given algorithm, for the i-th context we represent the accuracy (see Equation 4) of the algorithm for the first i contexts. From Figure 1, it is obvious that the most efficient is the

BA algorithm with the relation δ (the algorithm's accuracy grows with the number of contexts).

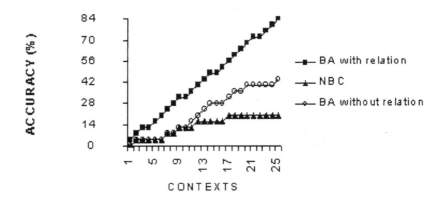

Fig. 1. The graphical representation of the experimental results described in Table 2

4.2 Second Experiment

Since for Romanian language does not exist neither a corpus nor something similar with WordNet, we make a WSD experiment using the BA algorithm, which requires only information that can be extracted from untagged corpus.

Our aim is to use the BA algorithm for the romanian language, to disambiguate the word *poarta* in some contexts obtained with an on-line corpus tool (at us *htdig* and a Romanian corpus).

We make the following specifications:

- the target word *poarta* has, in romanian language, four possible senses (two nouns and two verbs);
- we experiment our algorithm starting with 38 contexts for the target word;
- we start with four words as contextual features for the disambiguation (a single feature for each sense).

The accuracy of the BA algorithm in the proposed experiment is **60%**.

We make the following specifications:

- in the above experiment we grow the number of occurrences for each sense of the target word and we observe that: with two occurrences for each sense the algorithm's accuracy grows with **10%**, with three occurrences for each sense the accuracy grows with **15%**;

- we grow the number of input contexts (100) and we observe that the algorithm's accuracy grows with **15%**.

As a conclusion, if the number of words used as contextual features for the disambiguation and the number of contexts grows, the accuracy of the BA algorithm grows, too.

5   Conclusions and Further Work

If the Agent (described above) starts with a substantial initial knowledge (number of senses of the target word, set of words used as contextual attributes for the disambiguation) and if the environment consists in a big number of contexts, the the disambiguation (learning) algorithm works very well (the number of senses of the target word learned by the agent grows).

Further work is planned to be done in the following directions:

- For assuring a better efficiency of the disambiguation, we plain to retain in a database the results of the learning process;
- We plain to establish a better evaluation for our Agent, working with some standard ambiguous words and a more impressive amount of contexts from different corpora (as BNC http://sara.natcorp.ox.ac.uk/lookup.html);
- We will compare the results with those obtained with SENSEVAL's , two recent pilot applications in WSD;
- As input of our agent we plain to use SEMCOR [12], a manually sense tagged corpus, in which all words have been tagged with WordNet senses;
- At the University of Bucharest is in construction a WordNet for Romanian language, and we will use that as input for our Agent;
- We plain to study our approach in the context of combining labeled and unlabeled data with Co-Training as in [1];
- Our own goal is to solve with our method the disambiguation for a query in a future QA-system in Romanian which is now in construction;
- We also planned to improve the application using a subroutine which determines all the synonyms for the initial features of the target word (a hierarchical clustering algorithm, which is already implemented).

References

1. Blum, A., Mitchell, T. (1998) Combining Labeled and Unlabeled Data with C-Training. Proceedings of the 11th Annual Conference on Computational Learning Theory 92–100
2. Allen, J. (1995) Natural language understanding. 2nd edn. Benjamin/Cummings Publ.
3. Jurafsky, D., Martin, J. (2000) Speech and language processing. Prentice Hall

4. Manning, C., Schutze, H. (1999) Foundation of statistical natural language processing. MIT Press
5. Resnik, P., Yarowsky, D. (1998) Distinguishing Systems and Distinguishing sense: new evaluation methods fot WSD. Natural Language Engineering. **1**
6. Yarowsky, D. (1999) Hierarchical Decision Lists for WSD. Kluwer Acadmic Publishers
7. Sebastiani, F. (2001) A tutorial on Automated Text Categorization. ESSLLI, 1–25
8. Tatar, D. (2001) Inteligenta artificiala: demonstrare automata de teoreme, prelucrarea limbajului natural. Editura Microinformatica
9. Yarowsky, D. (1995) Unsupervised Word Sense Disambiguation Rivaling Supervised Methods. Proceedings of ACL'95, 189–196
10. Yarowsky, D.: WordSense Disambiguation Using Statistical Models of Roget's Categories Trained on Large Corpora. http:// citeseer.nj.nec.com
11. G. Serban, D. Tatar (2001) A new algorithm for WSD. Studia Univ. Babes-Bolyai, Informatica. **2**, 99–108
12. Kilgarriff, A.: Gold Standard Datasets for Evaluating Word Sense Disambiguation Programs. http:// citeseer.nj.nec.com
13. http://titania.cobuild.collins.co.uk/form.html
14. http://www.cogsci.princeton.edu/ wn/

Web Search Results Clustering in Polish: Experimental Evaluation of Carrot

Dawid Weiss and Jerzy Stefanowski

Institute of Computing Science, Poznań University of Technology,
ul. Piotrowo 3A, 60–965 Poznań, Poland,
E-mail: dawid.weiss@cs.put.poznan.pl, jerzy.stefanowski@cs.put.poznan.pl

Abstract. In this paper we consider the problem of web search results clustering in the Polish language, supporting our analysis with results acquired from an experimental system named Carrot. The algorithm we put into consideration – *Suffix Tree Clustering* has been acknowledged as being very efficient when applied to English. We present conclusions from its experimental application to Polish, demonstrating fragile areas of the algorithm related to rich inflection and certain properties of the input language. Our results indicate that the characteristics of produced clusters (number, distinctiveness), strongly depend on pre-processing phase. We also attempt to investigate the influence of two primary STC parameters: *merge threshold* and *minimum base cluster score* on the number and quality of results. Finally, we introduce two approaches to efficient, approximate conflation of Polish words: *quasi-stemmer* and an automaton-based lemmatization method.

1 Search Results Clustering Overview

Together with an exponential increase of the number of documents available in the Internet, comes the requirement for faster and more reliable tools for locating relevant information - the role currently played by search engines. While algorithms for indexing and querying large volumes of data have been substantially improved, the paradigm of searching for information based on providing query terms and retrieving a list of matching documents, remained almost the same since the very beginning. Considering solely the number of matching results for even very narrow topics, browsing such set is not feasible anymore. Besides, query terms are often ambiguous and spanning over multiple subjects, making it impossible to present the results as a linear ordering of relevance.

Searching and browsing are in very strong relationship to each other; it is natural that improving browsing techniques will also improve the overall performance of seeking for information. Search engines return a list of references to matching documents, each one usually comprising a title, URL and a short fragment of the source document, called a *snippet*. A snippet should contain enough information about the document it describes to give the user a clue what the entire document is about. This is the starting point for search results clustering which attempts to form *meaningful, thematic groups* of snippets. These groups are then presented instead, or in addition to the original document references. The user gets a much deeper insight into the subjects the query covered, simply by looking at the set of discovered groups.

2   STC Algorithm and Related Work

The problem of clustering in general is very well explored, thanks to the heritage drawn from statistics, economy and even fields of computer science such as data mining and information retrieval. In spite of this, several properties such as demand for linear complexity, limited amount of input data (a snippet instead of full body of a document) and processing of text features render search result clustering a research field worth investigating on its own.

Arguably the first query result visualization algorithm based on the paradigm of clustering was presented in the Scatter-Gather system [4], but it was not until *Suffix Tree Clustering* (STC) algorithm appeared [9], that the field of search results clustering, also called *ephemeral clustering* [5] had been given a substantial momentum.

STC has at least two distinguishing features: its time complexity is linear with respect to the number of clustered snippets, and it operates on phrases present in the text, in contrast to most previous efforts, built on top of standard IR measures of terms frequency distribution. STC attempts to cluster documents or search results according to shared *phrases* they contain, thus employing information about the proximity of terms, in addition to their frequencies.

STC is organized into two phases: discovering *base clusters* and combining (merging) them to form the final set of groups. In the first phase, all sentences in the search result (both in snippets and document titles) are inserted into a *generalized suffix tree*, where each symbol (node) in the tree represents a single term. Every node also holds references to the sentences (and documents) it occurred in. Thus, each path from some node to the root of the suffix tree denotes a phrase shared by all the documents that node holds references to.

For each node (phrase) referencing at least two documents we define a *base cluster score*: $s(m) = |m| \times f(|m_p|) \times \sum(tfidf(w_i))$, where $|m|$ is the number of words in phrase m not present in a stop list, $f(|m_p|)$ is a function penalizing short-phrases, $tfidf(w_i)$ is a standard Salton's *term frequency-inverse document frequency* term ranking measure. A *stop list* contains a set of *stop words* – terms which can be safely ignored because they are unlikely to form meaningful phrases on their own (conjunctions, prepositions or articles)[1]. A set of base clusters is then identified by selecting nodes with base cluster score higher than an arbitrarily chosen *minimal base cluster score threshold*. In the second stage of the algorithm, top-ranking base

---

[1] Ignoring stop words is a common practice in IR, but may sometimes lead to unfortunate results – consider Shakespeare's *"to be or not to be"*...

clusters are merged using a variation of single-link AHC algorithm, with binary merge criterion between base clusters a and b defined by the formula: $similarity(a, b) = 1 \Leftrightarrow \left(\frac{|a \cap b|}{|a|} > \alpha \right) \wedge \left(\frac{|a \cap b|}{|b|} > \alpha \right)$, where α in the above formula denotes an arbitrarily chosen *merge threshold* and by $|x|$ we understand a set of documents in cluster x.

Unfortunately due to space constraints, we are not able to give a broader insight into algorithms that followed STC (refer to [5] for a review of existing methods), but all of them, including STC, have been designed, implemented and evaluated for English only. This puts in question their applicability to other languages, because, as we are about to show, the properties of a language may significantly affect an algorithm's performance. To our best knowledge only Semantic Hierarchical Online Clustering (SHOC) algorithm [2], attempts to overcome the language issues. Instead of using words at the lowest level of granularity, it employs recurring sequences of characters, which makes it particularly suitable for oriental languages, where words are not graphically distinguished in the text.

Stunningly, with the abundance of new algorithms, there exists a significant shortage in evaluation techniques and measures of ephemeral clustering quality. Most algorithms are judged based on explicit user surveys or analysis of server logs aimed at comparing raw search interface to the clustered one [9], in spite of the fact that [6] clearly states such comparison can be misleading. A different approach is present in [5], where an information entropy – derived measure is used for evaluation. We chose to employ this measure as well, even though it has certain shortcomings discussed in Sect. 4.

3   Motivation

It seems that the issue of language-aware search results clustering has been to some point neglected. We decided to investigate the behavior of STC as a representative algorithm specific to the domain, when applied to Polish. In particular, we wanted to examine the impact on the quality of results when language with rich inflection is being processed. Our intention also was to determine any potential influence which data pre-processing activities such as ignoring meaningless terms (stop words) or various methods of term conflation (lemmatization or stemming) might have on the final clusters.

This was one of the main reasons of creating Carrot system. Carrot is an implementation of STC featuring Polish and English term conflation algorithms (ref. to Sect. 4.1) and stop words lists specific to both languages. Carrot was available online between Spring and Summer of 2001 before legal issues concerning automated querying of the background major search engine forced us to limit its public availability[2]. Another reason driving this experiment was to find out how the choice of STC's thresholds affect the quality

---

[2] More information can be found at *http://www.cs.put.poznan.pl/dweiss/carrot*

and stability of results. We also wanted to employ a mathematical measure of quality as opposed to user surveys.

4 The Experiment

In order to fulfill the goals given above, we prepared a small-scale experiment comparing clusters acquired from Carrot to a predefined, manually produced grouping. In order to keep the experiment realistic, we decided to utilize two sets of real search results downloaded from Vivisimo search engine in response to queries: *inteligencja* (intelligence) and *odkrywanie wiedzy* (knowledge discovery). Recently, the above set of queries was extended to include additional two in Polish and four in English.

The manual clustering of the two test queries had been performed by five experts in the field, independently. Original documents were not retrieved from the Web, so that humans had as much information about the clustered set, as the algorithm – a set of about 70 snippets per query. This again proved that clustering, even when done by hand, is unambiguous and problematic (ref. to [6]). Out of 80 snippet-to-cluster assignments, only 50% were fully consistent among all individuals. In the end there were about 14 manually created groups for each query.

Having a manual benchmark to relate to, we faced the problem of choosing possibly objective comparison function. The problems with empirical evaluation of ephemeral clustering systems have already been mentioned in Sect. 2. We decided to utilize Byron Dom's method [1], briefly presented in Def. 1. The manual clustering we produced was used as the *ground truth* set required by the measure.

Definition 1. Let X be a set of *feature vectors*, representing objects to be clustered, C be a set of *class labels*, representing the desired, optimal classification (also called a *ground truth set*), and K be a set of cluster labels assigned to elements of X as a result of an algorithm. Knowing X and K one can calculate a two-dimensional contingency matrix $H \equiv \{h(c, k)\}$, where $h(c, k)$ is the number of objects labeled class c that are assigned to cluster k. *Information-theoretic external cluster-validity measure* is defined by:

$$Q_0 = -\sum_{c=1}^{|C|} \sum_{k=1}^{|K|} \frac{h(c, k)}{n} \log \frac{h(c, k)}{h(k)} + \frac{1}{n} \sum_{k=1}^{|K|} \log \left(\frac{h(k) + |C| - 1}{|C| - 1} \right). \quad (1)$$

where $n = |X|$, $h(c) = \sum_k h(c, k)$, $h(k) = \sum_c h(c, k)$.

The experiment consisted of comparing results produced by STC to the ground truth set using the measure given in Def. 1. Full range of values for the key algorithm parameters – merge threshold and minimal base cluster score (see Sect. 2) – were taken into account. The experiment was repeated with and without term conflation algorithms and stop words removal in pre-processing phase.

For the experiment, we used a normalized version of the formula presented in Def. 1, where zero denotes no correspondence between ground truth set of clusters and the compared set, while one means the two are identical. It should be stressed that quality of results in search results clustering is hard to define. We express it as a singe-value relevance measure to the manual "perfect" clustering in order to analyze its relative changes and trends. Interpretation of specific values the measure can take is problematic since it is an aggregation of many different aspects.

Byron Dom's measure is defined for flat partitioning of an input set of objects, while STC produces flat, but overlapping clusters. In order to be able to apply the measure, we decided that snippets assigned to more than one cluster, would belong only to the one having the highest score assigned by STC.

4.1   Word Conflation Algorithms Used for Experiments

In many inflectional languages a word, depending on its function in a sentence, may appear in several graphical forms while maintaining roughly the same meaning. In most cases, there exists a constant part of a word called a *stem* and *affixes* which change (in Polish: pisz-*esz*, pisz-*emy*, pisz-*ecie*). While there is a great deal of confusion around the proper terminology, we shall call a set of all inflected forms of a word a *lemma* and lemma's representation as a symbol a *lexeme* (lexeme is not always identical to the stem, more than one stem may exist in one lemma). In Information Retrieval it is usually better to consider lexemes rather than exact word forms. *Lemmatization* is a process of mapping of a word form to its lexeme, but it usually requires significant linguistic resources (like dictionaries). Simplified (and approximate) algorithms called *stemming algorithms* work by replacing affixes iteratively, according to certain heuristic rules until a "stem" of a word is found.

A number of both commercial and public stemming and lemmatization algorithms exist for English. For Polish, certain commercial solutions exist [3], but according to our knowledge no freely available solutions have been proposed besides Krzysztof Szafran's SAM [7]. We initially made an attempt to utilize this free morphological analyzer, but it is a precompiled program and works only in batch-mode, practically making it impossible to use it in an online system. We therefore created two simple, but efficient, techniques for conflating words in Polish – one is an approximate stemming method called *quasi-stemmer*, the other is a basic lemmatization technique.

The quasi-stemmer is based on the use of over eight hundred thousand unique terms corpora of Polish texts, obtained by permission from Rzecz-pospolita newspaper archives. About five hundred most common suffixes from this corpora have been extracted and put in a lookup-table. The condition for determining whether two words α and β can be considered inflected forms of a common stem is given in Def. 2.

Definition 2. Two words α and β may be considered to originate from a common stem, if they contain a common prefix, which is at least n characters long, and the suffixes created by stripping that prefix exist in the predefined lookup table.

It is obvious that such simple technique has limited accuracy, hence the prefix *quasi-* in the name. Also, some caution must be undertaken since the equivalence relation implied by quasi-stemmer is not transitive. Nonetheless, the quasi-stemmer has a very appealing capability of comparing proper names and terms not present in the corpora. It performs very well on inflected names of places or people, for instance. Our experience shows that even such a simple technique brings an improvement in both cognitive and experimental results (refer to Sect. 5.1).

We also created another technique, this time based on an open source dictionary of Polish included with ispell[3]. This dictionary contains words and rules for deriving most of their inflected forms. We assumed (which is unfortunately not always the case), that a lemma of a word could be created from an entry in the dictionary and all the derived word forms. We then compressed this information into a finite state automaton, a very compact data structure used directly for lookups. If a given inflected term exists in the automaton, we can find its mapping to a "lexeme" – the dictionary form which generated it. For any term, the complexity of its stem-lookup operation is at most linear with the length of the term (this is the maximum time, if the lookup has been successful). Several problems still exist, like the mentioned fact that certain lemmas are represented in the dictionary as separate entries, or certain entries generate inflected forms representing more than one lemma – for example entry *betoniarz* (concrete mixer operator) and *betoniarka* (concrete mixer) end up with the same lexeme (*betoniarz*) and that should not be the case.

5   Results

5.1   Inflection and Stop Words

Polish inflection rules are much more complex than English. Words have different suffixes depending on the case, gender, number, person, degree, mode or tense. Oren Zamir in [9] claims that the importance of pre-processing is not a crucial factor of algorithm's performance. Our experiments and papers, such as [6], indicate it is not the case. We discovered that both cognitive and measured results improve when stop words are removed and term conflation

---

[3] *http://ispell-pl.sourceforge.net*

methods are used. In Fig. 1 one can observe the increase in quality measure value when the above techniques are used.

It was quite unexpected to us that removing stop words from the input had such high influence on results. This suggests that a more advanced technique of identifying meaningless terms could even further improve cluster quality. Maybe this could even go as far as discarding all non-nouns, or non-verbs.

Rich inflection in the input has a potential negative influence on the factor used for calculating cluster scores. The TFIDF (*term frequency inverse document frequency*) model used for scoring terms in phrases does not yield the actual importance of a given lexeme, because the score is distributed over all of its word forms.

No preferred term conflation methods was evident in our results. Both quasi-stemmer and the simple lemmatization yielded improvement over not altered input data, yet this improvement was

5.2   Phrases and Words

STC produces clusters based on common phrases occurring in snippets. The Polish language has a feature of not being as order-dependant as English. The meaning of a sentence can be as well carried solely by suffixes of the terms used. Compare this classic example from [8]: *John hit Paul* $\neq$ *Paul hit John*, while in Polish: *Jan uderzył Pawła* = *Pawła uderzył Jan*. Consequently, STC's primary assumption that common phrases form good clusters thus becomes questionable.

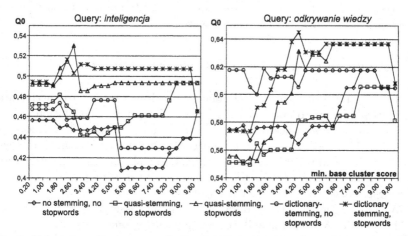

Fig. 1. Value of quality measure Q_0 in relation to changing base cluster score. Five different configurations of input pre-processing are shown. STC merge threshold is constant (0.6)

A potential solution here could be to utilize information about neighboring terms in the input, without strict assumption that they must form sequences. This approach has been partially studied for English in [5], where *lexical affinity* of pairs of terms was used instead of keywords. The use of word sets as opposed to sequences has been also studied in [9], where the author claimed phrases performed slightly better than sets (experiments in English).

5.3   Sensitiveness of STC Control Parameters

Oren Zamir in [9] claims that the influence of merge threshold over the results was unnoticeable and chooses an arbitrary value for his STC evaluation. We have found out that the results are in deed not very sensitive to the setting of the merge threshold, especially when the input data has been pre-processed. However, as seen in Fig. 2, low merge threshold for certain queries may lead to noticeably unstable quality measure indicator. This effect was gone (or rather smoothed-out) when any of the term conflation methods had been used.

Minimum base cluster score threshold expresses much stronger influence, especially on the number of discovered clusters (see Fig. 2). While this is obvious (because this parameter describes a cut-off threshold for phrases considered at later stages of STC), it is not at all clear what exact value this parameter should be set to. It seems that higher values stabilize the properties of merge threshold, also improving numerical measure of quality, but it also means that low-scoring clusters, perhaps interesting, are discarded in favor of large, strong ones (which are potentially obvious to the user). Also, as seen in Fig. 2, when pre-processing has been applied, there is a quick saturation of quality and a sudden drop in the number of clusters produced. The relation between the number of clusters, their quality and the minimum base cluster score should be a matter of further research.

6   Conclusions and Future Research Directions

In this paper we have presented conclusions drawn from an experimental application of STC to documents in Polish. The primary conclusion is that STC seems to be significantly sensitive to languages with rich inflection in the input. This affects both base cluster construction and scoring. Our experiments demonstrate that proper term conflation and stop words identification improved the stability of the algorithm and quality as a comparative measure of correspondence to a manual clustering.

We also argue that using phrases for finding base clusters may not be the best choice for inflectionally rich languages, giving Polish as an example. In this language word order may be tricky, and the same meaning may be expressed without using exactly the same sequence of terms. The most promising research direction in this area includes abandoning strict order of

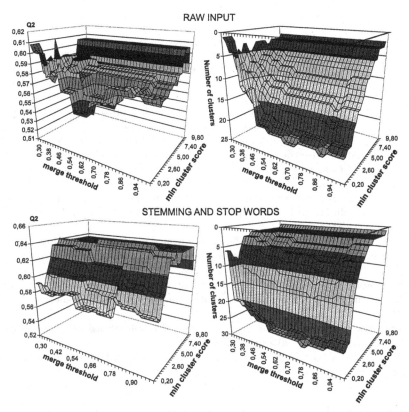

Fig. 2. A three-dimensional view of the relation between merge threshold and minimum base cluster score to the value of quality measure Q_2 (charts on the left side) and number of produced clusters (charts on the right side). Please note the axis of number of clusters is reversed for clarity

terms in the discovered base clusters, maybe replacing phrases with lexical affinities or frequent sets.

This paper also introduces two efficient methods for term conflation in Polish and shows how they improve the quality of results when applied to STC. Potential research directions in this area should be directed at developing methods capable of correctly processing proper names (users frequently look for places, people or things rather than abstract topics). Quasi-stemmer presented in this paper has this property to some extent, but an analysis of results if an advanced lemmatization method were used would be of great interest.

Acknowledgement. The authors would like to thank an anonymous reviewer for his insightful and helpful suggestions and Mr. Stanisław Osiński, Mr. Paweł Kowalik and Mr. Michał Wróblewski for their help in manual clustering. This research has been supported by grant 91-395/03-BW.

References

1. Byron E. Dom. An information-theoretic external cluster-validity measure. Technical Report IBM Research Report RJ 10219, IBM, 2001.
2. Zhang Dong. *Towards Web Information Clustering*. PhD thesis, Southeast University, Nanjing, China, 2002.
3. Elżbieta Hajnicz and Anna Kupść. Przegląd analizatorow morfologicznych dla języka polskiego. Technical Report 937, IPI PAN, 2001.
4. Marti A. Hearst and Jan O. Pedersen. Reexamining the cluster hypothesis: Scatter/gather on retrieval results. In *Proceedings of SIGIR-96, 19th ACM International Conference on Research and Development in Information Retrieval*, pages 76–84, Zürich, CH, 1996.
5. Yoëlle S. Maarek, Ronald Fagin, Israel Z. Ben-Shaul, and Dan Pelleg. Ephemeral document clustering for web applications. Technical Report RJ 10186, IBM Research, 2000.
6. Sofus A. Macskassy, Arunava Banerjee, Brian D. Davison, and Haym Hirsh. Human performance on clustering web pages: A preliminary study. In *Knowledge Discovery and Data Mining*, pages 264–268, 1998.
7. Krzasztof Szafran. Analizator morfologiczny SAM-95. Technical Report TR 96 226, Faculty of Mathematics, Informatics and Mechanics. Warsaw University, 1996.
8. Stanisław Szpakowicz. *Automatyczna analiza składniowa polskich zdań pisanych*. PhD thesis, Warsaw University, 1978.
9. Oren Zamir. *Clustering Web Documents: A Phrase-Based Method for Grouping Search Engine Results*. PhD thesis, University of Washington, 1999.

Part IV

Data Mining and Machine Learning Technologies

A Hybrid Genetic Algorithm - Decision Tree Classifier

Abdel-Badeeh M.Salem[1] and Abeer M.Mahmoud[2]

Faculty of Computer & Information Science, Ain Shams University, Abbassia,
Cairo, Egypt
e-mail: *absalem@asunet.shams.edu.eg*

Abstract. Studies of data mining classification algorithms have shown that these algorithms either cause a loss of quality or scalability aspects or cannot effectively uncover the data structure. This paper presents a new approach for developing two C4.5 based classifiers. The first, RFC4.5, uses the RainForest framework approach while the second, GARFC4.5, uses genetic algorithm. The two classifiers have been applied on medical database of 20MB size for thrombosis diseases, obtained from the discovery challenge competition of the 3rd European Conference on Principles and Practice of Knowledge Discovery in Database held in Prague, 1999. The results show that the two Classifiers give higher classification accuracy than traditional C4.5 classifier. For both classifiers, at a certain population size, it is found that the classification accuracy increases with sample size.

1 Introduction

Classification is an important data-mining problem that has a wide range of applications, including medical diagnosis, fraud detection, scientific experiments, credit approval and target marketing [1][4][5]. The input to a classification problem is a dataset of training records and each record has several attributes. Attributes whose domains are numerical are called numerical attributes (continuous attributes), whereas attributes whose domain is not are called categorical attributes (discrete attributes). There is one distinguished attribute, called class label, which is a categorical attribute with a very small domain. The remaining attributes are called predictor attributes; they are either continuous or discrete in nature. The goal of classification is to build a concise model of the distribution of the class label in terms of the predictor attributes. The resulting model is used to assign class labels to a database of testing records where the values of the predictor attributes are known but the value of the class label is unknown. Many classification models have been proposed in the literatures such as distributed algorithms, restricted search, data reduction algorithms, parallel algorithms, neural network and others[6][7]. Decision trees are especially attractive for a data-mining environment for several reasons. First, due to their intuitive representation, they are easy to assimilate by humans[2]. Second, they can be constructed relatively fast compared to other methods. Third, they are non-parametric and thus suited for exploratory knowledge discovery. Last, the accuracy of classification is comparable or superior to other models[6]. A variety of decision tree-based algorithms have been developed to extract valuable knowledge from very huge databases including ID3 and extensions, C4.5 and others [11][7]. However, these approaches either cause loss of accuracy or cannot effectively uncover the data structure[12]. C4.5[9] is a well-known decision tree algorithm that in most cases can generate near optimal decision trees but only when the training data are given all together. However when the training examples are given incrementally the C4.5 alone cannot be used efficiently[10]. Gehrke et al.[5] used the RainForest framework data access method for decision tree classifier that separates the scalability aspects of algorithms from the central features that determine the quality of the tree. Also genetic algorithms have been used for classification problems in data mining domain in variety of applications[12].

This paper presents a new hybrid classifier integrating the strengths of genetic algorithm, decision tree and RainForest approaches. The developed classifier GARFC4.5, has been applied to large medical database for thrombosis diseases. The data was made through the Discovery Challenge Competition , organized as part of the 3rd European Conference on Principles and Practice of Knowledge Discovery in Database in Prague[13]. The paper is organized as follows: section two introduces C4.5 decision tree algorithm, RainForest approach for fast decision tree construction and genetic algorithm. Section three explores two C4.5 based classifiers; namely; RFC4.5 and GARFC4.5. RFC4.5 uses the RainForest framework approach while GARFC4.5 is a hybrid classifier uses Genetic Algorithm. Section four presents medical database & data prepration. Section five shows the computational results. Section six ends up with conclusions.

2   Overview

2.1   The C4.5 Algorithm

The algorithm constructs a decision tree starting from a training set S , which is a set of cases , or tuples in database terminology. Each case specifies values for a collection of attributes and for a class. Each attribute may have either *discrete* or *continuous* values. Moreover, the special value *unknown* is allowed, to denote unspecified value. We denote with $C_1, \cdots, C_{Nclass}$ the value of the class.

A *decision tree* T is a model of the data that encodes the distribution of the class label in terms of the predictor attributes. It is a directed, a cyclic graph in form of a tree. The root of the tree does not have any incoming edges. Every other node has exactly one incoming edge and zero or more outgoing edges. If a node n has no outgoing edges we call a *leaf node*, otherwise we call an *internal node*. Each leaf node is labeled with one class label; each internal node is labeled with one predictor attribute called the *splitting attribute*. Each edge e originating from an internal node n has a predicate q associated with it where q involves only the splitting attribute of n. The set of predicates P on the outgoing edges of an internal node must be *non-overlapping* and *exhaustive*. A set of predicates P is *non-overlapping* if the conjunction of any two predicates in P evaluates to *false*. A set of predicates P is *exhaustive* if the disjunction of all predicates in P evaluates to *true*. We will call the set of predicates on the outgoing edges of an internal node n the *splitting predicates of n*; the combined information of splitting attribute and splitting predicates is called the *splitting criteria* of n and is denoted by $crit(n)$.

For an internal node n, let $E = \{e_1, e_2, \cdots, e_k\}$ be the set of outgoing edges and let $Q = q_1, q_2, \cdots, q_k$ be the set of predicates such that edge e_I is associated with predicate q_I. Let us define the notion of the *family* of tuples of a node with respect to database D. The family $F(r)$ of the root node r of decision tree T is the set of all tuples in D.

For a non-root node $n \in T, n \neq r$, let p be the parent of n in T and let $q_{p \to n}$ be the predicate on the edge $e_{p \to n}$ from p to n. The family of the node n is the set of tuples $F(n)$ such that for each tuple $t \in F(n), t \in F(p)$ and $q_{p \to n}(t)$ evaluates to true. Informally, the family of a node n is the set of tuples of the database that follows the path from the root to n when being classified by the tree. Each path W from the root r to a leaf node n corresponds to a classification rule $R = P \to c$, where P is the conjunction of the predicates along the edges in W and c is the class label of node n. Pruning, the method most widely used for obtaining right sized trees, which aim to reduce size of the tree and minimize misclassification rate as well.

C4.5 algorithm constructs the decision tree with divide and conquer algorithm (see Fig. 2.1), each node is associated with a set of cases and a test over the training cases to specify the discriminating attribute which splits the cases into subsets associated to children of the node. The splitting is repeated down the tree trying to produce leaves that contain only nodes belonging to the same class. The most popular split selection method for node tests depend on Entropy (S), Gain (S, A) and GainRatio (S, A) terms, where the attribute with highest GainRatio is chosen as the *splitting attribute*.

$$\mathbf{Entropy(S)} = \sum_{i=1}^{c} -p_i \log_2 p_i$$

Where P_i is the proportion of S to class i and c is the possible values of class (target) attribute.

$$\mathbf{Gain(S, A)} = \mathbf{Entropy(S)} - \sum_{\nu \in Values(A)} \frac{|S_\nu|}{|S|} \mathbf{Entropy(S_\nu)}$$

Where A is an attribute, and S_ν is the subset of S for which attribute A has value ν

$$\mathbf{SplitInformation(S, A)} \equiv -\sum \frac{S_i}{S} \log_2 \frac{S_i}{S}$$

```
ConstructTree(S)
(1) ComputeEntropy(S)
(2) If OneClass or FewCases
        return a leaf;
        create a decision node n;
(3) ForEach Feature A
        ComputeGain(S,A);
(4) n.test=AttributeWithBestGain;
(5) If n.test is continous
        Find threshold;
(6) ForEach S' in the splitting of S
(7) If S' is empty
        Child of n is a leaf
(8) Else
        Child of n=ConstructTree(S');
(9) return n
```

Fig. 1. Pseudo-code of the C4.5

$$\textbf{GainRatio(S, A)} \equiv \frac{\textbf{Gain(S, A)}}{\textbf{SplitInformation(S, A)}}$$

2.2   Rainforest Framework

Over the last few years, there has been a surge in the development of scalable data access methods for classification tree construction. An examination of the split selection methods in the literature reveals that the greedy schema can be refined to the generic RainForest Tree Induction Schema shown in Figure 2. Consider a node n of the decision tree, the split selection method has to make two decisions while examining the family of n : (i) It has to select the splitting attribute X, and (ii) it has to select the splitting predicates on X. Once decided on the splitting criterion, the algorithm is recursively applied to each of the children of n. Let SS denote a representative split selection method. Note that at a node n, the utility of a predictor attribute X as a possible splitting attribute is examined independent of the other predictor attributes: The sufficient statistics are the class label distributions for each distinct attribute value of X. We define the AVC-set of a predictor attribute X at node n to be the projection of F_n onto X and the class label where counts of the individual class labels are aggregated. The AVC-set of predictor attribute X at node n will be denoted by $AVCn(X)$. (The acronym AVC stands for Attribute Value Classlabel). To give a formal definition, we assumed without loss of generality that the domain of the class label is the set $1, \cdots, J$. Let a_n, X, x, i be the number of records t in F_n with attribute value $t \cdot X = x$ and class label $t \cdot C = i$. Formally,

$$a_n, X, x, i = |\{t \in F_n : t \cdot X = x \wedge t \cdot C = i\}|.$$

For a predictor attribute X, let $S = dom(X)xN^J$ where N denotes the set of natural numbers. Then

$$AVC_n(X) = (x; a_1, \cdots, a_J) \in S : \exists t \in F_n : (t \cdot X = x \wedge \forall i \in 1, \cdots, J : a_i = a_n, X, x, i)$$

We define the *AVC-group* of a node n to be the set of the *AVC-sets* of all predictor attributes at node n. Note that the size of the *AVC-set* of a predictor attribute X at node n depends only on the number of distinct attribute values of X and the number of class labels in F_n .

Input: node n, partition D, classification algorithm CL
Output: decision tree for D rooted at n
Top-Down Decision Tree Induction Schema:
BuildTree(Node n, datapartition D, algorithm CL)
(1) Apply LC to D to find crit(n)
(2) let K be the number of children of n
(3) if $(k > 0)$
(4)   Create K children $c_1, \cdots, c_k$ of n
(5)   Use best split to partition D into $D_1, \cdots, D_k$
(6)   for $(I = 1; I <= K; I + +)$
(7)   BuildTree(C_i, D_i)
(8)   endfor
(9) endif
RainForest Refinement:
(1a) for each predictor attribute P
(1b) Call CL.find-best-partitioning (AVC-set of P)
(1c) endfor
(2a)K= CL.decide-spliting-criterion ();

Fig. 2. Tree induction schema and refinement

It has been advisable for several algorithms to construct as many AVC-sets as possible in main memory while minimizing the number of scans over the training database. As an example of the simplest such algorithm, assume that the complete AVC-group of the root node fits into main memory. Then we can construct the tree according to the following simple schema: Read the training database D and construct the *AVC-group* of the root node n in memory.Then determine the splitting criterion from the AVC-sets through an in-memory computation. Then make a second pass over D and partition D into children partitions. This simple algorithm reads the complete training database twice and writes the training database once per level of the tree; more sophisticated algorithms are possible[5].

2.3   Genetic Algorithm

Genetic algorithm (GA) provides an approach to learning that is based loosely on simulated evolution[8]. The genetic algorithm methodology hinges on a population of potential solutions, and as such, exploits the mechanisms of natural selection well known in evolution[3]. Rather than searching from general to specific hypotheses, or from simple to complex, GA generates successor hypotheses by repeatedly mutating and recombining parts of the best currently known hypotheses[8]. Although different implementations of genetic algorithms vary in their details, they typically share the following structure: the algorithm operates by iteratively updating a pool of hypotheses, called the population. On each iteration, all members of the population are evaluated according to the fitness function. A new population is then generated by probabilistically selecting the fit individuals from the current population. Some of these selected individuals are carried forward into the next generation population intact. Others are used as the basis for creating new offspring individuals by applying genetic operations such crossover and mutation[8].

3   The Proposed Classifiers

3.1   RFC4.5

This section presents RFC4.5 algorithm, which combines C4.5 with RainForest. C4.5 has been widely implemented and tested. It constructs the decision tree with divide and conquer algorithm, each node is associated with a set of cases and a test over the training cases to specify the discriminating attribute which splits the cases into subsets associated to children of the node. The splitting is repeated down the tree trying to produce leaves that contain only nodes belonging to the same class. Pruning, the method most widely used for obtaining right sized trees, which aim to reduce size of the tree and minimize misclassification rate as well. Decision trees are usually simplified by discarding one or more subtrees and replacing them with leaves; as when building trees, the class associated with a leaf is found by examining the training cases covered by the leaf and choosing the most frequent class.

The RFC4.5 algorithm starts with reading the training record to constructs the AVC-group then determine the splitting criterion from the AVC-sets through an in-memory computation. Then make a second pass over training records and partition it according to AVC-set for attribute into children partitions. This simple algorithm reads the complete training database once and keep all the information it needs in its data structure instead of accessing database each time so it reduce access time as well. And again it repeats the process for each child of the pervious node until reaching all leaves. The algorithm starts with reading record set RS and for each feature (predicates attributes) it construct AVC-set by calling function create (). The algorithm then follow as C4.5 it calculate the Entropy () of RS and call BestSplit() function which calculate the GainRatio to find best feature and best split. For each child and after it distributes the original RS it recursive call the code again until it reach all leaves. It is noted that in most cases the part of pruning or simplifying the trees cause a loss of classification accuracy due to reducing the tree size or actually the number of nodes in the produced trees. For that reason a combination of pruning part in the Quinlan C4.5 algorithm with a simple pruning algorithm shown in Figure 3 is developed to give a simple tree, save the memory as well as keeping the classification accuracy high.

```
PrunTree(RootNode r)
(1) if (r.isLeaf)
            return;
(2) else
(3) if(r.AllChildern-Leaves)
(4)            if(r.AllSameClass)
(5)                r.Class=Childern-Class;
(6)                r.isLeaf=true;
(7)                Delete Childern;
(8)            endif
(9)            endif
(10) PrunTree(Childern[i]);
(11) endelse
```

Fig. 3. Simple tree-pruning algorithm

3.2   GARFC4.5

Integrating the strengths of *RainForest with* C4.5 and *Genetic Algorithm*, we implement GARFC4.5 algorithm in object-oriented manner to build better tree(s) for classification problems. There are three main steps to solve a problem using a genetic algorithm: Define a representation; define the genetic initialization, cross over, mutation and selection process and define the objective function. Each individual in the population represents a solution to classification problem in a single data structure, which is a decision tree. The initial population individuals (decision trees) are created by considering random sampling percentages of the training record sets. Crossover operation is the random exchange between subtree and subtree from two parent in the population only when crossover conditions are occurred. Mutation operation is implemented as random exchange between two subtree in the same tree also only if the mutation conditions are met. The objective function *(fitness)* is used to evaluate the individuals. Every genetic algorithm requires an objective function this is how the genetic algorithm determines which individuals are better than others. In classification problems it is preferred that the fitness function be the classification accuracy over the test set. So our objective function for the genetic algorithm is the percentage of classifying correct records from test records.

The methodology of object oriented GARFC4.5 classifier is shown in Figure 4. The classifier was implemented in Visual C++ 6.0 on windows 98 with 64.0MB RAM and 350 MHZ. The hybid classifier GARFC4.5 has two main components. The first one, RFC4.5 classifier, combines the previoue described advantage of RainForest with the traditional C4.5 algorithm. RFC4.5 takes a set of training records from the database and produce decision tree according to the randomly selected set of records. The second component, GA, performs the following operations : (a) taking , as input, the generated trees by RFC4.5, and (b) performs the GA operations. Crossover takes two generated trees from two different training data sets by RFC4.5, exchanges two or more subtree, and produces another two different trees with different fitness function value(classification accuracy). Mutation operation takes one generated tree by first component RFC4.5, exchanges two or more subtree of the same given tree or convert one or more subtree to a leaf, and produces a new tree with different fitness function value. The GA operation proceed until the best tree is obtained from all the decision trees generated from randomly selected set of training data by RFC4.5. Figure 5 shows a pseudo code for the Object Oriented GA.

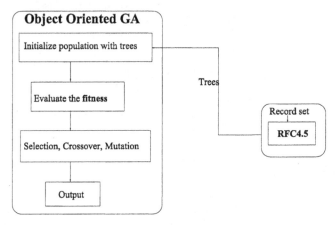

Fig. 4. The proposed GARFC4.5 methodology

```
(1)  GA.Initialize_ Population_Trees();
(2)  GA.Evaluate_Trees_Fitness();
(3)  While (!Termination_Condition)
(4)  GA.Select();
(5)  If (Crossover_Conditions)
(6)  GA.Crossover(parent1, parent2);
(7)  If (Mutation_Condition)
(8)  GA.Mutation(Parent);
(9)  GA.Evaluate_Individuals();
(10) EndWhile.
```

Fig. 5. Pseudo code for object oriented GA

4   Medical Database & Data Prepartion

As a case study for evaluating the developed GARFC4.5 classifier, a medical database has been used. The data was made through the Discovery Challenge Competition, organized as part of the 3rd European Conference on Principles and Practice of Knowledge Discovery in Database in Prague [13]. The database consists of three different tables, of size 20 MB. This database includes all patients (about 1000 records). The first table contains basic information about patients. Second table contains special laboratory examinations for collagen diseases. Table 1 shows the second database table attributes. The third table contains Laboratory examinations stored from 1980 to march 1999. The tests are not necessarily connected to thrombosis. The main goal was to discover some sensitive and specific patterns as well as achieving a high prediction rate for thrombosis disease. The target attribute thrombosis represents the degree of thrombosis 0:negative (no thrombosis), 1: positive (the most sever one), 2:positive (sever), 3: positive (mild).

TABLE 1 Schemata For The Thrombosis Data

Attr. name	Description	Data type
Sex	Male or Femal	Discrete
aCL_IgG	anti-Cardiolipin antibody(IgG) concentration	Continuous
aCL_IgM	anti-Cardiolipin antibody(IgM) concentration	Continuous
ANA	anti-nucleus antibody concentration	Continuous
ANA_Pattern	pattern observed in the sheet of ANA examination	Discrete
aCL_IgA	anti-Cardiolioin antibody(IgA) concentration	Continuous
Diagnosis	disease names	Discrete
KCT	measure of degree of coagulation	Discrete
RWT	measure of degree of coagulation	Discrete
LAC	measure of degree of coagulation	Discrete
Symptoms	other symptoms observed	Discrete

The original data tables have been imported and converted into access format. A straightforward analysis of temporal data was more problematic, since different patients had different numbers of tests, and missing data were more common so a lot of cleanup was necessary before mining. Some objects with duplicate identification numbers have been detected and a lot with missing values and some with unrealistic values. Concentrating on second table only for predicting thrombosis disease and the rest two tables for the other discovered knowledge.Examples for cleanup process were as follow:

- There exist four patients have duplicate values in IDs so they were removed.
- For description_date attribute three patients have wrong format so they were modified.
- For first_date attribute there exist 36 patient with wrong format, 96 times with bad values like (S32.6.9)
- For admission (+, -) there exist three patients have wrong values like ($\ddot{U}$H) and four patients with wrong data which can be updated like (+ve$\ddot{U}$ sh) which updated to (+).
- For diagnosis only one patient with wrong data exist and other three with miss spilling diagnosis name.

- For examination date attribute; one patient has the same examination in the same day with different results.
- For symptoms attribute there exist 6 patients with nonsense values, and there were same symptoms written in different way capital or small and as we know decision tree are case sensitive for attribute value.

After applying the data cleaning process we obtained 406 patients with reasonable data that we will apply our mining technique on. We let 70% of the data for training and 30evaluating the system.

5   Experiments & Computational Results

In our experiments, we set the probability of crossover to 0.9 and mutation rate to 0.03 after trying some other values and discovering that the previous values are the best (as known this value are problem specific). We used 20 generation as the stopping criterion as we notice that the fitness of the individuals doesn't improve anymore. The sampling percentages vary from 5% to 50% of the original dataset and the population size from 10 to 100.

Table 2 shows the computational results for GARFC4.5 from different combination of population size and the sampling percentages. The GARFC4.5 algorithm gives trees with higher classification rate than C4.5; 81.6%; alone. RFC4.5 algorithm with our Prune property also gives higher classification accuracy than traditional C4.5. The table also shows that RFC4.5 gives classification accuracy close to that of GARFC4.5 classifier. From the results we notice that when the sampling percentages become higher the fitness of the produced trees are improved but it's not a general phenomena as it could be wrong in different problem domain. The sampling percentages should be set according to the classification problem domain. GASC4.5 could have some limitations due to a lot of computational steps.

TABLE 2 Computational Results For GARFC4.5

Sample size	Best of initial Population/ by Object Oriented RFC4.5 PopulationSize							
	10	20	30	40	50	60	80	100
5%	0.861842	0.815789	0.861842	0.855263	0.865132	0.815789	0.861842	0.891447
10%	0.884868	0.832237	0.901316	0.911184	0.944079	0.927632	0.888158	0.911184
15%	0.911184	0.930921	0.878289	0.907895	0.884868	0.894737	0.901316	0.898026
20%	0.884868	0.930921	0.904605	0.898026	0.907895	0.927632	0.924342	0.930921
25%	0.944079	0.924342	0.907895	0.940789	0.953947	0.937500	0.940789	0.937500
30%	0.940789	0.940789	0.940789	0.940789	0.944079	0.963816	0.944079	0.940789
35%	0.973684	0.940789	0.944079	0.953947	0.960526	0.953947	0.953947	0.980263
40%	0.976974	0.980263	0.944079	0.953947	0.960526	0.963816	0.953947	0.976974
45%	0.983553	0.986842	0.970395	0.970395	0.976974	0.970395	0.986842	0.986842
50%	0.970395	0.970395	0.990132	0.976974	0.973684	0.990132	0.973684	0.986842
Sample size	Best of initial Population/ by Object Oriented RFC4.5 PopulationSize							
	10	20	30	40	50	60	80	100
5%	0.891447	0.891447	0.875000	0.881579	0.871711	0.865132	0.87553	0.901316
10%	0.917763	0.894737	0.930921	0.911184	0.871711	0.917763	0.914474	0.898026
15%	0.947368	0.930921	0.878289	0.907895	0.888158	0.894737	0.901316	0.950658
20%	0.950658	0.930921	0.904605	0.898026	0.907895	0.937500	0.953947	0.947368
25%	0.957237	0.924342	0.953947	0.953947	0.917763	0.937500	0.953947	0.940789
30%	0.960526	0.953947	0.940789	0.940789	0.944079	0.963816	0.980263	0.944079
35%	0.980263	0.940789	0.950658	0.963816	0.960526	0.953947	0.953947	0.980263
40%	0.983553	0.983553	0.963816	0.963816	0.986842	0.980263	0.980263	0.980263
45%	0.986842	0.990132	0.978974	0.970395	0.976974	0.970395	0.986842	0.986842
50%	0.990132	0.970395	0.990132	0.990132	0.980263	0.990132	0.973684	0.986842
ProgramRunTime/PopulationSize								
Time(s)	10	20	30	40	50	60	80	100
5%	7.228	6.027	7.316	9.169	8.711	10.383	13.280	12.865
10%	7.192	7.539	6.596	7.957	8.836	12.302	13.413	16.292
15%	6.796	8.215	9.176	10.210	10.975	13.122	14.019	16.356
20%	8.275	6.092	8.233	9.978	14.248	12.718	16.521	18.175
25%	7.418	8.175	9.625	12.349	11.984	15.499	16.316	21.138
30%	8.073	8.356	10.162	11.434	14.267	14.742	18.056	19.842
35%	8.026	8.369	10.367	13.328	13.703	16.210	19.438	23.273
40%	10.529	8.931	10.881	12.998	16.738	16.097	19.004	25.976
45%	9.272	8.225	12.440	13.807	16.862	17.525	22.509	26.764
50%	8.823	10.083	12.147	14.261	17.417	20.600	25.106	0.27647

6   Conclusion

This paper presents a new approach for developing two C4.5 based classifiers. The first classifier, RFC4.5, uses the RainForest framework approach while the second, GARFC4.5, is a hybrid classifier uses genetic algorithm. The role of C4.5 classifier is to construct a simple decision tree. The role of RainForest is to keep the scalability aspects in constructing the classifier. The role of Genetic algorithm is working as online or dynamic training. The two developed classifiers have been applied to large medical database for thrombosis diseases. When generate decision tree, it has been found that many branches can be modified to eliminate the redundancy and compress attributes that lead to the same class.

The computational results show that both GARFC4.5 and RFC4.5 classifiers give better decision trees with higher classification accuracy than the traditional C4.5 algorithm. The difference in classification accuracy between traditional C4.5 and RFC4.5 tends to the prune property, which is executed in different way, and also to the repeated values of predicting attributes in testing record set. RFC4.5, as traditional C4.5, involves in local greedy search in the data space beside it works as offline learning classifier. GARFC4.5 is better when the data set is given incrementally as online learning rather than completely at once. Also the results show that, both classifiers give high classification accuracy. Of course this result needs further study using other databases.

References

1. Agrawal, R., Imielinski, T., and Swami, A. (1993). Database mining: A performance perspective, IEEE TKDE.
2. Breiman, L., Friedman, J.H., Olshen, R.A., and Stone, P.J. (1984)Wadsworth International Group.
3. Cios, Krzysztof J., Witold, Pedrycz., and Roman, W.Swiniarski (1998). Data Mining Methods for Knowledge Discovery, Kluwer Academic publishers, boston/dordrecht/London.
4. Gehrke, J., Ramakrishnan, R., Ganti, V., and Yin Loh, W (1998). BOAT-Optimistic Decision Tree Construction.
5. Gehrke, J., Ramakrishnan, R., and Ganti, V (2000). Rainforest a framework for fast decision tree construction of large datasets. Data Mining and Knowledge Discovery, 4(2/3):127-162.
6. Lim, T.S., Loh, W.Y., and Shih, Y.S (2000). A comparison of prediction accuracy, complexity, and training time of thirty three old and new classification algorithms. machine learning, 48:203-228.
7. Michie, D., Spiegelhalter, D.J., and Taylor, C.C (1994). Machine Learining, Neural and Statistical Classification. Ellis HorWood.
8. Mitchell, Tom M (1997). Machine Learning, publisher MaGRAW-Hill-Internaltional editions, computer science series.
9. Quinlan, J.R. C4.5 (1993): Programming for Machine Learnining ,Morgan Kaufman Publishers.
10. Shin'ichi, Oka., and Qiangfu, Zhao (2001). Design of Decision Trees through Integration of C4.5 and GP, the university of Aizu, Aizu-Wakamatsu, Japan 965-8580.
11. Weiss, S.M., and Kulikowski, C.A (1991). computer systems that learn: classification and prediction methods from statistics, neural nets, machine learning, and expert systems.
12. Zhiwei, Fu (1999). An Innovative GA-Based Decision Tree Classifier in Large Scale Data Mining.
13. Zytkow, Jan.M., and Rauch, Jan (1999). Principles of data mining and knowledge discovery, third European conference, PKDD'99 Prague, Czech Republic, September 1999, proceedings, Springer, Lecture Notes in Artificial Intelligence 1704.

Optimization of the ABCD Formula Used for Melanoma Diagnosis

Alison Alvarez[1], Stanislaw Bajcar[2], Frank M. Brown[3], Jerzy W. Grzymala-Busse[3], and Zdzislaw S. Hippe[4]

[1] Department of Computer Science, George Washington University, Washington, DC, 20052, USA
[2] Regional Dermatology Center, 35-310 Rzeszow, Poland
[3] Department of Electrical Engineering and Computer Science, University of Kansas, Lawrence, KS 66045, USA
[4] Department of Expert Systems and Artificial Intelligence, University of Information Technology and Management, 35-225 Rzeszow, Poland

Abstract. The ABCD formula is used for computing a new attribute, called TDS, to help with melanoma diagnosis. In our research four discretization techniques were used, two of them never published before. We found four corresponding new ABCD formulas to compute TDS by applying more than 163 thousand experiments of variable ten-fold cross validation. Diagnosis of melanoma with each of these new ABCD formulas, when used with an appropriate discretization technique, is significantly more accurate (with the level of significance 5%) than diagnosis using the traditional ABCD formula. Finally, the rule sets, induced from data sets obtained using four new ABCD formulas and the traditional ABCD formula, were graded by an experienced melanoma diagnostician.

1 Introduction

Melanoma is an extremely dangerous skin cancer. Every improvement of melanoma diagnosis has a significant impact on saving human life. Research reported in this paper is a continuation of our previous research on improving diagnosis of melanoma based on data mining. Our previous results on optimization of the ABCD (Asymmetry, Border, Color, and Diversity) formula, used for melanoma diagnosis, were reported in [1], [5], [7] and [8]. In our current research the main objective was also the optimization of the ABCD formula. However, this time we used completely different methodology. First, in our previous research [8] we compared two discretization techniques, based on agglomerative and divisive cluster analysis. It turned out that the divisive cluster analysis technique was much better, most likely because the total number of final clusters using the divisive method was usually smaller than the total number of final clusters obtained by the agglomerative cluster method [13]. Therefore, this time we decided to use four different discretization schemes, two of them based on divisive cluster analysis. The remaining two discretization techniques are new, unpublished techniques. More importantly, in our previous research we compared discretization techniques used on melanoma data only for the traditional ABCD formula. This time we compared four different discretization techniques tested on six different melanoma data sets. Moreover, we computed more precisely the total number of errors by averaging results of 30 experiments of variable ten-fold cross validation, as oppose to a single execution of the fixed ten-fold cross validation, reported in our previous research.

Additionally, in our previous research we were searching for the optimal ABCD formula using fixed ten-fold cross validation (i.e., with fixed re-shuffling of the data set) and then validated our results using 30 experiments of variable ten-fold cross validation, with randomly changed re-shuffling of the data set. Results of 30 variable ten-fold cross validation were much worse since the best results obtained by fixed ten-fold cross validation were outliers. Therefore, this time we decided to use variable ten-fold cross validation to search for the optimal ABCD formula. Using 30 experiments for every candidate—ABCD formula—would result in enormous computational complexity. Our compromise was to apply five variable ten-fold cross validations for every candidate ABCD formula and then validation with 30 variable ten-fold cross validations. Thus in our current research we used only variable ten-fold cross validation instead of less reliable fixed ten-fold cross validation.

In some papers [10] and [14] claims were published that the ABCD formula does not improve diagnosis of melanoma. To validate these claims we ran experiments on melanoma data with a removed attribute called TDS, which is a result of the ABCD formula.

Finally, an experienced melanoma diagnostician graded rule sets induced from data sets obtained by using four new ABCD formulas and the fifth, traditional ABCD formula, each combined with the best appropriate discretization technique. Thus we obtained useful feedback: an expert's opinion on the usefulness of the final product of data mining.

2   ABCD formula

Our data on melanoma were collected at the Regional Dermatology Center in Rzeszow, Poland [9]. The data consisted of 410 cases. In diagnosis of melanoma an important indicator is TDS (total dermatoscopic score), computed on the basis of the ABCD formula [4] and [15], using four variables: *Asymmetry*, *Border*, *Color* and *Diversity*. The variable *Asymmetry* has three different values: *symmetric spot, one axial symmetry*, and *two axial symmetry*. *Border* is a numerical attribute, with values from 0 to 8. A lesion is partitioned into eight segments. The border of each segment is evaluated; the sharp border contributes 1 to *Border*, the gradual border contributes 0. *Color* has six possible values: *black, blue, dark brown, light brown, red* and *white*. Similarly, *Diversity* has five values: *pigment dots, pigment globules, pigment network, structureless areas* and *branched streaks*. In our data set *Color* and

Diversity were replaced by binary single-valued variables. The TDS is traditionally computed using the following formula (known as the ABCD formula):

$$TDS = 1.3 * Asymmetry + 0.1 * Border + 0.5 * \Sigma Colors + \Sigma Diversities,$$

where for *Asymmetry* the value *symmetric spot* counts as 0, *one axial symmetry* counts as 1, and *two axial symmetry* counts as 2, $\Sigma Colors$ represents the sum of all values of the six color attributes and $\Sigma Diversities$ represents the sum of all values of the five diversity attributes.

3   Discretization

The melanoma data set contained three numerical attributes: Asymmetry, Border and TDS. The numerical attributes should be discretized before rule induction. The data mining system LERS uses for discretization a number of discretization algorithms [2]. In our experiments a polythetic divisive method of cluster analysis [3] was used in two of the four used discretization techniques. *Polythetic* methods use all attributes while *divisive* methods begin with all cases being placed in one cluster. Our method was also *hierarchical,* i.e., the final structure of all formed clusters was a tree.

Initially all cases were placed in one cluster C_1. Next, for every case the average distance from all other cases was computed. The case with the largest average distance was identified, removed from C_1, and placed in a new cluster C_2. For all remaining cases from C_1 a case c with the largest average distance d_1 from all other cases in C_1 was selected and the average distance d_2 from c to all cases in C_2 was computed. If $d_1 - d_2 > 0$, c was removed from C_1 and put to C_2. Then the next case c with the largest average distance in C_1 was chosen and the same procedure was repeated. The process was terminated when $d_1 - d_2 \leq 0$. The partition defined by C_1 and C_2 was checked whether all cases from C_1 were labeled by the same decision value and, similarly, if all cases from C_2 were labeled by the same decision value (though the label for C_1 might be different than the label for C_2). For clusters that contain cases labeled by at least two distinct decision values the same procedure of splitting into two clusters was repeated until all final clusters were labeled by the same decision value.

Once clusters are formed the postprocessing starts. First all clusters are projected on all attributes. Then the resulting intervals are merged to reduce the number of intervals and, at the same time, preserving consistency. Merging of intervals begins from *safe merging*, where, for each attribute, neighboring intervals labeled by the same decision value are replaced by their union. The next step of merging intervals is based on checking every pair of neighboring intervals whether their merging will result in preserving consistency. If so, intervals are merged permanently. If not, they are marked as un-mergeable. Obviously, the order in which pairs of intervals are selected affects the final outcome. In our experiments we used two criteria:

- start from an attribute with the most intervals first,
- start from an attribute with the largest conditional entropy of the decision given attribute.

In our experiments four different discretization methods were used. These methods are named on the basis of two questions produced by LERS: first, whether to use divisive cluster analysis (0 means use cluster analysis, 1 means skip cluster analysis and project original, single cases on all attributes and continue with postprocessing), secondly, whether to process an attribute with the most intervals first (denoted by 0) or whether to process an attribute with the largest conditional entropy first (denoted by 1). Thus, our methods of discretization are denoted by 00, 01, 10, 11. For example, 00 denoted using cluster analysis and processing attributes with the most intervals first.

4   Rule Induction and Validation

In our experiments rules were induced by the algorithm LEM2 (Learning from Examples Module, version 2). LEM2 is a part of the system LERS (Learning from Examples based on Rough Sets) [6]. Rough set theory was initiated in 1982 [11], [12].

The most important performance criterion for methods of data mining is the total number of errors. For evaluation of an error number we used the ten-fold cross validation: all cases were randomly re-ordered, and then the set of all cases is divided into ten mutually disjoint subsets of approximately equal size. For each subset, all remaining cases are used for training, i.e., for rule induction, while the subset is used for testing. Thus, each case is used nine times for training and once for testing. Note that using different re-orderings of cases causes slightly different error numbers. LERS may use *constant ten-fold cross validation* by using the same way of case re-ordering for all experiments. Also, LERS may perform ten-fold cross validation using different case re-orderings for every experiment, called *variable ten-fold cross validation*.

5   Experiments

Our experiments were designed to find the optimal ABCD formula. We assumed that the optimal ABCD formula, for computing a new TDS, should be a linear combination of 13 attributes:

$$new\_TDS = c_1 * Asymmetry + c_2 * Border + c_3 * Color\_black +$$
$$c_4 * Color\_blue + c_5 * Color\_dark\_brown + c_6 * Color\_light\_brown +$$
$$c_7 * Color\_red + c_8 * Color\_white + c_9 * Diversity\_pigment\_dots +$$
$$c_{10} * Diversity\_pigment\_globules + c_{11} * Diversity\_pigment\_network +$$
$$c_{12} * Diversity\_structureless\_areas + c_{13} * Diversity\_branched\_streaks.$$

Our objective was to find optimal values for coefficients c_1, c_2,..., c_{13} for each discretization method. The criterion of optimality was the smallest total number of errors for variable ten-fold cross validation for data with 13 old, unchanged attributes and with a new fourteenth attribute, new_TDS, that replaced the original TDS attribute. In the first phase four series of experiments were performed, for different vectors $(c_1, c_2,..., c_{13})$, applying the same discretization method. For each vector $(c_1, c_2,..., c_{13})$ the corresponding new_TDS was computed and then the sequence of five variable ten-fold cross validations was used for the evaluation of the number of errors. The smallest error indicated the optimal choice of $(c_1, c_2,..., c_{13})$. In the sequel, such a data set, with TDS computed on the basis of the optimal $(c_1, c_2,..., c_{13})$ and with all remaining 13 attributes unchanged, will be called a data set obtained by the corresponding discretization method.

A special script was created to compute the new_TDS given ranges for all 13 coefficients c_1, c_2,..., c_{13}. Due to computational complexity, not all combinations of coefficients that are implied by Tables 1 – 4 were tested. Experiments were conducted in sets of a few thousand at a time. Some overlapping occurred between such sets of experiments. In phase 1 the total number of executed variable ten-fold cross validations was about 163 thousand. In phase 2 of our experiments, the four data sets obtained by the four discretization methods, plus the original data set and the original data set with removed attribute TDS, were discretized using all four discretization methods and then the total number of errors was computed using 30 experiments of variable ten-fold cross validations. In phase 2 the total number of executed variable ten-fold cross validations was about nine hundred.

Finally, five selected rule sets, induced from data sets obtained by the four discretization methods and the discretized using the same discretization method and, additionally, the original data set (with TDS computed using the original formula) and discretized using the 00 discretization method were graded by an experienced melanoma diagnostician. Every rule was graded on the scale from 0 to 5, with 0 meaning a useless rule, i.e., rule that is misleading, and 5 meaning an excellent rule.

TDS based on:	Discretization method			
	00	01	10	11
00 discretization method	12.17	12.73	13.20	13.03
01 discretization method	37.57	12.23	37.27	12.50
10 discretization method	12.67	12.80	12.23	12.73
11 discretization method	36.97	12.50	37.27	12.47
Original ABCD formula	13.97	14.43	14.17	14.07
No TDS	55.23	55.37	55.73	55.17

Table 1. Average number of errors

TDS based on:	Discretization method			
	00	01	10	11
00 discretization method	2.157	1.911	2.384	1.991
01 discretization method	3.137	2.553	4.920	1.943
10 discretization method	2.426	2.235	1.597	2.016
11 discretization method	4.335	2.610	3.403	2.640
Original ABCD formula	1.520	2.330	2.086	2.212
No TDS		4.666	4.672	5.212	3.403

Table 2. Standard deviation

6 Conclusions

Each entry from Table 1 is presented in the following analysis as a hyphen-ated expression. The initial part of the expression describes the discretization technique that was used to obtain a new TDS, the part that follows the hyphen describes the method used for discretization of the corresponding data set. *Orig* stands for the original data set, *NoTDS* stands for the original data set with the TDS attribute removed.

We present results of the standard statistical test on the difference between two means, used pairwise on 24 entries from Table 1, taking into account their corresponding standard deviation from Table 2, with the level of significance 5%, two tailed test. There are no significant differences in error numbers between any two methods from the set {00-00, 00-01, 00-10, 00-11, 01-01, 01-11, 10-00, 10-01, 10-10, 10-11, 11-01, 11-11}. Similarly, there are no significant differences in error numbers between any two methods from the following sets: {Orig-00, Orig-01, Orig-10, Orig-11}, {01-00, 01-10, 11-00, 11-10} and {NoTDS-00, NoTDS-01, NoTDS-10, NoTDS-11}, provided that both methods are taken from the same set. All methods from the set {00-00, 00-01, 01-01, 01-11, 10-00, 10-01, 10-10, 10-11, 11-01, 11-11} are better (the error numbers are smaller) than methods from the set {Orig-00, Orig-01, Orig-10, Orig-11}. Also, method 00-10 is better than any of the following methods: Orig-00, Orig-10, Orig-11; and method 00-11 is better than Orig-11. Finally, all methods from the set {Orig-00, Orig-01, Orig-10, Orig-11} are better than methods from the set {01-00, 01-10, 11-10, 11-11} as well as all methods from the set {01-00, 01-10, 11-00, 11-10} are better than methods from the set {NoTDS-00, NoTDS-01, NoTDS-10, NoTDS-11}.

The best discretization method is denoted by 00 (i.e., the discretization based on divisive cluster analysis and processing attributes with the most intervals first).

Among four discretization methods that were used for our experiments the best results were achieved when a discretization method used for searching for the optimal ABCD formula was the same as the discretization method used for discretization of the corresponding data set.

All four of our new ABCD formulas provide, when combined with appropriate discretization techniques, significantly better diagnosis of melanoma.

The choice among the two options used in discretization: processing attributes with the most intervals first or processing attributes with the largest conditional entropy first is more important than the choice for the remaining two possibilities: using or not using cluster analysis. The option of processing attributes with the largest entropy first is less stable (in Table 1, entries in rows denoted by 01 and 11 and columns 00 and 10 are much larger than other entries in the same rows).

Entry	Total weighted score per rule set	Average weighted score per rule
00-00	58.58	2.72
01-01	63.13	2.90
10-10	49.38	2.30
Orig-00	107.17	3.23

Table 3. Diagnostician's scores for rule sets

Another interesting observation can be made comparing Table 1 with Table 3, presenting the diagnostician's scores for five selected rule sets. Though all entries: 00-00, 01-01, 10-10, and 11-11 are significantly better (the number of errors during diagnosis of melanoma is significantly smaller than using Orig-00), the diagnostician definitely graded Orig-00 much higher. Most likely, the diagnostician was used to the original values of TDS and, consequently, the corresponding rules were praised higher. The intervals 1..4.9, 4.9..5.5, 5.5..8.7 for TDS, computed by the 00 discretization method, are very close to the intervals used by diagnosticians (1..4.75, 4.8..8.45, 5.5..8.7). This is another proof of the high quality of the 00 discretization method. As follows from Table 1, melanoma data with removed TDS are associated with much larger number of errors. Thus, the claim from [10] that the ABCD formula is not useful for diagnosis of melanoma seems to be unjustified. In the future we are planning to extract from our data cases with small lesions to check claim from [14].

Acknowledgment. This research has been partially supported by the State Committee for Research (KBN) of the Republic of Poland under the grant 7 T11E 030 21.

References

1. Bajcar, S., Grzymala-Busse, J. W., and Hippe, Z. S. (2003) A comparison of six discretization algorithms used for prediction of melanoma. Proceedings of the Eleventh International Symposium on Intelligent Information Systems, IIS'2002, Sopot, Poland, June 3-6, 2002, Physica-Verlag, 3–12.

2. Chmielewski, M. R. and Grzymala-Busse, J. W. (1996) Global discretization of continuous attributes as preprocessing for machine learning. Int. Journal of Approximate Reasoning **15**, 319–331.
3. Everitt, B. 1980) Cluster Analysis, Second Edition. Heinmann Educational Books, London, UK.
4. Friedman, R. J., Rigel, D. S., and Kopf, A. W. (1985) Early detection of malignant melanoma: the role of physician examination and self-examination of the skin. CA Cancer J. Clin. **35**, 130–151.
5. Grzymala-Busse, J. P., Grzymala-Busse, J. W., and Hippe, Z. S. (2001) Melanoma prediction using data mining system LERS. Proceeding of the 25th Anniversary Annual International Computer Software and Applications Conference COMPSAC 2001 Chicago, IL, October 8–12, 2001, IEEE Computer Society, 615–620.
6. Grzymala-Busse, J. W. (1997) A new version of the rule induction system LERS. Fundamenta Informaticae **31**, 27–39.
7. Grzymala-Busse, J. W. and Hippe, Z. S. (2002) Postprocessing of rule sets induced from a melanoma data set. Proceedings of the COMPSAC 2002, 26th Annual International Conference on Computer Software and Applications, Oxford, England, August 26–29, 2002, IEEE Computer Society, 1146–1151.
8. Grzymala-Busse, J. W. and Hippe, Z. S. (2002) A search for the best data mining method to predict melanoma. Proceedings of the RSCTC 2002, Third International Conference on Rough Sets and Current Trends in Computing, Malvern, PA, October 14–16, 2002, Springer-Verlag, 538–545.
9. Hippe, Z. S. (1999) Computer database NEVI on endargement by melanoma. Task Quarterly **4**, 483–488.
10. Lorentzen, H., Weismann, K., Secher, L., Peterson, C. S., Larsen, F. G. (1999) The dermatoscopic ABCD rule does not improve diagnostic accuracy of malignant melanoma. Acta Derm. Venereol. **79**, 469–472.
11. Pawlak, Z. (1982) Rough Sets. International Journal of Computer and Information Sciences, **11**, 341–356.
12. Pawlak, Z. (1991) Rough Sets. Theoretical Aspects of Reasoning about Data. Kluwer Academic Publishers, Dordrecht, Boston, London.
13. Peterson, N. (1993) Discretization using divisive cluster analysis and selected post-processing techniques. Department of Computer Science, University of Kansas, Internal Report.
14. Pizzichetta, M. A., Piccolo, D., Argenziano, G., Pagnanelli, G., Burgdorf, T., Lombardi, D., Trevisan, G., Veronesi, A., Soyer, H. P. (2001) The ABCD rule of dermatoscopy does not apply to small melanocytic skin lesions. Archives of Dermat. **137**, 1376-1378.
15. Stolz, W., Braun-Falco, O., Bilek, P., Landthaler, A. B., Cogneta, A. B. (1993) Color Atlas of Dermatology, Blackwell Science Inc., Cambridge, MA.

An Instance Reduction Algorithm for Supervised Learning

Ireneusz Czarnowski and Piotr Jędrzejowicz

Department of Information Systems
Gdynia Maritime University
Morska 83, 81-225 Gdynia
Poland
irek, pj@am.gdynia.pl

Abstract. The paper proposes a simple, heuristic, instance reduction algorithm (IRA), which can be used to increase efficiency of the supervised learning. IRA task is to select some instances from the original training set. A reduced training set consisting of selected instances is used as an input for the machine learning algorithm. This results in reducing time needed for learning or increasing learning quality or both. IRA is based on calculating for each instance in the original training set a value of its similarity coefficient and then grouping instances into clusters with identical values of the coefficient. Out of each cluster only a certain number of instances is selected to form a reduced training set. The approach has been validated by means of the computational experiment.

1 Introduction

As it has been observed in [12], the in supervised learning, a machine learning algorithm is shown a training set, which is a collection of training examples called instances. Each instance has an input vector and an output value. After learning from the training set, the learning algorithm is presented with additional input vectors, and the algorithm must generalize, that is to decide what the output value should be.

It is well known that in order to avoid excessive storage and time complexity, and possibly to improve generalization accuracy by avoiding noise and overfitting, it is often necessary or useful to reduce original training set by removing some instances before learning phase or to modify the instances using a new representation.

In the instance reduction approach a subset S of instances to keep from the original training set T can be obtained using incremental, decremental and batch search. An incremental search begins with an empty subset S and adds each instance in T to S if it fulfills some criteria (see for example [10]). The decremental search begins with $S = T$, and then searches for instances to remove from S (see for example [5,12]). Finally, in a batch search mode, all instances are evaluated using some removal criteria. Then all those that do meet the criteria are removed at a single step (see for example [11]).

The paper proposes a simple, heuristic, instance reduction algorithm (IRA). which belongs to the batch search mode class. The approach involves the following steps:

- Calculating for each instance from the original training set the value of its similarity coefficient.
- Grouping instances into clusters consisting of instances with identical values of this coefficient.
- Selecting the representation of instances for each cluster and removing remaining instances, thus producing the reduced training set.

Section 2 of the paper introduces the proposed similarity coefficient and presents a formal description of IRA. Section 3 describes computational experiment design and gives details on benchmark data sets used in the experiment. Section 4 discusses computational experiment results. Section 5 includes conclusions and suggestions for future research.

2   Instance Reduction Algorithm

Instance Reduction Algorithm (IRA) aims at removing a number of instances from the original training set T and thus producing the reduced training set S. It is assumed that a training set is a collection of training examples called instances. Each instance consists of an input vector and an output value. Let N denote the number of instances in T and n - the number of attributes, that is the number of elements in the input vector. Let also $X = \{x_{ij}\}$ $i = 1 \ldots N, j = 1 \ldots n$, denote a matrix of n columns and N rows containing all input vectors from T. IRA involves the following stages:

Stage 1. Transform X normalizing value of each x_{ij} into interval $< 0, 1 >$ and then rounding it to the nearest integer, that is 0 or 1.

Stage 2. Calculate for each instance from the original training set the value of its similarity coefficient I_i:

$$I_i = \sum_{j=1}^{n} x_{ij} s_j, i = 1 \ldots N, \tag{1}$$

where

$$s_j = \sum_{i=1}^{N} x_{ij}, j = 1 \ldots n. \tag{2}$$

Stage 3. Map input vectors (i.e. rows from X) into t clusters denoted as Y_v, $v = 1 \ldots t$. Each cluster containing input vectors with identical value of the similarity coefficient I_i, where t is a number of different values of I_i.

Stage 4. Set value of the representation level k, which denotes the maximum number of input vectors to be retained in each of t clusters defined in stage 3. Value of k is set arbitrarily by the user.

Stage 5. Select input vectors to be retained in each cluster. Let y_v denote a number of input vectors in cluster v, $v = 1 \dots t$. Then the following rules for selecting input vectors are applied:

- If $y_v \leq k$ then all input vectors in Y_v are kept.
- If $y_v > k$ then the order of input vectors in Y_v is randomized and the cluster partitioned into $d = \frac{y_v}{k}$ subsets denoted as D_{vu}, $u = 1 \dots d$. Generalization accuracy of each subset denoted as A_{vu} is calculated using the so called leave-$(d-1)$-out test with $X' = X - Y_v + D_{vu}$ as a training set. Subset of input vectors from cluster v maximizing value of A_{vu} is kept to be stored in the reduced training set.

Stage 6. The retained input vectors from all clusters are integrated into the reduced training set.

3   Computational Experiment

3.1   Data Sets Used in the Experiment

Benchmark problems used in the experiment include four well-known classification problems from Machine Learning Database Repository at the University of California [9] which are Wisconsin breast cancer, Cleveland heart disease, credit approval and thyroid disease. Additional problem is a customer intelligence in banking provided under the EUNITE World Competition 2002 [4]. All the above problems require classification based on input vectors with both - continuous and binary attributes.

Diagnosis of breast cancer involves classifying a tumor as either benign or malignant based on cell descriptions gathered by microscopic examination. Breast cancer databases was obtained from Dr. William H. Wolberg, University of Wisconsin Hospitals, Madison [8]. It includes 699 examples, 9 inputs and 2 outputs each.

Cleveland heart disease problem involves predicting heart disease, that is deciding whether at least one of four major vessels is reduced in diameter by more than 50%. The binary decision is made based on personal data. The data set includes 303 examples with 13 inputs and 2 outputs.

Credit card approval involves predicting the approval or rejection of a credit card to a customer. The data set consists of 690 examples with 15 inputs and two outputs with a good mix of attributes.

Thyroid disease problem involves predicting to determine whether a patient referred to the clinic hypothyroid. Therefore three classes are built: normal (not hypothyroid), hyperfunction and subnormal functioning. Because 92 percent of the patients are not hyperthyroid a good classifier must be significant better than 92%. The original data set consists of 3772 instances in training set with 21 inputs and three outputs.

The customer intelligence in banking problem requires classifying bank customers as either active or non-active. The data set consists of 12000 instances with 36 attributes each.

3.2   Learning Machine Used in the Experiment

To validate the proposed IRA it has been decided to evaluate generalization accuracy using a set of artificial neural networks as learning machines, presenting them with various reduced sets of instances during the supervised learning stage and comparing results with those obtained without reducing the respective training sets. All ANN used during the experiment have had the MLP structure with 3 layers - input, hidden and output. Number of neurons in layers 1, 2 and 3, respectively, has been set to the following values:

- Wisconsin breast cancer - 9, 9 and 1.
- Cleveland heart disease - 13, 13 and 1.
- Credit approval - 15, 15 and 1.
- Thyroid disease - 21, 21 and 3.
- Customer intelligence in banking (CI) - 36, 15 and 1.

The range of weights has been set to $[-10, 10]$ and the sigmoid activation (transfer) function has had the sigmoid gain value set to 1.0.

3.3   Training Algorithm

To train artificial neural networks an implementation of the population learning algorithm (PLA), originally proposed in [6], is used. Neural network trained using the PLA is further on referred to as the PLAANN. Possibility of applying population learning algorithms to train ANN has been investigated in earlier papers of the authors [1–3]. Several versions of the PLA have been designed, implemented and applied to solving variety of benchmark problems. Initial results were promising showing good or very good performance of the PLA as a tool for ANN training [3].

In order to increase efficiency of the approach it has been decided to use the parallel computing environment based on PVM (Parallel Virtual Machine). This allows for running parallel learning processes or groups of such processes and thus speeding up training of neural networks.

A neural network learns patterns by adjusting its weights. Learning process can be considered as a search for weights of connections between neurons such that a network can output the correct target pattern for each input pattern. Search processes aiming at finding the required weights are carried within the proposed tool as a population learning scheme in accordance with principles of the population learning algorithm.

The parallel PLA scheme is based on the co-operation between the master worker (server) whose task is to manage computations and a number of slave workers, who act in parallel, performing computations as requested by the master. The approach allows a lot of freedom in designing population-learning process. The master worker is managing communication flow during the population learning. It allocates computational tasks in terms of the required population size and number of iterations as well as oversees information exchange between slaves.

The following features characterize the proposed parallel implementation of PLA:

- Master worker defines number of slave workers and size of the initial population for each of them.
- Each slave worker uses identical learning and improvement procedures.
- Master worker activates parallel processing.
- After completing each stage workers inform master about the best solution found so far.
- Master worker compares the received values and sends out the best solution to all the workers replacing their current worst individual.
- Master worker can stop computations if the desired quality level of the objective function has been achieved. This level is defined at the beginning of computations through setting the desired value of the mean squared error on a given set of training patterns.
- Slave workers can also stop computations if the above condition has been met.
- Computation is carried for the predefined number of stages.

In case of all discussed implementations, the PLA code, which is run by each slave workers, is based on the following assumptions:

- An individual is a vector of real numbers from the predefined interval, each representing a value of weight of the respective link between neurons in the considered ANN.
- The initial population of individuals is generated randomly.
- There are five learning and improvement procedures used - standard mutation, local search, non-uniform mutation, gradient mutation and gradient adjustment.
- There is a common selection criterion for all stages; at each stage individuals with fitness below the current average for the population are rejected.

4   Computational Experiment Results

Instance reduction algorithm proposed in Section 2 has been used to generate training sets for all considered problems. For each problem five reduced instance sets have been prepared with a representation level set, respectively, to 1, 2, 3, 4, and 10. Number of thus obtained instances in each training set is shown in Table 1.

Table 1. Number of instances in the reduced training sets and the percentage of training examples retained

Problem	$k = 1$	$k = 2$	$k = 3$	$k = 4$	$k = 10$
Breast	223	304	367	415	530
	(32%)	(43%)	(52%)	(59%)	(76%)
Heart	132	166	187	213	241
	(44%)	(55%)	(62%)	(70%)	(80%)
Credit	147	178	199	203	240
	(21%)	(26%)	(28.8%)	(29.4%)	(35%)
CI	66	91	113	130	272
	(0.6%)	(0.8%)	(0.9%)	(1.1%)	(2.3%)
Thyroid	266	374	454	522	809
	(7%)	(10%)	(12%)	(14%)	(21%)

Applying IRA has clearly resulted in a substantial reduction of training set sizes as compared with original data sets. It remains to show that reducing training set size still preserves basic features of the analysed data. Intuitively, this can be seen, for example, in Fig. 1 and 2, where the initial distribution of values of attributes 35 and 36 in the customer intelligence in banking is compared with their distribution after IRA has been applied with a view to reduce a number of instances.

In the computational experiment the PLAANN classifier has been run 50 times for each representation level (i.e. 1, 2, 3, 4 and 10). Characteristics of thus obtained classifications averaged over 50 runs are shown in Table 2.

The column "original data set" in Table 2 shows results obtained by applying the PLAANN to the original, non-reduced training set using the "10 cross-validation" approach. The remaining accuracies shown in Table 2 have been obtained in a similar manner but for the reduced sets of instances as shown earlier in Table 1.

Overall performance of the PLAANN classifier seems quite satisfactory. It is also clear that increasing representation level leads to a better performance in terms of classifier quality at a cost of higher requirements in terms of computation time (see Fig. 3).

It might be worth noting that in case of the Customer Intelligent problem the PLAANN applied to original set of instances has not been able to find any satisfactory solution in a reasonable time and in fact classification process has been stopped. For the thyroid problem with original training set the PLAANN gives accuracy of classification at the level of about 92%, which is a standard performance. Accuracy of classification for the discussed problem has grown substantially with the reduction of the original dataset size.

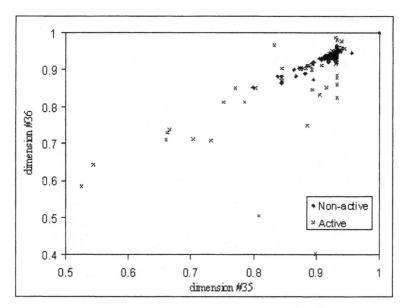

Fig. 1. Initial distribution of values of attributes 35 and 36 in the customer intelligence in banking

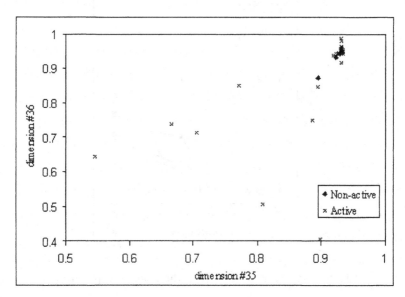

Fig. 2. Distribution after reducing number of instances from of 12000 to 272 (representation level = 10)

Table 2. Averaged accuracy (%) and its standard deviation as produced by PLAANN in the classification experiment

Problem	$k = 1$	$k = 2$	$k = 3$	$k = 4$	$k = 10$	Original data set
Breast	95.0	95.4	93.7	95.1	94.7	96.6
	± 1	± 1	± 4	± 1	± 2	± 1.4
Heart	77.7	79.5	85.7	83.1	86.2	85.7
	± 1.6	± 2.3	± 2	± 7	± 2	± 1.2
Credit	86.0	84.5	85.5	88.5	88.5	88.1
	± 1	± 6	± 6	± 1	± 1	± 1.2
CI	72.1	74.5	74	75.1	80.4	58.7
	± 2.8	± 1.2	± 1.7	± 1.7	± 2	± 3.4
Thyroid	94	94	96.4	99.1	98.9	93.1
	± 1.3	± 1	± 1.2	± 1	± 1	± 1

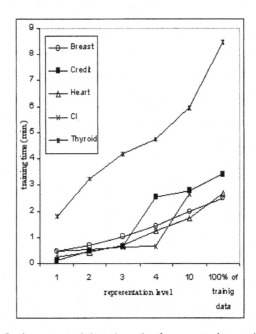

Fig. 3. Average training times in the reported experiment

Comparison of the proposed IRA with other approaches to instance reduction is shown in Table 3.

Table 3. Performance comparison of different instance reduction algorithms

Algorithm	Breast		Heart	(Cleveland)	Credit	
	Accuracy	Retained	Accuracy	Retained	Accuracy	Retained
IRA $(k = 1)$	93.04	37.34	74.00	33.66	85.34	28.38
IRA $(k = 10)$	95.84	81.12	82.70	67.33	85.10	69.36
kNN	96.28	100.00	81.19	100.00	84.78	100.00
CNN	95.71	7.09	73.95	30.84	77.68	24.22
SNN	93.85	8.35	76.25	33.88	81.31	28.38
IB2	95.71	7.09	73.96	30.29	78.26	24.15
IB3	96.57	3.47	81.16	11.11	85.22	4.78
DROP3	96.14	3.58	80.84	12.76	83.91	5.96

The "Retained" column in Table 3 shows what percent of the original training set has been retained by the respective reduction algorithm. All the results, except these produce by the proposed IRA, were reported in [12]. Results for IRA have been obtained under exactly the same experiment layout as described in [12]. Acronyms used in Table 3 stand for k Nearest Neighbour (kNN), Condensed Nearest Neighbour (CNN), Selective Nearest Neighbour (SNN), Instance Based (IB) and Decremental Reduction Optimization Procedure (DROP).

Computational experiment has been carried on a SGI Challenge R4400 workstation with 12 processors. A number of slave workers used by the master varied in different runs from 5 to 15 and in each run has been chosen randomly.

5   Conclusions

Main contribution of the paper is proposing and provisionally validating a simple, heuristic, instance reduction algorithm, which can be used to increase efficiency of the supervised learning. Computational experiment supports authors' claim that reducing training set size still preserves basic features of the analyzed data. The approach extends a range of available instance reduction algorithms. Moreover, it is shown that the proposed algorithm can be, for some problems, competitive in comparison with the existing techniques.

Properties of the proposed algorithm should be, however, further studied with a view of finding an efficient procedure of selecting a combination of instances to be retained from a cluster of instances with identical similarity coefficient.

Another direction of research should focus on establishing decision rules for finding a representation level suitable for each cluster allowing different representation levels for different clusters.

References

1. Czarnowski, I., Jędrzejowicz, P. (2002) Application of the Parallel Population Learning Algorithm to Training Feed-forward ANN. Proceedings of the Euro-International Symposium on Computational Intelligence (E-ISCI), Košice, Slovakia, June 16-19.
2. Czarnowski, I., Jędrzejowicz, P. (2002) Population Learning Metaheuristic for Neural Network Training. Proceedings of the Sixth International Conference on Neural Networks and Soft Computing (ICNNSC), Zakopane, Poland, June 6-10.
3. Czarnowski, I., Jędrzejowicz, P., Ratajczak, E. (2001) Population Learning Algorithm - Example Implementations and Experiments. Proceedings of the Fourth Metaheuristics International Conference, Porto, pp.607-612.
4. The European Network of Excellence on Intelligent Technologies for Smart Adaptive Systems (EUNITE) - EUNITE World competition in domain of Intelligent Technologies - http://neuron.tuke.sk/competition2/.
5. Gates, G.W. (1972) The Reduced Nearest Neighbor Rule. IEEE Transactions on Information Theory, IT-18-3, pp.431-433.
6. Jędrzejowicz, P. (1999) Social Learning Algorithm as a Tool for Solving Some Difficult Scheduling Problems. Foundation of Computing and Decision Sciences (24):51-66.
7. Li, J., Dong, G., Ramamohanarao, K. (2000) Instance-based classification by emerging patterns. Procidings of the Fourth European Conference on Principles and Practice of Knowledge Discovery in Database. Lyon, France, pp.191-200.
8. Mangasarian, O.L., Wolberg, W.H. (1990) Cancer diagnosis via linear programming, SIAM News, Vol. 23, Number 5, September, pp.1-18.
9. Merz, C.J., Murphy,P.M. (1998) UCI Repository of machine learning databases [http://www.ics.uci.edu/~mlearn/MLRepository.html]. Irvine, CA: University of California, Department of Information and Computer Science.
10. Salzberg, S. (1991) A Nearest Hyperrectangle Learning Method. Machine Learning, 6, pp.277-309.
11. Tomek, I. (1976) An Experiment with the Edited Nearest-Neighbor Rule . IEEE Transactions no Systems, Man, and Cybernetics, 6-6, pp.448-452.
12. Wilson, D.R., Martinez, T. R. (2000) Reduction techniques for instance-based learning algorithm. Machine Learning, Kluwer Academic Publishers, Boston, 33-3:257-286.

Dependence of Two Multiresponse Variables: Importance of The Counting Method

Guillermo Bali Ch.[1], Andrzej Matuszewski[2], and Mieczysław A. Kłopotek[2]

[1] Monterrey Institute of Technology, Mexico City Campus, gbali@itesm.mx
[2] Institute of Computer Science, Polish Academy of Sciences, Warsaw, Poland
{amat,klopotek}@ipipan.waw.pl

Abstract. The problem of testing statistical hypotheses of independence of two multiresponse variables is considered. The final aim of the investigation is the formation of methods that discover relevant information existing in a database with (given) population survey results. We show in this research that a formulation of null hypothesis has an impact on p-values of a given subset of data. It is possible therefore to establish: a significant dependence of responses in one sense and a lack of such dependence in another sense. Such a discovery can be meaningful for producers of questionnaire surveys. Specific algorithms for automated data mining can be formulated that consider specific null hypotheses on independency or dependency of survey questions.
Keywords: "pick any" questions, statistical dependency of database fields, bootstrap, questionnaire surveys, Yule-Pearson coefficient

1 Introduction

In a number of machine discovery applications dependence and/or independence between factors plays an important role, among others in Bayesian network learning, decision tree induction, construction of classifiers. For example automated construction of Bayesian network based classifiers heavily relies on conditional and marginal independence between variables.

While traditional Bayesian networks were handling discrete single-valued variables (Spirtes et al.,[13]), more attention is being paid currently to other types of variables, first of all to continuous ones (Olesen, 1993 e.g.). It seems, however, that much research is still needed in case of handling discrete multiresponse variables, in spite of some effort e.g. using Dempster-Shafer belief function approach (Klopotek [8], Wierzchon and Klopotek [15]).

The number of questionnaire population surveys that use multiresponse questions grows. This type of questions seems to be the most effective way to obtain reliable information from a given respondent. Clearly, they can be used for special ends and should be formulated in an appropriate manner.

In this paper we will assume that there are two questions, **A** and **B**, that ask respondent to pick any of the proposed answers according to his choice.

There are several aspects of assessing "correlation" (or the existence of mutual relations) between two multiresponses. We will discuss problems that are connected with testing the general dependency (or independency) between **A** and **B**. In section 2 we will give a brief overview of the literature on the subject of multiresponse independence analysis. Section 3 states our assumptions. In section 4 we present our working example. Sections 5, 6, 7 and 8 explain our methodology, while section 9 contains results of a simulation.

2   Review of literature

The main body of the existing literature refers to the case when only one variable is "genuine" multiresponse e.g.: Agresti and Liu [1], Bilder and Loughin [5], Bilder et.al. [3], Decady and Thomas [7], Loughin and Sherer [9].

One of the problems that has been researched and discussed (see: Bilder et. al. [3], Bilder and Loughin [4]) is the invariance of inference procedures under the different definitions of what is treated as "yes" or "pick". One might think therefore that also there are basically important (for respondent) those options, which were not chosen (by him). Such an opinion is absurd sometimes and therefore the invariance requirement seems to be exaggerated.

We consider here something similar to the invariance principle, but only in a psychological sense. We firmly believe and assume in our approach that only those chosen options characterize responder and/or are most informative.

Thomas and Decady [14] have introduced an elegant method that develops a "natural" form of statistic based on numbers: r_i and c_j defined below. This method must be considered:

1. in its actual form or with a small modification of the Agresti type
2. or as a basis of bootstrapping

in any theoretical and practical researches.

3   Number of answers limitations

In this paper we will discuss basically the following case. For both questions, **A** and **B**, there are three possible answers. We do not consider persons that gave no answer for at least one question. Such persons do not contribute in assessing the "correlation" between **A** and **B**.

The "traditional" multiresponse contingency table: $[n_{ij}]$ of 3×3 dimensions is formed in the following way. We call the method: a "traditional" because: it seems to be quite natural and it is implemented in many professional statistical software packages like SPSS and STATISTICA (documented in their respective Internet sites e.g.). If a responder chooses 1 answer (among 3 possibilities) to the question **A** and two answers to the **B**, then two ones are added to the contingency table. They appear in the row that corresponds to the answer to **A** in its two columns determined by the responders choices among the **B**'s items.

It is somewhat controversial what to do if a respondent chooses all 3 answers to one question. From one point of view, all those answers have certain influence on inferences connected with assessing the dependency between **A** and **B**. From other point, however, the basic (at least for this research) multi-aspect problem of defining "independency" would be even less intuitive if such respondents are taken into account.

In any case, persons with all possible answers chosen for at least one question are not modeled in our random-sampling schemes and therefore will not be considered in this research.

4   An example

Lets assume that managers in a certain organization decided to assess the capabilities of a group of employees in terms of activity and potential. The actual and future possibilities of an employee were categorized as follows:

- a_1 : "works hard and is active",
- a_2 : "has managerial capabilities",
- a_3 : "represents a high professional level".

According to the obvious fact mentioned in the previous section, persons that do not fulfill any of the above requirements can be treated as not interesting for the institution. Therefore, they do not enter in the analysis.

The same employees were asked the following question (as a second multiresponse **B**):

What is your favorite way of spending weekends?

The following options were offered as answers (named: b_1, b_2, b_3 respectively):

- "I always actively practice my hobby and/or physical-sporting activities",
- "My basic wish is to stay with the family",
- "I need many and long involving contacts with books, newspapers, electronic media".

Since the objective is to assess the pattern of correlation between at-work and weekend activities, the persons having assigned all options in **A** or **B** were treated as not "informative".

5   Votes vs. electors

There are two counting methods that will be considered here as bases for calculation the final contingency tables. First of them the traditional one (already mentioned) counts the votes that "received" a given pair of answers, $a_2\, b_3$ say.

Assume that 13 employees are both: 1) assessed as good candidates for the managerial post (a_2), and 2) self-classified as amateurs of information media (b_3). Some of them can "appear" in other cells of the contingency table, though. It should be noted that persons could exist among them, which appear in rows other than the second and columns other than the third. For our purposes it is important, nevertheless, to establish how many employees appear in the second row, i.e., how many of them have managerial possibilities. We will denote this number (being equal or higher than 13) as r2. Analogically, one can denote numbers of employees having a_1 or a_3. Numbers of persons corresponding to the columns will be denoted by c_j .

Attention should be paid to both sums of three components: r_i and c_j . They are greater or equal than N and not higher than 2N, where N is the number of employees that enter our "analysis of correlation".

Of course, the more options employee "has" the more times his/her "vote" enters contingency table. For our example it may seem that this is a quite appropriate solution. The more options an employee has, the more valuable he is for the institution - one can assume. Even for this specific example one can have some doubts, though. Our main aim is to check if there exists a correlation between **A** and **B**. Since employees, having nothing or everything (chosen) at one of the questions, dont count, should perhaps all others be treated equally?

There exists a simple method of counting that makes all respondents equal - we will call it democratic, therefore.

If an employee that has good managerial potential, and besides of being interested in media, he practices a hobby (i.e. he picked b_1), the democratic principle "counts" such employee in a following way. It adds 1 two times: in (2,1) and (2,3) positions of the contingency table.

Denote d_{ij} the "cells" of democratic contingency table. This table, contrary to the "traditional" multiresponse (called earlier as $[n_{ij}]$), has many properties of a simple "uniresponse" table. Total the sums of rows of $[d_{ij}]$ e.g. are equal to the number of employees N, which is not true for multiresponse table: $[n_{ij}]$.

6   Sampling schemes

As was said in section 3, we will be interested here in (at population) random ways of modeling how people is picking the options. There are many models that can be used for this purpose. We will apply firstly those with certain typical, psychological motivations.

POSITIVE scheme illustrates the following real practice that will be presented here for the variable **A** from the example. It may be consequence of either the bosss way of thinking or the actual employees activity at workplace that one option, of those responses proposed within the question A, is completely clear. Say employee obviously is at very high professional level

(a_3). The only problem remaining for his boss is whether there exists another option that can be appropriate.

The opposite way of thinking we will call NEGATIVE. Now it appears that one option becomes completely inapplicable as a characteristic of a given employee. Employee has obviously no chance at all to become a manager e.g. The problem is whether the other two options are valid or not.

The third scheme: PREDEFINED assumes that number of employees for a given pair of answers is given in advance. The sequence of 3 numbers must be predefined.

Another basic aspect of a sampling scheme is the INDEPENDENCE of sampling **A** and **B** responses. The aim of the proposed modeling is a null hypothesis: "a lack of correlation".

Taking into account the above principles the following actual sampling schemes are considered in this research e.g.:

POSITIVE INDEPENDENCE attempts to model marginals that fulfill a positive way of picking options.

NEGATIVE INDEPENDENCE purportedly models the negative choices. It appeared that the difference in the simulated behavior between the schemes is insignificant. There are important formal differences nevertheless in the process of bootstrapping of these schemes.

Both schemes in the proposed implementation have 7 analogous parameters that formally determine actual simulations. First 3 of them are strictly the same; next 2 (p_1, p_2) are technically absolutely the same but with a different rationale. Last 2: q_1, q_2 are different but have the same frequentist interpretation in one basic aspect.

These are the meanings and interpretations of the parameters. We will use the specific sample from our example to be clearer in explications.

N(1,1) - number of employees having (chosen) exactly one option in both questions.

N(1,2) - number of employees having one option in **A** and two options chosen for the question B.

N(2,2) - number of employees having two options in both questions.

It follows therefore that we will not discuss here both mixed-number-of-choices cases simulated together. Models are simplified therefore. Of course the fact that the one of mixed cases: (1,2), instead of the other: (2,1) is chosen is not significant because of the symmetry.

It is important how the marginals of **B** are generated for the (1,2) sub-sample. We show this for the Positive Independence first.

Let p_1, p_2 and $p_3 = (1 - p_1 - p_2)$ be the probabilities of choosing: a_1, a_2 and a_3, respectively. Parameters of generating columns are: q_1, q_2 and $q_3 = (1 - q_1 - q_2)$. For a given employee belonging the (1,2) sub-sample, a first **B** choice (a POSITIVE) is generated according these probabilities. The second choice (of the employee) is done with the probabilities that are proportional to those weekend activities, which have not been chosen as a first choice.

The negative scheme for **B** is based on parameters: ng_1, ng_2 and $ng_3 = 1 - ng_1 - ng_2$. Two weekend activities are chosen as a complement to the number generated with probabilities ng_i. If the number 3 is generated (with a probability: ng_3), then two first weekend activities are chosen.

Positive and negative schemes in the sub-sample (1,2) are equivalent in a sense that the expected occurrences of weekend activities for NEGATIVE scheme are matched with respective occurrences for POSITIVE. It is done through the calculation ng_j as appropriate functions of q_j.

Since the X^2 type statistics are going to be calculated for both schemes then marginals of (1,1) sub-sample have no influence as it was expected. It is true unless the marginal probabilities are very small. We choose 1/3 probability for all marginals in our simulation. It was less obvious that for the (2,2) sub-sample both: Positive and Negative, marginals do not influence neither. We put therefore equal marginal probabilities for this sub-sample, too. It does not mean that sub-samples of the (2,2) type are generated with the same mechanism for Positive and Negative as it is for the (1,1) sub-sample.

7   Review of correlation coefficients

We list the most important correlation coefficients between two variables. Value zero is always a necessary condition for independence between variables.

Continuous data
Coefficient (Pearsons r) belongs to the interval [-1,1]
Discrete, ordinal data
Coefficient (Kendalls tau) does not attain the limits of [-1,1] interval
One variable nominal, second at least ordinal
Coefficient eta belongs to the interval [0,1] – there disappears a distinction between the positive and negative correlation
Nominal data (assuming 3 categories for both varaiables)
Yule-Pearson coefficient (phi) belongs to [0,2] interval
Multiresponse data
Yule-Pearson coefficient or any of its equivalents belongs to the same interval as previous coefficient but it doesn't attain the upper limit

8   Statistics

The classical X^2 statistic for testing independence in two-way contingency table can be applied to both: $[n_{ij}]$ and $[d_{ij}]$ tables. Monographs oriented for the area of discrete statistics (like: Andersen [2] and Bishop et. al., [6]) offer different distributions as models of the data behavior. Once the data matrix fulfills these requirements under the null hypothesis, the chi square distribution for X^2 can be applied to test this hypothesis.

It follows from the descriptions of sampling schemes in the previous section that none of the classical sets of requirements can be fulfilled for tables generated through these schemes. This fact actually turned the area of multiple response researches into a special field, from the very beginning in the end last decade. In [10] an experiment is reported with $X^2([d_{ij}])$. It shows that this statistic has expectation equal to 2 under the hypothesis of independence, where number of responses is randomly chosen (for both: **A** and **B**) among 1 and 2. It is sharply lower value than that (equal 4) that follows from the classical requirements. $X^2([n_{ij}])$ tends to have slightly higher values.

The bootstrap-type procedures, oriented for calculation of a "coefficient of dependence", are considered here. We took into account the Yule-Pearson coefficient of "correlation". The theory behind this coefficient is described in [11].

Probably many practitioners still use X^2 for $[n_{ij}]$ assuming it has chi-square distribution with 4 degrees of freedom. The appropriate Yule-Pearson coefficient is our first statistic called TRADITIONAL.

Second chi-square-type statistic is named DEMOCRATIC.

Third statistic: PICK ANY is based on that introduced by Thomas and Decady [14] - already mentioned. Their statistic has $X^2([n_{ij}])$ form with the following 3 changes.

1. Instead of the sum of rows and columns - values that appear in components of the statistic - the values: r_i and c_j are put, respectively.
2. A correction factor multiplies the value of statistic.
3. The final statistic has 9 degrees of freedom chi-square distribution under the null hypothesis.

The Yule-Pearson coefficient was defined taking into account the above properties.

9   Simulation results

For a given set of 7 parameters, 10 000 samples were generated for each scheme of independence considered here. Actual sizes of the Yule-Pearson statistics were calculated through the estimation of its 95% and 99% percentiles. Besides the averages the sample estimators of standard deviations are reported in the second row for each counting method. First four tables correspond to equal probabilities (and numbers predefined) in (1,2) simulated table.

The last tables (5 and 6) correspond to the sequence: 5-3-1 of predefined sample sizes in columns of generated (1,2) tables. Probabilities in the NEGATIVE sampling scheme are matched accordingly.

Table 1. 95% percentiles of Yule-Pearson coefficients. N(1,1)=N(1,2)=9, N(2,2)=0

Sampling Counting	POSITIVE	NEGATIVE	PREDEFINED
TRADITIONAL	0.2412 0.0019	0.2402 0.0025	0.2334 0.0029
DEMOCRATIC	0.3152 0.0026	0.3151 0.0030	0.3111 0.0035
PICK ANY	0.3769 0.0017	0.3771 0.0039	0.3669 0.0031

Table 2. 99% percentiles of Yule-Pearson coefficients. N(1,1)=N(1,2)=9, N(2,2)=0

Sampling Counting	POSITIVE	NEGATIVE	PREDEFINED
TRADITIONAL	0.3234 0.0033	0.3222 0.0038	0.3049 0.0046
DEMOCRATIC	0.4200 0.0041	0.4232 0.0057	0.4103 0.0055
PICK ANY	0.4971 0.0083	0.5025 0.0091	0.4734 0.0051

Table 3. 95% percentiles of Yule-Pearson coefficients. N(1,1)=N(1,2)=N(2,2)=9

Sampling Counting	POSITIVE	NEGATIVE
TRADITIONAL	0.0674 0.0009	0.0673 0.0005
DEMOCRATIC	0.1445 0.0015	0.1455 0.0019
PICK ANY	0.1437 0.0011	0.1443 0.0011

Table 4. 99% percentiles of Yule-Pearson coefficients. N(1,1)=N(1,2)=N(2,2)=9

Sampling Counting	POSITIVE	NEGATIVE
TRADITIONAL	0.0946 0.0017	0.0938 0.0017
DEMOCRATIC	0.1949 0.0025	0.1978 0.0039
PICK ANY	0.1945 0.0044	0.1928 0.0029

Table 5. 95% percentiles of Yule-Pearson coefficients. N(1,1)=N(1,2)=9, N(2,2)=0

Sampling Counting	NEGATIVE	PREDEFINED
TRADITIONAL	0.2487	0.2443
	0.0026	0.0029
PICK ANY	0.3421	0.3795
	0.0048	0.0049

Table 6. 99% percentiles of Yule-Pearson coefficients. N(1,1)=N(1,2)=9, N(2,2)=0

Sampling Counting	NEGATIVE	PREDEFINED
TRADITIONAL	0.3418	0.3285
	0.0044	0.0040
PICK ANY	0.4651	0.5052
	0.0084	0.0059

10   Final remarks

As this study indicates, the issue of measuring dependence (through testing the independence) or correlation between two multiresponse questions (variables) is strongly related to ones understanding (or definition) of independence. It turns out that type of "voting" (traditional versus democratic) and type of sampling scheme influence the results of independence analysis. Though not investigated here, it may be concluded from our earlier investigations, that the types of independence described here differ strongly from those investigated in the Dempster-Shafer theory also. We can talk about shades of independence.

The issue of shades of independence in the context of multiresponse questions seems to be widely under-investigated in e.g. Bayesian network learning. Further studies are needed in order to find out whether or not one can allow for mixing of various types of independence understanding within a single Bayesian network and whether independence properties implicitly assumed within Bayesian network research hold also for independence types defined within multiresponse analysis.

References

1. Agresti A., Liu L-M, "Modelling a categorical variable allowing arbitrary many category choices", Biometrics 55, 936-943, 1999.
2. Andersen E. B., "The statistical analysis of categorical data", Springer Verlag, 1990.

3. Bilder C. R., Loughin T. M., Nettleton D., "Multiple marginal independence testing for pick any/C variables", Commun. Statist.-Simula., 29, 1285-1316, 2000.

4. Bilder C R., Loughin T. M., "On the first-order Rao-Scott correction of the Umesh-loughin-Sherer statistic", Biometrics 57, 1253-1255, 2001.

5. Bilder C. R., Loughin T. M., "Testing for conditional multiple marginal independence", Biometrics 58, pp. 200-208, 2002.

6. Bishop Y. M. M., Fienberg S. E., Holland P. W., "Discrete Multivariate Analysis: Theory and Practice, Cambridge-Mass. The MIT Press, 1975.

7. Decady Y. J., Thomas D. R., "A simple test of association for contingency tables with multiple column responses", Biometrics 56, 893-896, 2000.

8. Kłopotek M.A., "Methods of Identification and Interpretations of Belief Distributions in the Dempster-Shafer Theory" (in Polish), Publisher: Instytut Podstaw Informatyki PAN, Warszawa 1998.

9. Loughin T., Scherer P. N., "Testing association in contingency with multiple column responses", Biometrics 54, 630-637, 1998.

10. Matuszewski A., Trojanowski K., "Models of multiple response independence", in: M. A. Klopotek, M. Michalewicz, S. T. Wierzchon (ed.), "Intelligent Information Systems 2001", Physica-Verlag (Springer), pp.209-219, 2001.

11. Mirkin B., "Eleven ways to look at the chi-squared coefficient for contingency tables", American Statistician, 55, pp. 111-120, 2001.

12. Olesen K. G., Causal probabilistic networks with both discrete and continuous variables. IEEE Transactions on Pattern Analysis and Machine Intelligence, 3(15), 1993.

13. Spirtes P., Glymour C., Scheines R., "Causation, prediction and search", Lecture Notes in Statistics 81, 1993

14. Thomas D. R., Decady Y. J., "Analyzing categorical data with multiple responses per subject", SSC Annual Meeting, Proceedings of the Survey Methods Section, 121-130, 2000.

15. Wierzchoń S.T., Kłopotek M.A.: "Evidential Reasoning. An Interpretative Investigation." Publisher: Wydawnictwo Akademii Podlaskiej, Siedlce, 2002

On Effectiveness of Pre-processing by Clustering in Prediction of C.E. Technological Data with ANNs

Janusz Kasperkiewicz and Dariusz Alterman

Institute of Fundamental Technological Research Polish Academy of Sciences
Warsaw, Poland; jkasper@ippt.gov.pl, dalter@ippt.gov.pl

Abstract. Civil Engineering technological data are naturally clustered in a specific way. A black-box model of relation between concrete composition and concrete properties can be constructed using a suitable artificial neural network like Fuzzy ARTMAP that was implemented for the presented experiments. After training the system allows valuable prediction of technological data. It was expected that pretreatment of data by their clustering should enable improved prediction on testing examples. The clustering was realized in two different ways: with k - means algorithm and with GCA approach. The improvement of the precision of predictions was found rather limited, but the final efficiency was better, as more records have been positively recognized. The approach seems to be even more promising in case of data of particular internal structure and application of advanced procedures of clustering.

1 Civil engineering data; a HPC database example

Since more than two decades the PPO group at IFTR PAS is studying various aspects of brittle matrix composite materials, their composition, technology, properties, optimisation, etc. It became recently obvious that the amount of information that was previously analysed in a conventional way is rapidly increasing in time both in volume and in complexity, and that the only promising way of exploitation of such information is by application of various soft computing techniques, [3, 4, 5]. The paper is dedicated to results of a certain approach to improve quality of predictions of the property of strength of High Performance Concrete, (HPC), obtained on processing laboratory data collected from technical publications and internal reports during over a decade.

The HPC database had a simple structure: 6+1 of - respectively - the input and the output attributes. The input attributes were - per a cubic meter of concrete - quantities of cement, water, silica, superplasticizer and fine and coarse aggregates, respectively. The single attribute on the output side was the 28-days standard compressive strength of concrete. The collected database on HPC mixes consisted of 633 records concerning results from about 90 investigations, in which a single investigation produced from 1 to

45 records, and each record might represent one as well as many separate measurements.

Because many important attributes are lacking in the database, and also due to natural scatter of engineering data, the final database is expected to represent the knowledge on the composition and strength of concrete in a fuzzy, strongly imprecise way. However, this is the only information that is available in case of most technological results, and it seems to be quite a typical situation in civil engineering investigations.

2   Prediction of concrete strength with black-box models

It is probably a general observation that Civil Engineering data are highly inhomogeneous. In fact the attributes are often clustered. The effect is related to various technological reasons, but it is usually disregarded due to multidimensional description of technological observations.

The experiments on HPC data were performed with several artificial neural networks, (ANNs). Back propagation networks appeared to be ineffective in spite of thousands of training epochs. Algorithm of Fuzzy ARTMAP, however, gave much better results, as was reported in [5]. In view of this positive experience a special application has been prepared for a more user-friendly processing of the data. The program called **beton.exe** has an interface for selection of its control parameters (e.g. vigilance factors) by cross-validation and the results of each prediction is displayed graphically, in form of a diagram of predicted values of the attribute against their actual values, (cf. Fig.2).

As Fuzzy ARTMAP is based on the idea of similarities and dissimilarities between records it has an important feature of producing an escape answer in the testing mode, when the input attributes are too much different from the examples supplied during the training. This quality is absent in many other solutions of ANNs. The results of prediction of the network is therefore characterized not only by the accuracy of predictions, but also by the number of records recognized by the system, (a bi-criterial estimate). There is no indication which configuration of the control parameters is optimal and the best efficiency of the system should be found experimentally.

Another program, used for comparisons in prediction tests was **aiNet**. This is a pseudo ANNs algorithm - a self-organizing system, working on a principle of searching for an optimal estimator of the conditional average in the data distribution, [1]. This program was selected for its effortless operational effectiveness.

3 Clustering of data and its effectiveness

Because of the observed clustered structure of the HPC data it can be expected that models taking into account such clusterisation should better describe the reality, compared to results obtained on the original dataset. The idea therefore was to try a procedure as in the following.

during training

(1) normalize the input data

(2) define number N of disjoint clusters based on the input attributes values: $\{c_1, c_2, \ldots, c_N\}$

(3) train the network (e.g. Fuzzy ARTMAP) in each cluster separately, and store the respective memory modules: $\{mem_1, mem_2, \ldots, mem_N\}$

during testing

(4) allocate the testing records each to a cluster with the centre nearest to the record: $\{test_1, test_2, \ldots, test_N\}$ (a remark: an arbitrary threshold can be introduced here, to exclude certain testing records, being out-of-reach for the present experience of the system);

(5) find predictions of the testing records separately in respective clusters, (finding output values for $test_i$ with mem_i : i = 1, 2, ..., N).

To demonstrate essentials of the proposed idea a two-dimensional example has been prepared taking into account a simple dataset generated as a collection of pairs of points x, y, surrounding centres μ_i , i=1, 2, 3: (2, 2), (5, 6), and (13, 4), using MatLab function for random vectors of multivariate normal distribution - $mvnrnd(\mu, \sigma)$, σ being a covariance matrix, (symmetric, positive semi-definite), [6]. The three centres defined separate clusters of 100, 100 and 300 cases, respectively, Fig.1.

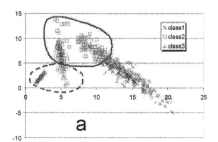 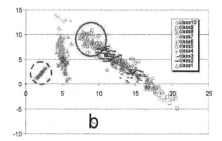

Fig. 1. Categorization of the 500 points dataset with k-means algorithm, (SPSS), into 3 and 10 clusters, respectively. At three clusters result (diagram 'a') records merge in rather unclear way, as indicated by encircling. At ten clusters (diagram 'b') the effect is better, especially the case of class 4 - inside the dashed circle

The database was processed using k-means cluster analysis algorithm of SPSS, with number of clusters specified by the operator, [7]. It can be seen in Fig.1-a that selecting a too low number of the clusters results in records from different subsets merging together, as indicated by encircling. On the other hand a too big number of clusters would make the whole treatment of the data complicated, and - which would be even worst - might result in lost of the expected ordering effect of the procedure.

It was found that predictions on split database were not much better, compared to the prediction on the whole dataset analysed together. The conclusion from the demonstration was, however, that the database may be decomposed into subsets of configurations more convenient for analysis. The best decomposition procedure should be determined experimentally.

In experiments on real, technological data, (HPC database), for clustering applied have been k-means routine available in SPSS, and the procedure realised using GCA (GradeStat), [2]. For training and predictions applied have been both Fuzzy ARTMAP and aiNet.

The result of locating 5 clusters in the whole HPC database was (after normalization of the input attributes) splitting it into classes of - respectively: 1, 1, 46, 570 and 15 records. It was decided to disregard the two smallest clusters as the outliers, and to process the remaining three groups of data separately, as suggested above.

Each cluster has been subdivided into training and testing subsets, in proportion 2/3 and 1/3 of records. After training and finding the separate predictions with aiNet the results have been re-combined and the coefficient of correlation have been calculated as r=0.881. When all the training subsets are combined the prediction on the testing subsets gives the correlation coefficient r = 0.877. There were no experiments with more clusters of the same data.

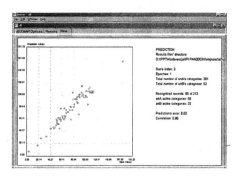

Fig. 2. Predictions of the strength of concrete against the actual values, obtained using beton.exe; 86 of 213 test records were recognized; coefficient of correlation was r=0.961.

In the second experiment the clusters have been obtained in another investigation using GradeStat package, [2]. Here, the total of 633 records has been divided into subsets RES, of 473 records, and OUT, of 160 records, more regular and less regular, respectively, from the point of view of grade correspondence analysis. Like before the predictions were done on 1/3 of the records, after training on approximately 2/3 of the records.

The observed correlations between the actual and the predicted values were: r=0.875 in case of combining separate predictions from the both clusters (182 of 213 records) and r=0.957 when the whole dataset was taken up together. In the latter case, however, only 86 records have been recognized out of 213, Fig. 2. This improved accuracy was obtained at price of the number of records identified: the clustering resulted in a more than two times increase of the number of positive predictions.

4   Conclusions

Experimental observations databases of the type similar to HPC database described in the paper can be used for training and prediction of the important properties of concrete, (like compressive strength). It was found that Fuzzy ARTMAP makes possible satisfactory predictions on records not available during training, with a possibility of increasing precision of the prediction at the payoff of the number of records recognized. aiNet almost always was giving predictions for all the records, but with a tendency to "flatten" the output, by predicting biased values, close to average values of respective attributes.

The procedure of clustering the training dataset into subsets, and - later on - of similar clustering of the testing dataset as well, e.g. by k-means algorithm, makes possible improvement in the number of recognized records, without diminishing the accuracy of prediction, (measured by the coefficient of correlation between the actual and the predicted values).

Acknowledgements. This paper was prepared with a support from the NATO SfP Project 97.1888 - Concrete Diagnosis, and partly with a help from the KBN grant: 8 T11F 01317.

References

1. aiNet - a neural network application for Windows environment. MS WORD documentation available at the Internet address: http://www.ainet-sp.si/
2. GradeStat   -   data   analysis   system;   IPI   PAN,   Warsaw, http://www.ipipan.waw.pl/
3. Kasperkiewicz J., Alterman D., Artificial intelligence in predicting propertiesof brittle matrix composites. In: *Brittle Matrix Composites 6 (BMC 6)*, Ed-s Brandt, Li and Marshall. Woodhead Publ. Ltd. and ZTUREK R-S. Inst., Cambridge and Warsaw 2000, 485-496.
4. Kasperkiewicz J., Artificial neural networks in engineering materials design. Proceedings of the International Conference *Challenges to Civil and Mechanical Engineering in 2000 and Beyond*, June 2-5, 1997, Wroclaw, vol.I, 103-126.

5. Kasperkiewicz J., Racz J., Dubrawski A., HPC strength prediction using artificial neural network. *Journal of Computing in Civil Engineering*, v.9, No.4, October 1995, 279-284.
6. MatLab - Statistics Toolbox; Internet address: http://www.matworks.com/
7. SPSS - a statistical analysis program; Internet address: http://www.spss.com/

The Development of the Inductive Database System VINLEN:
A Review of Current Research

Kenneth A. Kaufman[1] and Ryszard S. Michalski[1,2]

[1] Machine Learning and Inference Laboratory, George Mason University, Fairfax VA 22030, USA
[2] Institute of Computer Science, Polish Academy of Scinces, Warsaw, Poland

Abstract. Current research on the VINLEN inductive database system is briefly reviewed and illustrated by selected results. The goal of research on VINLEN is to develop a methodology for deeply integrating a wide range of *knowledge generation operators* with a relational database and a knowledge base. The current system has already integrated an AQ learning system for generating *attributional rules* in two modes: *theory formation*, in which generated rules are consistent and complete with regard to data, and *pattern discovery*, in which generated rules represent strong patterns, not necessarily consistent or complete. It also has integrated a conceptual clustering module for splitting data into conceptual classes, and providing descriptions of those classes. Preliminary data management and knowledge visualization operators, such as the *intelligent target data generator* (ITG) and *concept association graph* display, have also been integrated. To facilitate an easy interaction with the system, a user-oriented visual interface has been implemented. An example of results from applying VINLEN to a medical problem domain is presented to illustrate VINLEN knowledge discovery and representation capabilities.

1 Introduction

The field of databases is in the midst of an extraordinary growth, and databases are becoming omnipresent and globally connected. In this context, a new research direction has been recently proposed to deeply integrate database technology with modern methods for inductive knowledge generation and for storing and using the knowledge so created. Such integrated systems are called *inductive databases*. In contrast to conventional databases, inductive databases can answer not only queries for which answers are stored in the database, but also queries that require synthesizing and applying *plausible knowledge*, generated by inductive inference from the facts in the database and prior knowledge. Inductive databases can be viewed as a natural next step in the development of the database technology.

This paper presents a brief review of research on the inductive database system VINLEN, which is being developed at the Machine Learning and Inference Laboratory at George Mason University. In VINLEN, inductive inference capabilities, combined with standard relational database operators implemented through an SQL client, are implemented by developing a new type of database operators, called *knowledge generation operators (KGOs)*. KGOs operate on *knowledge segments*, consisting of a combination of one or more relational tables and related knowledge in the knowledge base. A KGO takes one or more input knowledge segments and generates an output knowledge segment.

Two important constraints have been imposed on knowledge generation operators in VINLEN: (1) that their results be in a form that is easy to understand and interpret by users, and (2) that knowledge generated can be expressed compactly and efficiently. These capabilities are achieved by applying ideas and methods for *natural induction*, in which inductive methods create data descriptions in the forms that appear natural to people, by employing *attributional calculus* as a representation language [8]. Attributional calculus is a logic system that combines elements of propositional, first order predicate, and multiple-valued logics. It serves both as an inference system and as a knowledge representation language. Attributional rules, the primary form of knowledge representation in VINLEN, are more expressive than conventional decision rules that use <attribute-relation-value> conditions.

2 VINLEN System

2.1 An Overview

Research on the VINLEN system grows out of our previous efforts on the development of INLEN, an early system for integrating databases and machine learning and inference mechanisms for the purpose of multistrategy learning, data mining, and decision support [12,9]. INLEN included multiple learning and discovery operators, the high-level knowledge generation language KGL-1 [4,10], and an advisory system. It did not, however, integrate an actual database system, instead being constrained to relatively small tables located in reserved files. VINLEN is an entirely new implementation that combines and extends INLEN with a wide range of new or improved knowledge generation operators and an advanced visual interface that provides an easy access to all system operators and components.

Researchers have started to acknowledge the advantages in integrating inductive inference capabilities in a database system. Among the approaches somewhat similar in philosophy to the one presented here are ones presented by Morik and Brockhausen [14], Mannila [7], and Roddick and Rice [15]. In general, VINLEN differs from these in terms of its much wider variety and complexity of knowledge generation operators, its knowledge storage mechanism, and its visual interface.

Fig. 1 presents a general schema for VINLEN. The top part presents a database, which can be local or distributed. VINLEN's database can be one of the widely used commercial databases, for example, ORACLE or ACCESS. The *Target Knowledge Specification* is a generalization of a database query;

specifically, it is a user's request for a knowledge segment to be created by the system.

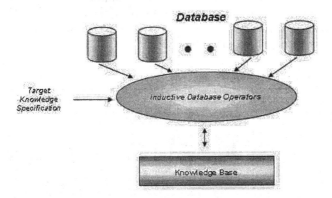

Fig. 1. A general schema of the VINLEN inductive database

The central part, *Inductive Database Operators*, contains a collection of operators that call upon various programs for knowledge generation (e.g., rule learning, conceptual clustering, target data generation, etc.) as well as SQL functions. These operators can be invoked by a user directly, or through a *knowledge query language (KQL)*.

The KQL is an extension of the SQL database query language that integrates SQL with the KGL-1 capabilities. In addition to conventional data management operators, it thus includes operators for conducting inductive and deductive inference, statistical analysis, and many supportive functions. KQL is quite different from traditional high-level languages for data exploration, which have generally been Prolog-based. Among the exceptions to the Prolog-based approach, M-SQL [3] is philosophically similar to KQL in that it builds upon the SQL data query language, integrating it with one inductive operator. KQML [2] allows the querying for specific pieces of knowledge, although it does not support the abstract templates and multiple discovery operators supported by KQL.

The VINLEN knowledge base contains definitions of the domains and types of attributes in the database, data constraints, value hierarchies of structured attributes, known relationships binding attributes, and any other background knowledge that users may have represented in the system. During the operation of an inductive database, the knowledge base is populated by newly generated data descriptions, hypothetical patterns, data classifications, statistical information, results from hypothesis testing, etc.

To provide a general overview and easy access to all VINLEN operators, we have developed a visual interface that consists of VINLEN views at different abstraction levels. Fig. 2 presents the most abstract view of the main panel of VINLEN. The central part contains icons for invoking databases (DB), knowledge bases (KB), and knowledge systems (KS) currently implemented in the system. The term knowledge system stands for a system integrating a database and a relevant knowledge base to support knowledge mining and knowledge use for a specific application problem. By clicking on DB, KB, or KS, the user can select and access available to VINLEN databases, knowledge bases, and knowledge systems, respectively.

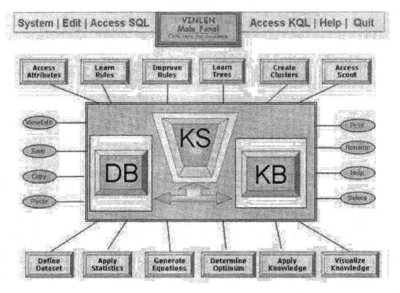

Fig. 2. VINLEN inductive database visual interface (main panel, abstract view)

The twelve rectangular buttons above and below the central region represent diverse classes of knowledge generation operators, and data and knowledge management operators. Each button invokes a pull-down menu whose options represent different programs for performing these classes of tasks, or different modes of operation of the same program. At the top of the interface is a set of pull-down menus consisting both of standard operators and mechanisms for direct access to an SQL client (for data management) and a KQL interpreter for invoking *knowledge scouts*, personal agents whose function is to synthesize and manage target knowledge [6].

The data in each VINLEN knowledge system are stored internally in relational tables, as are other system components. Prior knowledge and generated knowledge are stored in a hierarchy of relational tables. Parameter sets for individual operators are stored in system-specific tables. This storage methodology facilitates an efficient access to components by the system through a standard SQL interface.

2.2   Knowledge Representation

To illustrate the knowledge storage mechanism, consider a selection of attributional rules representing patterns discovered in experiments on a medical database containing information about men age 50-65, their diseases and their lifestyles (Fig. 3). These patterns were learned from 2063 cases describing patients with high blood pressure and 5288 patients without it (these total numbers of positive and negative examples are denoted P and N, respectively). The first pattern (A) represents a very simple and approximate regularity stating that patients with high or greater *rotundity* (a function of their weight and height) tend to have high blood pressure (the raw support of the rule, p, is 689 patients out of 2063, or 33%). The value $n = 1058$ indicates the number of patients, 20% of the 5288, that have high or higher rotundity, but do not have high blood pressure.

```
High_Blood_Pressure is present in patient-set if:
A   1   [Rotundity >= high]  (p:689, n:1058)
        (p:689, n:1058, q:0.23)
B   1   [Heart_Disease = present] (p:284, n:355)
    2   [Rotundity <= average]  (p:1328, n:4120)
        (p:205, n:282, q:0.14)
High_Blood_Pressure is absent if:
A   1   [education >= vocational]  (p:3691, n:1333)
    2   [Diabetes = absent]  (p:5089, n:1884)
    3   [Rectal_Polyps = absent]  (p:5072, n:1967)
        (p:3396, n:1167, q:0.26)
B   1   [Rotundity <= average]  (p:4120, n:1328)
    2   [Heart_Disease = absent]  (p:4933, n:1779)
    3   [Diabetes = absent]  (p:5089, n:1884)
    4   [Exercise >= medium]  (p:4219, n:1571)
    5   [Stroke = absent]  (p:5249, n:2018)
    6   [Kidney_Disease = absent]  (p:5228, n:2013)
    7   [Colon_Polyps = absent]  (p:5191, n:2025)
    8   [education <= hs_grad]  (p:1550, n:709)
        (p:851, n:238, q:0.18)
```

Fig. 3. Attributional rules indicating the likely presence or absence of high blood pressure

Thus, the above pattern indicates only a tendency, but not a very strong relationship. The parameter q denotes a measure of *pattern quality*, defined as the weighted product of the *completeness* (support percentage) and *consistency gain*, which is defined as the ratio:

$$q = (1 - C_p) / (1 - C_g)$$

where C_p is the consistency of the pattern $(p \,/\, (p + n))$, and C_g is the degree of confidence in randomly guessing a positive diagnosis $(P \,/\, (P + N))$. Thus, a rule with q=1 will be perfectly complete and consistent with regard to the training data, and a rule with q=0 will have a consistency equal to that of a random guess. Detailed explanations of the q measure can be found in [5,11].

The second pattern (B) found in the data indicates that presence of high blood pressure is associated with a history of heart disease and low to average rotundity. Of the cases that satisfied the heart disease condition, 284 had high blood pressure, and 355 did not. Of the cases that satisfied the rotundity condition, 1328 had high blood pressure, and 4120 did not. Of the cases that satisfied both conditions, 205 had high blood pressure (out of 2063, that is, 10%), and 282 did not (out of 5288, that is, 5%).

The above attributional rules are conjunctions of between one and eight *conditions*, which take the form [<attribute(s)> <relation> <value(s)>] The rules and conditions are annotated by weights indicating their positive and negative coverage. Each diagnosis/decision is represented by a *ruleset*; in this example, the two rulesets each consist of two rules. The two rulesets comprise what we call a *ruleset family*, a collection of knowledge relevant to the task of determining the value of some attribute(s), in this case high blood pressure. Thus, while the knowledge is stored in relational tables, the attributional calculus representation allows it to be organized in such a way that components can be accessed and analyzed by the system. Specifically, VINLEN organizes the knowledge in a hierarchical set of tables: **Ruleset**, which identifies the ruleset's consequent (decision) and the numbers of positive and negative examples from which the ruleset was generated; **Complex**, which includes information on specific rules, including their rank in their rulesets (ordered by support), and their positive and negative training example coverage; **Selector**, which contains the details of the individual conditions that comprise rules' antecedents, including the attribute, relation and value portions of the conditions, and their positive and negative training example coverage levels; and **Learn**, which archives the parameters under which the rulesets were learned.

Fig. 4 depicts this knowledge organization, which allows the user to employ knowledge scouts that query for and access components of the knowledge base at any or all levels of the ruleset family hierarchy. For example, the user can request the number of *rules* in a given *ruleset* that have exactly two *conditions* with at least 80% support and 80% confidence over the training data. Fig. 5 instantiates this knowledge schema for the two rules with decision High_Blood_Pressure is present.

2.3   Knowledge Visualization via Concept Association Graphs

Users often rely on visualization methods to provide a better understanding of the patterns in the data and knowledge. Visual representations provide more succinctly and comprehensibly the ideas that are important to the

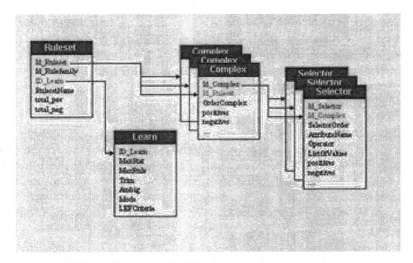

Fig. 4. Schema for the VINLEN inductive database visual interface

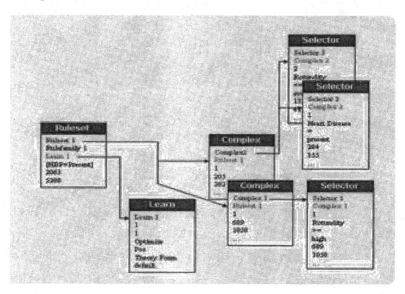

Fig. 5. Instantiated database structure of two rules for the decision High Blood Pressure is present

users. Hence, visualization technology is very much in the spirit of inductive databases. We have developed visualization operators based on the method of *diagrammatic visualization* [9], and on *association graphs* [1,6].

The latter method provides a novel approach to concept visualization. In it, the elements may represent attributes or high-level concepts, with annotated links showing details of the relationships. Fig. 6, for example, presents an association graph built from the rules listed in Fig. 3. The output attributes and decisions are represented by the darker ovals. The input attributes and values are represented by lighter ovals and rectangles, respectively. The square numbered boxes represent rules linking inputs to an output. Each link's width is an indication of the strength of the associated condition (if it connects a rule box to an input node) or of the rule as a whole (if it connects the rule box to a decision node). The strength is based on the consistency (C_p) of the condition or rule.

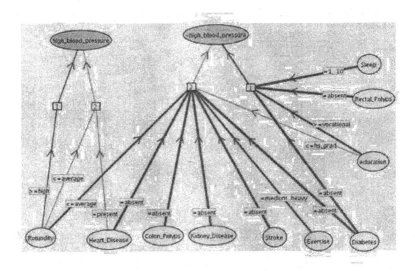

Fig. 6. An association graph representing rules for diagnosing high blood pressure

3   Conclusion

The VINLEN system is still under development, and is planned to include a very wide range of knowledge generation operators and related capabilities. Due to space limitations, this paper concentrated only on the rule generation and concept association graph visualization operators implemented in the current VINLEN system. Other major features already integrated in VINLEN include conceptual clustering, intelligent target data generation (selecting the most promising attributes and data records for completing a specified task in a given amount of time), database manipulation through an SQL client, and a mechanism for creating knowledge query language scripts to guide data exploration tasks.

Initial results indicate the great potential of this research for applications in many areas. Many questions and research issues remain, such as how to integrate symbolic and statistical reasoning capabilities in VINLEN, how to develop a knowledge query language that facilitates the creation of scripts for automated knowledge discovery, how to visualize complex regularities, and how to implement knowledge mining in temporal datasets. Potential approaches to this problem include, but are not limited to the construction of new temporal-oriented attributes that better characterize the target concepts, the characterization of sequential processes through the analysis of their components, and prediction methods based on previous events. To this end, we plan to conduct research on how to introduce the capability of *qualitative prediction* to a temporal database. Such a capability requires a method for inductively generating qualitative hypotheses about future events (e.g., [13]).

An inductive database can be used to build *knowledge scouts*, which are scripts for synthesis and management of target knowledge. During the course of its existence, a knowledge scout builds a model of interests and experiences of the user, and employs that model in synthesizing target knowledge. An important research issue for future study here is how to cope with changes in the users' interests (concept drift).

Summarizing, the main contribution of the VINLEN project is the development of a methodology for a tight integration of a data base, knowledge base, data management operators, knowledge generation operators, a general knowledge query language, and a user-oriented visual interface. Continuing research and development promises to achieve advances in the field of knowledge mining and solutions to real-world problems.

Acknowledgments

The authors thank Michal Draminski and Guido Cervone for their contributions to the implementation of VINLEN, the development of some of the component operators. This research was performed at the Machine Learning and Inference Laboratory at George Mason University. The Laboratory's research activities are supported in part by the National Science Foundation under Grants No. IIS-9906858 and IIS-0097476, and in part by the UMBC/LUCITE #32 grant. Any opinions, findings and conclusions or recommendations expressed in this material are those of the authors, and do not necessarily reflect the views of the National Science Foundation (NSF) or the UMBC/LUCITE sponsor.

References

1. Cervone, G., Michalski, R.S. (2003) Concept Association Graphs: CAG1. *Reports of the Machine Learning Laboratory*, George Mason University
2. Finan, T., Fritzson,R., McKay, D., McEntire, R. (1994) KQML as an Agent Communication Language. *Proceedings of the Third International Conference on Information and Knowledge Management (CIKM '94)*, ACM Press
3. Imielinski, T., Virmani, A., Abdulghani, A. (1996) DataMine: Application Programming Interface and Query Language for Data Mining. *Proceedings of the Second International Conference on Knowledge Discovery and Data Mining (KDD-96)*, pp. 256-261
4. Kaufman, K.A., Michalski, R.S. (1998) Multistrategy Data Mining via the KGL Metalanguage. *Proceedings of the Seventh Symposium on Intelligent Information Systems (IIS '98)*, Marlbork, Poland, pp. 39-48
5. Kaufman, K.A., Michalski, R.S. (1999) Learning From Inconsistent and Noisy Data: The AQ18 Approach. *Proceedings of the Eleventh International Symposium on Methodologies for Intelligent Systems (ISMIS '99)*, Warsaw, pp. 411-419
6. Kaufman, K.A., Michalski, R.S. (2000) A Knowledge Scout for Discovering Medical Patterns: Methodology and System SCAMP. *Proceedings of the Fourth International Conference on Flexible Query Answering Systems (FQAS'2000)*, Warsaw, pp. 485-496
7. Mannila, H. (1997) Inductive Databases and Condensed Representations for Data Mining. *Proceedings of the International Logic Programming Symposium (ILPS'97)*, pp. 21-30
8. Michalski, R.S. (2001) Attributional Calculus: A Representation System and Logic for Deriving Human Knowledge from Computer Data. *Reports of the Machine Learning and Inference Laboratory*, MLI 01-1, George Mason University
9. Michalski, R.S., Kaufman, K.A. (1998) Data Mining and Knowledge Discovery: A Review of Issues and a Multistrategy Approach. In Michalski, R.S., Bratko, I., Kubat, M. (eds.), *Machine Learning and Data Mining: Methods and Applications*, London, John Wiley & Sons, pp. 71-112
10. Michalski, R.S., Kaufman, K. (2000) Building Knowledge Scouts Using KGL Metalanguage. *Fundamenta Informaticae* **40**, pp. 433-447
11. Michalski, R.S., Kaufman, K. (2001) Learning Patterns in Noisy Data: The AQ Approach. In Paliouras, G., Karkaletsis, V., Spyropoulos, C. (eds.), *Machine Learning and its Applications*, Springer-Verlag, pp. 22-38
12. Michalski, R.S., Kerschberg, L., Kaufman, K.A., Ribeiro, J.S. (1992) Mining for Knowledge in Databases: The INLEN Architecture, Initial Implementation and first results. *Intelligent Information Systems: Integrating Artificial Intelligence and Database Technologies* **1**, pp. 85-113
13. Michalski, R.S., Ko, H., Chen, K. (1988) Qualitative Prediction: SPARC/G Methodology for Inductively Describing and Predicting Discrete Processes. In Van Lamsweerde, A., Dufour, O. (eds.), *Current Issues in Expert Systems*
14. Morik, K., Brockhausen, P. (1996) A Multistrategy Approach to Relational Knowledge Discovery in Databases. *Proceedings of the Third International Workshop on Multistrategy Learning (MSL-96)*, pp. 17-27
15. Rice, S.P., Roddick, J.F. (2000) Lattice-Structured Domains, Imperfect Data and Inductive Queries. *Proceedings of the Eleventh International Conference on Database and Expert Systems Applications (DEXA 2000)*

Automatic Classification of Executable Code for Computer Virus Detection

Paweł Kierski[2], Michał Okoniewski[1], and Piotr Gawrysiak[1]

[1] Institute of Computer Science, Warsaw University of Technology,
ul. Nowowiejska 15/19, 00-665 Warszawa
[2] MKS Sp. z o.o., ul. Fortuny 9, 01-339 Warszawa

Abstract. Automatic knowledge discovery methodologies has proved to be a very strong tool which is currently widely used for the analysis of large datasets, being produced by organizations worldwide. However, this analysis is mostly done for relatively simple and structured data, such as transactional or financial records. The real frontier for current KDD research seems to be analysis of unstructured data, such as freeform text, web pages, images etc. In this paper we present results of applying KDD methodology to such unstructured data - namely computer machine code. We show that it is possible to construct automatic classification system, that would be able to distinguish "good" computer code from malicious code - in our case code of computer viruses - and which therefore could act as an intelligent virus scanner. In our approach we use methods originating from text mining field, treating CPU instructions as a kind of natural language.

Keywords: classification, text mining, virus scanners, machine code analysis

1 Introduction

The main research problem described in this paper is knowledge discovery methods application for computer viruses identification. We treat executable code in a similar way as a document in text mining process. Thus instructions are treated as words. The goal is to find common qualities of a virus by setting up highly efficient classifier. This would be a step forward towards automatic discovery of new, previously unknown viruses - by exactly applying Piatetsky's [1] definition of data mining in the area of antiviral research.

The rest of the paper is organized as follows. Section 2 presents definitions on computer viruses detection and identification. In the section 3 are presented methods of input data preprocessing and training of classifiers, which are included in experimental system described in section 4 and evaluated in section 5. Conclussions and possible follw-up is described in section 6.

2 Computer viruses

Computer virus according to [2] is a fragment of executable program that fulfills following conditions:

- Is able to replicate itself.
- Replication is not a side-effect of the code execution.
- Viral code is attached to the host code in such a way that an attempt to execute a host process executes also a virus.
- At least one of the copies created is also a virus.

The **host**, to which it attaches a copy of viral code is a program written in executable form for a given system.

2.1   Detection and identification of viruses

Let us define the notions of detection and identification of computer viruses.

Detection is a process of checking if the executable file includes a virus, or a result of such process - a decision if the file is infected or clean.

Identification is a process of recognition of virus type precisely enough to "cure" the file - regain the form before infection or to state that such treatment is not possible.

The number of virus types may be estimated as tens of thousands - depending on method of estimation. The detection and identification of viruses are separate processes now. Usually the procedure has 2 or 3 levels of recognition. First step is a search for signatures - short fragments of viral code. Due to similarities between viruses the number of signatures is significantly smaller than number of virus types. Second step is the identification of fragment of suspected code using additional signatures check or checksums of viral area of code.

This method has a serious disadvantage. Virus scanner author has to possess a copy of every new virus type and publish updates of the scanning software frequently. This is why there is a demand for heuristic methods of detection of previously unknown viruses. Such methods should be based on useful general knowledge of general qualities of a virus.

2.2   False alarms

Virus detection without full identification or using heuristic methods is prone to two kinds of false alarm errors:

- false positive - identifying a virus in a clean file
- false negative - treating a virus-free file as infected one

To ensure the quality of a virus scanner, designers have to minimize the occurrences of both error types. Measures to estimate quality of detection are: detection rate $W = \frac{N_v'}{N_v}$, and relative ratio of false alarms $F = \frac{N_c v}{N_c}$, where N_v' is a number of well detected viral samples, N_v is a total number of samples, N_c number of virus-free samples, $N_c v$ number of virus-free samples recognized as infected.

Usually files are not infected, thus for a heuristic virus scanner value $F = 0.1\%$ means several "suspicious" files for a single PC-like system. The value of W should be as high as possible. "Not infected" should be preferred classification result - to minimize false alarms and save time for user intervention.

3   Construction of classifiers

To construct proper classifier of executable code, we have followed several iterative steps according to classic KDD-cycle [1].

3.1   Representation of executable code for classification

Executable code is written in CPU native language. Thus we may follow the rules commonly used in text mining. One of most popular preprocessing methods for text is representation in the vector form of words with occurrences count known as "bag of words". [3,4]. This representation of code for the classifier is prepared by counting the frequency of each instruction in executable code.

Of course there are differences between executable code and natural language. There are much less instruction types in the program than words (roots after stemming) in any language- usually the number ranges between 50 and 200. However, much more information is contained in instruction parameters than in grammar forms of words. For viruses - most information crucial for detection is contained in parameters. Number of instructions with possible parameters for x86 CPU's is over 10^{20}.

This is why in our system, after disassembling the program, some instructions were grouped together according to some intuitive rules. For example all logical operations (AND, OR, NOT, bitwise shifts) or MOV instructions (excluding MOV's to segment registers) were clustered.

At the same time, instructions or parameters specific for viruses are extracted from the program. For example the system counts separately numbers of DOS functions calls for file operations or changes of addresses of BIOS variables.

3.2   Discretization of attributes

Having prepared the vector representation one may use various discretization methods [5] to decrease the number of different values in the instruction vector.

Fragments of code analyzed by a virus scanner range from less than one hundred bytes to several kilobytes. When disassembled these result in several up to several thousand of instructions, so attribute values in the vector usually is smaller than 1000. However, many of them still need discretization.

Most often we experimented with function $n = |(log2x) + 1|$ which may be interpreted as the minimal number of digits in binary representation of number x.

3.3   Simplifying the representation

Next step of representation preparation is removal of less useful attributes - those with little informational value. Attributes that has the same value for all objects are obviously excluded. To assess informational value of the attribute one may use correlation coefficient between attribute and the decisive attribute

$$\rho_i = \frac{Cov(A_i, D)}{\sqrt{Var(A_i)Var(D)}} \tag{1}$$

where A_i is a random variable for attribute i and D for decisive attribute. The formula may be transformed as follows using expected values for attributes:

$$\rho_i = \frac{E[(A_i - EA_i)(D - ED)]}{\sqrt{(EA_i^2 - (EA_i)^2)(ED^2 - (ED)^2)}} \tag{2}$$

For ED, if we assume value 0 for "clean file" and 1 for "infected", we have $ED = \frac{n_w}{N}$ where n_w is number of infected files and N total number of files.

If the value of $|\rho_i|$ is close to 0, there is no significant linear correlation between A_i and D - so such attribute may be excluded from the representation.

Another method of assessing attribute usefulness originates from the theory of information. In the case of two information sources the entropy formula is:

$$H = -\sum_{i=1}^{n}\sum_{j=1}^{m} p(i.j)log_2 p(i, j) \tag{3}$$

thus

$$H = -\sum_{i=1}^{n} p(i)log_2 p(i) - \sum_{j=1}^{m} p(j)log_2 p(j) + \sum_{i=1}^{n}\sum_{j=1}^{m} p(i.j)log_2 \frac{p(i, j)}{p(i)p(j)} \tag{4}$$

The last addend is the measure of mutual information of i and j.

$$J = \sum_{i=1}^{n}\sum_{j=1}^{m} p(i.j)log_2 \frac{p(i, j)}{p(i)p(j)} \tag{5}$$

If sources are independent, for each (i, j) we get $p(i, j) = p(i)p(j)$ and accordingly $J = 0$.

Both methods were applied to reduce the number of attributes in the representation. Up to 100 attributes with highest ρ_i or J were selected.

3.4   Naive Bayes classifier

Based on the vector representation of a program code a Bayessian classifier was trained. Let us use the notation: N - number of objects, A - vector representation of an object, $D_i, i = 1..n$ - decisions (values of decisive attribute), n - number of classifying decisions, A_j - value of j attribute $(j = 1..m)$, m - number of attributes. Occurrence of specific value is treated as hypothesis to confirm.

The assumption of independence of $(A_j = k)$ and D for each of them results in formula:

$$P(A|D_i) = \prod_{j=1}^{m} P(A_j = k|D = D_i) \tag{6}$$

Thus, from Bayes theorem:

$$P(D_i|A) = \frac{(\prod_{j=1}^{m} P(A_j = k|D = D_i))P(D = D_i)}{P(A)} \tag{7}$$

$P(A)$ is independent from values of decisive attribute, so it is enough to compare slightly transformed values for classification:

$$k_i = \frac{|\{A : D = D_i\}| \prod_{j=1}^{m} |\{A : A_j = k \wedge D = D_i\}|}{N^{m+1}} \tag{8}$$

Bayesian classifier tends to prefer values D_i that occur often in the training set. Usually the classifier should be trained with similar distribution of D_i as in test dataset. Classifier may be also fine-tuned by multiplying values of k_i by constants K_i $(i = 1..m)$. To increase C times the number of samples for a given decisive attribute values K_i should be equal C^{m+1}

3.5   Decision rules approach

To compare with bayesian classifier some decision rule based approaches were applied. Parametrizing rule-based classifier is possible by choosing rules to be used. Especially, strong rules (rules with high support) should be useful to make a distinction between infected and clean files. Of course usually rule-based classifier is not complete. If there are no strong rules to make a decision - the object is not classified.

For this purpose, *Apriori Certain* was used [6,7]. This algorithm generates optimal in terms of length of conditional part of the rule using at most n data accesses. It allows to select strong rules as a result.

Fine-tuning of the classifier means here finding proper level of *minSupp* for rules with decisive attribute $D = 1$ ("infected file"). If we keep the set of rules with their supports, it would be possible to choose a subset of rules as needed. Choosing stronger rules decreases the number of false alarms, and increasing the subset is making the classifier more selective, but with more false alarms.

4   Design of experimental systems

The experiment was done with an archive of 401997 samples of binary code. Divided in 3 parts: infected samples $SV(|SV| = 249271)$, not infected samples used to train classifiers $SC1(|SC1| = 24779)$, and not infected samples used for testing classifiers $SC2(|SC2| = 127947)$. The number of samples in $SC2$ lets to test the percentage of false alarms with precision 0,001%. Elimination of false alarms during classification of test dataset means that false alarms should occur more seldom than once per 105 files. For a real virus scanner it is hardly acceptable, but during test phase it may be enough to assess the quality of classifiers for various parameters. This is why we assume that 0,003% of false alarms is acceptable. It means four false alarms in the test data set of clean, not infected files.

Generating of classifiers with over 800 attributes was impossible due to too much resources and time needed. This is why 5 to 25 attributes were selected according to measures of importance calculated according to criteria (2) and (5).

5   Experimental results

Classifiers based on 5, 10, 15, 20 and 25 best attributes were tested.

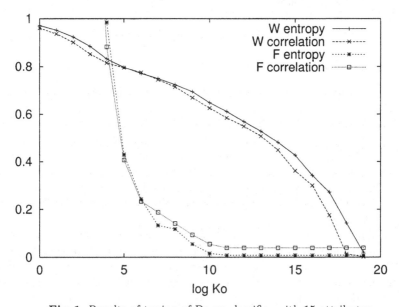

Fig. 1. Results of tuning of Bayes classifier with 15 attributes

For bayesian classifiers based on 5 and 10 attributes reduction of false alarms rate below required minimum caused too big loss in detection rate. Results in terms of F and W values were assessed by virus scanner experts. Best classification was obtained for 15 attributes (Fig. 1) chosen with mutual information measure (formula 5) and the value of $K_0 = 10^{11}$. The use of more than 15 attributes caused not acceptable growth of false alarm rate.

Also in the case of rule-based classifier, best results were obtained for 15 attributes (Fig. 2). However, the measure of correlation proved to be slightly better decrease false alarms - on the contrary to bayesian classifier.

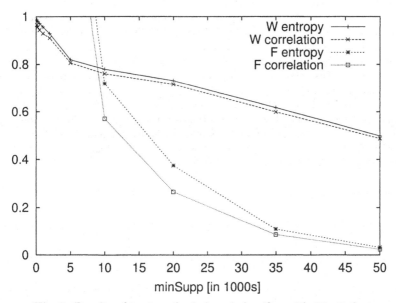

Fig. 2. Results of tuning of rule-based classifier with 15 attributes

6   Conclusions

In all experiments Bayes classifier turned out to be more effective than the rule-based classifier. It is however worth pointing out that with this classification method we can get false alarms even for samples that it was trained upon. On the other hand managing individual false alarm cases is relatively simple for decision rule-based classifier. This might have important practical consequences.

In this paper we show that in most cases it is enough to test frequency of relatively small number of instruction sequences to make a binary classification decision - virus code / safe code. This method is semi-automatic, as in most data mining applications, but enables us to create a robust virus scanner, that can perform more efficiently in the area of detection of previously unknown viral structures. It may also serve as a base for new heuristic virus scanners.

References

1. U.M. Fayyad, G. Piatetsky-Shapiro, P. Smyth, From Data Mining to Knowledge Discovery : An Overview, In: Advances in Knowledge Discovery and Data Mining, eds. Fayyad, Piatetsky-Shapiro et al. AAAI Press/ The MIT Press, 1996
2. Frisk Software International, F-Prot virus scanner description, 2001
3. A. McCallum, K. Nigam, A Comparison of Event Models for Naive Bayes Text Classification, AAAI-98 Workshop on Learning for Text Categorization, 1998
4. P. Gawrysiak, Automatyczna kategoryzacja dokumentow, PhD thesis, Warsaw University of Technology, 2001
5. S.H. Nguyen, Discretization of real value attributes - boolean reasoning approach, PhD Thesis, Warsaw University, 1997
6. M. Kryszkiewicz, Strong Rules in Large Databases, Proceedings of Seventh International Conference IPMU '98 Information Processing and Management of Uncertainty in Knowledge - based Systems, Paris, France 6-10 July 1998
7. M. Kryszkiewicz, H. Rybinski, Strong Rules in Large Databases, Proc. of EUFIT '98, Aachen, Germany, September 1998

Heuristic Search for Optimal Architecture of a Locally Reccurent Neural Network

Andrzej Obuchowicz and Krzysztof Patan

Institute of Control and Computation Engineering, University of Zielona Góra, Podgórna 50, 65-246 Zielona Góra, Poland

Abstract. The modelling of a dynamic non-linear system is the aim of this work. A neural model is constructed with dynamic artificial neurons, which contain inner feedbacks. Such a model consists of an adder module, a linear dynamic system and a non-linear activation function. Emphasis is put on the searching of the optimal, in the sense of a generalization capability, network architecture. On the basis of a graph of network structures and evaluation values assigned to them, heuristic search algorithms can be elaborated and implemented. The A^*-algorithm and Tabu Search are applied and compared.

One of the most interesting and promising solutions of dynamic system modelling problem is the application of the locally recurrent neural network, called the Dynamic Multi-Layer Perceptron (DMLP). The DMLP can have the same architecture as the well-known Multi-Layer Perceptron (MLP), but it is composed of Dynamic Neural Models (DNM) which consist of an adder module, a linear dynamic system - Infinite Impulse Response (IIR) filter, and a non-linear activation function ([1,2]). The calculated output error is propagated back to the input layer through hidden layers containing dynamic filters, similarly as in the standard back-propagation algorithm. As a result, the Extended Dynamic Back-Propagation algorithm may be defined [2]. This algorithm adjust connection weights as well as IIR filter parameters.

The selection of an optimal architecture of the MLP to be able to solve a given problem, is an important prerequisite problem for research workers. The MLP architecture influences on the rate of training procedure, network processing rate and quality. The crucial tradeoff one has to make is between the learning capability of the MLP and fluctuations due to the finite sample size [3]. The significance of the network architecture optimization increases when the dynamic network is taken into account. The number of free network parameters rapidly increases when one substitutes standard McCulloch-Pitt's neurons by the DNM units. Thus, apart from setting an appropriate number of hidden layers and the number of neurons in each of these layers, the dynamic order of each particular neuron have to be established in the DMLP.

The aim of this work is to design the optimal DMLP to model a given dynamic non-linear system. In order to solve this problem, the space of possible DMLP architectures has been organized in the digraph form [9]. The Tabu Search is proposed as a searching algorithm and compared with the A^* algorithm previously implemented [9].

In the following, the DMLP is introduced. Next, the graph of the DMLP architectures is given. Finally, the A*-algorithm and Tabu Search approaches to the DMLP architecture allocation is discussed and illustrated by an example.

1   Dynamic neural network

1.1   Dynamic neural model

The general structure of a neuron model proposed by Ayoubi [1] is considered. The dynamics is introduced to a neuron by adding a linear dynamic system (the IIR filter) to the neuron structure. Three main operations are performed in the DNM. First of all, the weighted sum of inputs is calculated according to the formula:

$$x(k) = \mathbf{w}^T \mathbf{u}(k) = \sum_{p=1}^{P} w_p u_p(k), \tag{1}$$

where $\mathbf{w} = [w_p \mid p = 1, 2, \ldots, P]^T$ denotes the input weights vector. The weights perform a similar role as in MLPs. Then this calculated membrane potential $x(k)$ is passed to the IIR filter. Here the filters under consideration are the linear dynamic systems of different orders, viz. the first, second order. The behavior of this linear system can be described by the following difference equation:

$$\tilde{y}(k) = \sum_{i=0}^{n} b_i x(k-i) - \sum_{i=1}^{n} a_i \tilde{y}(k-i), \tag{2}$$

where $\tilde{y}(k)$ is the filter output; $\mathbf{a} = [a_i \mid i = 1, 2, \ldots, n]^T$ and $\mathbf{b} = [b_i \mid i = 0, 1, \ldots, n]^T$ are feedback and feedforward paths' weights, respectively. Finally, the neuron output can be described by

$$y(k) = F(g\,\tilde{y}(k)), \tag{3}$$

where $F(\cdot)$ is a non-linear activation function that produces the neuron output $y(k)$ and g is the slope parameter of the activation function [6,11].

1.2   The DMLP

The DMLP presents an ordered pair $(A, \mathbf{v})$. A denotes the network architecture represented as follows:

$$A = ((V_m \mid m = 0, 1, \ldots, M), (o_s^m \mid m = 1, 2, \ldots, M; s = 1, 2, \ldots, s_m), \Lambda), \tag{4}$$

where $(V_m \mid m = 0, 1, \ldots, M)$ is a family of $M + 1$ sets of DNM units, called layers, including two disjunctive non-empty sets V_0, V_M that define the input (non-processing) and output (processing) units, respectively. $(o_s^m \mid$

$m = 1, 2, \ldots, M; s = 1, 2, \ldots, s_m)$ is a set of natural numbers, o_s^m denotes the filter order of the s-th DNM unit from the m-th layer, $s_m = \text{card}(V_m)$. $\Lambda = \bigcup_{m=0}^{M-1} V_m \times V_{m+1}$ is a set of edges that define the connections between units in the network. The vector of network parameters $\mathbf{v}$ can be expressed in the following way:

$$\mathbf{v} = \left(\mathbf{w}, (\mathbf{a}_s^m, \mathbf{b}_s^m, g_s^m \mid m = 1, 2, \ldots, M; s = 1, 2, \ldots, s_m)\right), \qquad (5)$$

where the set of weights $\mathbf{w}$ assigns a real value to each connection, $(\mathbf{a}_s^m, \mathbf{b}_s^m, g_s^m \mid m = 1, 2, \ldots, M; s = 1, 2, \ldots, s_m)$ describes feedback and feed-forward IIR synaptic vectors and the slope parameter, respectively, of the s-th DNM unit in the m-th layer.

1.3   Graph of the DMLP architectures

The graph $G(A)$ of the DMLP architectures [9] can be described by definition of the expansion operator $\Gamma(A)$, which generates all successors of a given architecture A.

The expansion operator $\Gamma(A)$ creates the following successors.

1. *Varying the number of hidden layers.* Assume the DMLP architecture A (4). The architecture $A^{(1)}$ with inserted hidden layer with one DNM unit v' of zero order is the successor of A.
2. *Varying the number of units in a hidden layer.* Assume the architecture A (4) that has at least one hidden layer ($M \geq 2$). Then, all architectures $A^{(2)}$ with an inserted unit v'' in a i-th layer for $i = 1, 2, \ldots, M - 1$ are successors of A.
3. *Varying the IIR order of the DNM unit.* Assume the architecture A (4). Then, all architectures $A^{(3)}$ with increased order of the IIR filter in the j-th DNM unit of the i-th layer $i = 1, 2, \ldots, M$, $j = 1, 2, \ldots, s_i$ are successors of A.

Thus, $\Gamma(A)$ maps an architecture A with $M - 1$ hidden layers and N processing DNM units onto $M + N$ successors.

2   DMLP architecture optimization approaches

2.1 $A^\star$ algorithm

The $A^\star$ algorithm, first described in [7,8], is a way to implement best-first search to a problem graph. The algorithm searches a directed graph in which each node n_i represents a point in the problem space. Each node will contain, in addition to a description of the problem state it represents, an indication of how promising it is, a parent link that points back to the best node from which it came, a list of the nodes that were generated from it. The parent link renders it possible to recover the path to the goal once the goal is found. The list of successors makes it possible, if a better path is found to already existing node, to propagate the improvement down to its successors.

A heuristic function $f(n_i)$ is needed to estimate the merits of each generated node. In the A^* algorithm this cost function is defined as a sum of two components:

$$f(n_i) = g(n_i) + h(n_i),\tag{6}$$

where $g(n_i)$ is the cost of the best path from the start node n_0 to the node n_i and it is known exactly to be the sum of the cost of each of the rules that were applied along the best path from n_0 to n_i, and $h(n_i)$ is the estimation of the addition cost getting from the node n_i to the nearest goal node. The function $h(n_i)$ contains the knowledge about the problem.

The outline of the A^* algorithm is described in many handbooks from the domain of Artificial Intelligence. In this work the algorithm included in [10] is implemented.

2.2   Tabu Search

The tabu search metaheuristic has been proposed by Glover [4]. This algorithms models processes existing in the human memory. This memory is implemented as a simple list of solutions explored recently. The algorithm starts from a given solution x_0, which is treated as actually the best solution $x^* \leftarrow x_0$. The tabu list is empty $T = \emptyset$. Next, the set of neighboring solutions is generated excluding solutions noted in the tabu list, and the best solution of this set is selected as a new base point. If x' is better than x^* then $x^* \leftarrow x'$. The actual base point x' is added to the tabu list. This process is iteratively repeated until a given criterion is satisfied.

There are many implementations of the Tabu Search idea, which differs between themselves in the method of the tabu list managing, e.g. Tabu Navigation Method (TNM), Cancellation Sequence Method (CSM), Reverse Elimination Method (REM). Particular description of these methods can be found in [5].

2.3   Implementations

In order to apply the A^* and Tabu Search algorithms to an architecture optimization of the DMLP we have to define:

- the optimization criterion — which is chosen in the form

$$J_T\big(\mathbf{y}_{A,\mathbf{v}^*}(k), \mathbf{y}(k) \mid k \in K\big) = \frac{\sum_{k \in K} \big(\mathbf{y}_{A,\mathbf{v}^*}(k) - \mathbf{y}(k)\big)^2}{\sum_{k \in K} \mathbf{y}^2(k)}\tag{7}$$

where $\mathbf{y}_{A,\mathbf{v}^*}(k)$ and $\mathbf{y}(k)$ are the output of the learned DMLP and desired output, respectively.

- an expansion operator $\Gamma(A) : \mathcal{A} \to 2^{\mathcal{A}}$ that maps any network architecture $A \in \mathcal{A}$ onto a set of successors (see Section 1.3).

Moreover, the following functions have to be defined for the A^* algorithm

- the cost function $g(A, A')$ assigned to each expansion operation:

$$g(A, A') = \begin{bmatrix} \gamma(A') - \gamma(A) \\ \delta(A') - \delta(A), \end{bmatrix} \tag{8}$$

where $\gamma(A)$ is the number of free parameters in the DMLP architecture A, and $\delta(A)$ is the number of hidden layers in A;
- the heuristic function $h(A)$

$$h(A) = \left(J_T\big(\mathbf{y}_{A_0, \mathbf{v}^*}(k), \mathbf{y}(k) \mid k \in K\big)\right)^{-1} \begin{bmatrix} J_T\big(\mathbf{y}_{A, \mathbf{v}^*}(k), \mathbf{y}(k) \mid k \in K\big) \\ 0 \end{bmatrix}, \tag{9}$$

where A_0 denotes the initial architecture of searching.

Because both $g(A, A')$ and $h(A)$ are vector functions, the relation $\mathbf{p} \dot{\le} \mathbf{q}$ should be defined

$$\mathbf{p} \dot{\le} \mathbf{q} \Leftrightarrow (p_1 \le q_1) \text{or}\big((p_1 = q_1) \text{and} (p_2 \le q_2)\big). \tag{10}$$

3   Modelling of the dynamic system – experiment

This section presents the experimental results achieved during selection of the optimal neural network structure using searching methods described in the previous sections. The neural network composed of DMNs is used here to identify the dynamic non-linear process represented by the following difference equation:

$$y(k + 1) = \frac{y(k)}{1 + y(k)^2} + u(k)^3 \tag{11}$$

where $u(k)$ and $y(k)$ are the input and output of the process at the instant k, respectively. The learning process is carried out off-line for 500 steps using the Extended Dynamic Back-Propagation algorithm and a pseudo-random input uniformly distributed in the interval $[-2, 2]$. The learning set consists of 200 patterns and the learning rate is equal to 0.01. The training procedure of each examined network structure is repeated four times in order to decrease a chance to get stuck in local minima of an error function. Furthermore, each neuron in the network has the hyperbolic tangent activation function.

The selection of the optimal neural network structure is performed using two searching methods: the A^* algorithm and the Tabu Search method. The second algorithm is tested with different number of structures memorized, in turn 3, 5 and 10. Results achieved during experiments are presented in Table 1, where the notation $N^n_{r,v,s}$ denotes n-th layer neural network with r

Table 1. Specification of the selected neural networks

CHARACTERISTICS	Method			
	Tabu Search – list length			$A^\star$
	3	5	10	
Network structure	$N^2_{1,1,1}$	$N^2_{1,4,1}$	$N^2_{1,4,1}$	$N^2_{1,5,1}$
1st layer filters orders	(2)	(2 1 1 0)	(2 1 1 0)	(2 1 0 2 1)
2nd layer filters orders	(0)	(1)	(1)	(2)
Modelling quality	0.145051	0.139015	0.139015	0.123431

inputs, v hidden neurons and s outputs, and $(or1\,or2\ldots orn)$ denotes that 1st neuron possesses $or1$ order filter, 2nd one – $or2$ order filter and n-th neuron – orn filter order.

Both searching methods start with the minimal network structure consisting the output neurons only. After that, the neural network is growing up. At each algorithm step one parameter can be changed: the number of hidden layers (maximum 2 hidden layers) or the filter order (maximum 2nd order) or the number of neurons in hidden layers. The Tabu Search algorithm makes is also possible to reduce the network size. As one can see in Table 1, the best results have been obtained using $A^\star$ algorithm. The optimal network structure selected with this method consists of two processing layers and five hidden neurons. The optimal path generated with the $A^\star$ algorithm is presented in Table 2. In order to find the optimal neural network, 558 structures have been tested. Each another network has a bigger size than the previous one.

Table 2. Optimal path generated with the $A_\star$ algorithm

Network No.	Network structure	Filters orders		Modelling quality
		1st layer (hidden)	2nd layer (output)	
0	$N^1_{1,1}$	–	(0)	0.267325
2	$N^1_{1,1}$	–	(1)	0.171869
3	$N^2_{1,1,1}$	(0)	(1)	0.144926
9	$N^2_{1,1,1}$	(1)	(1)	0.249667
12	$N^2_{1,1,1}$	(1)	(2)	0.239527
20	$N^2_{1,2,1}$	(1 0)	(2)	0.151304
28	$N^2_{1,3,1}$	(1 0 0)	(2)	0.152115
31	$N^2_{1,4,1}$	(1 0 0 0)	(2)	0.161490
118	$N^2_{1,4,1}$	(1 1 0 0)	(2)	0.168057
322	$N^2_{1,4,1}$	(1 1 0 1)	(2)	0.168039
426	$N^2_{1,5,1}$	(1 1 0 1 0)	(2)	0.208492
540	$N^2_{1,5,1}$	(1 1 0 1 1)	(2)	0.210782
543	$N^2_{1,5,1}$	(2 1 0 1 1)	(2)	0.220161
558	$N^2_{1,5,1}$	(2 1 0 2 1)	(2)	0.123431

Table 3. The optimal path generated with the Tabu Search (list length – 3)

Network No.	Network structure	Filters orders		Modelling quality
		1st layer (hidden)	2nd layer (output)	
0	$N^1_{1,1}$	–	(0)	0.267325
2	$N^1_{1,1}$	–	(1)	0.171863
3	$N^2_{1,1,1}$	(0)	(1)	0.163588
9	$N^2_{1,1,1}$	(1)	(1)	0.163725
28	$N^2_{1,1,1}$	(1)	(0)	0.147749
33	$N^2_{1,1,1}$	(2)	(0)	0.145051

In the case of the Tabu Search method, the results are also interesting. When a short tabu list has been used (length of 3), the algorithm have demonstrated the periodic behavior. After every 23 structures it generates the same optimal neural network of the $N^2_{1,1,1}$ class. To avoid such a periodic behavior, the longer tabu lists have been applied (length of 5 and 10). Table 1 clearly shows that using longer tabu lists, better results can be obtained. Moreover, in both cases the identical optimal path has been achieved. The conclusion is, that further increasing size of the tabu list does not yield better results. In Table 3 one can see the optimal path generated with the Tabu Search

Table 4. The optimal path generated with the Tabu Search (list length – 10)

Network No.	Network structure	Filters orders		Modelling quality
		1st layer (hidden)	2nd layer (output)	
0	$N^1_{1,1}$	–	(0)	0.267325
2	$N^1_{1,1}$	–	(1)	0.171863
3	$N^2_{1,1,1}$	(0)	(1)	0.163588
9	$N^2_{1,1,1}$	(1)	(1)	0.163725
38	$N^2_{1,1,1}$	(2)	(1)	0.152251
42	$N^2_{1,2,1}$	(2 0)	(1)	0.167976
49	$N^2_{1,2,1}$	(2 0)	(2)	0.167494
54	$N^2_{1,2,1}$	(2 1)	(2)	0.151649
61	$N^2_{1,2,1}$	(2 1)	(1)	0.151706
64	$N^2_{1,3,1}$	(2 1 0)	(1)	0.159347
74	$N^2_{1,3,1}$	(2 1 0)	(0)	0.159677
80	$N^2_{1,3,1}$	(2 0 0)	(0)	0.166961
84	$N^2_{1,3,1}$	(2 0 1)	(0)	0.150898
88	$N^2_{1,3,1}$	(2 0 1)	(1)	0.156903
93	$N^2_{1,3,1}$	(2 1 1)	(1)	0.161591
97	$N^2_{1,4,1}$	(2 1 1 0)	(1)	0.139015

method (list length equal to 3). First, the network is growing up, and for the network No. 28 the algorithm reduces the network size. This phenomenon is very attractive and can cause that Tabu Search may be more flexible method than the A^* algorithm. In turn, Table 4 shows the optimal path generated with the Tabu Search algorithm using the tabu list length of 10.

4   Conclusion

The paper deals with the optimization of the DMLP network. The space of DMLP architectures has been organized in the form of digraph. Two heuristic algorithms have been applied in order to find the optimal node of the digraph: the A^*-algorithm and Tabu Search. The experimental results show that effectiveness of both algorithms is comparable. Slightly better result were achieved using A^*-algorithm. The choice of the proper heuristic function, however, is not a trivial task. On the other hand the length of the Tabu Search list has a crucial importance to the effectiveness of the method.

References

1. Ayoubi, M. (1994) Fault diagnosis with dynamic neural structure and application to turbo–charger. In: Fault Detection Supervision and Safety for Technical Processes, Int. Symp. SAFEPROCESS'94, Espoo, Finland, Vol.2, pp.618–623.
2. Calado, J.M.F., Korbicz J., Patan K., Patton R.J. and Sa da Costa J.M.G. (2001) Soft computing approaches to fault diagnosis for dynamic systems. European Journal of Control, Vol.7, Nos. 2-3, pp. 248–286.
3. Geman, S., Bienenstack, E., and Doursat, R. (1992) Neural networks and the bias/variance dilemma. Neural Computation, Vol.4, No.1, pp.1–58.
4. Glover, F. (1986) Future Paths for Integer Programming and Links to Artificial Intelligence. Computers and Operations Research, Vol.5.
5. Glover, F., and Laguna, M. (1997) Tabu Search. Kluwer Academic Publishers.
6. Gupta, M.M., and Rao, D.H. (1993) Dynamic neural units with application to the control of unknown nonlinear systems. Journal of Intelligent and Fuzzy Systems, Vol.1, pp.73–92 .
7. Hart, P.E., Nilsson, N.J., and Raphael, B. (1968) A formal basis for the heuristic determination of a minimum cost paths. IEEE Trans. Systems, Man, Cybernetics, Vol. 4.
8. Nilsson, N.J. (1980) Principles of Artificial Intelligence. Springer-Verlag, New York.
9. Obuchowicz, A. (1999) Architecture optimization of a network of dynamic neurons using the A^* algorithm. In: Intelligent Techniques and Soft Computing, 7th European Congress EUFIT'99, Aachen, Germany, (published on CD-ROM).
10. Rich, E. (1983) Artificial Intelligence. McGraw-Hill Company, New York.
11. Żurada, J. (1993) Lambda learning rule for feedforward neural nerworks, In: Neural Networks, Int. Conf., San Francisco, USA, pp.1808–1811.

Discovering Extended Action-Rules (System DEAR)

Zbigniew W. Ras[1,2] and Li-Shiang Tsay[1]

[1] UNC-Charlotte, Computer Science Dept., Charlotte, NC 28223, USA
[2] Polish Academy of Sciences, Institute of Computer Science, Ordona 21, 01-237 Warsaw, Poland

Abstract. Action rules introduced in [3] and investigated further in [4] assume that attributes in a database are divided into two groups: stable and flexible. In general, an action rule can be constructed from two rules extracted earlier from the same database. Furthermore, we assume that these two rules describe two different decision classes and that our goal is to re-classify some objects from one of these decision classes to the other one. Flexible attributes provide a tool for making hints to a business user what changes within some values of flexible attributes are needed for a given object to re-classify this object to another decision class. In [3], we suggested what changes are needed to classification attributes listed in both rules but we did not consider situations when such an attribute is listed only in one of these rules. Also, neither in [3] nor [4] we provide a way to compute support and confidence of action rules. ! In this paper, we show how system $DEAR$ is discovering extended action rules which give better strategies for re-classifying objects than strategies provided by action rules. Also, the confidence of extended action rules is much higher than confidence of corresponding action rules. System $DEAR$, implemented in KDD Laboratory at UNC-Charlotte, requires Windows 95 or higher. It does not discretize numerical attributes which means some discretization algorithm has to applied before $DEAR$ is used

1 Introduction

Data mining became successful because knowledge extracted from data provides competitive advantage in support of decision making. In the banking industry, the most widespread use of data mining is in the area of customer and market modeling, risk optimization and fraud detection. In financial services, data mining is used in performing so called trend analysis, in analyzing profitability, helping in marketing campaigns, and in evaluating risk. There are many software applications on the market that use data mining for stock prediction.

In support of marketing, data mining helps us in predictive modeling of markets and customers. It can identify the traits of profitable customers and reveal the hidden traits. It helps to search for the sites that are most convenient for customers as well as trends in customer usage of products and services. Specific questions about profitability must be answered from a broader perspective of customer and market understanding. For instance, customer loyalty can be often as important as profitability. In addition to short term profitability the decision makers must also keep eye on the lifetime value of a customer. Also, a broad and detailed understanding of customers is needed to send the right offers to the right customers at the most appropriate time. Knowledge about customers can lead to ideas about future offers which will meet their needs.

In [3], the notion of an action rule was proposed. Any action rule provides hints to a business user what changes within so called flexible attributes are needed to re-classify some customers from a lower profitability class to a higher one. These customers form, so called, supporting class for that rule. Each action rule was constructed from two rules, extracted earlier, defining different profitability classes. It was assumed that values of stable attributes listed in both rules had to be the same. It was no constraint placed on values of flexible attributes listed in both rules. Also, no constraint was placed on values of attributes listed only in one of these rules.

In this paper, we show what constraints have to be placed on values of classification attributes if they are listed only in one of the rules from which an action rule is formed. The proposed constraints make the confidence of new rules, called extended action rules, much higher than the confidence of corresponding action rules. System $DEAR$, implemented in KDD Laboratory at UNC-Charlotte, discovering extended action rules requires Windows 95 or higher. Numerical attributes have to be discretized before system $DEAR$ is used.

2   Information Systems and Decision Tables

An information system is used for representing knowledge. Its definition, presented here, is due to Pawlak [2].

By an information system we mean a pair $S = (U, A)$, where:

- U is a nonempty, finite set called the universe,
- A is a nonempty, finite set of attributes i.e. $a : U \longrightarrow V_a$ is a function for $a \in A$, where V_a is called the domain of a.

Elements of U are called objects. In our paper objects are interpreted as customers. Attributes are interpreted as features, offers made by a bank, characteristic conditions etc.

In this paper we consider a special case of information systems called decision tables [2]. In any decision table together with the set of attributes a partition of that set into conditions and decisions is given. Additionally, we assume that the set of conditions is partitioned into stable conditions and flexible conditions. There is only one decision attribute to be seen as *profit ranking*. Its domain contains values being integers. This decision attribute classifies objects (customers) with respect to the profit for a bank. Date of

Birth is an example of a stable attribute. Interest rate on any customer account is an example of a flexible attribute (dependable on bank). We adopt the following definition of a decision table:

A decision table is any information system of the form $S = (U, A_1 \cup A_2 \cup \{d\})$, where $d \notin A_1 \cup A_2$ is a distinguished attribute called decision. The elements of A_1 are called stable conditions, whereas the elements of A_2 are called flexible conditions.

The cardinality of the image $d(U) = \{k : d(x) = k \text{ for some } x \in U\}$ is called the rank of d and is denoted by $r(d)$.

Let us observe that the decision d determines the partition $CLASS_S(d) = \{X_1, X_2, ... X_{r(d)}\}$ of the universe U, where $X_k = d^{-1}(\{k\})$ for $1 \leq k \leq r(d)$. $CLASS_S(d)$ is called the classification of objects in S determined by the decision attribute d.

In this paper, as we mentioned before, objects correspond to customers. Also, we assume that customers in $d^{-1}(\{k_1\})$ are more profitable for a bank than customers in $d^{-1}(\{k_2\})$ for any $k_2 \leq k_1$. The set $d^{-1}(\{r(d)\})$ represents the most profitable customers for a bank. Clearly the goal of any bank is to increase its profit. It can be achieved by shifting some customers from the group $d^{-1}(\{k_2\})$ to $d^{-1}(\{k_1\})$, for any $k_2 \leq k_1$. Namely, through special offers made by a bank, values of flexible attributes of some customers can be changed and the same all these customers can be moved from a group of a lower profit ranking to a group of a higher profit ranking.

3   Extended Action Rules

In this section we describe a method to construct extended action rules from a decision table containing both stable and flexible attributes.

Before we introduce several new definitions, assume that for any two collections of sets X, Y, we write, $X \sqsubseteq Y$ if $(\forall x \in X)(\exists y \in Y)[x \subseteq y]$. Let $S = (U, A_1 \cup A_2 \cup \{d\})$ be a decision table and $B \subseteq A_1 \cup A_2$. We say that attribute d depends on B if $CLASS_S(B) \sqsubseteq CLASS_S(d)$, where $CLASS_S(B)$ is a partition of U generated by B (see [2]). Assume now that attribute d depends on B where $B \subseteq A_1 \cup A_2$. The set B is called d-reduct in S if there is no proper subset C of B such that d depends on C. The concept of d-reduct in S was introduced to induce rules from S describing values of the attribute d depending on minimal subsets of $A_1 \cup A_2$. In order to induce rules in which the THEN part consists of the decision attribute d and the IF part consists of attributes belonging to $A_1 \cup A_2$, subtables $(U, B \cup \{d\}!)$ of S where B is a d-reduct in S should be used for rules extraction. By $L(r)$ we mean all attributes listed in the IF part of a rule r. For example, if $r = [(a_1, 3) * (a_2, 4) \longrightarrow (d, 3)]$ is a rule then $L(r) = \{a_1, a_2\}$. By $d(r)$ we denote the decision value of a rule. In our example $d(r) = 3$. If r_1, r_2 are rules and $B \subseteq A_1 \cup A_2$ is a set of attributes, then $r_1/B = r_2/B$ means that the conditional parts of rules r_1, r_2 restricted to attributes B are the same. For example if $r_1 = [(a_1, 3) \longrightarrow (d, 3)]$, then $r_1/\{a_1\} = r/\{a_1\}$.

Example 1. Assume that $S = (\{x_1, x_2, x_3, x_4, x_5, x_6, x_7, x_8\}, \{a, c\} \cup \{b\} \cup \{d\})$ be a decision table represented by Table 1. The set $\{a, c\}$ lists stable attributes, b is a flexible attribute and d is a decision attribute. Also, we assume that H denotes customers of a high profit ranking and L denotes customers of a low profit ranking.

..	.a.	.b.	.c.	.d.
x_1	0	S	0	L
x_2	0	R	1	L
x_3	0	S	0	L
x_4	0	R	1	L
x_5	2	P	2	L
x_6	2	P	2	L
x_7	2	S	2	H
x_8	2	S	2	H

Table 1. Decision System

In our example $r(d) = 2$, $r(c) = 3$, $r(a) = 2$,
$CLASS_S(d) = \{\{x_1, x_2, x_3, x_4, x_5, x_6\}, \{x_7, x_8\}\}$,
$CLASS_S(\{b\}) = \{\{x_1, x_3, x_7, x_8\}, \{x_2, x_4\}, \{x_5, x_6\}\}$,
$CLASS(\{a, b\}) = \{\{x_1, x_3\}, \{x_2, x_4\}, \{x_5, x_6\}, \{x_7, x_8\}\}$,
$CLASS_S(\{a\}) = \{\{x_1, x_2, x_3, x_4\}, \{x_5, x_6, x_7, x_8\}\}$,
$CLASS_S(\{c\}) = \{\{x_1, x_3\}, \{x_2, x_4\}, \{x_5, x_6, x_7, x_8\}\}$,
$CLASS(\{b, c\}) = \{\{x_1, x_3\}, \{x_2, x_4\}, \{x_5, x_6\}, \{x_7, x_8\}\}$.

So, $CLASS(\{a, b\}) \subseteq CLASS_S(d)$ and $CLASS(\{b, c\}) \subseteq CLASS_S(d)$. It can be easily checked that both $\{b, c\}$ and $\{a, b\}$ are d-reducts in S.

Rules can be directly derived from d-reducts and the decision table S. In our example, we get the following optimal rules:
$$(a, 0) \longrightarrow (d, L), \ (c, 0) \longrightarrow (d, L),$$
$$(b, R) \longrightarrow (d, L), \ (c, 1) \longrightarrow (d, L),$$
$$(b, P) \longrightarrow (d, L), \ (a, 2) * (b, S) \longrightarrow (d, H), \ (b, S) * (c, 2) \longrightarrow (d, H).$$

Now, let us assume that $(a, v \longrightarrow w)$ denotes the fact that the value of attribute a has been changed from v to w. Similarly, the term $(a, v \longrightarrow w)(x)$ means that $a(x) = v$ has been changed to $a(x) = w$. Saying another words, the property (a, v) of a customer x has been changed to property (a, w).

Assume now that $S = (U, A_1 \cup A_2 \cup \{d\})$ is a decision table, where A_1 is the set of stable attributes and A_2 is the set of flexible attributes. Assume that rules r_1, r_2 have been extracted from S, B_1 is a maximal subset of A_1 such that $r_1/B_1 = r_2/B_1$, $d(r_1) = k_1$, $d(r_2) = k_2$ and $k_1 \leq k_2$. Also, assume that $(b_1, b_2, ..., b_p)$ is a list of all attributes in $L(r_1) \cap L(r_2) \cap A_2$ on which r_1, r_2 differ and $r_1(b_1) = v_1$, $r_1(b_2) = v_2,..., r_1(b_p) = v_p$, $r_2(b_1) = w_1$, $r_2(b_2) = w_2,..., r_2(b_p) = w_p$.

By (r_1, r_2)-action rule on $x \in U$ we mean a statement:
$$[(b_1, v_1 \longrightarrow w_1) \wedge (b_2, v_2 \longrightarrow w_2) \wedge ... \wedge (b_p, v_p \longrightarrow w_p)](x) \Rightarrow$$
$$[(d, k_1 \longrightarrow k_2)](x).$$

If the value of the rule on x is true then the rule is valid. Otherwise it is false.

Let us denote by $U^{<r_1>}$ the set of all customers in U supporting the rule r_1. If (r_1, r_2)-action rule is valid on $x \in U^{<r_1>}$ then we say that the action rule supports the new profit ranking k_2 for x.

To define an extended action rule, let us assume that two rules are considered. We will present them in a table to clarify the process of constructing an extended action rule. Here, St means stable attribute and Fl means flexible one.

$A - (St)$	$B - (Fl)$	$C - (St)$	$E - (Fl)$	$G - (St)$	$H - (Fl)$	D
a_1	b_1	c_1	e_1			d_1
a_1	b_2			g_2	h_2	d_2

Table 2. Two Decision Rules

The classical representation of these two rules will be:
$$r_1 = [a_1 * b_1 * c_1 * e_1 \longrightarrow d_1], \ r_2 = [a_1 * b_2 * g_2 * h_2 \longrightarrow d_2].$$

Assume now that object x supports rule r_1 which means that it is classified as d_1. In order to re-classify x to class d_2, we not only need to change its value B from b_1 to b_2 but also we have to require that $G(x) = g_2$ and the value H for object x has to be changed to h_2. This is the meaning of the extended (r_1, r_2)-action rule given below:

$$[(B, b_1 \longrightarrow b_2) \wedge (G = g_2) \wedge (H, \longrightarrow h_2)](x) \Rightarrow (D, d_1 \longrightarrow d_2)(x).$$

Assume now that by $NT(t)$ we mean the number of tuples having property t.

By support of the extended (r_1, r_2)-action rule (given above) we mean:
$$NT[(A = a_1) * (B = b_1) * (C = c_1) * (E = e_1) * (G = g_2) * (D = d_1)]$$

By the confidence of the extended (r_1, r_2)-action rule (given above) we mean:
$$\{NT[(A = a_1) * (B = b_1) * (C = c_1) * (E = e_1) * (G = g_2) * (D = d_1)]/$$
$$NT[(A = a_1) * (B = b_1) * (C = c_1) * (E = e_1) * (G = g_2)]\} \cdot$$
$$\{NT[(A = a_1) * (B = b_2) * (C = c_1) * (G = g_2) * (H = h_2) * (D = d_2)]/$$
$$NT[(A = a_1) * (B = b_2) * (C = c_1) * (G = g_2) * (H = h_2)]\}.$$

Clearly, for any extended (r_1, r_2)-action rule support and confidence is defined in a similar way.

Example 2. Assume that $S = (U, A_1 \cup A_2 \cup \{d\})$ is a decision table from the Example 1, $A_2 = \{b\}$, $A_1 = \{a, c\}$. It can be checked that rules $r_1 = [(b, P) \longrightarrow (d, L)]$, $r_2 = [(a, 2) * (b, S) \longrightarrow (d, H)]$, $r_3 = [(b, S) * (c, 2) \longrightarrow (d, H)]$ can be extracted from S. Clearly x_5, $x_6 \in U^{<r_1>}$. Now, we can construct (r_1, r_2)-action rule executed on x:
$$[(b, P \longrightarrow S)](x) \Rightarrow [(d, L \longrightarrow H)](x).$$

The extended (r_1, r_2)-action rule executed on x has the form:
$$[(a = 2) \wedge (b, P \longrightarrow S)](x) \Rightarrow [(d, L \longrightarrow H)](x).$$
Clearly objects x_5, x_6 support both rules.

Example 3. Assume that $S = (U, A_1 \cup A_2 \cup \{d\})$ is a decision table represented by Table 2. Assume that $A_1 = \{c, b\}$, $A_2 = \{a\}$.

..	.c.	.a.	.b.	.d.
x_1	2	1	1	L
x_2	1	2	2	L
x_3	2	2	1	H
x_4	1	1	1	L

Table 3. Decision Table

Clearly $r_1 = [(a, 1) * (b, 1) \longrightarrow (d, L)]$, $r_2 = [(c, 2) * (a, 2) \longrightarrow (d, H)]$ are optimal rules which can be extracted from S. Also, $U^{<r_1>} = \{x_1, x_4\}$.
If we construct an extended (r_1, r_2)-action rule
$$[(a, 1 \longrightarrow 2) \wedge (c = 2)](x) \Rightarrow [(d, L \longrightarrow H)](x)$$
then only x_1 supports that rule.

The corresponding (r_1, r_2)-action rule $[(a, 1 \longrightarrow 2)](x) \Rightarrow [(d, L \longrightarrow H)](x)$ is supported by x_1 and x_4.

Algorithm to Construct Extended Action Rules

Input:

Decision table $S = (U, A_1 \cup A_2 \cup \{d\})$,

A_1 - stable attributes, A_2 - flexible attributes,

λ_1, λ_2 - weights.

Output:

R - set of extended action rules.

Step 0.

$R := \emptyset$.

Step 1.

Find all d-reducts $\{D_1, D_2, ..., D_m\}$ in S which satisfy the property

$card[D_i \cap A_1]/card[A_1 \cup A_2] \leq \lambda_1$

(reducts with a relatively small number of stable attributes)

Step 2.

FOR EACH pair (D_i, D_j) of d-reducts (found in step 1) satisfying

the property $card(D_i \cap D_j)/card(D_i \cup D_j) \leq \lambda_2$ DO

find set R_i of optimal rules in S using d-reduct D_i,

find set R_j of optimal rules in S using d-reduct D_j.

Step 3.

FOR EACH pair of rules (r_1, r_2) in $R_i \times R_j$ having

different THEN parts DO

if B_1 is a maximal subset of A_1 such that $r_1/B_1 = r_2/B_1$, $d(r_1) = k_1$,

$d(r_2) = k_2$ and $k_1 \leq k_2$, then

if $(b_1, b_2, ..., b_p)$ is a list of all attributes in $Dom(r_1) \cap Dom(r_2) \cap A_2$

on which r_1, r_2 differ and

$r_1(b_1) = v_1, r_1(b_2) = v_2,..., r_1(b_p) = v_p$,

$r_2(b_1) = w_1, r_2(b_2) = w_2,..., r_2(b_p) = w_p$

and if $(A_1 - B_1) \cap Dom(r_2) = \{a_1, a_2, ..., a_q\}$, and

$r_2(a_1) = u_1, r_2(a_2) = u_2,..., r_2(a_q) = u_q$

and if $[Dom(r_2) \cap A_2] - \{b_1, b_2, ..., b_p\} = \{c_1, c_2, ..., c_r\}$

and $r_2(c_1) = t_1, r_2(c_2) = t_2,..., r_2(c_r) = t_r$,

then the following extended (r_1, r_2)-action rule add to R:

if $[(a_1 = u_1) \wedge (a_2 = u_2) \wedge ... \wedge (a_q = u_q) \wedge (b_1, v1 \longrightarrow w1) \wedge (b_2, v_2 \longrightarrow w_2) \wedge ... \wedge (b_p, v_p \longrightarrow w_p) \wedge (c_1, \longrightarrow t_1) \wedge (c_2, \longrightarrow t_2) \wedge ... \wedge (c_r, \longrightarrow t_r)](x)$

then $(d, k_1 \longrightarrow k_2)(x)$

The extended (r_1, r_2)-action rule says that if the change of values of attributes of customer x match the left-hand side of this rule, then the ranking profit of customer x should change from k_1 to k_2.

This algorithm was implemented as part of the system $DEAR$ (Discovering Extended Action Rules) under Windows 98 and tested on several banking databases. The quality of action rules discovered by $DEAR$ is much higher than by DAR (Discovering Action Rules). System DAR is the implemented version of the algorithm proposed in [3] by Ras and Wieczorkowska.

References

1. Chmielewski M. R., Grzymala-Busse J. W., Peterson N. W., Than S., (1993), The Rule Induction System LERS - a version for personal computers in *Foundations of Computing and Decision Sciences*, Vol. 18, No. 3-4, 1993, Institute of Computing Science, Technical University of Poznan, Poland, 181-212
2. Pawlak Z., (1985), *Rough Ssets and decision tables*, in Lecture Notes in Co Computer Science 208, Springer-Verlag, 1985, 186-196.
3. Ras, Z., Wieczorkowska, A., (2000), Action Rules: how to increase profit of a company, in *Principles of Data Mining and Knowledge Discovery*, (Eds. D.A. Zighed, J. Komorowski, J. Zytkow), Proceedings of PKDD'00, Lyon, France, LNCS/LNAI, No. 1910, Springer-Verlag, 2000, 587-592
4. Ras, Z., Gupta, S., (2002), Global action rules in distributed knowledge systems, in *Fundamenta Informaticae Journal*, IOS Press, Vol. 51, No. 1-2, 2002, 175-184
5. Skowron A., Grzymala-Busse J., (1991), From the Rough Set Theory to the Evidence Theory, in *ICS Research Reports*, 8/91, Warsaw University of Technology, October, 1991

Discovering Semantic Inconsistencies to Improve Action Rules Mining

Zbigniew W. Ras[1,2] and Angelina A. Tzacheva[1]

[1] UNC-Charlotte, Computer Science Dept., Charlotte, NC 28223, USA
[2] Polish Academy of Sciences, Institute of Computer Science, Ordona 21, 01-237
Warsaw, Poland

Abstract. A new class of rules, called action rules, show what actions should
be taken to improve the profitability of customers. Action rules introduced in [3]
and investigated further in [7] assume that attributes in a database are divided
into two groups: stable and flexible. These reflect the ability of a business user to
influence and control their change for a given consumer. In this paper, we introduce
a new classification of attributes partitioning them into stable, semi-stable, and
flexible. Values of stable attributes can not be changed for a given consumer (for
instance *maiden name* is an example of such an attribute). So, stable attributes
have only one interpretation. If values of an attribute change in a deterministic
way as a function of time (for instance values of the attribute *age*) we call them
semi-stable. All remaining attributes are called flexible. Clearly, in the process of
action rule extraction, stable attributes are highly undesirable. What about semi-
stable attributes? Although, they seem to be quite similar to stable attributes, the
difference between them is quite essential. Semi-stable attribute may have many
different interpretations but among them only one interpretation is natural and it
is called standard. All its other interpretations are called non-standard. In a non-
standard interpretation, a semi-stable attribute can be classified as flexible (business
user may control its change). In a single database we may easily fail to identify
attributes which have non-standard interpretation. Query answering system based
on distributed knowledge mining, introduced in [4,5], will be used in this paper
as a tool to identify which semi-stable attributes have non-standard interpretation
so they can be classified as flexible. This way, by decreasing the number of stable
attributes in a database we may discover action rules which would not be discovered
otherwise.

1 Introduction

Ras and Wieczorkowska [3] introduced the notion of action rules. Special
type of rules can be constructed from classification rules to suggest a way
to re-classify customers (objects) to a desired state. In e-commerce applications, this re-classification may mean that a consumer not interested in a
certain product, now may buy it, and therefore may fall into a group of more
profitable customers.

These groups are described by values of classification attributes in a decision table schema. In [3], all attributes are divided into stable and flexible. In

this paper, a new subclass of attributes called semi-stable attributes is introduced. Semi-stable attributes are typically a function of time, and undergo deterministic changes (for instance the attribute *age*). Different interpretations, called non-standard, of such attributes may exist, and in such cases all these attributes can be treated the same way as flexible attributes. In the algorithm of action rule extraction, presented in [3], attributes which are not flexible are highly undesirable. By identifying which semi-stable attributes have non-standard interpretation, we increase the number of flexible attributes and the same increase the chance to generate more precise action rules.

Assuming that attribute is flexible, we may find a way to change its value for a given object. However, quite often, such a change cannot be done directly to a chosen attribute (for instance to the attribute *profit*). In that situation, definitions of such an attribute in terms of other attributes have to be learned. These new definitions are used to construct action rules showing what changes in values of some attributes, for a given consumer, are needed in order to re-classify this consumer the way business user wants. We may search for definitions of these flexible attributes looking at either local or remote sites for help.

The application of semi-stable attributes to the process of action rules mining involves detection of nonstandard interpretations of semi-stable attributes. At local system level detection is possible, but limited to dependencies existing between local attributes. At distributed information systems level the detection of nonstandard interpretations involves discovering semantic inconsistencies, addressed by Ras and Dardzinska [8].

2   Information Systems and Decision Tables

An information system is used for representing business knowledge. Pawlak [2] gives the following definition:

By an information system we mean a pair $S = (U, A)$, where:

- U is a nonempty, finite set of objects (called customer identifiers),
- A is a nonempty, finite set of attributes i.e. $a : U \longrightarrow V_a$ for $a \in A$, where V_a is called the domain of a.

Information systems can be seen as generalizations of decision tables [2]. Partition of the set of attributes into conditions and decisions is given in any decision table. We assume that the set of conditions is partitioned into stable, semi-stable, and flexible conditions. Attribute $a \in A$ is called stable for the set U if its values assigned to objects from U can not be changed by a business user. An attribute is called semi-stable, if it is a function of time and it is changing in its standard interpretation in a deterministic way. Otherwise, it is called flexible. *Date of birth* is an example of a stable attribute.

Age is an example of semi-stable attribute (its value *young* in a non-standard interpretation may mean behaving as a young person). Interest rate on any customer account is an example of a flexible attribute. For simplicity reason, we will consider decision tables with only one decision. We adopt the following definition of a decision table:

By a decision table we mean any information system of the form $S = (U, A_1 \cup A_2 \cup A_3 \cup \{d\})$, where $d \notin A_1 \cup A_2 \cup A_3$ is a distinguished attribute called the decision. Elements of A_1 are called stable conditions, the elements of A_3 are called semi-stable, and $A_2 \cup \{d\}$ are called flexible conditions.

The assumption that attribute d is flexible is quite essential. Otherwise we would be unable to re-classify objects in U from the point of view of attribute d. So, if d is flexible and we want to change its value for a given object, values of some attributes from $A_2 \cup A_3$ have to be changed as well.

Before we proceed, certain relationships between values of attributes from A_2 and A_3 and values of the attribute d have to be presented first.

3   Action Rules

Ras and Wieczorkowska [3] propose a method to construct action rules from a decision table containing both stable and flexible attributes. In this section, let us assume that for each attribute in A_3 we know its semantics and also we know if it is stable or flexible. So $A_3 = \emptyset$, which means some semi-stable attributes are moved to A_1, some to A_2.

Assume now that for any two collections of sets X, Y, we write, $X \subseteq Y$ if $(\forall x \in X)(\exists y \in Y)[x \subseteq y]$. Let $S = (U, A_1 \cup A_2 \cup \{d\})$ be a decision table and $B \subseteq A_1 \cup A_2$. We say that attribute d depends on B if $CLASS_S(B) \subseteq CLASS_S(d)$, where $CLASS_S(B)$ is a partition of U generated by B (see [2]).

Assume now that attribute d depends on B where $B \subseteq A_1 \cup A_2$. The set B is called d-reduct in S if there is no proper subset C of B such that d depends on C. The concept of d-reduct in S was introduced, in rough sets theory (see [2], [6]), to identify minimal subsets of $A_1 \cup A_2$ such that rules describing the attribute d in terms of these subsets are the same as rules describing d in terms of $A_1 \cup A_2$. It was shown that in order to induce rules in which THEN part consists of the decision attribute d and IF part consists of attributes belonging to $A_1 \cup A_2$, only subtables $(U, B \cup \{d\})$ of S where B is a d-reduct in S can be used for rules extraction.

By $L(r)$ we mean all attributes listed in IF part of rule r. For example, if $r = [(a_1, 3) * (a_2, 4) \longrightarrow (d, 3)]$ is a rule then $L(r) = \{a_1, a_2\}$. By $d(r)$ we denote the decision value of a rule r. In our example $d(r) = 3$. Similarly, $a_1(r) = 3$.

If r_1, r_2 are rules and $B \subseteq A_1 \cup A_2$ is a set of attributes, then $r_1/B = r_2/B$ means that the conditional parts of rules r_1, r_2 restricted to attributes B are the same. For example if $r_1 = [(a_1, 3) \longrightarrow (d, 3)]$, then $r_1/\{a_1\} = r/\{a_1\}$.

Algorithm for constructing action rules, implemented as system DAR was given in [3].

For each pair of rules (r_1, r_2) satisfying the conditions $r_1/A_1 = r_2/A_1$, $d(r_1) = k_1$, $d(r_2) = k_2$ where $k_1 < k_2$, if $(b_1, b_2, ..., b_p)$ was a list of all attributes in $L(r_1) \cap L(r_2) \cap A_2$ on which r_1, r_2 differ and

$$r1(b1) = v1, r1(b2) = v2, ..., r1(bp) = vp, \qquad (1)$$
$$r2(b1) = w1, r2(b2) = w2, ..., r2(bp) = wp \qquad (2)$$

then the algorithm DAR generates the following (r_1, r_2)-action rule:

if $[(b_1, v_1 \longrightarrow w_1) \wedge (b_2, v_2 \longrightarrow w_2) \wedge ... \wedge (b_p, v_p \longrightarrow w_p)](x)$, then the ranking profit of customer x is expected to change from k_1 to k_2.

Object x supports this action rule if it satisfies the properties: $b_1(x) = v_1$, $b_2(x) = v_2, ..., b_p(x) = v_p$, $d(x) = k_1$.

By the support of an action rule r we mean all the objects supporting that rule. It is denoted by $Sup(r)$.

If the change of values of attributes of object $x \in Sup(r)$ will match the term
$[(b_1, v_1 \longrightarrow w_1) \wedge (b_2, v_2 \longrightarrow w_2) \wedge ... \wedge (b_p, v_p \longrightarrow w_p)](x)$
and the resulting object is either classified in S as k_2 or not classified at all, then the action rule r successfully supports x. The set of objects successfully supported by r is denoted by $SSup(r)$.

By the confidence of rule r we mean $Conf(r) = SSup(r)/Sup(r)$.

Example 1. Assume that $S = (\{x_1, x_2, x_3, x_4, x_5, x_6, x_7, x_8\}, \{a, c\} \cup \{b\} \cup \{d\})$ is a decision table represented as **Table 1**. The set $\{a, c\}$ contains stable attributes, b is a flexible attribute and d is a decision attribute.

It can be easily checked that $\{b, c\}$, $\{a, b\}$ are the only two d-reducts in S.

Applying for instance $LERS$ discovery system (see [1]), the following definitions are extracted from S:

$(a, 0) \longrightarrow (d, L)$, $(c, 0) \longrightarrow (d, L)$, $(b, R) \longrightarrow (d, L)$, $(c, 1) \longrightarrow (d, L)$,
$(b, P) \longrightarrow (d, L)$, $(a, 2) \wedge (b, S) \longrightarrow (d, H)$, $(b, S) \wedge (c, 2) \longrightarrow (d, H)$.

Now, let us assume that $(a, v \longrightarrow w)$ denotes the fact that the value of attribute a has been changed from v to w. Similarly, the term $(a, v \longrightarrow w)(x)$ means that $a(x) = v$ has been changed to $a(x) = w$. Saying another words, the property (a, v) of object x has been changed to property (a, w).

If we take $r = [(b, R \longrightarrow S) \Longrightarrow (d, L \longrightarrow H)]$ as the action rule, then $Sup(r) = \{x_2, x_4\}$, $Conf(r) = 1$.

	a	b	c	d
x_1	0	S	0	L
x_2	0	R	1	L
x_3	0	S	0	L
x_4	0	R	1	L
x_5	2	P	2	L
x_6	2	P	2	L
x_7	2	S	2	H
x_8	2	S	2	H

Table 1. Decision System

4   Semi-Stable Attributes

The notion of action rules introduced in [3] divides attributes into two groups: stable and flexible. These reflect the ability of a business user to influence and control their change for a given consumer. In the process of action rule extraction stable attributes are highly undesirable. We introduce a new classification of attributes into stable, semi-stable, and flexible taking into consideration semantics of attributes which clearly may differ from database to database.

Value of a stable attribute a for a given object cannot be changed by a business user in any interpretation of a. All such interpretations are called standard.

An example of such an attribute is *date of birth*. Standard interpretations of this attribute may differ in a granularity level. It is possible that one stable attribute implies another one.

There is a special subset of attributes called semi-stable, which at first impression may look stable, but they are a function of time and undergo changes in a deterministic way. Therefore, they cannot be called stable. The change is not necessarily in a linear fashion, where the attribute may be stable for a period of time, and then begin changing in certain direction as shown on Figure 1.

Semi-stable attributes may have many interpretations, some of which might be nonstandard. We denote by $M_s(a)$ the set of standard interpretations of attribute a and by $M_n(a)$ the set of non-standard interpretations of a. If the attribute a has a nonstandard interpretation, $I(a) \in M_n(a)$, then it can be changed, and thus it may be seen as a flexible attribute in action rule extraction.

For instance, if $a = $ "age" and $Dom(a) = \{young, middle - aged, old\}$, the author of the database may indeed input *young* for a person who behaves as young when their actual age is *middle-aged*. Then the interpretation is nonstandard. The business user can therefore influence this attribute. For

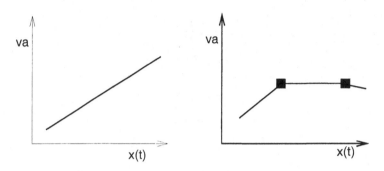

Fig. 1. Semi-stable attributes

example, if the following action rule was mined for object x

$$r_1 = [[(a, young \longrightarrow middle - aged)](x) \Longrightarrow [(d, L \longrightarrow H)](x)]$$

with respect to decision attribute d *(ex. loyalty)* the business user would like to change the attribute value *young* to *middle-aged* for object x. Since the database author interpretation is nonstandard related to the behavior associated with certain age, if the object is put into special conditions that can affect its behavior, such as top university, the attribute value can be changed, and the same object x might be re-classified from *low loyalty* to *high loyalty*.

Many cases of nonstandard interpretations could be found in databases. It is particularly important for those to be detected when mining for global rules in distributed knowledge systems. An example is the attribute *height*. Consider the following situation: Chinese people living in the mountains are generally taller than majority of Chinese population.

If $Dom(a) = \{short, medium, tall\}$ for attribute *height* and a system S_1, contains data about Chinese population in the mountains, the author of the database may consider a certain Chinese person living in the mountains *medium* height in relation to the rest. Now, assume another system S_2 containing data about Chinese people living in popular urban area. In global action rule extraction, if S_2 is mined for rules, the interpretation would regard the height value *medium* from S_1 as *tall*. Therefore, the interpretation in S_1 is nonstandard.

Numeric attributes may possess nonstandard interpretations as well. Consider for instance the attribute *number of children*. When one is asked about the number of children one has, that person may count step-children, or children who have died. In such a case, the interpretation is nonstandard.

A flexible attribute is an attribute which value for a given object varies with time, and can be changed by a business user. Also, flexible attributes may have many interpretations. *Interest rate* is an example of a flexible attribute.

Assume that $S = (U, A)$ is an information system which represents one of the sites of a distributed knowledge system (DKS). Also, let us assume that each attribute in A is either stable or flexible but we may not have sufficient temporal information about semi-stable attributes in S to decide which one is stable and which one is flexible. In such cases we will search for additional information, usually at remote sites for S, to classify uniquely these semi-stable attributes either as stable or flexible.

5   Discovering Semantic Inconsistencies

Different interpretations of flexible and semi-stable attributes may exist. Semi-stable attributes, in a non-standard interpretation, can be classified as flexible attributes and therefore can be used in action rule extraction. We discuss a detection method of nonstandard interpretations of a semi-stable attribute at local knowledge system level, and next at distributed knowledge systems level. Detection of a nonstandard interpretation at local level is limited to the dependency of one semi-stable attribute to another semi-stable attribute for which it is known that its interpretation is standard. Attribute related to time must be available in the information system, such as the attribute *age*. Furthermore, information about certain breakpoints in attribute behavior is required, such as the break points shown in Graph 2. This information can be placed in the knowledge system ontology.

Assume that $S = (X, A)$ is an information system and I is the interpretation used for attributes in S. Also, assume that both $a, b \in A$ are semi-stable attributes, $I(a) \in M_s$ and the relation $\prod_{\{a,b\}}(S) \subseteq \{(v_a, v_b) : v_a \in Dom(a),\ v_b \in Dom(b)\}$ is obtained by taking projection of the system S on $\{a, b\}$. The ontology information about break points for attributes a and b in S, represented in the next section as relation $R_{I(a),I(b)}$, is assuming that the interpretation I is standard for attribute b. It is possible that some tuples in $\prod_{\{a,b\}}(S)$ do not satisfy the break points requirement given. In such a case the interpretation of B is nonstandard, $I(B) \in M_n$.

Consider the following situation:

Example 3. Assume that $S_1 = (U_1, \{a\} \cup \{h, j\} \cup \{b\})$ is an information system represented by Table 2, where $\{a\}$ is a set of stable attributes, $\{h, j\}$ is a set of semi-stable attributes, and $\{b\}$ is a set of flexible attributes, where h is *height* and j is *the number of of aunts and uncles*. The interpretation of j is known to be standard, $I(j) \in M_s$. The system represents a local site.

Graph 2 shows the break points defined by the system's ontology for attributes h and j as a function of time t. The number of aunts and uncles grows as the height grows, since the person is young, and the parents brothers and

	a	h	b	j
x_1	0	a	S	m
x_2	0	b	R	m
x_3	0	b	S	n
x_4	0	c	R	m
x_5	2	b	P	n
x_6	2	b	P	n
x_7	2	a	S	m
x_8	2	c	S	m

Table 2. Information System

sisters have newborn children. The number of aunts and uncles decreases, as the height becomes constant or shrinks, since the person is middle-aged or old, and the number of aunts and uncles decreases as they die. Therefore,

if $I(h) \in M_s$ and $I(j) \in M_s$, then $R_{I(h),I(j)} = \{(a,m),(b,m),(b,n),(c,n)\}$ is placed in the ontology layer for system S_1.

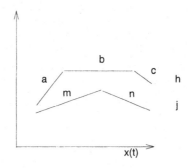

Fig. 2. Graph 2

From Graph 2, we see that relation instance $(c,m) \in \prod_{\{h,b\}}(S_1)$ representing objects x_4, x_8 does not belong to $R_{I(h),I(j)}$. Therefore, $I(h) \in M_n$.

In other words, objects x_4, x_8 do not satisfy the break point requirement given on Graph 2, thus the interpretation of attribute h*height* is nonstandard.

Local knowledge systems provide limited capability of detecting nonstandard semantics. Distributed knowledge systems supply greater ability to detect nonstandard semantics. They also give the opportunity to business users to seek alternative solution at remote sites. This is particularly important in a situation when they are either not willing or not able to undertake the suggested actions from their local site. In a distributed knowledge systems (DKS) scenario semantic inconsistencies can be detected even if temporal information is not available. With large number of sites containing similar attributes in DKS, certain trends can be observed, such as association rules with high confidence and support common for all the sights. In such a situation, it is also possible that a small number of sites do not support those common rules, or even contradict them. This case presents a hint for non-standard attribute interpretation, and semantic inconsistencies.

Assume S_1, S_2,..., S_n are large information systems which are parts of DKS. Association rule mining is performed on all systems in DKS. Rules satisfying the minimum support min_{sup} and minimum confidence min_{conf} thresholds are mined. If a rule

$$[w_a \star w_b \star ... \star \star w_c] \longrightarrow w_d \, [min_{sup}, min_{conf}],$$

where $d \in A$ is a semi-stable attribute in $S = (U, A)$ is extracted from S and supported by many sites in DKS, it is called a trade, or common rule. If the common rule is not supported, or it is contradicted (2) by only a small number of sites in DKS

$$\sim ([w_a \star w_b \star ... \star \star w_c] \longrightarrow w_d) \, [min_{sup}, min_{conf}],$$

then the attribute d has nonstandard interpretation in S.

Assume that we do not know if the interpretation of a semi-stable attribute $d \in A_1$ at site $S_1 = (U_1, A_1)$ is standard. We have discussed the case when attribute d is defined in S_1 in terms of a semi-stable attribute b, which has standard interpretation in S_1. Namely, if we identify an object in S_1 which description contradicts information about attributes d, b stored as a part of ontology of $S1$, then attribute d has non-standard interpretation in S_1. However, it can happen that we do not have any information about the interpretation of b in S_1. We can either look for a definition of d in terms of another semi-stable attribute in S_1 or look for a definition of d in terms of attribute b at another site of DKS. If we cannot find any attribute other than d, which is semi-stable and has non-standard interpretation in S_1, we contact another site.

Let A_1^{ss} be the set of all semi-stable attributes in S_1. We search for sites S_i such that $d \in A_i$ and $A_i^{ss} \cap A_1^{ss} \neq \emptyset$. Let I_d be the collection of such sites and $b \in A_1^{ss} \cap A_i^{ss}$, where $i \in I_d$.

In the case where, the interpretation of both attributes d, b is standard, if $I \in I_d$ satisfies the property that any $b \in A_1^{ss} \cap A_i^{ss}$ has standard interpretation in S_i, then i is not considered. Thus, we need to observe another site from I_d.

Now let us assume that $I_{\{d,b\}}$ is the set of sites in I_d such that $b \in A_1^{ss} \cap A_i^{ss}$. We extract rules at sites $I_{\{d,b\}}$ describing d and having b on their left side. Either association rules discovered at site S_1 will support association rules discovered at majority of sites $I_{\{d,b\}}$ or conflict many of them. We claim that the interpretation of attribute d is standard in the first case. In the second case, it is non-standard.

This algorithm was tested on a DKS consisting of thirteen sites representing thirteen Insurance Company Datasets, with a total of 5000 tuples in all DKS sites. Semi-stable attributes with non-standard interpretation have been detected and used jointly with flexible attributes in action rules mining. The confidence of these action rules is usually higher than the confidence of the corresponding action rules based only on flexible attributes.

References

1. Chmielewski M. R., Grzymala-Busse J. W., Peterson N. W., Than S., (1993), The Rule Induction System LERS - a version for personal computers in *Foundations of Computing and Decision Sciences*, Vol. 18, No. 3-4, 1993, Institute of Computing Science, Technical University of Poznan, Poland, 181-212

2. Pawlak Z., (1985), *Rough Ssets and decision tables* , in Lecture Notes in Co Computer Science 208, Springer-Verlag, 1985, 186-196.

3. Ras, Z., Wieczorkowska, A., A(2000), Action Rules: how to increase profit of a company, in *Principles of Data Mining and Knowledge Discovery*, (Eds. D.A. Zighed, J. Komorowski, J. Zytkow), Proceedings of PKDD'00, Lyon, France, LNCS/LNAI, No. 1910, Springer-Verlag, 2000, 587-592

4. Ras, Z., Zytkow, J.M., (2000), Mining for attribute definitions in a distributed two-layered DB system, in *Journal of Intelligent Information Systems*, Kluwer, Vol. 14, No. 2/3, 2000, 115-130

5. Ras, Z., (2001), Query Aaswering based on distributed knowledge mining, in *Proceedings of IAT2001,*, Maebashi City, Japan, October 23-266

6. Skowron A., Grzymala-Busse J., (2001), From the Rough Set Theory to the Evidence Theory, in *ICS Research Reports*, 8/91, Warsaw University of Technology, October, 1991

7. Ras, Z., Gupta, S., (2002), Global action rules in distributed knowledge systems, in *Fundamenta Informaticae Journal*, IOS Press, Vol. 51, No. 1-2, 2002, 175-184

8. Ras, Z., Dardziska A., (2002), Handling semantic inconsistencies in distributed knowledge systems using ontologies, in *Foundations of Intelligent Systems*, Proceedings of ISMIS'02 Symposium, Lyon, France, LNCS/LNAI, Vol. 2366, Springer-Verlag, 2002, 66-74

Incremental Rule Induction for Multicriteria and Multiattribute Classification

Jerzy Stefanowski and Marcin Żurawski

Institute of Computing Science, Poznań University of Technology, 60-965 Poznań, Poland. E-mails *jerzy.stefanowski@cs.put.poznan.pl, mzurawski@mikrob.pl*

Abstract. Incremental induction of decision rules within the dominance-based rough set approach to the multicriteria and multiattribute classification is discussed. We introduce an algorithm, called Glance, that incrementally induces a set of all rules from examples. We experimentally evaluate it and compare with two other algorithms, previously proposed for this kind of classification problem.

1 Introduction

Induction of decision rules from provided learning examples is one of the main problems considered in machine learning and knowledge discovery. Several algorithms for rule induction have already been proposed for various types of *attributes* (nominal, numerical, etc.) describing examples, see e.g. reviews in [4,7]. However, in some problems one can meet yet another type of attributes with *preference order* in their domains. For example, when considering buying a car, an attribute like "fuel consumption" has a clear preference ordered scale - the less, the better. The attributes with preference ordered domains are called *criteria*. Let us notice that this kind of semantic information is often present in data related to cost, gain or quality of objects like in economical data; however it is neglected by typical knowledge discovery tools. On the other hand, consideration of criteria is the key aspect of Multiple-Criteria Decision Analysis (MCDA) [1,6].

In this paper we consider one of the major MCDA problems called *multiple-criteria classification* (or sorting) *problem*. It concerns an assignment of objects evaluated by a set of criteria into pre-defined, and also preference-ordered, decision classes. When solving such a problem, any reasonable regularities to be discovered from data have to take into account a *dominance principle*. This principle requires that an object x having better, or at least the same, evaluations on a set of criteria than object y cannot be assigned to the worse decision class than object y. Some data may contain objects violating this principle. This consitutes a kind of *inconsistency* that should be handled in a proper way from MCDA point of view. Greco, Matarazzo and Slowinski [1] have proposed an extension of the rough sets theory [5] that is able to deal with this kind of inconsistency. It is called Dominance-based Rough Set Approach (DRSA). Decision rules, which are induced from examples within DRSA framework, have a special syntax, that requires a new type of algorithms for their induction.

First, an algorithm DOMLEM was introduced [3]. It induces a minimal set of DRSA rules covering all examples in the input data. On the other hand, sometimes it is also interesting to induce other sets of the rule, i.e. a set of all rules which can be induced from the given data or a subset of all rules which satisfy user predefined requirements concerning, e.g. minimal number of supporting objects or maximal length of the condition part. In [7] there have been discussed two algorithms for inducing such sets of rules, algorithm *AllRules* to generate the set of all rules and algorithm *DomExplore* to produce the satisfactory subset of rules. However, these algorithms require all examples to be read into memory before induction. Experiments have also showed that they are efficient rather on smaller data sets. Hence, these algorithms are not suitable for handling large amounts of examples or for situations where only the part of data is available at the begining and other parts are provided later. Thus, we are interested in an incremental processing of examples while inducing rules.

The aim of this paper is to present a new algorithm for inducing dominance-based decision rules for the multicriteria and multiattribute classification problem. It is called "Glance" and induces all rules in an incremental way. Moreover, we experimentally evaluate it on several data sets and compare its efficiency with two previously proposed, algorithms: *AllRules* and *Explore*.

The next section of this paper gives basic notations of the dominance based rough set approach and decision rules. In section 3, the *Glance* algorithm is introduced. Then, its experimental evaluation is described in section 4. Discussion of the experimental results and final remarks are contained in the last section.

2   Dominance based decision rules

Selected concepts of DRSA are presented, for more details see e.g. [1–3]. Let us assume that learning examples are represented in *decision table DT* = $(U, A \cup \{d\})$, where U is a set of examples (objects), A is a set of *condition attributes* describing examples, V_a is a domain of a. Let $f(x, a)$ denote the value of attribute $a \in A$ taken by object $x \in U$. The domains of attributes may be preference–ordered or not - in the first case the attributes are called *criteria* while in the latter case we call them *regular attributes*. Preference scale of each criterion a induces a *complete preorder* (i.e. a strongly complete and transitive binary relation) $\succeq$ in set U. For objects x and y, $\succeq$ means that object x is "*at least as good*" as y with respect to criterion a. The asymmetric part of $\succeq$ is denoted by $\succ$. An object x is preferred to object y on criterion a if $a(x) \succ a(y)$ and x is indiscernible to y if $a(x) = a(y)$.

$d \notin A$ is a decision attribute that partitions a set of examples into k decision classes $\{Cl_t : t = 1, \ldots, k\}$. These classes are preference ordered, i.e. for all r, s such that $s > r$, objects from Cl_s are prefered to objects from Cl_r. In multicriteria classfication problems it is typical to consider *upward union* and *downward union* of classes instead of single decision classes. The upward and downward unions are defined, respectively, as:

$$Cl_g^\geq = \bigcup_{h \geq g} Cl_h, \quad Cl_g^\leq = \bigcup_{f \leq g} Cl_f, \quad g = 1, \ldots k.$$

The statement $x \in Cl_g^\geq$ means "x belongs to at least class Cl_g", while $x \in Cl_g^\leq$ means "x belongs to at most class Cl_g.

The rough set theory requires determining relation between objects. For each criterion the *dominance relation* is defined in the following way. Let $P^\succeq$ be a subset of criteria from A; the object x *dominates* object y with respect to subset $P^\succeq$ iff $a(x) \succeq a(y)$ for all $a \in P^\succeq$. For each regular attribute from $P^= \subset A$ there exists *indiscernible relation*, which is an equivalence relation as in the original rough sets theory [5].

For any subset $P \subseteq A$ two sets of objects defined: $R_P^+(x)$ and $R_P^-(x)$. Given object $x \in U$, $R_P^+(x)$ is the set of all objects $y \in U$ which dominates x with respect to $P^\succeq$ (criteria of P) and are indiscernible with x with respect to $P^=$ (attributes of P). Analogously, $R_P^-(x)$ is the set of all objects $y \in U$ which are dominated by x with respect to $P^\succeq$ and are indiscernible with x with respects to $P^=$. If the decision table contains *inconsistent* objects, the sets $R_P^+(x)$ and $R_P^-(x)$ can be used to define lower and upper approximations of upward unions $Cl_g^\geq$ and downward unions $Cl_g^\leq$, respectively. For instance, the lower approximation of $Cl_g^\geq, g \leq k$ is the set $\{x \in U : R_P^+(x) \subseteq Cl_q^\geq\}$, i.e. it contains all objects belonging to $Cl_q^\geq$ without ambiguity with respect to their descriptions.

Examples belonging to rough approximations of upward and downward unions of classes are used for induction of "$if \ldots then \ldots$" decision rules. The two kinds of these rules are distinguished:

$D_\succeq$–decision rules with the following syntax: $if\ (f(x, a_1) \succeq r_{a_1}) \wedge \ldots \wedge (f(x, a_m) \succeq r_{a_m}) \wedge f(x, a_{m+1}) = r_{a_{m+1}} \wedge \ldots \wedge (f(x, a_p) = r_{a_p})\ then\ x \in Cl_g^\geq$,

$D_\preceq$–decision rules with the following syntax: $if\ (f(x, a_1) \preceq r_{a_1}) \wedge \ldots \wedge (f(x, a_m) \preceq r_{a_m}) \wedge f(x, a_{m+1}) = r_{a_{m+1}} \wedge \ldots \wedge (f(x, a_p) = r_{a_p})\ then\ x \in Cl_g^\leq$,

where $P = \{a_1, \ldots, a_p\} \subseteq A, P^\succeq = \{a_1, \ldots, a_m\}, P^= = \{a_{m+1}, \ldots, a_p\}$, $(r_1, \ldots, r_p) \in V_{a1} \times V_{a2} \times V_{ap}$ and $g = 1, \ldots k$.

The $D_\succeq$–rule says: "if an evaluation of object x on criteria a_i is at least as good as a threshold value r_i $(i = 1, \ldots, m)$ and on attribute a_j object x is indiscernible with value r_j $(i = m+1, \ldots, p)$ then object x belongs to a least class Cl_g". Similarly the $D_\preceq$–decision rule means that an object is evaluated as "at most good as a value" and belongs to at most a given class.

Each dominance based decision rule has to be *minimal*. Since a decision rule is an implication, by a minimal decision rule we understand such an implication that there is no other implication with an antecedent of at least the same weakness (in other words, rule using a subset of elementary conditions or/and weaker elementary conditions) and a consequent of at least the same strength (in other words, rule assigning objects to the same union or sub-union of classes).

Consider a $D_{\succeq}$–decision rule $if\,(f(x, a_1) \succeq r_{a_1}) \wedge \ldots \wedge (f(x, a_m) \succeq r_{a_m}) \wedge f(x, a_{m+1}) = r_{a_{m+1}}) \wedge \ldots \wedge (f(x, a_p) = r_{a_p})\,then\,x \in Cl_g^{\geq}$. If there exists an object y such that $f(y, a_1) = r_{a_1} \wedge \ldots \wedge (f(y, a_m) = r_{a_m} \wedge \ldots \wedge f(y, a_p) = r_{a_p}$, then y is called basis of the rule. Each $D_{\succeq}$–rule having a basis is called *robust* because it is "founded" on an object existing in the data table. Analogous definition of robust decision rules holds for the other types of rules.

We say that an object supports a decision rule if it matches both condition and decision parts of the rule. On the other hand, an object is covered by a decision rule if it matches the condition part of the rule. The rule can be described by a parameter *strength*, which is the number of supporting objects.

The set of induced rules is *complete*, if is covers all objects from the data table in such a way that objects belonging to lower approximations of unions are re-assigned to their original class while inconsistent objects are assigned to cluster of classes referring to their inconsistency [1]. The set of rules is called *minimal* if it is a set of minimal rules that is complete and non-redundant, i.e. exclusion of any rule from it makes it incomplete.

3   "Glance" algorithm

The minimal set of rules can be induced from examples of multicriteria and multiattribute classification problem by means of the *DOMLEM* algorithm [3,7]. Such a set of rules is usually sufficient for aims of *predicting classification* of new (or testing) objects. However, if the aim of the induction process is *descriptive*, i.e. discovered rules should help in explaining or better understanding of circumstances in which decisions were made, another kind of rule sets are also useful [7]. One possibility is to create all decision rules, which could be generated from the given decision table, while another option is to discover the satisfactory subset of all rules, which satisfy user's requirements e.g. sufficiently high strength and confidence, contain limited number of elementary conditions. The algorithms for inducing such sets of dominance based rules were discussed in [7,8]. Let us notice however, that their computational costs are higher than for DOMLEM; in particular it is true in case of looking for all rules, which is a problem characterized by an exponential complexity with respect to a number of attributes. Moreover, these algorithms require all examples to be read into memory before induction – what is a serious drawback if knowledge is to be discovered from real databases. Let us also remind that some learning problems are characterized by an incremental processing of information, i.e. descriptions of examples are available sequentially in steps. To overcome all of these problems we have decided to consider *incremental learning*, i.e. the learning algorithm has to learn rules from provided examples and then refine knowledge representation when new examples become available. The paradigm of incremental learning has been already considered in the field of machine learning or knowledge discovery from databases, however we have to skip the review of existing approaches due to a limited size of the paper.

We introduce an algorithm called *Glance*, which is incremental and stores in memory only rules and not descriptions of learning examples. In its basic version, it induces the set of all rules from examples of multiattribute and multicriteria classification problems. These rules are not necessarily based on some particular objects, i.e. they are non-robust. The general scheme of the algorithm for the case of inducing certain rules is presented below.

Procedure *Glance*
(**input** U -set of examples; P - set of criteria and regular attributes;
$Cl^{\geq \leq} = \{Cl_1^{\leq}, \ldots, Cl_{t-1}^{\leq}, Cl_2^{\geq}, \ldots, Cl_t^{\geq}\}$ - family of unions of decision classes;
output R - set of decision rules);
begin
    for each union $u \in Cl^{\geq \leq}$ **do**
    begin
        Let r be a rule with empty condition part;   {it covers each example}
        $R_u \leftarrow r$;   {add to the set of rules for union u}
    end{for}
    for each example $x \in U$ **do**
        for each u such that $f(x, d) \notin u$ **do** {d is a decision attribute}
            for $r \in R_u$ **do**
                if (r covers x) **then**   {r does not discriminate properly}
                begin
                    $R_u \leftarrow R_u \setminus r$;
                    for each $a \in P$ **do**
                    begin   {specialize rules}
                        Let s be a condition on attr. a excluding x;
                        $r_{new} \leftarrow r \cup s$;
                        if (r_{new} is minimal) **then** $R_u \leftarrow R_u \cup r_{new}$;
                    end; {for each $a \in P$}
                end; {if r covers x}
    R $\leftarrow \bigcup_u R_u$;
end{procedure}

The main idea of *Glance* is concordant with inductive learning principle, according to which an algorithm starts from the most general description and then updates it (specializes) so that it is consistent with available training data (also considering approximations). The most general description are rules with empty condition part, which means that such rules cover all examples. At the beginning of calculations, a set of rules R_u containing only one rule with an empty condition part is assigned to every union u. When the new object, e.g. $x \in Cl_g$ is available, it is determined for which unions u this object should be treated as a negative example (these unions include all $Cl_f^{\leq}$ such that $f < g$ for downward unions and $Cl_h^{\geq}$ such that $h > g$ for upward unions). For all such unions their rule sets R_u are checked. If any rule $r \in R_u$

covers x, then it has to be updated as it does not discriminate all negative examples properly. It is removed from R_u and specialized by adding on each attribute or criterion an elementary condition that it is not satisfied by description of x. For criteria, elementary conditions are in forms $(f(x, a) \succ v_a)$ or $(f(x, a) \prec v_a)$, depending on the direction of preference and unions; where v_a is the value of criterion a for object x. For regular attributes, conditions are in form $(f(x, a) \notin \{v_1, \ldots, v_l\})$ and v_a is added to this list. After constructing a new rule its minimality has to be checked. While inducing rules incrementally, minimality is checked only between rules concerning the same union of classes. Verification of minimality between unions is performed only at the very end of calculations, when it is certain that no more learning examples will become available. Furthermore, at the end of the induction process, these conditions in all rules are transformed to representation using operators $(\preceq, \succeq, =)$ as in the syntax of dominance based rules.

The computational complexity of the basic version of *Glance* is exponential with respect to a number of attributes. The user can define the maximal accepted number of elementary conditions in the syntax of the rule.

There is also an extended version of *Glance* [8], which allows to induce a satisfactory set of rules with other pre-defined constraint expressing minimum accepted (relative) strength of the rule. However, it requires to maintain in memory additional information about the number of processed examples, which are covered by already generated rules satisfying the constraint – for more details see [8].

4   Experiments

The main aim of an experimental evaluation of the *Glance* algorithm is check how the computational time and the number of rules are changing with increasing number of objects and attributes/criteria.

Moreover, we want to compare its performance with previously known algorithms *DomExplore* and *AllRules* [8]. Both of these algorithms are dedicated for multicrieria sorting problems, but they work in an non-incremental way. *DomExplore* induces the satisfactory set of *non-robust* rules and it is inspired by data mining algorithms, which search multiattribute (only) data for all "strong" rules with strength and confidence not less than predefined thresholds [7]. The *AllRules* algorithm has been developed by Greco and Stefanowski and it induces the set of *all robust* rules, which are based on some particular (non-dominated) objects belonging to approximations of upward or downward unions of classes. It is strongly tied to dominance based rough sets approach and uses its specific properties to reduce descriptions of objects being basis for robust rules.

The experiments were performed on family of artificial data sets. They were randomly generated according to chosen probability distribution (for details see [8]) and differ by the number of objects. Moreover for each series

Table 1. The computation time (in seconds) for compared algorithms while changing the number of objects

Algorithm	No. of attrib.	Number of objects							
		500	1000	1500	2000	3000	4000	5000	6000
DomExplore	3	0.10	0.33	0.72	1.15	2.69	4.9	8.3	14.99
	6	3.68	21.4	73.7	178.4	704.9	–	–	–
AllRules	3	0.10	0.16	1.41	3.13	7.69	18.4	34.3	53.01
	6	1.43	5.5	13.1	27.2	108.1	281.9	502.9	712.7
Glance	3	0.05	0.05	0.05	0.06	0.06	0.06	0.1	0.12
	6	1.26	3.13	5.5	7.86	15.38	21.2	26.31	38.7

of data sets we could change the number of attributes and the proportion of criteria to regular attributes. Firstly, all three algorithms were compared on the same data sets with respect to time of computation. In two series of data, with number of attributes 3 (2 criteria and 1 regular attribute) and 6 (2 criteria and 4 regular attributes), we systematically changed the number of objects from 500 till 6000. The results are presented in Table 1; symbol "–" means that the algorithm exceeded the accepted resources. Let us remark, that the number of rules induced by algorithms may be different as *Allrules* generates robust rules. Then, in Table 2 we present the change of the number of rules induced by *Glance* algorithm while incrementally increasing the number of objects (with 6 attributes).

Table 2. The number of rules induced by Glance while processing the different number of objects

	Number of objects										
	500	1000	1500	2000	2500	3000	3500	4000	4500	5000	6000
no. rules	1024	1580	1961	2153	2271	2495	2629	2705	2791	2962	3072

Table 3. The computational time (in seconds) of algorithms while changing the number of regular attributes and criteria in data containing 100 objects

Algorithm	No. of criteria and regular attributes					
	3	6	9	12	15	18
DomExplore	0	0.11	0.77	9	41.14	284.79
AllRules	0	0.06	0.82	7.58	68.66	624.83
Glance	0	0.22	44.9	439.41	–	–

Then, we have examined the influence of changing the number of attributes for the time of computation (with fixed number of objects equal to 100), see Table 3. For *Allrules* the results are growing with the number of attributes. For *DomExplore* the time also grows exponentially with the number of attributes, but respective results are around 10 times longer than *AllRules*. However, for *Glance* time exceeded the accepted limit when number of attributes was greater than 12.

Let us comment that, we also examined the impact of varying the proportion of regular attributes to criteria in the data sets on the time of computation and the number of rules for all compared algorithms. Due to the limitation on the paper size, there results will be published elsewhere.

5   Discussion of results and final remarks

Discovering knowledge from data containing criteria requires new tools. In this paper, we have considered the dominance-based rough set approach. Within this framework, we introduced a new algorithm *Glance*, which induces the set of all rules from provided examples. Unlike the previous algorithms, the new one works in an incremental way, i.e. every time new examples are available it updates so far obtained rules without having to generate new rules from scratch. It stores only rules in memory but not processed examples. The user can also define the constraint for maximal number of elementary condition to be used in a rule. Moreover, the algorithm can be extended to allow the user specifying constraints on minimal (relative) strength of a rule - for details see [8].

The algorithm *Glance* has been experimentally evaluated and compared with algorithms *AllRules* and *DomExplore*. The results given in Table 1 show that the computational time of *Glance* increases "linearly" with the number of objects. Furthermore it is always the best, while *AllRules* is the second and *DomExplore* is the worst. On the other hand, additional experiments indicate that if the number of attributes/ criteria is higher than 9, the winner is *AllRules*. Probably it is due to the fact that for all three algorithms the number of rules increases with the number of attributes, and *Glance* needs to check all generated rules when a new example is being processed. This aspect could be somehow reduced by using in *Glance* the minimal rule length constrain. Analyzing the number of rules from Table 2, one can notice that the highest increase occurs at the beginning of learning, then next parts of examples (≥ 3500) cause smaller changes. Additional results indicate that for *AllRules* it is better to process rather criteria than regular attributes, while *DomExplore* is more efficient on regular attributes only. *Glance* seems to be located between these two extremes.

The above observations obtained on artificial data, have been also confirmed in other experiments on some real data coming from UCI ML Repository - we skip them due to limited size of this paper.

In future research, we also want to examine classification accuracy of the *Glance* algorithm. However, predicting classification of a new object is a different problem than in only attribute case, because rules indicate unions not single classes. So, new strategies should be developed for determining final decision when the new objects matches conditions of many rules.

Acknowledgments. The research of the first author has been supported from the DS grant.

References

1. Greco S., Matarazzo B., Slowinski R., (1999) The use of rough sets and fuzzy sets in MCDM. In: Gal T., Stewart T., Hanne T. (Eds) Advances in Multiple Criteria Decision Making, Kluwer, chapter 14, pp. 14.1-14.59.
2. Greco S., Matarazzo B., Slowinski R. (2001), Rough sets theory for multicriteria decision analysis. European Journal of Operational Research, **129** (1), 1–47.
3. Greco S., Matarazzo B., Slowinski R., Stefanowski J. (2000), An algorithm for induction decision rules consistent with the dominance principle. In: Ziarko W., Yao Y. (eds), Proc. 2nd Int. Conference on Rough Sets and Current Trends in Computing, Banff, October 16- 19, 2000, 266-275.
4. Grzymala-Busse J.W. (1992) LERS - a system for learning from examples based on rough sets. In: Slowinski R. (Ed.), Intelligent Decision Support, Kluwer Academic Publishers, 3–18.
5. Pawlak Z. (1991), Rough sets. Theoretical aspects of reasoning about data. Kluwer Academic Publishers.
6. Roy B. (1995), Multicriteria Methodology for Decision Aiding. Kluwer Academic Publishers (Orignal French edition 1985).
7. Stefanowski J. (2001), Algorithims of rule induction for knowledge discovery. (In Polish), Habilitation Thesis published as Series Rozprawy no. 361, Poznan Univeristy of Technology Press, Poznan.
8. Zurawski M. (2001), Algorithms of induction decision rules for multricriteria decision problems. M.Sc. Thesis, Poznań University of Technology. Supervisor Stefanowski J., 110 pages.

Statistical and Neural Network Forecasts of Apparel Sales

Les M. Sztandera[1], Celia Frank[2], Ashish Garg[2], and Amar Raheja[3]

[1] Computer Information Systems Department, Philadelphia University,
   Philadelphia, PA 19144, USA
   SztanderaL@PhilaU.edu
[2] School of Textiles and Materials Technology, Philadelphia University,
   Philadelphia, PA 19144, USA
   FrankC@PhilaU.edu, Garg2@PhilaU.edu
[3] Department of Computer Science, California State Polytechnic,
   Pomona, CA 91768, USA
   Raheja@csupomona.edu

Abstract. In this paper, we are investigating both statistical and soft computing (e.g., neural networks) forecasting approaches. Using sales data from 1997 - 1999 to train our model, we forecasted sales for the year 2000. We found an average correlation of 90% between forecast and actual sales using statistical time series analysis, but only 70% correlation for the model based on neural networks. We are now working to convert standard input parameters into fuzzified inputs. We believe that fuzzy rules would help neural networks learn more efficiently and provide better forecasts.

1 Introduction

In today's world, all industries need to be adaptable to a changing business environment in the context of a competitive global market. To comply with higher versatility and disposability of products for consumers, firms have adopted new forms of production behaviour with names such as "just-in-time" and "quick response". To react to worldwide competition, managers are often required to make wise decisions rapidly. In fact, they must often anticipate events that may affect their industry. More than 80% of the US textile and apparel businesses have indicated an interest in time-based forecasting systems and have incorporated one or more of these technologies into their operations. To make decisions related to the conception and the driving of any logistic structures, industrial managers must rely on efficient and accurate forecasting systems. Better forecasting of production, predicting in due time a sufficient quantity to produce, is one of the most important factors for the success of a lean production. Forecasting garment sales is a challenging task because many endogenous as well as exogenous variables, e.g., size, price, colour, climatic data, effect of media, etc. are involved. Our approach is to forecast apparel sales in the absence of most of the above-mentioned factors and then use principles of fuzzy logic to incorporate the various parameters that affect sales. The forecasting modelling investigates the use of two statistical time series models, Seasonal Single Exponential Smoothing (SSES) and Winters' Three-Parameter Model. It then investigates soft computing models using Artificial Neural Networks (ANNs). A foundation has been made for multivariate fuzzy logic based model by building an expandable database and a rule base. After a substantial amount of data is collected, this model can be used to make predictions for sales specific to a store, color, or size of garment.

2   Databases

A U.S.-based apparel company has been providing sales data for various types, or classes, of apparel for previous years. From January 1997 until February 2001, sales data was collected for each day for every class for every store. From March 2001 onward, more detailed information about each garment is available including its size and colour. The data for the previous years (1997-2000) has much fewer independent variables as compared to the data for the year 2001. Using the data from January 1997 till March 2000, time series analysis was performed. A data file for each class was made that contains information about its total sales (in dollars) each day. For example, input file for class X looks as shown in Table 1: Total sales in terms of num-

Table 1. Format of Input Data File for Class X for Time Series Analysis

Date	Total Sales ($)
6/28/1997	2378
9/05/1998	1405
3/14/1999	3546
1/08/2000	5983

ber of units would have been a better option but it was not possible since the number of units sold was missing from many rows in the raw data. From March 2001 onwards, aggregation was done in several steps. Initially, from daily sales files, a database was made containing information about total sales for a particular garment-class for each color, each size, and each store. The total number of rows in this format was too large to be used as an input file for the multivariate model. In order to reduce the dimensionality of the database, two compressions were done. First, daily sales data was converted into monthly sales data. Secondly, color information was compressed by aggregating sales of similar colors. For example, Color Code 23 represents light green, 25 dark green, and 26 medium green. Instead of having different rows for different tones of green, we compressed this information into a single row by assigning Color Code 2 for all greens.

3   Pre-Analysis

Data was analysed for trends and seasonality. This analysis helps in choosing an appropriate statistical model, although this kind of preparation is not necessary for soft computing based models.

For pre-analysis, three classes were chosen, one each from the Spring, Fall and Non-seasonal garment categories. Analysis was spread among all the categories to remove any bias due to type of class.

Sales data often showed a weekly trend with sales volume increasing during weekends as compared to weekdays. This trend was evident from the daily sales graphs for all the classes. It was noticed that sales volume peaks on 4th, 11th, 18th, and 25th day of January, which are all Saturdays, and generally second highest sales are on Sundays.

This observation was further supported by qualitative means by calculating Auto Correlations Functions (ACFs). ACF is an important tool for discerning time series patterns. ACF were calculated for 350 observations for all the 3 classes. ACF for a given lag k is given by equation 1:

$$ACF(k) = \frac{\sum_{t=1+k}^{n} (Y_t - \overline{Y})(Y_{t-k} - \overline{Y})}{\sum_{t=1}^{n}(Y_t - \overline{Y})^2} \tag{1}$$

Upon further analysis of the data, it was observed that fraction contributions towards total sales in any week of the year and for any class remain significantly constant. Hence, information about fraction contribution can be used to forecast daily sales after forecasting weekly sales.

Garment sales are generally seasonal with demand increasing for a particular type in one season and for a different type in another season. To investigate seasonality, the same methodology was used as was used to establish weekly trend. Both graphically as well as using ACF, it was shown that sales of all three classes under consideration show strong and distinct seasonal trend. Interestingly, class CN has been categorized as non-seasonal; still it showed increase in seasonality although not as distinctively as shown in other two classes. While investigating the annual seasonal trend, data was aggregated into weekly increments. The sinusoidal pattern in ACFs is typical of many seasonal time series.

Currently we are investigating elimination of possible trends and the use of partial ACFs. This could, perhaps, lead to applying later an additive model of seasonality, and possibly a better approach than an application of the multiplicative model.

4   Methodology and Results

Evident from the format of the database for years 1997-2001 only sales information with respect to time for various garments is available. Hence, only univariate time series and soft computing models were investigated using these data. From March 2001 onwards, much more elaborate sales data is available. Using this data set, a multivariate forecasting model was implemented that could prove useful for inventory maintenance. Six classes, two each from the Spring (AS and BS), the Fall (AF and BF), and the Non-seasonal (CN and DN) categories, were chosen for each model. They were built using four years sales data, and the next two months data was forecasted. The forecasted data was then compared with actual sales to estimate the forecasting quality of the model.

As more data are collected in the future, a long-term trend will be investigated in the modelling process.

4.1   Seasonal Single Exponential Smoothing

Exponential Smoothing (EXPOS) is one of the most widely used forecasting methods. As discussed previously regarding seasonality in sales data, a Single Exponential Smoothing (SES) model with a factor of seasonality was investigated. A seasonal SES model requires four pieces of data: the most recent forecast, the most recent actual data, a smoothing constant, and length of the seasonal cycle. The smoothing constant (a) determines the weight given to the most recent past observations and therefore controls the rate of smoothing or averaging. The value of a is commonly constrained to be in the range of zero to one. The equation for Seasonal SES is:

$$F_t = \alpha A_{t-s} + (1 - \alpha) F_{t-s} \qquad (2)$$

where:
F_t = Exponentially smoothed forecast for period t
s = Length of the seasonal cycle
A_{t-s} = Actual in the period t-s
F_{t-s} = Exponentially smoothed forecast of the period t-s
α = Smoothing constant, alpha

Weekly sales data was used for the forecast model given by the above equation. Hence, s was chosen to be 52 (number of weeks in a year). There are many ways of determining alpha. Method chosen in the present work was based on Minimum Squared Error (MSE). Different alpha values were tried for modelling sales of each class and the alpha that achieved the lowest SE was chosen. After choosing the best alpha value, a forecast model was built for each class using four years weekly data. Using the model, a weekly sales forecast was done for January and February of 2001. In order to forecast daily sales, the fraction contribution of each day was multiplied by total forecasted sales of each week. The data for the remaining five classes are available on the project website. Table 2 gives the alpha value, R^2 of the model, and correlation coefficient between actual and forecasted daily sales for January 3 2001-February 27 2001 for all classes. It can been seen that even with single

Table 2. Values of Alpha, R^2, and Correlation Coefficients for Seasonal SES Model

Class	AS	BS	AF	BF	CN	DN
α	1.4	1.4	0.9	1.3	1.3	1.0
R^2	0.738	0.832	0.766	0.872	0.762	0.831
Corr.[1]	0.906	0.893	0.862	0.910	0.892	0.722

parameter Seasonal SES, R^2 is on an average more than 0.75 implying that the model is able to explain 75% of the variation in the data. Correlation coefficients between actual and forecasted sales of January 3 '01-February 27 '01 are also quite high except for class DN.

Winters' Three Parameter Exponential Smoothing Winters' powerful method models trend, seasonality, and randomness using an efficient exponential smoothing process. The underlying structure of additive trend and multiplicative seasonality of Winters' model assumes that:

$$Y_{t+m} = (S_t + b_t)I_{t-L+m} \tag{3}$$

S_t = smoothed nonseasonal level of the series at end of t
b_t = smoothed trend in period t
m = horizon length of the forecasts of Y_{t+m}
I_{t-L+m} = smoothed seasonal index for period $t + m$

That is, Y_{t+m} the actual value of a series, equals a smoothed level value S_t plus an estimate of trend b_t times a seasonal index I_{t-L+m}. These three components of demand are each exponentially smoothed values available at the end of period t. The equations used to estimate these smoothed values are:

$$S_t = \alpha(\frac{Y_t}{I_{t-L}}) + (1 - \alpha)(S_{t-1} + b_{t-1}) \tag{4}$$

$$b_t = \beta(S_t - S_{t-1}) + (1 - \beta)b_{t-1} \tag{5}$$

$$I_t = \gamma(\frac{Y_t}{S_t}) + (1 - \gamma)I_{t-L+m} \tag{6}$$

$$Y_{t+m} = (S_t + b_t m)I_{t-1+m} \tag{7}$$

Y_t = value of actual demand at end of period t

α = smoothing constant used for S_t

S_t = smoothed value at end of t after adjusting for seasonality

β = smoothing constant used to calculate the trend (bt)

b_t = smoothed value of trend through period t

I_{t-L} = smoothed seasonal index L periods ago

L = length of the seasonal cycle (e.g., 12 months or 52 weeks)

γ = smoothing constant, gamma for calculating the seasonal index in period t

I_t = smoothed seasonal index at end of period t

m = horizon length of the forecasts of Y_{t+m}

Equation 4 calculates the overall level of the series. S_t in equation 5 is the trend-adjusted, deseasonalized level at the end of period t. St is used in equation 7 to generate forecasts, Y_{t+m}. Equation 5 estimates the trend by smoothing the difference between the smoothed values S_t and S_{t-1}. This estimates the period-to-period change (trend) in the level of Y_t. Equation 6 illustrates the calculation of the smoothed seasonal index, It. This seasonal factor is calculated for the next cycle of forecasting and used to forecast values for one or more seasonal cycles ahead. For choosing α (alpha), β (beta), and γ (gamma) Minimum Squared Error (MSE) was used as criterion. Different combinations of alpha, beta, and gamma were tried for modelling sales of each class and the combination that achieved the lowest RSE was chosen. After choosing the best alpha, beta, and gamma values, the forecast model was built for each class using four years of weekly sales data. Using the model, a weekly sales forecast was done for Jan and Feb of 2001. In order to forecast daily sales afterwards, the fractional contribution of each day was multiplied by total forecasted sales of each week. Higher R^2 values and the ability of Winters' model to better define the model are due to the additional parameter beta utilized for trend smoothing. Although curve fitting is very good, correlation coefficients between actual and forecasted sales are not as high. On observing the graphs of actual versus forecasted values for all the classes, it can be observed that trend (growth or decay) has always been over estimated and, hence, forecasted values are too high or too low. As in any multiplicative model, the division by very small numbers or multiplication by extremely large values is a problem with equation 3 which could have resulted in overestimation of the trend. Table 3 gives the alpha, beta, gamma, R^2 of the model, and the correlation coefficient between actual and forecasted daily sales for Jan 3 2001-Feb 27 2001.

4.2   Soft Computing Methods

Artificial Neural Network (ANN) Model. Neural networks mimic some of the parallel processing capabilities of the human brain as models of simple and complex forecasting applications. These models are capable of identifying non-linear and interactive relationships and hence can provide good forecasts. In our research, one of the most versatile ANNs, the feed forward, back propagation architecture was implemented. The hidden layers are the regions in which several input combinations from the input layer are fed and the resulting output is finally fed to the output layer. $\{x_1,, \ldots, x_M\}$ is the training vector and $\{z_1,, \ldots, z_M\}$ is output vector. The error E of the network is computed as the difference between the actual and the desired output of the training vectors and is given in equation 8:

$$E = \frac{1}{P.N_L} \sum_{n=1}^{N_L} \sum_{p=1}^{P} (t_n^{(p)} - s_n^{(p)})^2 \tag{8}$$

Table 3. Values for α (alpha), β (Beta), γ (Gamma), R^2, and Correlation Coefficients for Winters' Model

Class	AS	BS	AF	BF	CN	DN
α	0.60	0.50	0.50	0.80	0.60	0.50
β	0.01	0.01	0.01	0.01	0.01	0.01
γ	1.00	0.47	0.91	0.72	1.00	0.82
R^2	0.923	0.969	0.951	0.685	0.941	0.933
Corr.[1]	0.903	0.920	0.869	0.667	0.927	0.777

[1] Correlation coefficient between actual and forecasted sales of Jan 3 '01-Feb 27 '01

where $t_n^{(p)}$ is the desired output for the training data vector p and $s_n^{(p)}$ is the calculated output for the same vector. The update equation for the weights of individual nodes in the different layers is defined using the first derivative of the error E as given in equation 9:

$$w_i^{(n)}(l) = w_i^{(n)}(l) + \Delta w_i^{(n)}(l)$$

$$where \quad \Delta w_i^{(n)}(l) = -\eta \frac{\partial E}{\partial w_i^{(n)}(l)} \tag{9}$$

ANN consisted of three layers viz. input layer, hidden layer and output layer with 10, 30 and 1 neuron respectively. 217 weeks sales data was divided into three parts. First part consisted of 198 weeks, which was used to train the network. The second part with 10 weeks data was used to test the network for its performance. The third part with 9 weeks data was used to compare forecasting ability of the network by comparing forecasted data with the actual sales data. In order to forecast daily sales afterwards, the fraction contribution of each day was multiplied by total forecasted sales of each week. Table 4 gives the R^2 of the model, and correlation coefficient between actual and forecasted daily sales for Jan 3 2001-Feb 27 2001

Table 4. Values of R^2, and Correlation Coefficients for ANN Model

Class	AS	BS	AF	BF	CN	DN
R^2	0.963	0.941	0.953	0.906	0.953	0.916
Corr.[1]	0.878	0.906	0.704	0.793	0.914	0.845

[1] Correlation coefficient between actual and forecasted sales of Jan 3 '01-Feb 27 '01

R^2 values for all the classes are much higher than those obtained from Seasonal SES, and Winters' Model. High R^2 values and the strength of the ANN model are due to the ability of ANNs to learn non-linear patterns. Although curve fitting is very good, correlation coefficients between actual and forecasted sales are not that good. This might be due to over learning of the network. A potential problem when working with noisy data is the so-called over-fitting. Since ANN models can approximate essentially any function, they can also over fit all kinds of noise perfectly. Typically, sales data have a high noise level. The problem is intensified by a number of outliers (exceptionally high or low values). Unfortunately, all three conditions that increase the risk of over-fitting are fulfilled in our domain and have impacted correlations.

5   Conclusion

Time series analysis seemed to be quite effective in forecasting sales. In all of the three models, the R^2 and the correlation coefficient were significantly high. The three parameter Winters' model outperformed Seasonal SES in both explaining variance in the sales data (in terms of R^2) and forecasting sales (in terms of correlation coefficient). ANN model performed best in terms of R^2 among three models. But correlations between actual and forecasted sales were not satisfactory. A potential problem when working with noisy data, a large number of inputs, and small training sets is the so-called over-fitting. Since big ANN models can approximate essentially any function they can also over fit all kinds of noise perfectly. Unfortunately, all three conditions that increase the risk of over-fitting are fulfilled in our domain. Typically, sales data have a high noise level. The problem is intensified by a number of outliers (exceptionally high or low values). A multivariate fuzzy logic based model could model the sales very well, as it would take into account many more influence factors in addition to time. This naturally leads to the first extension of this work. Extensions of the concept of discovery learning are of current interest and are being investigated.

Acknowledgment. The United States Department of Commerce/National Textile Center Grant S01-PH10 has supported this research. Mothers Work, Inc has provided additional research support.

References

1. Armstrong J. S. (2001) Principles of Forecasting. Kluwer, Boston

2. Garg A. (2002) Forecasting Women's Apparel Sales Using Mathematical Modeling, M.S. Thesis, Philadelphia Univ., Philadelphia.
3. Frank C., Garg A., Raheja A., Sztandera L. (2003) Forecasting Women's Apparel Sales Using Mathematical Modeling. Int. J. Clothing Science Technology, in press.
4. Sztandera L., Frank C.,Garg A., Raheja A. (2003) A Computational Intelligence Forecasting Model for Apparel Sales. Proc. 29th Int. Aachen Textile Conf., Aachen, in press.

Mining Knowledge About Process Control in Industrial Databases

Robert Szulim[1] and Wojciech Moczulski[2]

[1] University of Zielona Góra, Department of Electrical Metrology, Podgórna 50, 65-246 Zielona Góra, Poland
[2] Silesian University of Technology, Department of Fundamentals of Machinery Design, Konarskiego 18a, 44-100 Gliwice, Poland

Abstract. In the paper concept of using industrial databases to discover knowledge about process control is shown. A simplified description of a copper furnace and its internal processes is presented. A concept of building data warehouse based on archive data that came from industrial SCADA systems is discussed. Some examples of techniques used for processing very large databases are presented. A concept of knowledge discovery about dynamic industrial processes that are of strong stochastic nature is shown. Finally, a concept of knowledge-based expert system designed to help in operation and control of the industrial object is described.

1 Introduction

There are many complex industrial objects monitored by computer systems. Those systems are called SCADA - Supervisory Control and Data Acquisition. Many of those systems must control hundreds of sensors installed at the object. A human being – the process operator - must assess results coming from sensors. Results are usually shown on computer screens as so-called synoptic screens – Fig. 1.

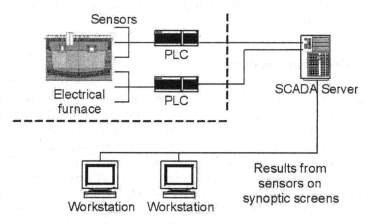

Fig. 1. SCADA system

Most decisions concerning operation of such objects must be taken using results obtained from sensors. Many times the decision making requires watching results from previously finished working periods as well. In such a complex environment it is easy to miss some symptoms of a bad condition of the object. Wrong decision or taken too late may cause a malfunction of the object or degradation of the overall performance of the plant. We believe that using knowledge-based expert computer system could help in maintaining work of the object. The system should be able to watch all of the sensors of the object. Present results might be compared with results from previous working periods and some pieces of advice for personnel could be generated.

2   Simplified description of the object and processes it carries out

As the object we consider the electrical copper furnace [1,2]. The electric furnace is one of the elements used in the process of copper production. Slag containing 15% of copper is loaded into furnace together with other technological supplements. The main task of the furnace is the so-called slag reduction. The process has periodical character. Optimal duration of one period is about 8 hours. Each operation period consists of three characteristic stages. The first stage (**load**) starts from loading coke and solidified slag. During this stage a large amount of electric energy is fed by using high values of phase current and deep immersion of electrodes in slag in order to cause turbulent motion of the charge inside the furnace. This stage takes usually 90 minutes. The second stage (**reduction**) continues until required fraction of copper in slag is achieved and takes about 300-400 minutes. Electric energy is dispatched to achieve and hold high temperature inside the furnace – about 1350 – 1400 degrees centigrade. Control chemical analyses and tapings are made during this stage. The last stage (**tapping**) starts when slag inside furnace contains desired percentage of the cooper (about 0,6 %). Its duration is about 90 minutes.

3   Problem description

The analyzed object and the technology are very complex and energy consuming. The electric furnace is an important subsystem of the factory. It is also critical for long-term work planning of the complete factory. Disturbances in work periodicity influence the operation of the complete plant.

Work of the object is maintained by specialized personnel. People that work there must possess extensive knowledge and experience about process of slag reduction. Improper decisions concerning operation of the object, or the ones taken too late, might result in very serious consequences. Moreover, the object can operate in one of three states: normal work, overload and overmelt. Both two last states are very dangerous and disadvantageous for the furnace. They may lead to fast overmelt, damage or emission of the poisonous elements to atmosphere, increase process duration, decrease overall performance, increase consumption of electric energy. Recovering from this state is a difficult and time-consuming process. Sometimes it is necessary to conduct a few smelts to achieve stable work of the furnace.

To keep normal state of the process, the personnel must take care about present and past parameters and indices. The process is very complex and has dynamic character, with strong stochastic nature that is caused by random properties of initial load and additives, and also by open-loop control by human personnel.

4   Attempts to solve the problem

Several attempts were taken to solve the problem of maintaining early mentioned object. An expert system was created, whose knowledge base contained rules that were acquired from domain experts by means of questionnaires and interviews [3,4]. Knowledge used in this system was acquired from experts based on their experience. There was a lot of problems with extracting knowledge from experts. They had different opinions about the same aspects of work of the object. Results of operation of the expert system were unsatisfactory. Another attempt was to build a software-based energy balance module. Work of that module was determined by many indices and parameters of the process. A few of that parameters, as some temperatures of the furnace and percentages of different elements in slag, were impossible to estimate with sufficient quality. It was most likely the cause of unsatisfactory performance of this module. Furthermore, there were problems with stability as well. Both shortly described attempts of solving problems with maintaining object were not good enought to support personnel decisions.

5   Knowledge-based approach to modelling of industrial processes

Work of many complex industrial object is maintained only by human operators. Operators must decide about work of the object by themselves. They usually use their own experience and are supported by SCADA systems to watch parameters of the process. However, there are known some attempts to support them by knowledge-based systems. Numerous techniques are used.

The so-called 'model-based' approach may be based on mathematical, physical, or rough input–output model of the considered process. Many examples of practical applications of neural networks, fuzzy and neuro-fuzzy systems are known [11].

In many cases it is impossible or difficult to find a mathematical model describing the process. In such a case other approaches might be used like *regression analysis* [5,8], *fuzzy sets* [10,17], *neural networks* [12] and *artificial intelligence, knowledge engineering* and *other* [11,13,6,7,16,9].

Each approach has its limitations and advantages. Using neural networks to identify models of nonlinear, dynamic processes is popular. However, neural networks are considered as a 'black–box' because knowledge representation cannot be easily understood by human operators. In many cases this is not acceptable. Although this may be overcome by using fuzzy neural networks that represent knowledge as a set of rules, the disadvantage consists in number of rules that may increase rapidly with increased number of input signals and fuzzy sets for each input. Knowledge represented by means of rules may be also difficult to understand, especially in the case if many rules are contained in the system.

Rough sets are often used for knowledge discovery on dynamic objects as well. As an example of a similar system to support work of rotary kiln stoker by using rough set and decision rules might be used [15,14]. The system may predict condition of the kiln by watching its present parameters. However, this object is much simpler than electrical furnace and processes carried out by the kiln are easier to model.

6   Objective of the research

Work of many complex objects is supported by computer SCADA systems. Those systems are capable to store acquired data in text files in mass storage memory. In the most cases those archive data are not used by companies anymore. Usually those data cannot be used directly because of format of the files. In order to make them usable to systems other than SCADA ones, they must be converted and stored in known database format. Special kinds of databases called *data warehouses* are used for such purposes.

By applying AI methods to preprocessed data stored in data warehouse it is possible to find some unknown and potentially useful patterns inside the data. Those patterns may carry new knowledge about the object.

Our long-term goal is to build an expert system to help in maintaining work of the electrical furnace. Knowledge for such a system is going to be discovered in archive measurement data coming from SCADA systems. Archive data require preprocessing, since formats of the original data make application of data mining and knowledge discovery methods difficult. Therefore, archive data was converted and stored in the data warehouse. Special software modules were developed to carry out this task.

7   Data warehouse

SCADA systems have the ability to backup its data. Backup files consist of thousand of files of a few different formats. Those archive data coming from such a system together with some other files with notes made by the personnel will be used as a source of raw data. Formats of those data are non-standard in most cases. There are a few kinds of files. Some of them contain data from PLC controllers read out in one-minute interval. Other files have an event character, i.e. they store data only in the case of occurrence of an event like alarm, loading od additives and temperature read-outs. There are also some files processed manually by operators available, which contain valuable information as duration of subsequent periods of the process that are not stored by SCADA systems. Some files have text format suitable for printing them by operators. Monitoring of each process results in a lot of such files. In order to use all these data it is purposeful to build *data warehouse*. A s a data warehouse we understand a database with the ability to store large quantity of data coming from different sources. The data in the warehouse will be easy to use because they will be spread out into separated tables stored usually in one place.

To create the data warehouse all backup files were preprocessed using special software import modules. Each group of files (of the same format) has its own import module. Data from files within one group are stored into one table of the data warehouse. As a platform for the data warehouse the Microsoft SQL 2000 was used [18].

7.1   Browsing process data

Special programs should be used to browse such a large database. Data warehouse is not equipped with browsing capabilities. In order to conveniently browse data in the data warehouse, a special application was developed using Microsoft Access as a platform. The application simplifies browsing process data in a few ways. Table views, forms, charts and pivot tables are used to facilitate browsing data in various forms.

7.2   Using large volume of data

Each database containing many records requires special mechanism to process data. In the client – server environment two approaches are possible. In classic approach data are stored in the database server and all operations on them are made by using a special application. To process data the application connects to the server through ODBC link. Every time when large amount of data must be processed they must be dispatched by network link to application. This situation may cause heavy network traffic and lower performance of the whole operation. Another reason of low performance is ODBC mechanism, which is not effective enough to transfer large quantities of data.

All database servers have special internal procedures for fast data processing called *stored procedures*. Their functionality is usually similar to programming languages. Very often JAVA and C languages are used to control data flow, and SQL to manipulate on data. By now the fastest way to process even very large quantities of data is only by using stored procedures that operate on data in various ways. Complex operations might be performed by SQL language structures. To enable special access to each record one by one, cursor mechanism might be used. Some event-based operations may be carried out using special procedures – *triggers*.

To operate on data, a client application is responsible only to call proper procedure on the server by network link. Then all operations are carried out by the server. No data are sent to the client by the network link. Only status of the operation is returned. This reduces network traffic and time to access the data, hence the most efficient data processing in databases is achieved.

However, there are some operations that are difficult to implement in this way. This includes complex algorithms of dealing with large quantity of memory or special mathematical operations. In that case some part of program is running at a workstation and only some part of algorithm is launched on the database server, as e.g. complex SQL statements.

8   Concept of knowledge discovery process

The reduction of slag in the electric furnace is going to be considered as a dynamic process. During analyzes both the actual and historical parameters are taken into account. Discovered knowledge should refer to actual and historical data. Building expert system based on this knowledge should be possible. This system should analyze actual process data together with archival data. Conclusions obtained by the expert system would support the personnel in controlling operation of the electric furnace. It is likely that predictions of the consumption of the electric energy and duration of the cycle might be calculated, basing on input parameters of the process. Both parameters are extremely important and have great influence on costs of the copper production. The concept of knowledge discovery consists of the following parts.

8.1   Extraction and transformation of data

Process of extraction and transformation of data to the warehouse is the very first step. Data from many files must be transferred to corresponding tables in the data warehouse. The format of data is often very specific for a corresponding application and must be converted to the database standard. A special program must be written to import data to every table.

8.2   Data cleaning, filtering, and validating

Data imported from archive files must be cleaned. A particular attention should be paid to personnel notes whose format might be wrong because of human mistakes, as e.g. date and time fields must contain proper characters. Values stored in tables should be validated. To validate such a large volume of data stored procedures of data warehouse should be used.

8.3 Data decomposition

Process of slag reduction consists of three parts. It is possible to decompose data assigning them to subsequent parts of the complete process. Duration of each part of the process entered manually by operators was used for this purpose. Each table includes records with time stamp field. Special procedures were developed to partition all available data series in all SCADA tables by parts of the process. Because of large volume of data it is very important to do that in an automated way by using special mechanisms.

8.4 Quantizing continuous signals

Quantizing will be used to change quants of values of the table fields to their corresponding labels. Large volume of data in the tables requires to develop special mechanism to accomplish this part in a fast and convenient way. Labels and ranges of the quants will be possible to define and change in easy and flexible way. Special forms will be developed to simplify quantization. To accelerate labeling tables and field, this mechanism has to be implemented as stored procedures on the data warehouse server.

8.5 Clustering similar processes, periods and events

Decomposition of realizations of process variables into characteristic, repeating events, which represent operation of the object, makes easier comparing process data between realizations. Those groups could be used to create a model to represent them. Models are going to represent one of possible scenarios of the operation of the furnace. Actual process data could be matched to models in order to predict the object's operation. Predictions made in this way should help the personnel of the object to take right decisions that satisfy such criteria, as e.g.: proper condition of the furnace, required quality of product, regular operation of the complete plant and low costs.

Linguistic representation of continuous process parameters. The control goal of the personnel consists in dispatching equal amount of energy to each electrode, which is obtained by controlling voltage (common for all electrodes) and immersion of individual electrodes. Both these controls affect current in each electrode. However, individual currents depend also from condition inside the furnace, as condition of individual electrodes, amount of slag inside furnace etc. Therefore, currents in electrodes may be asymmetric in many periods of operating time – see Fig. 2.

It is very difficult to group and classify such events like that from electrodes. The current has dynamic character, changes during the process and depends on often unknown conditions. We decided to use other quantity to classify events from electrodes – the *overall power consumption* whose course

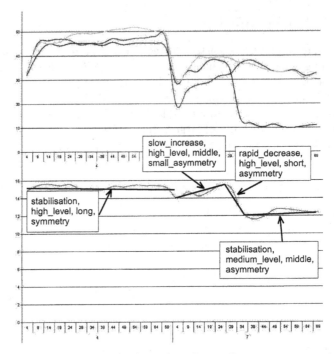

Fig. 2. Current in the electrodes (upper) and overall power consumption course (lower)

is easier to classify. We propose a *linguistic event description* for electrode control events. Our method consists in describing power events series by fuzzy descriptions. We want to describe character of changes of power consumption or other measurement as well. An example of classification of two different realizations of the process is shown in Fig. 2 where power consumption is presented. First one is quite stabilized during its duration. In the second realization four parts may be detected. Each part is assigned its linguistic description. This description is represented by the following generalized dynamic linguistic statement:

$$\langle [object,]\ [attribute,]\ type, level, duration, symmetry\_of\_currents\rangle \quad (1)$$

where:

type describes kind of event and may take values: *rapid_decrease, slow_decrease, stabilisation, slow_increase, rapid_increase*;

level represents level of power at the beginning of the analyzed part taking values from *low, medium,* and *high*;

duration may be expressed as *short, middle, long*;

symmetry_of_currents is an important feature of the part that carries information about the process, and may be expressed as *symmetry, small_asymmetry, asymmetry*.

Fig. 2 shows an example of event classification. To generate such descriptions a program was developed using Borland Delphi with access to data warehouse. The events in each realizations are described using the statement (1). Membership functions for each attribute are stored in the database and may be optimized with respect to some defined quality of classification.

Measure of similarity of process realizations. To compare two linguistic event descriptions a special measure of similarity is defined. Two realizations are analyzed taking into account consecutive events. If numbers of events in realizations are different, special *dummy* events are introduced into the shorter description. Single events are compared using fuzzy similarity relations [10] π_t, π_l, π_d defined for *type*, *level*, and *duration*, respectively. For two individual corresponding events e_{1k}, e_{2k} where $k = 1, \ldots, K$ and K is the number of events after equaling lengths of both linguistic descriptions, the similarity measure is defined as:

$$\pi[k] = \pi(e_{1k}, e_{2k}) = w_t \cdot \pi_t(e_{1k}, e_{2k}) + w_l \cdot \pi_l(e_{1k}, e_{2k}) + w_d \cdot \pi_d(e_{1k}, e_{2k}) \quad (2)$$

where w_t, w_l, w_d are corresponding weights for similarities π_t, π_l, π_d. Finally, similarity of two processes is calculated by combining values $\pi[k]$.

At this stage of research we do not take advantage of the last linguistic component *symmetry_of_currents* of the statement (1) that will be used later.

9   Expert system

Data warehouse will be used to collect data coming from SCADA files as live process data. This approach simplifies connection to live data. The expert system will access data about the actual process from the data warehouse. The expert system will use its knowledge base to forecast a scenario of the successive operation of the object and recommended controls. Conclusions and advices for personnel will be generated. Personnel will watch them on their computers together with SCADA results. This should help them in better control of the process. Some symptoms difficult to observe will be possible to track by such a system easier and faster then before, too.

10   Conclusions

In the paper the concept of using industrial measurement data of electrical furnace to build knowledge-based expert system was presented. The process of transformation of sets of raw data into a data warehouse together with some techniques of processing large databases was shown. A concept of knowledge discovery process was outlined. Stages of creation of an expert system for aiding personnel in taking decisions on operation of the electric arc furnace were presented as well.

References

1. Technological instruction of copper reduction in furnace slag, technological chart of the copper production (in Polish)– technological materials.
2. Estimating optimal technological parameters of reduction furnace slag in modernized electric furnace (in Polish), IMN Gliwice, 1997.
3. J. Bolikowski, J. Donizak, E. Michta, W. Miczulski, R. Szulim, Report on 1st and final part of Research Project "Project and implementation of the computer system to store measurement data required to process of slag reduction in electric furnace and to examine technical state of the devices" (in Polish), Zielona Góra 1999.
4. J. Bolikowski, M. Cepowski, E. Michta, W. Miczulski, R. Szulim, Expert system for supporting control of the process of slag reduction (in Polish), 4th Scientific Conference "Knowledge Engineering and Expert Systems", Wrocław, Poland, Wrocław University of Technology, 2000, vol. 2, p.229–236.
5. S. Brandt, Statistical and Computational Methods in Data Analysis, Springer Verlag, New York 1997
6. W. Cholewa, W. Moczulski, Technical Diagnostic with Hidden Model, Methods and Models in Automation and Robotics, Poland, Szczecin 2002.
7. P. Cichosz, Self-learning systems (in Polish). WNT, Warszawa 2000.
8. P. Czop, Diagnostic models of the rotary machine in temporal condition of work, PhD dissertation, Silesian University of Technology, Department of Fundamentals of Machinery Design, Gliwice 2001.
9. G. Drwal, Analyzes of classification issues in uncertain information condition, PhD dissertation, Silesian University of Technology, Faculty of Automation, Electronics and Computer Science, Gliwice 1999.
10. J. Kacprzyk, Fuzzy sets in System Analysis. PWN, Warszawa 1986.
11. J. Korbicz, J.M. Kościelny, Z. Kowalczuk and W. Cholewa, editors. Diagnostics of Processes. Models, Artificial Intelligence Methods, Applications (in Polish), PWN, Warszawa 2002.
12. J. Korbicz, A. Obuchowicz, D. Uciński, Artificial Neural Networks (in Polish), Akademicka Oficyna Wydawnicza, Warszawa 1994.
13. B. Kuipers, Qualitative Simulation, http://www.cs.utexas.edu/users/qr/papers-QR.html, internet
14. A. Mrózek, Use of Rough Sets and Decision Tables for Implementing Rule-Based Control of Industrial Processes, Bulletin of the Polish Academy of Sciences, vol 34, No. 5–6, p. 358–371, 1986.
15. A. Mrózek, L. Płonka, Data analyzes by using rough sets (in Polish), AOW, Warszawa 1999.
16. A. Niederliński, Rule-based expert systems (in Polish). WPJKS, Gliwice 2000.
17. D. Rutkowska, M. Piliński, Neural networks, genetic algorithms and fuzzy systems. (in Polish). PWN, Warszawa 1999.
18. http://www.microsoft.com/sql, internet

Acquisition of Vehicle Control Algorithms

Shang Fulian and Wojciech Ziarko

Department of Computer Science, University of Regina, Regina, SK, S4S 0A2
Canada

Abstract. In this paper, we present a multi-input multi-output (MIMO) data-acquired controller using system of hierarchical decision tables for a simulated vehicle driving control problem. The simulator incorporates dynamic mathematical model of a vehicle driving on a track. Sensor readings and expert driver control actions are accumulated to derive the vehicle control model. Sensor readings include random error to reflect realistic data acquisition conditions. The methodology of rough sets is being used to process the data and to automatically derive the control algorithm. In the experiments, the automated vehicle control algorithms derived from different driving patterns and with substantial sensor error consistently demonstrated astonishing robustness in their ability to properly drive the vehicle.

1 Introduction

The derivation of control algorithms from operation data has been of interest since early days of AI-related research. The objective was to substitute the complex mathematical modelling steps, which are often impossible to accomplish, with automated generation of control algorithm from operation data acquired through sampling actual operation processes conducted by an experienced human operator or operators. This empirical approach is primarily motivated by the lack of mathematical models of complex non-linear systems and the observed ability of human operators to learn, after some training, to effectively control such systems. In particular, known working examples of trainable control systems developed within the framework of rough set theory include experimental applications such as cement kiln production control [10], balancing inverted pendulum [7], and others [11],[12],[13],[14], [16], [17].

One of the major challenges in automated control systems design is the development of a vehicle control algorithm to substitute for a human driver. The problem is very complex in general due to large number of factors affecting the driver's control decisions during normal driving process. Many of the factors, such as visual or audio inputs are not easy to measure and interpret. However, for some special applications, the number of control inputs can be reduced to just a few to provide sufficient quality control of a moving vehicle. In those applications, the control information is obtained, for example, by reading predefined signals beamed by properly located transponders or by taking frequent measurements of vehicle speed and position relative to the track boundary. Possible applications of these simplified controllers may include, for example, automated trucks following a specially prepared route or automated farming tractors following a particular work pattern on a field.

In this article, we report our experiences with automated controller of a moving vehicle simulator. The simulator incorporates dynamic characteristics of a real car in the form of a mathematical model. The model car can be controlled by a software-simulated "expert driver" robot, or by any other controller implemented externally. To implement a new controller for the car, the simulator provides a number of sensor inputs which can be sampled with high frequency. In addition, the expert driver can provide information as to what control actions it would take in different car states. The control actions involve setting the car speed and the steering wheel angle. Improper control actions may cause a "crash" resulting in the car getting out of the track.

The objective of our experiments was to investigate the feasibility of applying decision tables [2], [5] acquired from car runs' training data to produce a working controller, referred to as *rough controller*, for the car. The methodology based on constructing a linear hierarchy of decision tables was applied [15]. In the process of passing from data to decision tables, the algorithms of rough set model were adapted (see, for example, [2], [6], [4], [9]).

In what follows, the basics of the design of the simulator are presented first. Then we talk about the simulation process and data acquisition from the simulated runs. This is followed by the introduction of the mathematical fundamentals of the rough set approach to the derivation of the linear hierarchy of control decision tables. In this part, we present the decision table hierarchy generation algorithm referred to as HDTL. We conclude with the discussion of experimental car runs controlled by the decision tables.

2   The Vehicle Model

In order to simulate a quasi-realistic car driving environment, we used a pre-existing racing car simulator RARS (Robot Auto Racing Simulator) [18]. The simulator adopts a simplified yet still powerful physical model of a car driving on a track. Traction force and skidding, etc. are implemented in the simulator. In addition, it is possible to expand the simulator by providing a control software for it or to introduce a random error to sensor readings to better simulate real driving scenarios.

In figure 1 the basic elements of the car model in two dimensional space are presented. In addition, the following variables and relations among them are used:

$L = V + W$, where V is a car's velocity vector, W is a velocity vector of tire bottom surface relative to the car and L is a slip vector or velocity of tire bottom relative to the track surface;

$L = V - vc * P$, where vc is a desired velocity of the car and P is a car's pointing vector, a unit vector in the direction of the car;

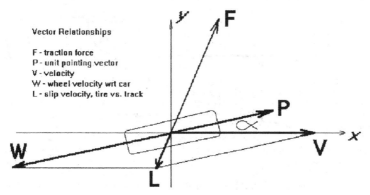

Fig. 1. Moving Vehicle Model

$Lt = V - vc * \cos(\alpha)$, where Lt is an x-component of L and α is the front tire angle;

$Ln = -vc * \sin(\alpha)$, where Ln is an y-component of L;

$F = -u(l) * \frac{L}{l}$, where F is a force vector pushing the car in the direction opposite to L, l is the scalar value of L and $u(l)$ is a friction function given by: $u(l) = FMAX * \frac{l}{K+l}$ with $FMAX$ and K being constant values;

$Ft = -f * \frac{Lt}{l}$, where Ft is an x-component of F and f is a scalar value of F;

$Fn = -f * \frac{Ln}{l}$, where Fn is an y-component of F;

$pwr = vc * (Ft * \cos(\alpha) + Fn * \sin(\alpha))$, where pwr is the car engine power;

There are several additional assumptions incorporated in the simplified physical model such as: no air-drag is considered, no road surface condition is taken into account with respect to friction, no gear changing is assumed, controller only outputs α and vc and no maximum torque is assumed.

3   Simulation Process

In principle, the simulator can run several cars. There are several state variables for each car in the simulator to represent the current status of each car, such as speed, direction and position on the track. At each tick of the clock, the simulator generates a situation vector of state variables for each car. The simulator then calls a car controller function using the current state vector. The car expert controller (robot) decides how to control the car for the next step and returns control decisions to the simulator. The simulator verifies the control decisions with the physical model and decides if it is necessary to modify the control decision speed or direction subject to the limits of the physical model. Finally the simulator calls the OpenGL graphics engine to render the racing process for each step.

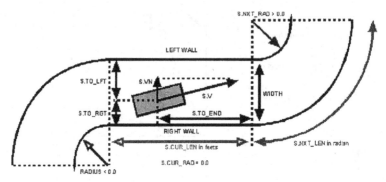

Fig. 2. Moving Vehicle State Variables

4   Acquisition of Training Data

Training samples were acquired during a training simulation process. We run the car by the "robot" driver on a predefined track. During this data acquisition process, several state variables were sampled with high frequency (see figure 2).

There are altogether eight such variables, divided into two groups: the *condition variables* and *decision variables*. The condition variables represent input information to be used in the development of the rough controller, such as the current position of the car. The following condition variables have been used:

   to_lft to mean the distance of car to the left wall of the track;
   to_end to denote the distance to the end of the straight or curve;
   cur_rad to represent the radius of the inner edge of the track;
   nex_rad representing the radius of the next segment;
   v for the current speed of the car;
   vn for the current speed of the car relative to the edge of the track;

The decision variables represent the control actions taken by the robot while driving the car. They are respectively: the front tire angle α and the tire speed *vc*. From each sample run, we collected about 15000 training samples, each being a snapshot of these eight variables. We used a predefined training data acquisition protocol that would not only record the working data of the robot, but would also get some extra information from the robot as to what to do in slightly different states than the current state the vehicle is in.

5   Derivation of Control Algorithm from Data

The process of deriving control algorithm from data consists of several major stages. In this process, we were trying to construct a linear hierarchy of decision tables. To acquire the decision tables, the original sensor readings data have to be discretized and subjected to rough set theory-based analysis and optimization. In this section, we comment in more detail about these major stages of control algorithm derivation.

5.1   Data Discretization

According to rough set methodology, prior to forming decision tables, both the condition and the decision variables need to be discretized into relatively few ranges to reduce the number of possible combinations and to increase the coverage of the space of observations.

A well known problem associated with the discretization of decision variables is the stability of the resulting controller. Generally, it is difficult to predict if the resulting decision table-based controller is stable or not, given a robot controller data to derive decision tables from. To deal with this problem, we intentionally made the original robot controller output less precise than it really was by mapping it into a few discrete levels. The experimental results indicated that the simulated vehicle has some tolerance to control errors. As a result, the wheel angle α was ground down into 32 levels, representing -1 to 1 radius and the speed vc was cut down to 13 levels, representing 20 - 150 feet/sec. speed range. With this discretization, experiments had shown that the robot controller is still stable, although the control process is noticeably not as smooth as with the original robot controller.

We discretized the condition variables by adopting a simple binary split approach. In this approach, each variable's range was split into two approximately even sub-ranges, resulting in a new, binary-valued representation in the decision table. When constructing the hierarchy of decision tables, each table in the hierarchy was derived using progressively finer discretization than its parent by doubling the number of discrete levels. Typically, the highest number of discrete levels required to discretize condition variables in our experiments was 16.

5.2   Decision Tables Acquired from Data

In this subsection, we briefly summarize the relevant aspects of the rough set model [2] as they relate to the acquisition of multi-valued decision tables from data for control applications.

Let U be a finite set of objects called a *universe*. The universe normally is a set of objects about which observations are acquired by sampling sensor readings or by some other means. The observations are expressed through values of a finite set of functions $C \cup D$ on U, referred to as *attributes*. The functions belonging to the set C are called *condition attributes*, whereas functions in D are referred to as *decision attributes*. In more detail, each attribute a belonging to $C \cup D$ is a mapping $a : U \to V_a$, where V_a is a finite set of values called the *domain* of the attribute a. In the context of the control application, condition attributes are functions obtained by discretizing values of condition variables and the decision attribute is a function obtained by

to_lft	cur_rad	nex_rad	v	vn	to_end	vc
0	8	0	12	4	0	4
0	8	0	12	8	0	4
0	12	0	8	4	8	3
0	12	0	8	8	8	3
0	12	0	12	4	8	4
0	12	0	12	8	8	4
4	8	0	12	4	12	6
4	8	0	12	8	12	6
...	...	...	...	...	...	...

Table 1. Partial Decision Table

discretizing values of one of the decision variables (in our application, different decision variables are treated independently).

The set of condition attributes C defines a mapping denoted as $\mathbf{C} : U \to \mathbf{C}(U) \subseteq \otimes_{a \in C} V_a$, where $\otimes$ denotes Cartesian product operator of all attribute domains of attributes in C. Similarly, the set of decision attributes D corresponds to a mapping $\mathbf{D} : U \to \mathbf{D}(U) \subseteq \otimes_{a \in D} V_a$ and both condition and decision attributes jointly define a mapping denoted as $\mathbf{C} \cup \mathbf{D} : U \to \mathbf{C} \cup \mathbf{D}(U) \subseteq \otimes_{a \in C \cup D} V_a$. For each combination of condition attribute values $x \in \mathbf{C}(U)$, the set $E_x = \mathbf{C}^{-1}(x) = \{e \in U : \mathbf{C}(e) = x\}$ is called the an *elementary* set. In other words, each elementary set is a collection of objects with identical values of the attributes belonging to the set C. The collection of all elementary sets forms a partition $R_C = \{E_x\}_{x \in \mathbf{C}(U)}$ of the universe U. In practice, the partition R_C is a representation of the limits of our ability to distinguish individual objects of the universe. The pair $\mathbb{A} = (U, R_C)$ is called an *approximation space*.

Similar to condition attributes, the decision attributes also define a partition of U. This partition consists of a collection of *decision classes* $R_D = \{F_x\}_{x \in \mathbf{D}(U)}$ corresponding to different combinations of values of the decision attributes in D, that is: $F_x = \mathbf{D}^{-1}(x) = \{e \in U : \mathbf{D}(e) = x\}$.

We define a *decision table* as a relation $DT_{C,D}(U) \subseteq \otimes_{a \in C \cup D} V_a$ between combinations of values of condition and decision attributes such that $DT_{C,D}(U) = \mathbf{C} \cup \mathbf{D}(U)$. Since any decision table with multiple decision attributes can be decomposed into several tables with singular decision attribute, in the rest of the paper we will assume that D contains only one attribute d, that is $D = \{d\}$. An example of partial decision table acquired from the vehicle simulator data with a single decision attribute vc is shown in table 1 .

Our interest here is in the analysis of the relation between condition attributes and decision attributes in decision tables. We would like to find out whether this relation is functional, or if not, which part of it is not functional. The *positive region* $POS_{C,D}(U)$ of the partition R_D in the approximation space $\mathbb{A}$ is introduced as $POS_{C,D}(U) = \cup \{E \in R_C : \exists F \in R_D (E \subseteq F)\}$. The decision table corresponding to the positive region $DT_{C,D}(POS_{C,D}(U))$, called *positive region decision table*, will be denoted as $DT_{C,D}^+(U)$. This decision table is deterministic as it represents the functional part of the decision table $DT_{C,D}(U)$. The complement of the positive region, $BND_{C,D}(U) = U - POS_{C,D}(U)$ is called the the *boundary region* of the partition R_D in the approximation space $\mathbb{A}$.

The positive region of the partition R_D is a union of those elementary sets which are entirely included in one of the blocks of the partition R_D defined by the decision attributes. All other blocks are in the boundary region $BND_{C,D}(U)$. In terms of decision tables, the positive region of the classification represented by the values of the decision attribute can be identified by a union of all those elementary sets whose attribute value combinations in the decision table are associated with a single value of a decision attribute only. This gives a simple method of identifying both the boundary and the positive regions, or functional and non-functional parts of the relation respectively, by finding "inconsistent" rows in the decision table.

One more aspect, which is not discussed here due to space limitations, is the optimization of decision tables. It is possible to eliminate redundant columns from the table by employing the rough set theory concept of *attribute reduct* (see, for example [2], [3]). This kind of optimization is being done during derivation of control tables from data for the rough controller described here.

5.3 Linear Hierarchy of Decision Tables

In the process of deriving control algorithm for vehicle control problem, we were aiming, for obvious reasons, at producing an uncertainty-free controller in which every individual decision would be taken with full certainty. Translating this requirement into decision tables would mean that every decision table used for control would have to be boundary area free. Although it is possible, in theory, to discretize the condition variables fine enough to ensure boundary-free classification, that would lead to excessive number of possible attribute combinations leading to relatively narrow domain coverage. In practice, that would translate into many situations in which there would be no matching attribute-value pattern in the decision table. Another approach, which was originally proposed in [15], suggests building a linear hierarchy of linked decision tables. In the hierarchy, the top table would be build based on the original condition attributes and the universe U, while all other tables would be derived from their parent tables by considering the parent's boundary as the universe for the current level and redefining attributes.

To present the details of the algorithm, referred to as *HDTL algorithm*, let U', C', D' denote initial universe (input data), initial condition attributes and initial decision attributes on U', respectively. In addition, let U, C, D

be variables representing current universe (current data set), condition and decision attributes on U, respectively.

Algorithm HDTL

> *Initialization:*
> 1. $U \longleftarrow U'$, $C \longleftarrow C'$, $D \longleftarrow D'$;
> 2. **Compute** $POS_{C,D}(U)$;
> *Iteration:*
> 3. **repeat**
>   {
> 4. **while** $(POS_{C,D}(U) = \emptyset)$
>   {
> 5. $C \longleftarrow$ **new** (C,U); *define new condition attributes*
> 6.   **Compute** $POS_{C,D}(U)$;
>   }
> 7. **Output** $DT^+_{C,D}(U)$; *output positive region decision table*
> 8. **if** $(POS_{C,D}(U) = U)$ **then exit.**
> 9. $U \longleftarrow U - POS_{C,D}(U)$;
> 10. $C \longleftarrow$ **new** (C,U); *define new condition attributes*
> 11. $D \longleftarrow D|_U$; *modify decision attributes*
> 12. **Compute** $POS_{C,D}(U)$;
>   }

The algorithm HDTL produces a series of decision tables, creating after each pass a positive region decision table obtained from the current universe corresponding to prior level boundary region. The tables are subsequently linked into a hierarchical structure allowing for rapid determination of a value of the decision attribute in a new case by passing from top level table down to the table in which a matching combination of condition attributes in the positive region is found.

The algorithm starts with initial data set and initial definitions of condition and decision attributes obtained after discretization (line 1) and computation of initial positive region (line 2). During each pass through the iterative block, the current set of condition attributes is eventually refined, for example by increasing discretization precision, or redefined until positive region is non-empty (lines 4-6). Following that stage, the positive region decision table is output and the current boundary region becomes a new universe (lines 7, 8). The condition attributes are defined again (line 10) and the definition of the decision attributes is changed by restricting them to current universe (line 11). The process continues until current universe is boundary-free, that is until all original input cases are uniquely classified into one of the decision classes.

The algorithm, in general, may not be able to produce a deterministic classifier, for example if the decision attribute is not deterministically definable in terms of values of original condition variables, that is condition variables prior to discretization. In our application, the original operation control data were all numeric and were always verified prior to running HDTL process to make sure that the the decision attribute, that is after discretization of the original decision variable, is functionally dependent on original condition variables.

6 Experiments

Training samples were acquired during numerous training simulated runs. We run the car by a robot controller on a certain track and recorded the simulation process. We used a training protocol that not only recorded the working data of the robot controller, but also asked the robot controller what to do slightly different situations than the current situation. During data acquisition process, we applied an error compensation technique by forcing the robot controller to see the condition variables through the same discrete ranges as the rough controller and to act on discrete values for decision variables as the rough controller would do. One of the reasons for doing this was to reduce random errors cause by the discretization. The other reason was to eliminate the problem of inconsistent decisions for a given discretization method. Random noise was introduced to condition variables and decision variables in our experiments to simulate sensors measurement errors. We were able to introduce up to 10 % of evenly distributed noise while while maintaining the stability of the rough controller. The rough controller was able to complete the whole track in 100 % of test runs when sensor errors stayed within 10 % limit. In general, the performance of the rough controller exceeded our expectations. In some instances, it succeeded in properly driving the car with sensor errors reaching 40 %. The rough controller can run on different tracks if it is trained to do so and if the discretization of condition variables is adjusted to suit each track.

7 Concluding Remarks

The approach adapted in our simulated vehicle control experiments can be used to capture the underlining physical models of complex control problems or natural processes where the physical nature of the process is too complex to map it into mathematical equations. It appears that the hierarchical decision table method, as summarized in the HDTL algorithm, provides a practical way of capturing process model from a large number of sampled data.

We are planning to continue our work towards acquisition of control algorithms from expert operator control data. The next step in this direction is going to be the inclusion of obstacles and other vehicles on the the track. Another extension to investigate is making control decisions with uncertain input information. The application of the variable precision rough set model extension (see, for example [4], [6], [13], [1]) in this context will be researched.

Acknowledgments: The research reported in this article was partially supported by a research grant awarded to the second author by the Natural Sciences and Engineering Research Council of Canada.

References

1. Beynon, M.: An Investigation of β reduct selection within the variable precision rough set model. In Ziarko, W. and Yao, Y. (eds). Rough Sets and Current Trends in Computing, Lecture Notes in AI 2005, Springer Verlag, 2001, 30-45.
2. Pawlak, Z.: Rough sets: Theoretical aspects of reasoning about data. Kluwer, 1991.
3. Ślęzak, D.: Approximate decision reducts. Ph.D. thesis, Institute of Mathematics, Warsaw University, 2001.
4. Ziarko, W.: Variable precision rough sets model. Journal of Computer and Systems Sciences, vol. 46. no. 1, 1993, 39-59.
5. Ziarko, W.: Probabilistic decision tables in the variable precision rough set model. Computational Intelligence, vol. 17, no 3, 2001, 593-603.
6. Yao, Y.Y., Wong, S.K.M.: A decision theoretic framework for approximating concepts, Intl. Journal of Man-Machine Studies, 37, 1992, 793-809.
7. Plonka, L. and Mrozek, A. Rule-based stabilization of the inverted pendulum. Computational Intelligence, vol. 11, no. 2, 1995, 348-356.
8. Slowinski, R. (ed.) Intelligent Decision Support: Handbook of Applications and Advances of Rough Sets Theory, Kluwer, 1992.
9. Grzymala-Busse, J. W. LERS - a System Learning from Examples Based on Rough Sets. In [8], 3-18.
10. Mrozek, A.: Rough sets in computer implementation of rule-based industrial processes. In [8], 19-31.
11. Khasnabis, S. Arciszewski, T. Hoda, S. Ziarko, W. Urban rail corridor control through machine learning. Transportation Research Record, no. 1453, 1994, 91-97.
12. Szladow, A. Ziarko, W. Adaptive process control using rough sets. Proc. of the Intl. Conf. of Instrument Society of America, ISA/93, Chicago, 1993, 1421-1430.
13. Ziarko, W. Katzberg, J. Rough sets approach to system modelling and control algorithm acquisition. Proc. of IEEE WESCANEX 93 Conference, Saskatoon, 1993, 154-163.
14. Ziarko, W. Generation of control algorithms for computerized controllers by operator supervised training. Proc. of the 11th IASTED Intl. Conf. on Modelling, Identification and Control, Innsbruck, 1992, 510-513.
15. Ziarko, W. Acquisition of Hierarchy-Structured Probabilistic Decision Tables and Rules from Data. Proc. of the World Congress on Computational Intelligence, IEEE International Conference on Fuzzy Systems, Honolulu, 2002, 779-785.
16. Munakata, T. Rough control: a perspective. In Lin, T.Y. Cercone, N. (eds.) Rough sets and data mining: analysis for imprecise data. Kluwer, 1997, 77-88.
17. Skowron, A. Peters, J. Suraj, Z. An application of rough set methods in control design. Proc. of the Workshop on Concurrency, Warsaw, 1999, 214-235.
18. http://rars.sourceforge.net/

Part V

Logics for Artificial Intelligence

On Algebraic Operations on Fuzzy Numbers

Witold Kosiński[1], Piotr Prokopowicz[2], and Dominik Ślęzak[3,1]

[1] Polish–Japanese Institute of Information Technology
Research Center
ul. Koszykowa 86, 02-008 Warsaw, Poland
[2] The University of Bydgoszcz
Institute of Environmental Mechanics and Applied Computer Science
ul. Chodkiewicza 30, 85-064 Bydgoszcz, Poland
[3] The University of Regina
Department of Computer Science
Regina, SK, S4S 0A2 Canada
email: *wkos@pjwstk.edu.pl, piotrekp@ab-byd.edu.pl, slezak@pjwstk.edu.pl*

Abstract New definition of the fuzzy counterpart of real number is presented. An extra feature, called the orientation of the membership curve is introduced. It leads to a novel concept of an ordered fuzzy number, represented by the ordered pair of real continuous functions. Four algebraic operations on ordered fuzzy numbers are defined; they enable to avoid some drawbacks of the classical approach.

Keywords: fuzzy numbers, quasi-convexity, orientation, algebraic operations

1 Introduction

In 1978 in their fundamental paper on operations on fuzzy numbers Dubois and Prade [3] proposed a restricted class of membership functions, called (L, R)–numbers. The essence of their representation is that the membership function is of a particular form that is generated by two so-called shape (or spread) functions: L and R. In this context (L, R)–numbers became quite popular, because of their good interpretability and relatively easy handling for simple operation, i.e. for the fuzzy addition. However, if one wants to stay within this representation while following the extension principle, approximations of fuzzy functions and operations are needed. They may lead to large computational errors that cannot be further controlled when applying them repeatedly, which was stated in several places, e.g. [14,15].

On the other hand operations on the so-called convex fuzzy numbers (compare [2,3,4,14]) following the Zadeh's extension principle, are similar to the interval analysis. The operations on intervals have several inconsistencies, which then appeared in the fuzzy calculation on convex fuzzy numbers. Moreover, if one wants to stay within this class further drawbacks appear. It turns that, for a fuzzy number A, the difference $A - A$ is usually a fuzzy zero, not the crisp zero. Furthermore, even for two fuzzy numbers A and B with simple triangular membership functions, the equation $A + X = B$ may be not solvable with respect to X. Consequently, if the fuzzy number C is the result of the addition $A + B$, then in general the difference $C - B$ is not equal to the fuzzy number A.

The goal of our paper is to overcome the drawbacks by constructing a revised definition of the fuzzy number enlarged by new elements. The new concepts make possible a simple utilizing the fuzzy arithmetic by constructing an Abelian group of fuzzy numbers. The other aim of the paper is to lay down foundations for the full set of four operations on fuzzy numbers in a way that makes of them a field of numbers with properties similar to those of the field of (crisp) real numbers. The convex fuzzy numbers become a subset of of the new model, called the field of ordered fuzzy numbers. That subset, however, is not closed even under arithmetic operation. At the present state of development only arithmetic operations are algorithmized and implemented in the Delphi environment. In the further paper the next step will be done.

2   Fuzzy number as a relation in the plane

Doing the present development, we would like to refer to one of the very first representations of a fuzzy set defined on a universe X (the real axis $\mathbb{R}$, say) of discourse, i.e. on the set of all feasible numerical values (observations, say) of a fuzzy concept (say: variable or physical measurement). In that representation (cf. [4,16]) a fuzzy set (read here: a fuzzy number) A is defined as a set of ordered pairs $\{(x, \mu_x)\}$, where $x \in X$ and $\mu_x \in [0, 1]$ has been called the grade (or level) of membership of x in A. At that stage, no other assumptions concerning μ_x have been made. Later on, one assumed that μ_x is (or must be) a function of x. However, originally, A was just a relation in a product space $X \times [0, 1]$. We know that not every relation must be a functional one. It is just a commonly adopted point of view, that such a kind of relation between μ_x and x should exist, which leads to a membership function $\mu_A : X \to [0, 1]$ with $\mu_x = \mu_A(x)$. In our opinion such a point of view may be too restrictive and here most of the above and further quoted problems have their origin. We summarize this in the form of the following primitive representations of a fuzzy number:

$$A = \{(x, y) \in \mathbb{R} \times [0, 1] : y = \mu_A(x), \text{where } \mu_A \text{ membership function}\} \quad (1)$$

or

$$A = \{(x, y) \in \mathbb{R} \times [0, 1] : y \text{ membership level of } x \} . \quad (2)$$

In our previous paper [9], we claimed that an extra feature of fuzzy number should be added in order to distinguish between two kinds of fuzzy numbers: mirror images of positive numbers $A \in \mathcal{FN}^+$ (the subset of fuzzy numbers $\mathcal{FN}$ corresponding to (quasi)convex membership functions supported within $\mathbb{R}^+$ – the positive real axis) and negative numbers $B \in \mathcal{FN}^-$ defined directly on $\mathbb{R}^-$. To solve this problem we have introduced in [9,10] the concept of the *orientation of the membership curve* of A.

A number of attempts to introduce non-standard operations on fuzzy numbers have been made [1,2,5,13,14]. It was noticed that in order to construct suitable operations on fuzzy numbers a kind of invertibility of their membership functions is required. In this way the $L - P$ was proposed as well as the more general approach to fuzzy numbers regarded as fuzzy sets [16] defined over the real axis, that fulfill some conditions, e.g. they are normal, compactly supported and in some sense convex (cf. [2,3,4,14]). In [7,11] the idea of modeling fuzzy numbers by means of quasi-convex functions (cf. [12]) is discussed. There the membership functions of fuzzy numbers were searched in the class of quasi-convex functions. Namely, if χ_r is a characteristic function of the one-element set $\{r\}$ with $r \in \mathbb{R}$, then we have specified two types of fuzzy numbers $A = (\mathbb{R}, \mu_A)$:

- **crisp**, where $\mu_A = \chi_r$, $r \in \mathbb{R}$.
- **genuinely fuzzy**, where
    (1) μ_A is normal, i.e. $1 \in \mu_A(\mathbb{R})$,
    (2) the support of μ_A is an interval (l_A, p_A), $l_A, p_A \in \mathbb{R}$,
    (3) $-\mu_A$ is strictly quasi–convex.

Then due to the fundamental theorem for strictly quasi–convex functions (Martos, 1975) we could formulate the following property:

Proposition 1. *Let genuinely fuzzy $A = (\mathbb{R}, \mu_A)$ be given. Then there exist $1_A^-, 1_A^+ \in (l_A, p_A) = \operatorname{supp}A$ such that μ_A is increasing on $(l_A, 1_A^-)$, decreasing on $(1_A^+, p_A)$, and constantly equal to 1 on $[1_A^-, 1_A^+]$.*

Thanks to this property, which is a kind of convexity assumption, made by other authors, we have got a quasi-invertibility of the membership function μ_A. The quasi-invertibility has enabled to state quite an efficient calculus on fuzzy reals, similar to the interval calculus. However, some unpleasant situations could occur (cf. Fig. 1). Even starting from the most popular trapezoidal membership functions, algebraic operations could lead outside this family, towards such generalized quasi-convex objects.

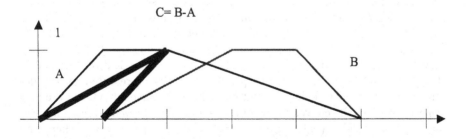

Figure1. Subtraction of genuinely fuzzy numbers with help of the interval calculus.

Our more general definition overcame the main drawback of other approaches, namely that the difference $A - A$ is usually a fuzzy zero – not the crisp zero. Moreover, it seems to provide a solution the problem of finding a solution of the equation $A + X = C$. In our previous papers [8,9] we have already tried to answer these questions in terms of the so-called *improper parts* of fuzzy numbers (cf. the object C on Fig. 1). Namely, given "proper" fuzzy numbers A and C, we can always specify possibly "improper" X such that $A + X = C$. Moreover, as shown in the foregoing sections, we can do it in a very easy (one could say: obvious) way, much clearer and efficient than in case of other approaches.

Here we should mention that Klir was the first, who in [5] has revised fuzzy arithmetic to take relevant requisite constraint (the equality constraint, exactly) into account and obtained $A - A = 0$ as well as the existence of inverse fuzzy numbers for the arithmetic operations. Some partial results of the similar importance were obtained by Sanchez in [13] by introducing an extended operation of a very complex structure. Our approach, however, is much simpler from mathematical point of view, since it does not use the extension principle but refers to the functional representation of fuzzy numbers in a more direct way.

3   Ordered fuzzy numbers

In [11] the idea of treating fuzzy numbers as pairs of functions defined on interval $(0, 1]$ has occurred. Those two functions, introduced with the help of α−sections, played the role of generalized inverses of monotonic parts of fuzzy membership functions. We were following this idea by providing in our previous paper [8] much clearer, a novel interpretation of fuzzy numbers.

Namely, considering a continuous function μ_A with its support on the interval $[a, b]$, where $a \leq b$, $\mu_A(a) = \mu_A(b) = 0$ we have added in $\mathbb{R}^2$ the segment $[a, b] \times \{0\}$ to the plot of μ_A, then the resulting set

$$\mathcal{C}_A = \{(x, y) \in \mathbb{R}^2 : y = \mu_A(x), x \in [a, b]\} \cup [a, b] \times \{0\} \qquad (3)$$

became a plane closed curve. Such a curve can possess two orientations: *positive* and *negative*. To have a common approach to any crisp (singleton) fuzzy number $r \in \mathbb{R}$, given by a characteristic function χ_r of the one–element set $\{r\}$, and to genuinely (i.e. non–singleton) fuzzy numbers, in terms of closed curves, we have added the interval $[0, 1] \times r$ to the graphical representation of a crisp number χ_r. Then the crisp number r was represented by the degenerated closed curve being that interval. Then we have introduced the first generalization of the convex fuzzy numbers, namely an *oriented fuzzy number* as a triple $A = (\mathbb{R}, \mu_A, \mathsf{s}_A)$, where $\mu_A : \mathbb{R} \to [0, 1]$ can represent either crisp or genuinely fuzzy number (cf. Proposition 1), and $\mathsf{s}_A \in \{-1, 0, 1\}$ denotes orientation of A.

We have distinguished two types of oriented fuzzy numbers. For a crisp number A we have $\mathsf{s}_A = 0$, while for a genuinely fuzzy one A, there is $\mathsf{s}_A = 1$ or $\mathsf{s}_A = -1$, depending on the orientation of curve $\mathcal{C}_A$. The notion of a fuzzy number can be redefined in terms of tuple $A = (\mathcal{C}_A, \mathsf{s}_A)$. Consider closed curve $\mathcal{C}_A \subseteq \mathbb{R}^2$ written as

$$\mathcal{C}_A = [l_A, p_A] \times \{0\} \cup [1_A^-, 1_A^+] \times \{1\} \cup \text{up } A \cup \text{down } A \qquad (4)$$

where up A, down $A \subseteq \mathbb{R}^2$ are the plots of some monotonic functions. Abbreviations up and down do not correspond to the fact whether a given function is increasing or decreasing but to the orientation of $\mathcal{C}_A$. They simply label the ascending and descending parts of the oriented curve. We call such interpreted subsets $\mathcal{C}_A \subseteq \mathbb{R}^2$ *membership curves* of oriented fuzzy numbers.

What is important in this new representation of a fuzzy number is the fact that the curve does not need to be (a part of) a graph of any function defined on real axis. Moreover, having the concept of the orientation introduced we can look once more on the representation of a fuzzy set (number) from Introduction (cf. (1), (2) and equipped with an extra feature:

$$A = \{\{(x, y) \in \mathbb{R} \times [0, 1] : y \text{ membership level of } x\} \cup \{\text{orientation}\}\}. \quad (5)$$

Due to (1) and (5) we may now redefine the oriented fuzzy number, identified by its membership curve as a continuous function $\boldsymbol{f}_A : \mathbb{R} \to \mathbb{R} \times [0, 1]$ with some properties and an interval $[t_0, t_1]$ such that

$$A = \{(x, y) \in \mathbb{R} \times [0, 1] : \exists t \in [t_0, t_1](x, y) = \boldsymbol{f}(t)\} \subseteq \mathbb{R}^2.$$

The function $\boldsymbol{f}$ has been called in [10] the **fuzzy observation**. Due to the main properties of quasi-convex real valued functions (cf. [12]) we may formulate the required properties of the function $\boldsymbol{f}$.

Proposition 2. *Let $\boldsymbol{f} : \mathbb{R} \to \mathbb{R} \times [0, 1]$ and $\boldsymbol{f}(t) = (x_f(t), \mu_f(t))$. It satisfies the following properties:*

1. *$\mu_f(t) = 1$ for some $t \in \mathbb{R}$*
2. *$\{t \in \mathbb{R} : \mu_f(t) > 0\} = (t_0, t_1)$, for some $t_0, t_1 \in \mathbb{R}$*
3. *$-\mu_f : \mathbb{R} \to [-1, 0]$ is strictly quasi-convex*

if and only if there exist $t^-, t^+ \in (t_0, t_1)$ such that μ_f is equal to 1 on (t^-, t^+), increasing on (t_0, t^-), and decreasing on (t^+, t_1).

Proposition 3. *Let $\boldsymbol{f} : \mathbb{R} \to \mathbb{R} \times [0, 1]$ and $\boldsymbol{f}(t) = (x_f(t), \mu_f(t))$. It has the following properties:*

1. *$\mu_f(t) = 1$ for some $t \in \mathbb{R}$*
2. *$\{t \in \mathbb{R} : \mu_f(t) > 0\} = (t_0, t_1)$, for some $t_0, t_1 \in \mathbb{R}$*
3. *$-\mu_f : \mathbb{R} \to [-1, 0]$ is strictly quasi-convex*

if and only if there exist continuous functions $\mu_f^\uparrow, \mu_f^\downarrow : [0,1] \to \mathbb{R}$ *such that:*

$$f(t_0) = (\mu_f^\uparrow(0), 0), \ f(t^-) = (\mu_f^\uparrow(1), 1), \ f(t_1) = (\mu_f^\downarrow(0), 0), \ f(t^+) = (\mu_f^\downarrow(1), 1) \tag{6}$$

and:

$$\forall_{t \in (t_0, t^-)} \exists!_{y \in (0,1)} f(t) = (\mu_f^\uparrow(y), y) \qquad \forall_{t \in (t^+, t_1)} \exists!_{y \in (0,1)} f(t) = (\mu_f^\downarrow(y), y) \tag{7}$$

We introduce the **up part** of f_A and **down part** of f_A and their projections on $\mathbb{R}$ as a pair $(\mu_A^\uparrow, \mu_A^\downarrow)$, where $\mu_A^\uparrow, \mu_A^\downarrow : [0,1] \to \mathbb{R}$ are continuous functions (cf. Fig.2).

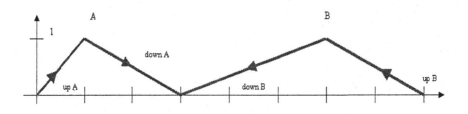

Figure2. Up-part and down-part of two fuzzy numbers.

Membership curves provide graphical representation of ordered fuzzy numbers as well. Functions $\mu_A^\uparrow$ and $\mu_A^\downarrow$ correspond to up A, down $A \subseteq \mathbb{R}^2$ as follows:

$$\text{up } A = \{(\mu_A^\uparrow(y), y) : y \in [0,1]\} \qquad \text{down } A = \{(\mu_A^\downarrow(y), y) : y \in [0,1]\} \tag{8}$$

Now we can introduce the new definition of fuzzy number.

Definition 1. By *ordered fuzzy real* we mean an ordered pair $A = (\mu_A^\uparrow, \mu_A^\downarrow)$, where $\mu_A^\uparrow, \mu_A^\downarrow : [0,1] \to \mathbb{R}$ are continuous functions. We call the corresponding (due to Propositions 1 and 2) function $f : \mathbb{R} \to \mathbb{R} \times [0,1]$ a *fuzzy A-observation*, denoted by $f_A = (x_A, \mu_A)$, where $x_A : \mathbb{R} \to \mathbb{R}$ and $\mu_A : \mathbb{R} \to [0,1]$.

One can see that by defining $x_A : \mathbb{R} \to \mathbb{R}$ as $x_A(t) = t$ or $x_A(t) = -t$ we obtain the notion of the genuinely fuzzy number. Functions $(\mu_A^\uparrow, \mu_A^\downarrow)$ correspond then to inverses of monotonic parts of μ_A, if μ_A is the membership function of convex (or quasi–convex) function.

As an example, let us consider Figure 3, which illustrates intuition of defining the reverse of ordered (oriented) fuzzy real. According to the idea mentioned above such a reverse should obtain an opposite orientation. By performing operation of addition on the pairs of ascending and descending parts of A and $-A$, we are then able to obtain crisp zero.

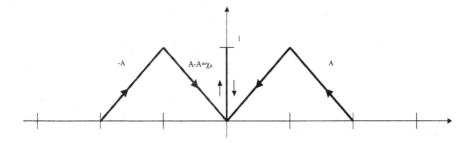

Figure3. Addition of reverse ordered fuzzy numbers.

4   Algebraic operations on ordered fuzzy numbers

Let us formulate operations of addition and subtraction on ordered fuzzy numbers. The following definition coincides with that presented for oriented fuzzy numbers in [8] and [9]. It is also related to the idea of adding and subtracting fuzzy numbers by means of inverses of their monotonic parts [11]. However, characteristics of the ordered fuzzy number enables much clearer definition.

Definition 2. Let three ordered fuzzy numbers $A = (\mu_A^\uparrow, \mu_A^\downarrow)$, $B = (\mu_B^\uparrow, \mu_B^\downarrow)$ and $C = (\mu_C^\uparrow, \mu_C^\downarrow)$, be given. We say that:

1. C is the sum of A and B, denoted by $C = A + B$, iff

$$\forall_{y \in [0,1]} \left[\mu_A^\uparrow(y) + \mu_B^\uparrow(y) = \mu_C^\uparrow(y) \wedge \mu_A^\downarrow(y) + \mu_B^\downarrow(y) = \mu_C^\downarrow(y) \right] \quad (9)$$

2. C is the subtraction of B from A, denoted by $C = A - B$, iff

$$\forall_{y \in [0,1]} \left[\mu_A^\uparrow(y) - \mu_B^\uparrow(y) = \mu_C^\uparrow(y) \wedge \mu_A^\downarrow(y) - \mu_B^\downarrow(y) = \mu_C^\downarrow(y) \right] \quad (10)$$

3. C is the product of A times B, denoted by $C = A \cdot B$, iff

$$\forall_{y \in [0,1]} \left[\mu_A^\uparrow(y) \cdot \mu_B^\uparrow(y) = \mu_C^\uparrow(y) \wedge \mu_A^\downarrow(y) \cdot \mu_B^\downarrow(y) = \mu_C^\downarrow(y) \right] \quad (11)$$

4. Assume additionally that

$$\forall_{y \in [0,1]} \left[\mu_B^\uparrow(y) \neq 0 \wedge \mu_B^\downarrow(y) \neq 0 \right] \quad (12)$$

Then we say that C is division of A by B, denoted by $C = A/B$, iff

$$\forall_{y \in [0,1]} \left[\mu_A^\uparrow(y)/\mu_B^\uparrow(y) = \mu_C^\uparrow(y) \wedge \mu_A^\downarrow(y)/\mu_B^\downarrow(y) = \mu_C^\downarrow(y) \right] \quad (13)$$

If $x_A(t) = -t$, then $\mu_A^\uparrow$ and $\mu_A^\downarrow$ are invertible functions of variable x. Inverses

$$(\mu_A^\uparrow)^{-1} = \mu_A|(l_A, 1_A^-) \qquad (\mu_A^\downarrow)^{-1} = \mu_A|(1_A^+, p_A) \qquad (14)$$

form decreasing and increasing parts of membership function μ_A, respectively. The result of addition of numbers A and B, where $x_A(t) = x_B(t) = -t$, is defined exactly as in case of convex fuzzy numbers, for which the interval arithmetic is used (cf. [2,7]). Such numbers obtain positive orientation, which is also the case of their sum.

Subtraction of A from A is the same operation as addition to A its reverse, i.e. the number $-A = (-\mu_A^\uparrow, -\mu_A^\downarrow)$. Then we get $C = (\mu_C^\uparrow, \mu_C^\downarrow)$, where

$$\mu_C^\uparrow(y) = \mu_A^\uparrow(y) - \mu_A^\uparrow(y) = 0 \qquad \mu_C^\downarrow(y) = \mu_A^\downarrow(y) - \mu_A^\downarrow(y) = 0 \qquad (15)$$

Since the crisp number $r \in \mathbb{R}$ (in other words – characteristic function χ_r) can be represented by the ordered pair $(\mu_r^\uparrow, \mu_r^\downarrow)$, where

$$\forall_{y \in [0,1]} \left[\mu_r^\uparrow(y) = \mu_r^\downarrow(y) = r\right] \qquad (16)$$

the result of such operation is just the crisp number $r = 0$, like in Figure 3.

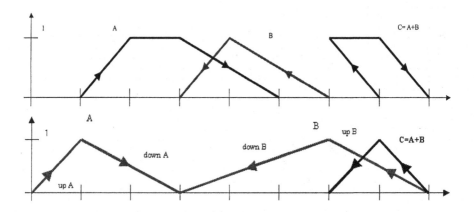

Figure4. Addition of ordered fuzzy numbers with opposite orientation.

Addition of oppositely oriented fuzzy numbers is an ordered fuzzy real, but may not be a genuinely (classical) fuzzy number. In general, one can expect the results not necessarily interpretable in terms of classical approach but still valuable as representing some fuzzy membership information, like in Figure 4. Obviously, this is not a problem at the level of ordered fuzzy numbers, where orientation of the membership curve just illustrates the position of $\mu_C^\uparrow$ with respect to $\mu_C^\downarrow$.

It is obvious that the set of ordered fuzzy numbers equipped with two arithmetic operations: addition and subtraction form an Abelian group. Even more, if crisp numbers, represented by characteristic functions of one-element sets, are regarded as a field with the well-defined operations between pairs crisp and fuzzy numbers (compare eqs (13)–(17) in [8]) then the set of ordered fuzzy numbers is a linear space.

As far as two other algebraic operations are concerned, we can see that division by fuzzy numbers is restricted to those which do not possess in their membership curves pairs of the form $(0, y)$ with $y \neq 0$. It is obvious, because 1 cannot be the result of the product of 0 times finite number. In Fig. 5 product of two ordered triangular numbers is presented.

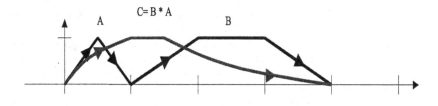

Figure5. Product of two ordered fuzzy numbers.

5   Conclusions

We introduced the notion of an ordered fuzzy real and explained its correspondence with fuzzy numbers defined by quasi-convex membership functions, as well as with oriented fuzzy numbers known from our previous research. We stated definitions of operations of addition, subtraction, multiplication and division on ordered fuzzy numbers and provided some examples of their results. In this way we have constructed a field of ordered fuzzy numbers as well as a linear space. In the nearest future we are going to extend this framework to functions of fuzzy variables.

Acknowledgements The authors thank Dr. Z. Kulpa from IPPT PAN, who after completing the paper has pointed out some connections of the oriented membership functions with *directed intervals* and the so-called Kaucher arithmetic. The work on the paper has been conducted in the framework of the two projects of the Research Center of Polish-Japanese Institute of Information Technology and of the IMSiIS of the University of Bydgoszcz, both supported by the Polish National Committee for Scientific Research (KBN).

References

1. E. Czogała, W. Pedrycz, *Elements and methods of fuzzy set theory* (in Polish), PWN, Warszawa, Poland (1985).

2. J. Drewniak, *Fuzzy numbers* (in Polish), in: Fuzzy sets and their applications, J. Chojcan, J. Łęski (eds), WPŚ, Gliwice, Poland (2001) 103–129.

3. D. Dubois, H. Prade, *Operations on fuzzy numbers*, Int. J. System Science, **9** (1978) 576–578.

4. J. Kacprzyk, *Fuzzy Sets in System Analysis* (in Polish) PWN, Warszawa, Poland (1986).

5. G.J. Klir, *Fuzzy arithmetic with requisite constraints*, Fuzzy Sets and Systems, **91** (1997) 165–175.

6. W. Kosiński, *On algebraic operations on fuzzy numbers*, Invited lecture at IC-NNSC'2002, June 11–15, Zakopane, Poland (2002).

7. W. Kosiński, K. Piechór, P. Prokopowicz, K. Tyburek, *On algorithmic approach to operations on fuzzy numbers*, in: Methods of Artificial Intelligence in Mechanics and Mechanical Engineering, T. Burczyński, W. Cholewa (eds), PACM, Gliwice, Poland (2001) 95–98.

8. W. Kosiński, P. Prokopowicz, D. Ślęzak, *Fuzzy numbers with algebraic operations: algorithmic approach*, in: Advances in Soft Computing, Proc. of IIS'2002 Sopot, Poland, June 3-6, 2002, M. Kopotek, S.T. Wierzcho, M. Michalewicz (eds.) , Physica Verlag, 2002, pp. 311-320.

9. W. Kosiński, P. Prokopowicz, D. Ślęzak, *On algebraic operations on fuzzy numbers*, in: Advances in Soft Computing, Proc. of the Sixth International Conference on Neural Network and Soft Computing, Zakopane, Poland, June 11-15, 2002, L. Rutkowski, J. Kacprzyk (eds.), Physica-Verlag, 2003, in print.

10. W. Kosiński, P. Prokopowicz, D. Ślęzak, *Ordered fuzzy numbers*, submitted for publication in Bulletin of the Polish Academy of Sciences, Sér. Sci. Math., (2002).

11. W. Kosiński, P. Słysz, *Fuzzy numbers and their quotient space with algebraic operations*, Bull. Polish Acad. Scien., **41/3** (1993) 285–295.

12. B. Martos, *Nonlinear Programming – Theory and methods*, PWN, Warszawa, Poland (1983) (Polish translation of the English original published by Akadémiai Kiadó, Budapest, 1975).

13. E. Sanchez, *Solutions of fuzzy equations with extended operations*, Fuzzy Sets and Systems, **12** (1984) 237–248.

14. M. Wagenknecht, *On the approximate treatment of fuzzy arithmetics by inclusion, linear regression and information content estimation*, in: Fuzzy sets and their applications, J. Chojcan, J. Łęski (eds), WPŚ, Gliwice, Poland (2001) 291–310.

15. M. Wagenknecht, R. Hampel, V. Schneider, *Computational aspects of fuzzy arithmetic based on Archimedean t-norms*, Fuzzy sets and Systems, **123/1** (2001) 49–62.

16. L.A. Zadeh, *Fuzzy sets*, Information and Control, **8** (1965) 338–353.

17. L.A. Zadeh, *The concept of a linguistic variable and its application to approximate reasoning, Part I*, Information Sciences, **8** (1975) 199–249.

18. L.A. Zadeh, *The role of fuzzy logic in the management of uncertainty in expert systems*, Fuzzy Sets and Systems, **11** (1983) 199–227.

Dual Resolution for Logical Reduction of Granular Tables

Antoni Ligęza

Institute of Automatics AGH, al. Mickiewicza 30, 30-059 Kraków, Poland
e-mail: ligeza@agh.edu.pl

Abstract. Dual resolution is a universal logical inference method for automated theorem proving. Contrary to classical Robinson's resolution it operates on formulae of conjunctive form and the direction of logical consequence is inverse with respect to resolvent generation. This method is convenient especially for proving completeness and formulae reduction. In the paper it is shown how to use dual resolution to reduce attributive decision tables preserving logical equivalence. As atomic values of attributes are to be glued together, granular tables are used to express the results of reduction.

Keywords: *granular sets, granular relations, dual resolution, granular attributive decision tables*

1   Introduction

Dual resolution is a universal logical inference method for automated theorem proving. Contrary to classical Robinson's resolution it operates on formulae of conjunctive form. Two such formulae can be resolved in a way analogous to the classical Robinson's resolution. The resulting formula is called *a dual resolvent* and the inference method is called *backward dual resolution* or *bd-resolution* for short. This is so because the direction of logical consequence is inverse with respect to resolvent generation, i.e. the disjunction of the parent formulae are logical consequence of their be-resolvent.

The method of bd-resolution is especially convenient for proving completeness and for formulae reduction. In fact, disjunction of formulae constituting a tautology (or covering the whole universe) can be reduced to *true*; in case the set is incomplete, it can be reduced to some minimal form.

In the paper it is shown how to use dual resolution to reduce attributive decision tables preserving logical equivalence. As atomic values of attributes are to be glued together, granular tables are used to express the results of reduction. Such extended attributive tables represent granular relations, where the values of attributes can be non-atomic ones: sets, intervals or lattice elements are admitted. Contrary to the inductive generation of rules from examples, the proposed logical reduction preserves logical equivalence, i.e. the resulting reduced table is logically equivalent to the initial one.

2 Granular sets, relations and tables

Consider a set V and several subsets of it, say $V_1, V_2, \ldots, V_k$. A *semi-partition* of V [3] is any collection of its subsets $V_1, V_2, \ldots, V_k$. A semi-partition is normalized (in normal form) iff $V_i \cap V_j = \emptyset$ for all $i \neq j$. A semi-partition is also called an s-partition or sigma-partition (σ-partition) for short. An s-partition of V (or maybe in V) will be denoted as $\sigma(V)$. If not stated explicitly, all the considerations will concern normalized s-partitions.

Now, for a given set S a granular set over S can be defined as follows.

Definition 1 *A granular set $G(S)$ is a pair $G(S) = \{S, \sigma(S)\}$, where $\sigma(S)$ is any s-partition defined on S. If the s-partition $\sigma(V)$ is unnormalized, then the granular set is also determined as an unnormalized one.*

The set S is called the domain *of the granular set, while the s-partition $\sigma(S)$ defines the so-called* signature *of granularity.*

Granular sets can be used to analyze and manipulate certain domains with a variable degree of detail. Using the idea of granular set a granular relation can be defined in a straightforward way. Consider some collection of sets $D_1, D_2, \ldots, D_n$. Let there be defined some granular sets on them, i.e. $G_1 = (D_1, \sigma_1(D_1)), G_2 = (D_2, \sigma_2(D_2)), \ldots, G_n = (D_n, \sigma_n(D_n))$.

Definition 2 *A granular relation $R(G_1, G_2, \ldots, G_n)$ is any set R_G such that $R_G \subseteq U_G$ where*

$$U_G = \sigma_1(D_1) \times \sigma_2(D_2) \times \ldots \times \sigma_n(D_n). \tag{1}$$

The set U_G will be referred to as granular universe *or* granular space. *If at least one of the granular subsets was unnormalized, the relation is also said to be unnormalized one.*

The elements (rows) of a granular relation will be called *boxes*. Note that in fact a granular relation defines a kind of meta-relation, i.e. one based on sets instead of single elements. In fact, if R is a relation defined as $R \subseteq D_1 \times D_2 \times \ldots \times D_n$, then any tuple of R is like a thread in comparison to elements of R_G which are like a cord or a pipe. A granular relation can be presented in the form of extended attributive decision table, i.e. such table, where non-atomic values of attributes cane be used. More details can be found in [3].

3 The idea of dual resolution

The bd-resolution inference method [1] is a general theorem proving method which in fact is dual to classical Robinson's resolution; furthermore, it works backwards, in the sense that normally the disjunction of the parent formulae is a logical consequence of the generated bd-resolvent – the direction of generation of new formulae is inverse with respect to the one of logical entailment. Detailed presentation of the theory of backward dual resolution can be found in [1] or in [2].

In order to prove that a formula is tautology it must be first transformed into Disjunctive Normal Form (DNF); so, all the formulae have the form of conjunctions of literals. The basic inference step consists in combining two such conjuncts into a new formula (the so-called bd-resolvent or dual resolvent), such that the disjunction of the parent formulae is a logical consequence of it. In case eventually an empty formula is obtained (i.e. one always true, obtained from a disjunction of the form $p \vee \neg p$), the initial formula in DNF is proved to be a tautology.

Let us examine a single step of dual resolution. For simplicity, let us consider two propositional logic formulae, $\psi_1 \wedge \omega$ and $\psi_2 \wedge \neg \omega$, where ω is a propositional symbol while ψ_1 and ψ_2 are arbitrary conjunctions of literals. The basic form of the bd-resolution rule is as follows:

$$\frac{\psi_1 \wedge \omega, \psi_2 \wedge \neg \omega}{\psi_1 \wedge \psi_2}$$

where $\psi_1 \wedge \psi_2$ is the *bd-resolvent* of the parent formulae. Contrary to classical resolution, the disjunction of the parent formulae is a logical consequence of their bd-resolvent. To see that it is enough to assume that if $\psi_1 \wedge \psi_2$ is assumed to be true, than both of the formulae must be true as well; hence, since either ω or its negation must be true, one of the initial formulae must be true, and so the initial disjunction must be true as well.

A logical definition of bd-resolution for further use is given below. For simplicity the problem of substitutions in First Order Predicate Calculus is omitted (it can be handled by appropriate use of the assumed (partial) interpretation **I**).

Definition 3 (Generalized bd-resolution) *Let $\psi_1 = \psi^1 \wedge \omega^1, \psi_2 = \psi^2 \wedge \omega^2, \ldots, \psi_j = \psi^j \wedge \omega^j$ be some conjunctive formulae, such that $\omega^1, \omega^2, \ldots, \omega^j$ are certain formulae satisfying the so-called* completeness condition *of the form $\omega \models_{\mathbf{I}} \omega^1 \vee \omega^2 \vee \ldots \vee \omega^j$. Then formula*

$$\psi = \psi^1 \wedge \psi^2 \wedge \ldots \wedge \psi^j \wedge \omega \tag{2}$$

will be called a bd-resolvent *of $\psi_1, \psi_2, \ldots, \psi_j$.*

As before, bd-resolvent is logical consequence of disjunction of the parent formulae. Moreover note that if ω is logically equivalent to $\omega^1 \vee \omega^2 \vee \ldots \vee \omega^j$ and $\psi^1 = \psi^2 = \ldots = \psi^j$ equals to ψ, then the bd-resolvent is logically equivalent to the disjunction of parent formulae. Obviously, assuming that the parent disjunction is true, at least one of the formulae $\psi \wedge \omega^i$ must be true. Hence, both ω^i and and ψ must be true, and so must be ω. Below it will be shown how bd-resolution can be applied to rule reduction.

The idea of reduction of a set of rules is simple: two (or more) rules with the same conclusion can be glued according to the following principle. Let rule r_1 be of the form $r_1 : \psi_1 \longrightarrow h$ and let rule r_2 be of the form $r_2 : \psi_2 \longrightarrow h$. Logical conjunctions of such rules (implications) can be replaced by logically equivalent rule r of the form $r : \psi_1 \vee \psi_2 \longrightarrow h$. Now, if $\psi_1 \vee \psi_2$ can be replaced by its dual resolvent logically equivalent but of simpler form, reduction can take place.

In order to provide an intuitive example consider two rules of the form:

$$r_1 : [color = white] \wedge [shape = circle] \longrightarrow [class = wheel]$$

and

$$r_2 : [color = black] \wedge [shape = circle] \longrightarrow [class = wheel]$$

Obviously, these rules can be replaced with logically equivalent rule r of the form:

$$r : [color = white \vee black] \wedge [shape = circle] \longrightarrow [class = wheel]$$

Further, if there are only two colors, i.e. black and white, then the rule can be simplified to $r : [shape = circle] \longrightarrow [class = wheel]$.

In general, consider k rules with the same conclusion, such that their preconditions differ only with respect to ω_i, $i = 1, 2, \ldots, k$, where $\omega_i = (A_j = t_{ij})$ defines the value of the same single attribute A_j. Assume that the following completeness condition holds, i.e. $\models \omega_1 \vee \omega_2 \vee \ldots \vee \omega_k$. Then the following reduction scheme can be applied:

$$
\begin{array}{l}
r^1 : \phi \wedge \omega_1 \longrightarrow h \\
r^2 : \phi \wedge \omega_2 \longrightarrow h \\
\quad \vdots \quad \vdots \qquad \vdots \quad \vdots \\
r^k : \phi \wedge \omega_k \longrightarrow h \\
\hline
r : \quad \phi \qquad \longrightarrow h
\end{array}
$$

For intuition, the preconditions of the formulae are replaced by a joint condition representing the disjunction of them; roughly speaking, the sets described with the preconditions are "glued" together into a single set. The resulting rule is logically equivalent to the set of initial rules.

In certain cases a complete reduction as shown above may turn out to be inapplicable; however, it may still be possible to simplify the set of rules if only the sub-formulae ω_i, $i = 1, 2, \ldots, k$ can be replaced with a single equivalent formula, as in the intuitive example provided above. In our case, a collection of certain elements can be always replaced by a subset containing all of them (and nothing more), while a collection of intervals can be replaced with their sum (which may be a single, convex interval). In general, let us assume that $\omega_1 \vee \omega_2 \vee \ldots \vee \omega_k \models \omega$, and $\omega \models \omega_1 \vee \omega_2 \vee \ldots \vee \omega_k$ The reduction can take the following logical form:

$$r^1 : \phi \wedge \omega_1 \longrightarrow h$$
$$r^2 : \phi \wedge \omega_2 \longrightarrow h$$
$$\vdots \quad \vdots \qquad \vdots \quad \vdots$$
$$\frac{r^k : \phi \wedge \omega_k \longrightarrow h}{r : \quad \phi \wedge \omega \longrightarrow h}$$

Formula ω must be expressible within the accepted language. In case of a single attribute the internal disjunction can be applied just by specifying the appropriate subset.

4   Granular tables reduction

Reduction of rules is an operation similar to finding minimal representation for propositional calculus formulae or boolean combinatorial circuits. The main idea of reduction of rules is to minimize the number of rules (records, rows) while preserving logical equivalence. An interesting possibility consists in replacing a number of rules having the same conclusions with a single equivalent rule. Using the tabular knowledge representation, reduction takes the following form:

rule	A_1	A_2	...	A_j	...	A_n	H
r^1	t_1	t_2	...	t_{1j}	...	t_n	h
r^2	t_1	t_2	...	t_{2j}	...	t_n	h
$\vdots$	$\vdots$	$\vdots$		$\vdots$		$\vdots$	$\vdots$
r^k	t_1	t_2	...	t_{kj}	...	t_n	h
rule	A_1	A_2	...	A_j	...	A_n	H
r	t_1	t_2	...	_	...	t_n	h

provided that $t_{1j} \cup t_{2j} \cup \ldots \cup t_{kj} = D_j$. Of course, the rules $r^1, r^2, \ldots, r^k$ are just some selected rows of the original table containing all the rules.

In a more general case the complete reduction (resulting in obtaining D_j) may be impossible. In such cases partial reduction leading to granular values may still be useful. It takes the following form::

rule	A_1	A_2	...	A_j	...	A_n	H
r^1	t_1	t_2	...	t_{1j}	...	t_n	h
r^2	t_1	t_2	...	t_{2j}	...	t_n	h
$\vdots$	$\vdots$	$\vdots$		$\vdots$		$\vdots$	$\vdots$
r^k	t_1	t_2	...	t_{kj}	...	t_n	h
rule	A_1	A_2	...	A_j	...	A_n	H
r	t_1	t_2	...	t	...	t_n	h

provided that $t_{1j} \cup t_{2j} \cup \ldots \cup t_{kj} = t$. As above, the rules $r^1, r^2, \ldots, r^k$ are just some selected rows of the original table containing all the rules.

5   An example: how the reduction works

Consider the following Optician's Decision Table [5], being a perfect example of a tabular system.

Number	Age	Spectacle	Astigmatic	Tear p.r.	Decision
1	y	m	y	n	H
2	y	n	y	n	H
3	p	m	y	n	H
4	q	m	y	n	H
5	y	m	n	n	S
6	y	n	n	n	S
7	p	m	n	n	S
8	p	n	n	n	S
9	q	n	n	n	S
10	y	m	n	r	N
11	y	m	y	r	N
12	y	n	n	r	N
13	y	n	y	r	N
14	p	m	n	r	N
15	p	m	y	r	N
16	p	n	n	r	N
17	p	n	y	r	N
18	p	n	y	n	N
19	q	m	n	r	N
20	q	m	n	n	N
21	q	m	y	r	N
22	q	n	n	r	N
23	q	n	y	r	N
24	q	n	y	n	N

This table can be regarded as a complete decision table (ready for use) or as a specification of some tabular rule based system. A brief analysis assures us that there are no redundant or subsumed rules. The system is deterministic and complete.

In order to perform reduction one has to search for certain rows identical apart values of a single attribute. For example, rows no. 1 and 2 of the initial table differs only on the second position, and the given values are the only ones, i.e. the completeness condition is satisfied. The two rows can be reduced to a single row of the table below (row no. 1). Reduction to complete domains (total reduction in which no subsets of the domains are admissible) leads to:

Number	Age	Spectacle	Astigmatic	Tear p.r.	Decision
1	y	–	y	n	H
2	–	m	y	n	H
3	y	–	n	n	S
4	p	–	n	n	S
5	–	n	n	n	S
6	–	–	–	r	N
7	p	n	y	–	N
8	q	m	n	–	N
9	q	n	y	–	N

If granular tables are admitted, a further total reduction is possible; as the result the following table is obtained:

Number	Age	Spectacle	Astigmatic	Tear p.r.	Decision
1	y	–	y	n	H
2	–	m	y	n	H
3-4	{ y, p }	–	n	n	S
5	–	n	n	n	S
6	–	–	–	r	N
7-9	{ p, q }	n	y	–	N
8	q	m	n	–	N

The output table is in its minimal form (maximally reduced one). It is still complete (logically equivalent to its input form) and deterministic.

6   An example: a mushroom table

Below a more extensive example of reduction is presented. It is taken from [6] and it uses a part of the mushroom identification table[2]. As the original table is too large (it contains 8124 rows and uses 22 attributes) the presented example is an extract of 36 rows and 5 attributes. The initial table (before reduction) is presented in Fig. 1.

After some first stage of reduction the table obtained in Fig. 2 containing 12 rows is obtained. As one can see, the attribute T takes now granular values.

After further reduction one obtains the table of 6 rows presented in Fig. 3; now also attribute W takes granular values.

The final reduction leads to a maximally reduced form and the granular table presented in Fig. 4. To conclude, the initial table was reduced to two logically equivalent rules.

---

[2] ftp://ftp.ics.uci.edu/pub/machine-learning-databases/mushroom/

No.	K	T	Z	W	P	O
1	r	b	z	l	s	BE
2	r	s	z	l	s	BE
3	r	o	z	l	s	BE
4	r	b	z	l	n	BE
5	r	s	z	l	n	BE
6	r	o	z	l	n	BE
7	r	b	z	k	s	BE
8	r	s	z	k	s	BE
9	r	o	z	k	s	BE
10	r	b	z	k	n	BE
11	r	s	z	k	n	BE
12	r	o	z	k	n	BE
13	o	s	g	k	l	SG
14	s	s	g	k	l	SG
15	o	b	g	k	l	SG
16	s	b	g	k	l	SG
17	o	o	g	k	l	SG
18	s	o	g	k	l	SG
19	o	s	g	t	l	SG
20	s	s	g	t	l	SG
21	o	b	g	t	l	SG
22	s	b	g	t	l	SG
23	o	o	g	t	l	SG
24	s	o	g	t	l	SG
25	o	s	g	k	o	SG
26	s	s	g	k	o	SG
27	o	b	g	k	o	SG
28	s	b	g	k	o	SG
29	o	o	g	k	o	SG
30	s	o	g	k	o	SG
31	o	s	g	t	o	SG
32	s	s	g	t	o	SG
33	o	b	g	t	o	SG
34	s	b	g	t	o	SG
35	o	o	g	t	o	SG
36	s	o	g	t	o	SG

Fig. 1. The initial table

7   Related work

In *Machine Learning* (ML) several rules somewhat related to dual resolution
are known. A number of such rules is discussed in the work of Michalski,
e.g. [4]. Note, however, that the principal difference is that in ML all the
rules are over-generalizing ones, e.g. the *dropping condition* rule, the *closing*

No.	K	T	Z	W	P	O
1	r	s,b,o	z	l	s	BE
2	r	s,b,o	z	l	n	BE
3	r	s,b,o	z	k	s	BE
4	r	s,b,o	z	k	n	BE
5	o	s,b,o	g	k	l	SG
6	s	s,b,o	g	k	l	SG
7	o	s,b,o	g	t	l	SG
8	s	s,b,o	g	t	l	SG
9	o	s,b,o	g	k	o	SG
10	s	s,b,o	g	k	o	SG
11	o	s,b,o	g	t	o	SG
12	s	s,b,o	g	t	o	SG

Fig. 2. The table after first reduction.

No.	K	T	Z	W	P	O
1	r	s,b,o	z	l,k	s	BE
2	r	s,b,o	z	l,k	n	BE
3	o	s,b,o	g	k,t	l	SG
4	s	s,b,o	g	k,t	l	SG
5	o	s,b,o	g	k,t	o	SG
6	s	s,b,o	g	k,t	o	SG

Fig. 3. The table after second reduction.

No.	K	T	Z	W	P	O
1	r	s,b,o	z	l,k	s,n	BE
2	o,s	s,b,o	g	k,t	l,o	SG

Fig. 4. The maximally reduced form.

interval rule or the *climbing generalization tree* rule where the result is a logical consequence of the parent items, while in case of dual resolution the inverse is true. Moreover, in case of rule reduction with bd-resolution *logical equivalence* is preserved.

This is also the case of the so-called *inductive resolution* rule [4]. Taking the logical point of view, the rule is of the form:

$$\frac{P \wedge F_1 \longrightarrow K, \neg P \wedge F_2 \longrightarrow K}{F_1 \vee F_2 \longrightarrow K}.$$

The rule leads to over-generalization; in case of resolution of rules (the classical one) one would have $F_1 \wedge F_2 \longrightarrow K$ as the logical consequence. In case of dual resolution of the preconditions the result is the same. However, if by chance $F_1 = F_2$ the rules of classical resolution, dual resolution and inductive resolution give the same result.

8   Conclusions

The presented dual resolution rule allows for reduction of rule based systems and tabular systems to some minimal form. The bd-resolution rule can be considered to be an extension of the minimization algorithms known reduction of combinatorial logical circuits (ones based on use of Karnaugh Tables or Quine-McCluskey approach) over attributive and first-order logic. An important feature is that the resulting system is logically equivalent to the initial one. This means that the proposed approach is safe – no over-generalization takes place. Simultaneously, one cannot expect to obtain reduction comparable to the one in case of rule induction.

Taking the above into account, the potential applications are considered to cover rule reduction in case safety is the principal interest, i.e. the obtained set of rules should be equivalent to its specification given in the form of extensively specified rules. This may apply to some technical systems, such as control and monitoring systems or as decision support ones. Some examples can be found on the WWW page [7].

References

1. Ligęza, A. (1993) Logical foundations for knowledge-based control systems – Knowledge representation reasoning and theoretical properties, *Scientific Bulletins of AGH*, No.: 1529, *Automatics 63*, Kraków, Poland
2. Ligęza, A. (1994) Backward dual resolution. Direct proving of generalization. In: H. Jaakkola *et al* (Eds.) *Information Modelling and Knowledge Bases V: Principles and Formal Techniques*, IOS Press, Amsterdam, 1994, 336-349
3. Ligęza A. (2002) Granular sets and granular relations. An algebraic approach to knowledge representation and reasoning. In: M.A. Kłopotek, S.A. Wierzchoń and M. Michalewicz (Eds.) Intelligent Information Systems 2002, *Advances in Soft Computing*, Physica-Verlag, 331-340
4. Michalski, R.S.(1983) A theory and methodology of inductive learning. In: Michalski R.S., Carbonell J.G. and Mitchell T.M. (Eds.) *Machine Learning: An Artificial Intelligence Approach*, Tioga, Palo Alto 1983, 83-133
5. Pawlak, Z. (1991) *Rough Sets. Theoretical Aspects of Reasoning about Data.* Kluwer Academic Publishers, London
6. Potempa T. (2002) Selected issues of data analysis and verification using Oracle. An M.Sc Thesis, AGH, Cracow, (in Polish)
7. The page of *Regulus* projects. http://spock.ia.agh.edu.pl/index/index.html.

Computer-Oriented Sequent Inferring without Preliminary Skolemization

Alexander Lyaletski

Faculty of Cybernetics, Kiev National Taras Shevchenko University,
2, Glushkov avenue, building 6, 03022 Kyiv, Ukraine

Abstract. In this paper[1] sequent formalism used in theorem-proving technique of a system of automated deduction, SAD[2], is described. The specific feature of SAD is that a mathematical text under consideration is formalized using a certain formal language, which is close to a "natural" one and can be translated into a certain first-order language in order to apply sequent-based methods having the following features: goal-oriented reducing an assertion to be proven to a number of auxiliary assertions, quantifier-handling technique admitting efficient logical inferring in a signature of an initial theory without skolemization, separating deduction from equation solving. One of these methods is expressed here in the form of a special sequent-type calculus. Some results about its soundness and completeness are given.

1 Introduction

A programme of automated theorem proving in mathematics called Evidence Algorithm, EA, was advanced by Academician V.Glushkov in his paper [1]. Its main objective is to help working mathematicians in mathematical text processing, i.e. in computer-aided constructing and verifying long, but, in some sense, "evident" proofs. In particular, V.Glushkov proposed to make investigation simultaneously into formalized languages for presenting mathematical texts in the form most appropriated for a user and formalization and evolutionary development of computer-maid proof step.

As the result of modern view on EA, the first version of a system of automated deduction called SAD [2] has been implemented by now. The system SAD corresponds well to contemporary trends of computer mathematical services construction and based on special sequent formalism.

All the existent computer-oriented calculi proposed for first-order logic can be divided into two parts: the calculi which rely upon results of Gentzen [3] and Kanger [4] (such as different goal-oriented sequent calculi, tableaux methods, etc.) and methods which rely upon results of Skolem [5] and Herbrand [6] (such as the resolution method [7], Maslov's inverse method [8], etc.). In this connection, we note that Gentzen-Kanger calculi significantly yield proof search efficiency to resolution-type methods. As a result, studies

---

[1] Supported by the INTAS 2000-447.
[2] See Web-cite "ea.unicyb.kiev.ua".

on automated theorem proving were later concentrated mainly on improving the potentialities of the Skolem-Herbrand approach.

The lower efficiency of Gentzen's approach (as compared with the Skolem-Herbrand approach) can be explained by the fact that even Kanger-type calculi, which reject obligatory preliminary skolemization, result in arising superfluous enumeration caused by the possibility of different orders of logical rule applications and by the necessity of principal formula duplication when some of those rules are applied. At the same time, machine-oriented Gentzen-Kanger type methods reflect proof techniques that are more "natural" for a human and, therefore, and are more appropriate for EA. That is why deduction in the EA-style has been developed in the direction of improving sequent calculi since appearance of the EA programme.

The first attempt of construction of EA-style technique was made in investigations of formal application of some proof search methods used in mathematical papers. As a result, a certain approach to inference search was developed for Group Theory and Set Theory. It found its final completion as a sequent calculus for auxiliary goal search [9], which was intended for ascertainment of the deducibility of 1st-order classical logic formulas. (This calculus has close relation to Kanger's calculus from [4] and possesses all its deficiencies.)

Attempts to overcome these deficiencies gave rise to an original notion of admissible substitutions investigated firstly in [10] from the point of view of obtaining new variants of Herbrand's theorem. Later, this notion has been incorporated in usual Gentzen calculi [11] in order to show how additional efforts connected with the possibilities of different orders of quantifier rule applications can be optimised by means of use of a special form formalism.

The first representative of use of this formalism for EA was a sequent calculus proposed in [12]. Subsequent attempts to improve the EA-style approach to sequent inferring in the EA-style gave arise to a family of a-sequent calculi (see, for example, [13,14]). That is why a deductive technique of the modern version of SAD is based on a subsequent modification of the sequent formalism, and one of its possible modification is described below as a calculus gD_2. The soundness and completeness of gD_2 are proven.

2   Preliminaries

We consider the sequent form of first-order classical logic with the universal and existential quantifiers and with the propositional connectives of implication ($\supset$), disjunction ($\vee$), conjunction ($\wedge$), and negation ($\neg$).

Sequents have the form $\Delta \to \Gamma$, where Δ and Γ are sequences of formulas. Formulas from Γ are said to be *premises*, and formulas from Δ are said to be *goals*. The calculus G from [15] is denoted by Gal here.

We understand *positive* $(P\lfloor F^+\rfloor)$ *and negative* $(P\lfloor F^-\rfloor)$ *occurrences* of a formula F in a formula P in the usual sense. If S is a sequent of the form $F_1, \ldots, F_m \rightarrow G_1, \ldots, G_n$, and F is a formula, then $S \lfloor F^+ \rfloor$ $(S \lfloor F^- \rfloor)$ denotes the same as $\phi(S) \lfloor F^+ \rfloor$ $(\phi(S) \lfloor F^- \rfloor)$, where $F_1, \ldots, F_m$, $G_1, \ldots, G_n$ are formulas, and $\phi(S)$ is $(F_1 \wedge \ldots \wedge F_m) \supset (G_1 \vee \ldots \vee G_n)$. Therefore, we can talk about a *positive (negative) occurrence of F in S*.

The expression $F^\neg$ denotes the result of one-step carrying of the negation into a formula F. If x is a variable, t is a term, and F is a formula, then $F|_t^x$ denotes the result of simultaneously replacing x by t.

Unknown, fixed, and indexed variables. In what follows, W denotes a set of formulas or of sequents.

We assume any formula F, sequent S, or set W does not contain the same quantifier more then one time. The notions of bound variables and scopes of quantifiers are assumed to be known to a reader.

For a formula G (for a sequent S), a variable v is said to be *unknown w.r.t.* G *(w.r.t. S)* if and only if there exists a formula F such that $G \lfloor (\forall v F)^- \rfloor$ or $G \lfloor (\exists v F)^+ \rfloor$ $(S \lfloor (\forall v F)^- \rfloor$ or $S \lfloor (\exists v F)^+ \rfloor)$ holds. Accordingly, v is said to be *fixed w.r.t.* G *(w.r.t. S)* if and only if there exists F such that $G \lfloor (\forall v F)^+ \rfloor$ or $G \lfloor (\exists v F)^- \rfloor$ $(S \lfloor (\forall v F)^+ \rfloor$ or $S \lfloor (\exists v F)^- \rfloor)$ holds. Obviously, any bound variable in G (in S) is either an unknown or fixed variable (either a "dummy" or a "parameter" in the terminology of [4]). It is clear what the phrase "a variable is unknown (fixed) w.r.t. W" means for a set W of formulas or of sequents.

For every variable v we introduce a countable set of new variables of the form ${}^k v$ $(k = 0, 1, 2, \ldots)$ called *indexed*, where k is an index.

If F is a formula containing, possibly, indexed variables, then for any index k the *expression ${}^k F$* denotes the result obtained from F by (i) deleting all indexes from bound variables and (ii) subsequently replacing every bound variable v by ${}^k v$.

Admissible substitutions. We treat the notion of a substitution as in [7]. Any substitution component is considered to be of the form t/x, where x is a variable (denominator), and t is a term (numerator) of a substitution. A reader is assumed to be familiar with the notion of a (simultaneous) unifier of sets of expressions.

An *equation* is a pair of terms s and t written as $s \approx t$. Let us assume that L is a literal of the form $R(t_1, \ldots, t_n)$ $(\neg R(t_1, \ldots, t_n))$ and M is a literal of the form $R(s_1, \ldots, s_n)$ $(\neg R(s_1, \ldots, s_n))$, where R is a predicate symbol. Then $\Sigma(L, M)$ denotes the set $\{t_1 \approx s_1, \ldots, t_n \approx s_n\}$. In this case, L and M are said to be *equal modulo $\Sigma(L, M)$* $(L \approx M$ *modulo $\Sigma(L, M)$*).

Let W be a set of formulas (sequents), σ be a substitution, and $t/u \in \sigma$, where u is an unknown variable w.r.t. W. Let the term t contain a variable v fixed w.r.t. W. In this case, we write $v \ll_W^\sigma u$, defining a relation $\ll_W^\sigma$. Obviously, the relation $\ll_W^\sigma$ is antisymmetric.

A formula F induces an *antisymmetric relation* $\preceq_F$ on a set of bound variables from F by the following: $u \preceq_F w$ holds if and only if a quantifier from F containing w occurs in the scope of a quantifier from F containing u.

Let $\preceq_V$ be an antisymmetric relation on a set of variables V, and W be a set of formulas or sequents. A substitution σ is said to be *admissible* for W w.r.t. $\preceq_V$ if and only if (i) denominators of σ are unknown variables w.r.t. W, and (ii) the transitive closure of $\preceq_V \cup \ll^\sigma_W$ is an antisymmetric relation.

In what follows, we only consider *sequents containing exactly one goal in their succedents*.

3   Special Calculus of Sequents

This section is devoted to a Gentzen-type calculus gD_1 containing sequents of gD_1 of the form $\Gamma \to G, [\Lambda]$, where Γ is a sequence of formulas, G is a formula, and Λ is a sequence of literals.

We assume that all variables in any sequent of gD_1 are indexed. Also, note that rules of gD_1 can be applied in any *arbitrary order*. (The notations L^+, M^+, and M^- play an auxiliary role and use only for underlining close relation of gD_1 to a calculus from the next section.)

Axioms. Axioms of gD_1 have the following form: $\Gamma \to \#, [\Lambda]$

Goal-Splitting Rules. All rules are applied "from top to bottom".

$(\to \supset_1)_{gD_1}$-rule:

$$\frac{\Gamma \to G \supset G_1, [\Lambda]}{\Gamma \to G_1, [\Lambda]}$$

$(\to \supset_2)_{gD_1}$-rule:

$$\frac{\Gamma \to G \supset G_1, [\Lambda]}{\Gamma \to \neg G, [\Lambda]}$$

$(\to \vee_1)_{gD_1}$-rule:

$$\frac{\Gamma \to G \vee G_1, [\Lambda]}{\Gamma \to G, [\Lambda]}$$

$(\to \vee_2)_{gD_1}$-rule:

$$\frac{\Gamma \to G \vee G_1, [\Lambda]}{\Gamma \to G_1, [\Lambda]}$$

$(\to \wedge)_{gD_1}$-rule:

$$\frac{\Gamma \to G \wedge G_1, [\Lambda]}{\Gamma \to G, [\Lambda] \quad \Gamma \to G_1, [\Lambda]}$$

$(\to \neg)_{gD_1}$-rule:

$$\frac{\Gamma \to \neg G, [\Lambda]}{\Gamma \to G^\neg, [\Lambda]}$$

$(\to \forall)_{gD_1}$-rule:

$$\frac{\Gamma \to \forall^k x G, [\Lambda]}{\Gamma \to G, [\Lambda]}$$

$$\frac{\Gamma \to \exists^k x G, [\Lambda]}{\Gamma, \neg^l(\exists^k x G) \to G|_t^{k x}, [\Lambda]}$$

where l is a new index and the term t is required to be free for the variable $^k x$ in the formula G. Moreover, t contains only functional symbols from sequents.

Premise-Duplicating Rule. This rule is applied only if F is a variant of a formula from an initial sequent:

$$\frac{\Gamma_1, F\lfloor M^+ \rfloor, \Gamma_2 \to G\lfloor L^+ \rfloor, [\Lambda]}{\Gamma_1, {}^l F, F\lfloor M^+ \rfloor, \Gamma_2 \to G\lfloor L^+ \rfloor, [\Lambda]}$$

where G is a formula and l is a new index.

Auxiliary-Goals Rules. All rules are applied from "top to bottom".

$(\supset_1 \to)_{gD_1}$-rule:

$$\frac{\Gamma_1, F \supset F_1 \lfloor M^+ \rfloor, \Gamma_2 \to G \lfloor L^+ \rfloor, [\Lambda]}{\Gamma_1, F_1 \lfloor M^+ \rfloor, \Gamma_2 \to G \lfloor L^+ \rfloor, [\Lambda] \quad \Gamma_1, \Gamma_2 \to F, [L, \Lambda]}$$

$(\supset_2 \to)_{gD_1}$-rule:

$$\frac{\Gamma_1, F \lfloor M^- \rfloor \supset F_1, \Gamma_2 \to G \lfloor L^+ \rfloor, [\Lambda]}{\Gamma_1, (\neg F) \lfloor M^+ \rfloor, \Gamma_2 \to G \lfloor L^+ \rfloor, [\Lambda] \quad \Gamma_1, \Gamma_2 \to \neg F_1, [L, \Lambda]}$$

$(\vee_1 \to)_{gD_1}$-rule:

$$\frac{\Gamma_1, F \lfloor M^+ \rfloor \vee F_1, \Gamma_2 \to G \lfloor L^+ \rfloor, [\Lambda]}{\Gamma_1, F \lfloor M^+ \rfloor, \Gamma_2 \to G \lfloor L^+ \rfloor, [\Lambda] \quad \Gamma_1, \to \neg F_1, [L, \Lambda]}$$

$(\vee_2 \to)_{gD_1}$-rule:

$$\frac{\Gamma_1, F \vee F_1 \lfloor M^+ \rfloor, \Gamma_2 \to G \lfloor L^+ \rfloor, [\Lambda]}{\Gamma_1, F_1 \lfloor M^+ \rfloor, \Gamma_2 \to G \lfloor L^+ \rfloor, [\Lambda] \quad \Gamma_1, \Gamma_2 \to \neg F, [L, \Lambda]}$$

$(\wedge_1 \to)_{gD_1}$-rule:

$$\frac{\Gamma_1, F \lfloor M^+ \rfloor \wedge F_1, \Gamma_2 \to G \lfloor L^+ \rfloor, [\Lambda]}{\Gamma_1, F \lfloor M^+ \rfloor, \Gamma_2 \to G \lfloor L^+ \rfloor, [\Lambda]}$$

$(\wedge_2 \to)_{gD_1}$-rule:

$$\frac{\Gamma_1, F \wedge F_1 \lfloor M^+ \rfloor, \Gamma_2 \to G \lfloor L^+ \rfloor, [\Lambda]}{\Gamma_1, F_1 \lfloor M^+ \rfloor, \Gamma_2 \to G \lfloor L^+ \rfloor, [\Lambda]}$$

$(\neg \to)_{gD_1}$-rule:

$$\frac{\Gamma_1, \neg(F \lfloor M^- \rfloor), \Gamma_2 \to G \lfloor L^+ \rfloor, [\Lambda]}{\Gamma_1, F^\neg \lfloor M^+ \rfloor, \Gamma_2 \to G \lfloor L^+ \rfloor, [\Lambda]}$$

$(\forall \to)_{gD_1}$-rule:

$$\frac{\Gamma_1, \forall^k x F \lfloor M^+ \rfloor, \Gamma_2 \to G \lfloor L^+ \rfloor, [\Lambda]}{\Gamma_1, {}^l(\forall^k x F), (F \lfloor M^+ \rfloor)|_t^{k x}, \Gamma_2 \to G \lfloor L^+ \rfloor, [\Lambda]}$$

$(\exists \to)_{gD_1}$-rule:

$$\frac{\Gamma_1, \exists^k x F \lfloor M^+ \rfloor, \Gamma_2 \to G \lfloor L^+ \rfloor, [\Lambda]}{\Gamma_1, F \lfloor M^+ \rfloor, \Gamma_2 \to G \lfloor L^+ \rfloor, [\Lambda]}$$

where l is a new parameter and the term t is required to be free for the variable $^k x$ in the formula F.

Termination Rules. The calculus gD_1 contains two termination riles.

$(\rightarrow \#_1)_{gD_1}$-rule: $(\rightarrow \#_2)_{gD_1}$-rule:

$$\frac{\Gamma_1, L, \Gamma_2 \rightarrow L, [\Lambda]}{\Gamma_1, L, \Gamma_2 \rightarrow \#, [\Lambda]} \qquad\qquad \frac{\Gamma \rightarrow L, [\Lambda_1, \widetilde{L}, \Lambda_2]}{\Gamma \rightarrow \#, [\Lambda_1, \widetilde{L}, \Lambda_2]}$$

Let S be a usual sequent $F_1, \ldots, F_n \rightarrow G$, where $F_1, \ldots, F_n$, and G are formulas. Then the sequent S^{gD_1} of the form $^1F_1, \ldots, {}^nF_n, \neg^{(n+1)}G \rightarrow {}^0G, [\,]$ is called an *initial sequent w.r.t. S* for gD_1.

When searching a proof of an initial a-sequent S^{gD_1}, an *inference tree* Tr w.r.t. S^{gD_1} is constructed. At the beginning of searching, Tr contains only S^{gD_1}. The subsequent nodes are generated by means of rules of gD_2. Inference trees grow "from top to bottom" in accordance with the order of inference rule applications.

An inference tree Tr w.r.t. S^{gD_1} is considered to be a *proof tree w.r.t.* S^{gD_1} if and only if every leaf of Tr is an axiom of gD_1.

Proposition 1. *Let formulas* $F_1, \ldots, F_n$ *form a consistent set of formulas,* G *be a formula, and* S *be the sequent* $F_1, \ldots, F_n \rightarrow G$. *The sequent* S *is inferred in the calculus Gal if and only if there exists a proof tree w.r.t.* S^{gD_1} *in the calculus* gD_1.

Proof Draft. Since gD_1 contains the usual quantifier rules, it is sufficient to prove Prop. 1 at the propositional level (see [16]). Q.E.D.

4   Goal-Driven Calculus of A-Sequents

A-sequents have the form $\{\Pi\}, \Gamma \rightarrow G, [\Lambda], \langle E \rangle$, where Π is an antisymmetric relation determined on a set of variables, Γ is a sequence of formulas, G is a formula, Λ is a sequence of literals, and E is a set of equations.

A calculus of a-sequents, gD_2, is defined by the following way.

Axioms. Axioms have the following form: $\{\Pi\}, \Gamma \rightarrow \#, [\Lambda], \langle E \rangle$

Goal-Splitting Rules. All rules are applied from "top to bottom".

$(\rightarrow \supset_1)_{gD_2}$-rule: $(\rightarrow \supset_2)_{gD_2}$-rule:

$$\frac{\{\Pi\}, \Gamma \rightarrow G \supset G_1, [\Lambda], \langle E \rangle}{\{\Pi\}, \Gamma \rightarrow G_1, [\Lambda], \langle E \rangle} \qquad\qquad \frac{\{\Pi\}, \Gamma \rightarrow G \supset G_1, [\Lambda], \langle E \rangle}{\{\Pi\}, \Gamma \rightarrow (\neg G), [\Lambda], \langle E \rangle}$$

$(\rightarrow \vee_1)_{gD_1}$-rule: $(\rightarrow \vee_2)_{gD_1}$-rule:

$$\frac{\{\Pi\}, \Gamma \rightarrow G \vee G_1, [\Lambda], \langle E \rangle}{\{\Pi\}, \Gamma \rightarrow G, [\Lambda], \langle E \rangle} \qquad\qquad \frac{\{\Pi\}, \Gamma \rightarrow G \vee G_1, [\Lambda], \langle E \rangle}{\{\Pi\}, \Gamma \rightarrow G_1, [\Lambda], \langle E \rangle}$$

$(\rightarrow \wedge)_{gD_2}$-rule: $(\rightarrow \neg)_{gD_2}$-rule:

$$\frac{\{\Pi\}, \Gamma \rightarrow G \wedge G_1, [\Lambda], \langle E \rangle}{\{\Pi\}, \Gamma \rightarrow G, [\Lambda], \langle E \rangle \quad \{\Pi\}, \Gamma \rightarrow G_1, [\Lambda], \langle E \rangle} \qquad \frac{\{\Pi\}, \Gamma \rightarrow (\neg G), [\Lambda], \langle E \rangle}{\{\Pi\}, \Gamma \rightarrow G^\neg, [\Lambda], \langle E \rangle}$$

$(\rightarrow \forall)_{gD_2}$-rule: $\qquad\qquad\qquad\qquad$ $(\rightarrow \exists)_{gD_2}$-rule:

$$\frac{\{\varPi\}, \varGamma \rightarrow \forall^k xG, [\varLambda], \langle E \rangle}{\{\varPi\}, \varGamma \rightarrow G, [\varLambda], \langle E \rangle} \qquad\qquad \frac{\{\varPi\}, \varGamma \rightarrow \exists^k xG, [\varLambda], \langle E \rangle}{\{\varPi'\}, \varGamma, \neg G' \rightarrow G, [\varLambda], \langle E \rangle}$$

where G' is $^l(\exists^k xG)$ for a new index l and $\varPi'$ is $\varPi \cup \{< {}^k y, {}^l z >: < {}^k y, {}^k z > \in \varPi$ & "it is not true that $(< {}^k z, {}^k x > \in \varPi$ & $< {}^k x, {}^k y > \in \varPi)$"$\} \cup \preceq_{G'}$.

Premise-Duplicating Rule. This rule is applied only if F is a variant of a formula from an initial a-sequent:

$$\frac{\{\varPi\}, \varGamma_1, F\lfloor M^+ \rfloor, \varGamma_2 \rightarrow L, [\varLambda], \langle E \rangle}{\{\varPi\}, \varGamma_1, {}^l F, F\lfloor M^+ \rfloor, \varGamma_2 \rightarrow L, [\varLambda], \langle E \rangle}$$

where L is a literal and l is a new index.

Auxiliary-Goals Rules. All rules are applied "from top to bottom".

$(\supset_1 \rightarrow)_{gD_2}$-rule:

$$\frac{\{\varPi\}, \varGamma_1, F \supset F_1 \lfloor M^+ \rfloor, \varGamma_2 \rightarrow L, [\varLambda], \langle E \rangle}{\{\varPi\}, \varGamma_1, F_1 \lfloor M^+ \rfloor, \varGamma_2 \rightarrow L, [\varLambda], \langle E \rangle \quad \{\varPi\}, \varGamma_1, \varGamma_2 \rightarrow F, [L, \varLambda], \langle E \rangle}$$

$(\supset_2 \rightarrow)_{gD_2}$-rule:

$$\frac{\{\varPi\}, \varGamma_1, F \lfloor M^- \rfloor \supset F_1, \varGamma_2 \rightarrow L, [\varLambda], \langle E \rangle}{\{\varPi\}, \varGamma_1, (\neg F) \lfloor M^+ \rfloor, \varGamma_2 \rightarrow L, [\varLambda], \langle E \rangle \quad \{\varPi\}, \varGamma_1, \varGamma_2 \rightarrow \neg F_1, [L, \varLambda], \langle E \rangle}$$

$(\vee_1 \rightarrow)_{gD_2}$-rule:

$$\frac{\{\varPi\}, \varGamma_1, F \lfloor M^+ \rfloor \vee F_1, \varGamma_2 \rightarrow L, [\varLambda], \langle E \rangle}{\{\varPi\}, \varGamma_1, F \lfloor M^+ \rfloor \varGamma_2 \rightarrow L, [\varLambda], \langle E \rangle \quad \{\varPi\}, \varGamma_1, \varGamma_2 \rightarrow \neg F_1, [L, \varLambda], \langle E \rangle}$$

$(\vee_2 \rightarrow)_{gD_2}$-rule:

$$\frac{\{\varPi\}, \varGamma_1, F \vee F_1 \lfloor M^+ \rfloor, \varGamma_2 \rightarrow L, [\varLambda], \langle E \rangle}{\{\varPi\}, \varGamma_1, F_1 \lfloor M^+ \rfloor, \varGamma_2 \rightarrow L, [\varLambda], \langle E \rangle \quad \{\varPi\}, \varGamma_1, \varGamma_2 \rightarrow \neg F, [L, \varLambda], \langle E \rangle}$$

$(\wedge_1 \rightarrow)_{gD_2}$-rule:

$$\frac{\{\varPi\}, \varGamma_1, F \lfloor M^+ \rfloor \wedge F_1, \varGamma_2 \rightarrow L, [\varLambda], \langle E \rangle}{\{\varPi\}, \varGamma_1, F \lfloor M^+ \rfloor, \varGamma_2 \rightarrow L, [\varLambda], \langle E \rangle}$$

$(\wedge_2 \rightarrow)_{gD_2}$-rule:

$$\frac{\{\varPi\}, \varGamma_1, F \wedge F_1 \lfloor M^+ \rfloor, \varGamma_2 \rightarrow L, [\varLambda], \langle E \rangle}{\{\varPi\}, \varGamma_1, F_1 \lfloor M^+ \rfloor, \varGamma_2 \rightarrow L, [\varLambda], \langle E \rangle}$$

$(\neg \rightarrow)_{gD_2}$-rule:

$$\frac{\{\varPi\}, \varGamma_1, \neg(F \lfloor M^- \rfloor), \varGamma_2 \rightarrow L, [\varLambda], \langle E \rangle}{\{\varPi\}, \varGamma_1, F^{\neg} \lfloor M^+ \rfloor, \varGamma_2 \rightarrow L, [\varLambda], \langle E \rangle}$$

$(\exists \rightarrow)_{gD_2}$-rule:

$$\frac{\{\Pi\}, \Gamma_1, \exists^k x F \lfloor M^+ \rfloor, \Gamma_2 \rightarrow L, [\Lambda], \langle E \rangle}{\{\Pi\}, \Gamma_1, F \lfloor M^+ \rfloor \Gamma_2 \rightarrow L, [\Lambda], \langle E \rangle}$$

$(\forall \rightarrow)_{gD_2}$-rule:

$$\frac{\{\Pi\}, \Gamma_1, \forall^k x F \lfloor M^+ \rfloor, \Gamma_2 \rightarrow L, [\Lambda], \langle E \rangle}{\{\Pi'\}, \Gamma_1, F', F \lfloor M^+ \rfloor, \Gamma_2 \rightarrow L, [\Lambda], \langle E \rangle}$$

where $L \approx M$ *modulo* $\Sigma(L, M)$, F' is ${}^l(\forall^k x F)$ for a new index l, and Π' is $\Pi \cup \{< {}^k y, {}^l z >: < {}^k y, {}^k z > \in \Pi$ & "it is not true that $(< {}^k z, {}^k x > \in \Pi$ & $< {}^k x, {}^k y > \in \Pi)"\} \cup \preceq_{F'}$.

Premise-Adding Rule. After every application of the $(\forall \rightarrow)_{gD_2}$-rule, a new premise $F \lfloor M^+ \rfloor$ can be added to the antecedent of any a-sequent containing a formula, which includes the marked occurrence of M^+.

Termination Rules. The calculus gD_2 contains two termination rules.

$(\rightarrow \#_1)_{gD_2}$-rule:

$$\frac{\{\Pi\}, \Gamma_1, M, \Gamma_2 \rightarrow L, [\Lambda], \langle E \rangle}{\{\Pi\}, \Gamma_1, M, \Gamma_2 \rightarrow, \#, [\Lambda], \langle E \cup \Sigma(L, M) \rangle}$$

where L and M are literals and $L \approx M$ (*modulo* $\Sigma(L, M)$).

$(\rightarrow \#_2)_{gD_2}$-rule:

$$\frac{\{\Pi\}, \Gamma \rightarrow L, [\Lambda_1, L', \Lambda_2], \langle E \rangle}{\{\Pi\}, \Gamma \rightarrow, \#, [\Lambda_1, L', \Lambda_2], \langle E \cup \Sigma(\widetilde{L}, L') \rangle}$$

where L and L' are literals and $\widetilde{L} \approx L'$ (*modulo* $\Sigma(\widetilde{L}, L')$).

Main Results. Let S be a usual sequent $F_1, \ldots, F_n \rightarrow G$, where $F_1, \ldots, F_n$, and G are formulas. Then an a-sequent S^{gD_2} of the form $\{\Pi_0\}, {}^1 F_1, \ldots, {}^n F_n, \neg^{(n+1)} G \rightarrow {}^0 G, [\,], \langle \ \rangle$ is called an *initial a-sequent w.r.t.* S for gD_2, where Π_0 is $\preceq_{1_{F_1}} \cup \ldots \cup \preceq_{n_{F_n}} \cup \preceq_{(n+1)G} \cup \preceq_{0G}$.

When searching a proof of an initial a-sequent S^{gD_2}, an *inference tree Tr* w.r.t. S^{gD_2} is constructed. At the beginning of searching, Tr contains only S^{gD_2}. Subsequent nodes are generated by means of the rules of gD_2.

Note that any possible application of goal-splitting rules precedes to any application of auxiliary-goals rules always.

Let Tr be a tree of a-sequents, W be a set of all the a-sequents from Tr, and $\preceq\preceq_{Tr}$ be a union of all the antisymmetric relations of Tr. A substitution σ is called *admissible for Tr* if and only if σ is admissible for W w.r.t. $\preceq\preceq_{Tr}$.

An inference tree Tr w.r.t. S^{gD_2} is considered to be a *proof tree w.r.t.* S^{gD_2} if and only if the following conditions are satisfied: (i) every leaf of Tr is an axiom and (ii) if E is a union of sets of equations from of all the leaves of Tr, then there exists a substitution σ such that σ is a simultaneous unifier of all the equations from E and σ is an admissible substitution for Tr.

Proposition 2. *Let $F_1, \ldots, F_n$ form a consistent set of formulas, G be a formula, and S be the sequent $F_1, \ldots, F_n \to G$. A proof tree w.r.t. S^{gD_1} exists in gD_1 if and only if there exists a proof tree w.r.t. S^{gD_2} in gD_2.*

Proof Draft. *Sufficiency.* Let Tr be a proof tree w.r.t S^{gD_2} in gD_2, s be a unifier of all the equations of Tr, and s be admissible for Tr. Obviously, we can assume that the terms of s do not contain unknown variables and that for every unknown variable x from Tr there exists a term t such that $t/x \in s$.

Let $\prec\prec_{Tr}$ be a union of all the antisymmetric relations of Tr. Since s is admissible for Tr, it is possible to give a linear order $\trianglelefteq_V$ on a set V of all the variables from Tr such that the transitive closure of $\prec\prec_{Tr} \cup \ll_W^\sigma$ will be included in $\trianglelefteq_V$. Moreover, s will be admissible for W w.r.t. $\trianglelefteq_V$.

The tree Tr can be transformed into a proof tree Tr' w.r.t. S^{gD_2} (in gD_2) satisfying the following conditions: All rules in Tr' are applied in such a way that the first quantifier rule application is determined by the first variable of w when looking through w from left to right, the second quantifier rule application is determined by the second variable of w, and so on.

Using the admissibility of s for W w.r.t. $\trianglelefteq_V$, it is easy to convert the tree Tr' into a proof tree Tr_1 w.r.t. S^{gD_1} in the calculus gD_1. To do this, it is necessary to repeat process of the construction of Tr' in the order defined by $\trianglelefteq_V$ and to replace of rules of gD_2 by corresponding rules of gD_1.

Necessity. Let Tr_1 be a proof tree w.r.t. S^{gD_1} in gD_1. Repeating process of the construction of Tr_1, replacing in Tr_1 every rule application by its analog in gD_2, and subsequently generating substitution components of a substitution s in accordance with quantifier rules of gD_1, we can transform Tr_1 into such a tree Tr w.r.t. S^{gD_2} in gD_2 that Tr contains leafs labeled only axioms of S^{gD_2}, s unifies all the equations from Tr, and s is admissible for Tr. (The last condition follows from the fact that s utilizes Gentzen's notion of admissibility.) This means that Tr is a proof tree in gD_2. Q.E.D.

As corollaries of Prop. 1 and Prop. 2, we have the following assertions.

Proposition 3. *Let formulas $F_1, \ldots, F_n$ form a consistent set of formulas, G be a formula, and S be the sequent $F_1, \ldots, F_n \to G$. The sequent S is inferred in Gal if and only if there exists a proof tree w.r.t. S^{gD_2} in gD_2.*

Proposition 4. *A formula G is valid if and only if there exists a proof tree w.r.t. the initial a-sequent $\{\preceq_{0G} \cup \preceq_{1G}\}, \neg^1 G \to {}^0 G, [\ ] \langle\ \rangle$ in gD_2.*

Proposition 5. *The calculus gS from [13] is sound and complete.*

5 Conclusion

The sequent formalism described above may be useful if necessity appears to construct such an inference search technique that admits: search for deduction in the signature of an initial theory (i.e. without skolemization), reduction of an assertion to be proven to a number of new auxiliary assertions, separation of finding a solution for "equations" from the deductive process, application of such "natural" methods as definitions and auxiliary-propositions applications, use of specific equality handling rules and various tools of computer algebra systems, and construction of a flexible interactive search mode.

In particular, this kind of a-sequent formalism was exploited when implementing the first version of SAD.

References

1. Glushkov V. M. (1970): Some problems of automata theory and artificial intelligence (in Russian). Kibernetika **2**, 3-13.
2. Verchinine K., Degtyarev A., Lyaletski A., Paskevich A. (2002): A system for automated deduction, SAD: a current state. Proceedings of the Intern. Workshop Automath'2002, Edinburgh, Great Britain.
3. Gentzen G. (1934): Untersuchungen uber das Logische Schliessen. Math. Zeit. **39**, 176-210.
4. Kanger S. (1963): Simplified proof method for elementary logic. Comp. Program. and Form. Sys.: Stud. in Logic. North-Holl., Publ. Co.
5. Skolem T. (1920): Logisch-kombinatorische untersuchungen uber die erfullbarkeit oder beweisbarkeit mathematischer satze. Skriftner utgit ar Videnskapsselskaper i Kristiania **4**, 4-36.
6. Herbrand J. (1930): Recherches sur la theorie de la demonstration. Travaux de la Societe des Sciences et de Lettres de Varsovie,Class III, Sciences Mathematiques et Physiques **33**.
7. Robinson J. (1965): A machine-oriented logic based on resolution principle. J. of the ACM **12**, No 1, 23-41.
8. Maslov S. Y. (1964): Inverse method for ascertainment of deducibility in classical predicate calculus. Doklady Akademiyi Nauk SSSR **159**.
9. Anufriyev F. V. (1969): An algorithm of theorem proof search in logical calculi (in Russian). Teoriya avtomatov, Institute of Cybernetics, Kiev, **1**.
10. Lyaletski A. V. (1981): A variant of Herbrand's theorem for formulas in the prefix form (in Russian). Kibernetika **1**, 112-116.
11. Lyaletski A. V. (1991): Gentzen calculi and admissible substitutions. Actes Preliminaieres, du Symposium Franco-Sovietique "Informatika-91", Grenoble, France, 99-111.
12. Degtyarev A. I., Lyaletski A. V. (1981): Logical inference in SAD (in Russian). Matematicheskie Osnovy Sistem Iskusstvennogo Intellekta, Institute of Cybernetics, Kiev, 3-11.
13. Degtyarev A., Lyaletski A., Morokhovets M. (1999): Evidence Algorithm and sequent logical inference search. LNAI **1705**, 44-61.
14. Degtyarev A., Lyaletski A., Morokhovets M. (2001): On the EA-style integrated processing of self-contained mathematical texts. Symbolic Computation and Automated Reasoning (the book devoted to the CALCULEMUS-2000 Symposium: edited by M. Kerber and M. Kohlhase), A K Peters, USA, 126-141.
15. Gallier J. (1986): Logic for computer science: foundations of automatic theorem proving. Harper and Row, Inc., New York, 513 pp.
16. Lyaletski A., Paskevich A. (2001): Goal-driven inference search in classical propositional logic. Proceedings of the Inter. Workshop STRATEGIES'2001. Siena, Italy, 65-74.

Converting Association Rules into Natural Language – an Attempt[*]

Petr Strossa[1] and Jan Rauch[2]

[1] kizips@vse.cz,
[2] rauch@vse.cz,
   Department of Information and Knowledge Engineering,
   Faculty of Informatics and Statistics, University of Economics,
   W. Churchill Sq. 4, 130 67 Prague, Czech Republic

Abstract. An attempt to convert association rules concerning a medical data set is described. It is shown that the association rules can be formulated in reasonable sentences of a natural language. A limited language model for formulating association rules in English and Czech is described. This model mainly consists of a set of formulation patterns, which only slightly depend on the subject domain, and tables of expressions (verb phrases, noun phrases etc.). The morphological problems and their solutions suitable for the limited subject domain, both for English and Czech, are described.

1   Introduction

This paper presents an attempt to convert association rules (**ARs**) into a natural language (**NL**). The goal is to improve a presentation level of results of a large data mining tasks to users – non specialists in data mining. We deal with a data set concerning the STULONG project – a longitudinal study of the risk factors of the atherosclerosis in the population of 1 419 middle aged men. The supposed users of the data mining results are clinicians. Most of them are not familiar with data mining terminology and thus the formulation of results in natural language can remarkably help them in the first contact of the results.

We use the ARs to describe interesting patterns concerning analysed data in the first stage. There is a large number of attributes describing the observed men. It results in a large number of ARs. Ranking the whole set of resulting rules is not sufficient. The resulting ARs must be grouped in a natural way corresponding to the structure and properties of the attributes. A system of ARs satisfying this requirement is outlined in section 2.

The general aspects of the formulation of ARs in NL are discussed in section 3. Data structures and algorithm for converting ARs into NL are suggested in section 4. Various aspects of conversion both into English and

---

[*] The work described here has been supported by the European project **TARSKI** – **Cost Action 274** and by the project **ZA471011** of the Ministry of Education of the Czech Republic.

into Czech language are also discussed. Some concluding remarks are given in section 5.

2   The System of Association Rules for STULONG

There are two data matrices describing patients. The ENTRY data matrix contains results of observation of 219 attributes of entry examinations of each patient. The CONTROL data matrix contains results of observation of 66 attributes at 10 610 examinations made in the years 1976–1999. We concentrate on the ENTRY data matrix.

The 219 attributes of the ENTRY data matrix are divided into 10 groups – social characteristics, physical activity, smoking etc. (see http://euromise. vse.cz/stulong-en/a-otazky/vv/). We can ask various analytic questions concerning pairs of attribute groups. One question corresponds to one such pair. The response to a particular analytic question is the set of all true ARs concerning the corresponding pair of attribute groups. We are interested in particular groups of patients defined by the attribute[1] Studied group of patients with values Normal group, Risk group, etc.

We search for ARs of the form $\varphi \Rightarrow_{p,s} \psi$. Here φ is a Boolean attribute (**BA**) derived from atributes of one group (φ is called *antecedent*) and ψ is a BA derived from atributes of another group (ψ is called *succedent*).

The association rule $\varphi \Rightarrow_{p,s} \psi$ is true in the Normal group of patients if the condition $\frac{a}{a+b} \geq p \wedge a \geq s$ is true where a is the number of patients from the Normal group satisfying both φ and ψ and b is the number of patients from the Normal group satisfying φ but not satisfyingψ. In other words, at least $100p$ per cent of the patients from the Normal group satisfying φ also satisfy ψ, and there are at least s patients from the Normal group satisfying both φ and ψ. The symbol $\Rightarrow_{p,s}$ is called *founded implication 4ft-quantifier*.

We use the 4FT-MINER procedure that is a part of the LISP-MINER system (see http://lispminer.vse.cz/). Let us remark that the 4FT-MINER mines not only for simple ARs of the form $\varphi \Rightarrow_{p,s} \psi$ but also for ARs corresponding to statistical hypoteheses test, for equvivalency ARs and for conditional ARs.

We give an example of AR that is a part of analytical question *Activities* $\Rightarrow_{p,s}$ *Body*. This question is specified as a task for the 4FT-MINER procedure to find all ARs such that:

- the antecedent is derived from attributes (columns of data matrix ENTRY) Physical activity in a job, Physical activity after a job, Way how he gets to work and Duration of the way to work;

---

[1] We mostly use simple NL labels for the data columns and their categorial values throughout this paper, instead of arbitrary formal identifiers and numerical values coding the native categories.

- the succedent is generated from attributes Height, Weight and BMI (Body Mass Index);
- We use the 4ft-quantifier $\Rightarrow_{p,s}$ with parameters $p = 0.6$ and $s = 20$.

The resulting set of ARs contains 161 elements. Most of them concern the normal group of patients. The strongest AR concerning the normal group of patients is:

Physical activity after a job(great activity) $\land$ Physical activity in a job(he mainly sits) $\Rightarrow_{0.77,20}$ BMI(22;26)

It means that both (A) and (B) are satisfied:

(A) 77 per cent ($= 0.77 \times 100$) of patients satisfying conditions
   – 'great activity after job' and
   – 'mainly sits in the job'
   also satisfy the condition
   – 'Body Mass Index is in the interval (22;26)'.
(B) There are 20 patients satisfying all the mentioned conditions.

This AR can be expressed in a given NL, e.g., in the following way:

"20, i.e. 77 % of the patients with great activity after job that mainly sit in the job have BMI = 22 through 26."

The whole system of analytic question is presented at http://euromise.vse.cz/stulong-en/a-otazky/vv/.

3   Association Rules Formulated in Natural Language

3.1   Introductory Examples

Let us start now with an extremely simple form of an AR – an implication with a conjunction of two *single value* basic Boolean attributes (**SVBBAs** – i.e. basic BAs used with one-value subsets during the mining process) as antecedent and one SVBBA as succedent. Here is a trivial example of such an AR as a formal expression (using only NL labels of the data columns and their values):[2]

Physical activity after a job(great activity) $\land$ Physical activity in a job(he mainly sits) $\Rightarrow_{percentage,count}$ BMI(normal)

Here is a number of examples how such an AR can be formulated in (more or less) current, informal English:

---

[2] Just for the purpose of the following description we suppose a formally changed domain of the attribute BMI so that we could sensibly work with *single values* here. Subset and interval values are dealt with elsewhere [4].

- X patients confirm this dependence: if a patient has great activity after job and a sedentary job, then he has a normal value of BMI.
- A combination of great activity after job and a sedentary job implies a normal value of BMI. This fact is confirmed by X patients.
- X patients with great activity after job and a sedentary job have no problem with obesity.
- X patients that make sports intensively and mainly sit in their job have a normal value of BMI.
- X patients highly physically active after their job and mainly sitting in their job have a normal value of BMI.

In all these examples the symbol X represents an arbitrary partial formulation like 'N (i.e. P %) of T', 'P % (as a matter of fact, N from a total of T) of the', ...

3.2   A Generalization

What general phenomena can be seen in the NL formulation examples given above? First of all, every SVBBA – let us use the symbol B for it – can be expressed by some (usually not all!) of the following forms, which we shall call **single value NL expressions (SVNLEs):**

- **a verb phrase** B^V – e.g. 'makes sports intensively', 'mainly sits in his job', 'has reached university education', 'has finished a university';
- **a participial phrase** B^P – e.g. 'making sports intensively', 'mainly sitting in his job', 'having reached university education';
- **a noun phrase** B^N – e.g. 'great activity after job', 'a sedentary job', 'university education';
- **an adjectival phrase** – e.g. 'university-educated'; a preliminary study in this point has shown that it is useful to distinguish between two kinds of adjectival phrases both in Czech and in English – one, like 'university-educated', being used as a *left complementation* of a noun, the other, like 'born in 1920' or 'highly physically active after his job', being used as a *right complementation*; we shall use the symbolic representation B^L for the former and B^R for the latter.

There is a regular morphological relation between B^V and B^P both in Czech and in English. (In English it can be represented by a set of rules for creating the *-ing-forms* from finite verb forms. A similar, only a little larger set of rules can be given for the formation of participles in Czech.) The relations between B^V, B^N, B^L and B^R do not show any general regularity either in English or in Czech – although some partially useful (?) regularities could be found here, too: e.g., a plausible (but not necessarily 'the best') B^V is often constructed as 'has B^N'...

The use of all SVNLEs in the formulations of our ARs has the following **syntactic limitations:**

- A B^V has always the word 'patient' or 'patients' as its subject.
- A B^N may be used:
  - as the subject or direct object of the verb 'imply' (or the like), or as a complementation in the phrase 'combination of ...' (or the like);
  - as the object of an appropriate verb form generally corresponding to the data column (e.g. 'have' or 'have reached' for the column Reached education);
  - in a prepositional phrase complementing the word 'patient(s)' – in this case the *usual* preposition is 'with', but sometimes there may be another preposition more appropriate for a specific B^N, e.g. 'in' for B^N = 'a managerial position', and sometimes (like at the attribute Way how he gets to work) it seems that no such prepositional phrase is usable.
- A B^P, B^L or B^R always modifies the word 'patient(s)'.

In accordance with these syntactic functions *some words* in the SVNLEs are subject to some **morphological transformations.** Only one type of these transformations seems to be relevant both for English and for Czech: the **finite verb** in B^V must be in 3rd person **singular or plural form** according to the form of the subject ('patient' or 'patients').[3] The same may hold for some other words in B^V, as e.g. 'his' vs. 'their'. If there is such a word in B^V, then the same transformation is also relevant for the derived B^P.

There are two other situations requiring specific morphological transformations if we want to generate correct formulations in Czech:

- The syntactic head of a B^N (i.e. the most 'central' noun in it) together with all adjectives and/or participles modifying it directly must be in the appropriate **case** – depending on
  - the verb having the B^N as its object (usually verbs require B^N in accusative, but exceptions exist), or
  - the noun modified by the B^N (e.g. 'kombinace' ['combination'] – such situations usually require genitive case, but even here, at least theoretically, exceptions *might* exist), or
  - the preposition making a prepositional phrase modifying a noun (such a prepositional phrase is usually called *non-congruent attribute* in Czech grammars; different Czech prepositions require B^N in different cases, e.g. 's(e)' ['with'] requires instrumental case, 'v(e)' ['in'] requires locative case).

---

[3] The tense and other morphological characteristics of a B^V do not change throughout the formulations of ARs. In most B^Vs the finite verb is always in present tense, e.g. 'sit(s)'. In some English B^Vs it is constantly in present perfect tense ('have/has reached') or perhaps in simple past tense ('entered'). Both these situations correspond to a constant use of past tense in Czech.

- The syntactic head of a B^P, B^L or B^R (i.e. the participle or adjective in it) must be in the same **number and case** as the noun modified by it. (This is the reason why adjectival and participial phrases are called *congruent attributes* in Czech grammars. In fact this required *congruence* concerns not only number and case, but also *grammatical gender*. However, in our formulations the adjectival and participial phrases always modify the word 'pacient' ['patient'], so their gender is always *masculine animate*.)

4 Conversion of Association Rules into Natural Language

4.1 Formulation Patterns

For an AR in the general form $A_1 \wedge A_2 \Rightarrow_{percentage,count} S$ the following (and many more similar) *formulation patterns (FPs)* can be used (approximately equivalent constructions in English and Czech are given; small capital expressions in parentheses denote the necessary morphological transformations of the SVNLEs; the expression 'PP' in parentheses means that an appropriate prepositional phrase shall be formed from a B^N; 'PP2' means the same but without repeating the preposition if it is the same as at the previous B^N with a PP directive):

- E: X patients confirm this dependence: if a patient has A_1^N and A_2^N, then he has S^N.
  C: X pacientů potvrzuje tuto závislost: má-li pacient A_1^N(ACCUSATIVE) a A_2^N(ACCUSATIVE), pak má S^N(ACCUSATIVE).
- E: A combination of A_1^N and A_2^N implies S^N. This fact is confirmed by X patients.
  C: Kombinace A_1^N(GENITIVE) a A_2^N(GENITIVE) implikuje S^N(ACCUSATIVE). Tento fakt potvrzuje X pacientů.
- E: X patients A_1^N(PP) and A_2^N(PP2) S^V(PLURAL).
  C: X pacientů A_1^N(PP) a A_2^N(PP2) S^V.
- E: X patients that A_1^V(PLURAL) and A_2^V(PLURAL) S^V(PLURAL).
  C: X pacientů, kteří A_1^V(PLURAL) a A_2^V(PLURAL), S^V.
- E: X patients A_1^R(PLURAL) and A_2^P(PLURAL) [4] S^V(PLURAL).
  C: X pacientů A_1^R(GENITIVE PLURAL) a A_2^P(GENITIVE PLURAL) S^V.

4.2 The SVNLE Generator

Although it could seem that every SVNLE corresponding to a SVBBA can be put together from a part corresponding to the data column and another part

---

[4] This might look odd, since English adjectives and participles do not distinguish singular and plural forms, but the PLURAL directive here must arrange for making, e.g., the form 'mainly sitting in <u>their</u> job' from 'mailny sits in his job'!

corresponding to the specific value, the data explored so far have shown us that such a division has a practical sense only when constructing a verb (or participial) phrase – and even then this cannot be applied universally: e.g., for the data column Reached education with values basic school, apprentice school, secondary school, university and not stated, most of the cases can be expressed by a verb form 'has reached' and a noun phrase identical with the specific value, but the last value requires a different construction of the verb phrase, e.g. 'has not stated the reached education'. As for the noun and adjectival phrases, it turns out as appropriate that these SVNLEs be associated immediately with the specific pairs ⟨data column, value⟩.

Hence it follows that the SVNLE generator should be provided with two tables:

- **T1,** assigning a **basic verb phrase V0** (possibly empty) to every **data column.** V0 can be either a single verb form or a complex phrase, e.g. 'gets to work'. A non-empty V0 should be a verb phrase that can be used – together with various complementations – to express various (though not necessarily all) values of the data column.
- **T2,** assigning a following set of parameters (some of them possibly empty) to every pair ⟨**data column, value**⟩:[5]
    - **AL** – an adjectival phrase usable as left complementation of the word 'patient' (e.g. 'university-educated');
    - **AR** – an adjectival phrase usable as right complementation of the word 'patient' (e.g. 'born in Y' – where Y stands for the actual value in the Year of birth column);
    - **C0** – a phrase usable as a complementation of V0 (e.g. 'on foot' or 'by public means of transport');
    - **N0** – a noun phrase usable in constructions without any verb or (at least sometimes) with such universal verbs as 'have' (e.g. 'basic education');
    - **NAF** (this is an abbreviation of *Non-congruent Attribute Formation,* a term from Czech grammar, but it could be read *Prepositional Attribute Formation* for English) – a code denoting how a (postpositive) complementation of the word 'patient' can be made from N0; for English, this will regularly be just a preposition (mostly 'with'); for Czech, a preposition plus a case directive must usually be given (e.g. 've+LOCATIVE'), but theoretically a case directive alone might sometimes be sufficient;
    - **VP** – a specific verb phrase (this should be given especially when no V0 is available or when no C0 can be given to complement the V0; a good example can be the already mentioned 'has not stated the reached education').

---

[5] However, for the attributes with natively numerical values, as e.g. Height or BMI (in its original conception), it is sufficient to assign one set of these parameters – with a variable symbol standing for the actual value – to the whole data column.

It is supposed that the values of C0 will be given directly in their necessary forms, so that no morphological transformations must be made with them. In the values of all the other parameters given above, the words possibly subject to any morphological transformation(s) must be marked with a special symbol. We have decided to use the '_' character before a word, complemented by a *morphological pattern symbol*, for this purpose. Thus, e.g., a real value of V0 may be '_vx:gets to work' and a real value of VP may be 'mainly _vx:sits in _pron:his job'.

If the SVNLE generator is called to make some B^V for B representing a value Z in a data column Y, it may proceed in two ways:

- it may combine V0 assigned to Y (if it is not empty) with C0 assigned to the pair $\langle Y, Z \rangle$ (if this is not empty), or
- it may use the VP value assigned to the pair $\langle Y, Z \rangle$ (if it is not empty).

If the SVNLE generator is called to make some B^N, B^L or B^R for B representing a value Z in a data column Y, it must simply find the value of N0, AL or AR assigned to the pair $\langle Y, Z \rangle$. If this value is empty, then the SVNLE generator must return a special value (e.g. empty string) meaning that the FP requiring the particular SVNLE cannot be used.

If the expression required has the form B^N(PP) or B^N(PP2), the code in NAF must be applied to N0 (and again, if NAF is empty, the FP must be rejected).

Generally the SVNLE generator must call a **morphological module (MM)** to make the forms required by the FP. A special case of this also arises when a B^P is required: in this case the generator will first construct B^V in a way described above, and then call the MM to transform the 3rd person singular verb form (marked with '_') into the corresponding participle. (Thus, e.g., 'getting to work on foot' will be obtained from V0 = '_vx:gets to work' and C0 = 'on foot'.) Some details of the MM implementation are dealt with elsewhere [4], but the mechanism is still being optimized, especially for Czech.[6]

5   Concluding remarks

The data structures and corresponding algorithm are now being implemented in the frame of the STULONG project (see http://euromise.vse.cz/ challenge/en/projekt/). During this process, several interesting points have been revealed:

---

[6] One must take into account, e.g., that a Czech adjective may require various gender forms in various phrases – but once it is used in a particular phrase (such as N0 = 'částečná invalidita' ['partial invalidity']), it may undergo further changes only in case and number.

- A recursive rewriting rule system (corresponding to a very simple context-free grammar) will probably be much more efficient than a simple set of FPs. As an example of a particular useful 'higher-level non-terminal symbol' in the FP system we can introduce the symbol P_A for any permissible NL representation of '*patient satisfying both antecedents*'. While this symbol can be further rewritten by 'A_1^L _n1:patient A_2^P', 'A_1^L _n1:patient A_2^N(PP)', 'A_1^L _n1:patient that A_2^V', '_n1:patient that A_1^V and A_2^V', '_n1:patient A_1^N(PP) and A_2^N(PP2)' and in many other ways, the FPs for the whole implication can have the following forms using the P_A symbol:

  - X P_A(PLURAL) also S^V(PLURAL).
  - A P_A usually also S^V. This rule is confirmed by X patients.
  - S^N is characteristic for X P_A(PLURAL).
  - ...

- There is a specific problem certainly worth a further study. While the two antecedents in a formal implication are interchangeable, they do *not* seem universally interchangeable in a NL formulation – or at least the formulation does not seem equally *natural* with any order of the antecedents. Moreover, the most natural conjunction corresponding to the logical symbol ∧ is sometimes not 'and' – it may be 'but'. What criteria should be taken into account for the best ordering of the antecedents and choice between 'and' and 'but'? We can consider these two variants of expressing (from the formal-logic point of view) the same P_A (as introduced above):

  - 'patient that makes sports intensively and mainly sits in his job'
  - 'patient that mainly sits in his job but makes sports intensively'

  The conjunction 'but' is appropriate between NL expressions that are understood as somehow contradictory. It seems that a good help for detecting this in the context of medical data could be assigning a value from the set {healthy, neutral, unhealthy} to every pair ⟨data column, value⟩ in the T2 table (see 4.2). And, of course, a 'really natural' ordering of the data columns should be provided for.

- Special attention should be paid to the questions of *negation*. The most serious problem of the negation seems to be a kind of *incompatibility* between its meanings in the formal world of the data columns and in informal NL interpretation. The problem can be illustrated by the following example. A formal negation of the BA Reached education(university) means in fact nothing more than that the patient has a value different from university in the column Reached education – i.e. his value may also be not stated! A NL negative formulation like 'has not reached university education' can never be interpreted as 'has not stated the reached education'. If the patients with the value not stated in a tested column are to be taken into account, there is probably only one (and quite a clumsy one!) form of *NL negation* remaining: a positive formulation 'patients that v' must be replaced with 'patients having not stated that they v'...

- Another problem that still has to be studied is how to formulate **a double implication or an equivalence** in NL. So far it seems that NLs do not offer many elegant ways for expressing such logical relations between statements. It looks more as if these constructions of formal logic were – from the point of view of intuitive description of the world – something unnatural! At least there is a question coming out of this whether it would not be the most natural to present a double implication simply as two implications, only joined together by means of a phrase like 'And the other way round:'...

We hope that the implementation of our limited language model is a viable way of automating the formulation of ARs in NL – albeit still with some limitations concerning some of the logical operators as they were described in this section. (After all, *if such limitations remain for objective reasons*, they may simply show that various types of ARs are not equally easy to handle for a human being without mathematical education.)

References

1. Aggraval, R. et al.: Fast Discovery of Association Rules. In Fayyad, U. M. et al.: *Advances in Knowledge Discovery and Data Mining*, AAAI Press / The MIT Press, 1996, 307–328.
2. Matheus, J. et al.: Selecting and Reporting What is Interesting: The KEFIR Application to Healthcare Data. In Fayyad, U. M. et al.: *Advances in Knowledge Discovery and Data Mining*, AAAI Press / The MIT Press, 1996, 495–515.
3. Rauch, J., Šimůnek, M.: Mining for 4ft Association Rules. In Arikawa, S., Morishita, S. (eds.): *Discovery Science 2000*, Springer-Verlag 2000, 268–272.
4. Strossa, P., Rauch, J.: Association Rules in STULONG and Natural Language. In Berka, P. (ed.): *ECML/PKDD-2002 Workshop Proceedings: Discovery Challenge Workshop Notes*, Report B-2002-8, Helsinki, Universitas Helsingiensis 2002, ISBN 952-10-0639-0, [13 p.]. See also `http://lisp.vse.cz/challenge/ecmlpkdd2002/`.

Proof Searching Algorithm for the Logic of Plausible Reasoning

Bartłomiej Śnieżyński

Institute of Computer Science
University of Mining and Metallurgy
Kraków, Poland
e-mail: sniezyn@agh.edu.pl

Abstract. Logic of plausible reasoning (LPR) is a knowledge representation and inference theory which is based on human reasoning techniques. Formalism can be described as a labeled deductive system; therefore reasoning can be considered as looking for proofs of given formulas. The aim of the paper is to present a proof searching algorithm for the LPR. The algorithm would be a core element of any information system using LPR.

Keywords: Logic of plausible reasoning, uncertain knowledge representation, proof searching algorithm.

1 Introduction

Logic of plausible reasoning (LPR) was developed by Collins and Michalski [2] in 1989. This formalism is a knowledge representation and inference technique, which can be used to represent human reasoning [1]. There are two important features of LPR: several inference patterns are defined (not only Modus Ponens) and many parameters estimate certainty. Therefore LPR differs from others theories used in AI to represent uncertain knowledge.

The aim of this paper is to present an algorithm which is able to find a proof of a formula from a given set of formulas. It would be a core element of any information system using LPR as a knowledge representation system.

In the following sections we define LPR language and a proof system. Next we present the proof algorithm and we show its properties.

2 Logic of Plausible Reasoning

LPR can be defined as a labeled deductive system [3]. *Language* consists of a finite set of constant symbols C, four relational symbols and logical connectives: $\rightarrow, \wedge$. The relational symbols are: V, H, S, E. They are used to represent: statements, hierarchy, similarity and dependency respectively.

Statements are represented as object-attribute-value triples. If $o, a, v \in C$, relation $V(o, a, v)$ represents a fact that object o has an attribute a equal v.

Relation $H(o_1, o, c)$, where $o_1, o, c \in C$, means that o_1 is o in a context c. Context is used for specification of the range of inheritance. o_1 and o have the same value for all attributes which depend on attribute c of object o.

Relation $S(o_1, o_2, c)$ represents a fact, that o_1 is similar to o_2 ($o_1, o_2, c \in C$). Context, as above, specifies the range of similarity. Only these attributes of o_1 and o_2 have the same value which depends on attribute c.

Dependency relation $E(o_1, a_1, o_2, a_2)$, where $o_1, a_1, o_2, a_2 \in C$, means that values of attribute a_1 of object o_1 depend on attribute a_2 of the second object. To represent mutual dependency [2] we need a pair of such expressions.

In object-attribute-value triples, value should be below an attribute in a hierarchy: if $V(o, a, v)$ is in a knowledge base, there should be also $H(v, a, c)$ for any $c \in C$.

Now we are able to define *formulas* of LPR. If $o, o_1, ..., o_n, a, a_1, ..., a_n, v, v_1, ..., v_n, c \in C$ then $V(o, a, v)$, $H(o_1, o, c)$, $S(o_1, o_2, o, a)$, $E(o_1, a_1, o_2, a_2)$, $V(o_1, a_1, v_1) \wedge ... \wedge V(o_n, a_n, v_n) \rightarrow V(o, a, v)$ are formulas of LPR.

To deal with uncertainty we use labels. Hence we need a *label algebra* $\mathcal{A} = (A, \{f_{r_i}\})$.

A is a set of labels which estimate uncertainty of formulas. *Labeled formula* is a pair $f : l$ where f is a formula and $l \in A$ is a label. A set of labeled formulas can be considered as a *knowledge base*.

All inference patterns are defined as proof rules (see below). Every proof rule r_i has a sequence of premises (of length p_{r_i}) and a conclusion. $\{f_{r_i}\}$ is a set of functions which are used to generate a label of a conclusion: for every proof rule r_i an appropriate function $f_{r_i} : A^{p_{r_i}} \rightarrow A$ should be defined. For rule r_i with premises $p_1 : l_1, ..., p_n : l_n$ the plausible label of its conclusion is equal $f_{r_i}(l_1, ..., l_n)$. Examples of plausible algebras can be found in [7,8].

Having the language we can define a proof system. We consider a subset of inference patterns described in [2]. Omitted patterns have minor role in inference.

There are 4 types of proof rules: *GEN*, *SPEC*, *SIM* and *MP*. They correspond to the following inference patterns: generalization, specialization, similarity transformation and modus ponens. Some transformations can be applied to different types of formulas; therefore indexes are used to distinguish different versions of rules. Rules can be divided into several groups according to type of formulas on which rules operate and types of inference patterns.

First is group of *statement proof rules*. It consist of 6 rules presented in table 1. They are used to perform reasoning on statements, hence first premises and the conclusions are object-attribute-value triples. Rules indexed by o transform object argument, what correspond to generalization and specialization of the statement. Rules indexed by v operate on values. Applying these rules changes the detail level of a description and corresponds to abstraction and concretion.

Second is group of *dependency proof rules*, which is shown in table 2. They operate on dependency formulas and allow to use generalization, spe-

Table 1. Statement proof rules.

$$GEN_o \; \frac{\begin{array}{c} H(o_1, o, c) \\ E(o, a, o, c) \\ V(o_1, a, v) \end{array}}{V(o, a, v)} \quad SPEC_o \; \frac{\begin{array}{c} H(o_1, o, c) \\ E(o, a, o, c) \\ V(o, a, v) \end{array}}{V(o_1, a, v)} \quad SIM_o \; \frac{\begin{array}{c} S(o_1, o_2, c) \\ E(o_1, a, o_1, c) \\ E(o_2, a, o_2, c) \\ V(o_2, a, v) \end{array}}{V(o_1, a, v)}$$

$$GEN_v \; \frac{\begin{array}{c} H(v_1, v, c) \\ H(v, a, c_1) \\ H(o_1, o, c_2) \\ V(o_1, a, v_1) \end{array}}{V(o_1, a, v)} \quad SPEC_v \; \frac{\begin{array}{c} H(v_1, v, o) \\ H(o_1, o, c_1) \\ V(o_1, a, v) \end{array}}{V(o_1, a, v_1)} \quad SIM_v \; \frac{\begin{array}{c} S(v_1, v_2, o) \\ H(o_1, o, c_1) \\ V(o_1, a, v_2) \end{array}}{V(o_1, a, v_1)}$$

Table 2. Proof rules based on dependencies.

$$GEN_E \; \frac{\begin{array}{c} E(o_1, a_1, o_1, a_2) \\ H(o_1, o, c) \\ E(o, a_1, o, c) \end{array}}{E(o, a_1, o, a_2)} \quad SPEC_E \; \frac{\begin{array}{c} E(o, a_1, o, a_2) \\ H(o_1, o, c) \\ E(o, a_1, o, c) \end{array}}{E(o_1, a_1, o_1, a_2)} \quad SIM_E \; \frac{\begin{array}{c} E(o_1, a_1, o_1, a_2) \\ S(o_1, o_2, c) \\ E(o_1, a_1, o_1, c) \end{array}}{E(o_2, a_1, o_2, a_2)}$$

Table 3. Proof rules using implications.

$$SPEC_{o\rightarrow} \; \frac{\begin{array}{c} H(o_1, o, c) \\ V(o, a_1, v_1) \wedge \ldots \wedge V(o, a_n, v_n) \rightarrow V(o, a, v) \end{array}}{V(o_1, a_1, v_1) \wedge \ldots \wedge V(o_1, a_n, v_n) \rightarrow V(o_1, a, v)}$$

$$MP \; \frac{\begin{array}{c} V(o_1, a_1, v_1) \wedge \ldots \wedge V(o_n, a_n, v_n) \rightarrow V(o, a, v) \\ V(o_1, a_1, v_1) \\ \vdots \\ V(o_n, a_n, v_n) \end{array}}{V(o, a, v)}$$

cialization and analogy inference patterns to change objects in dependency relations.

Third group presented in table 3 consist of two *rules using implications*. $SPEC_{o\rightarrow}$ is used to transform implication formulas and can be used to represent quantification over all objects which are below of given object in a hierarchy. Second rule is modus ponens, which is well known inference pattern.

Proof can be defined as a tree. Tree P is a proof of labeled formula φ from a set of labeled formulas KB if a root node of P is equal φ and for every node ψ: if ψ is a leaf, then $\psi \in KB$; else, there are nodes $(\psi_1, ..., \psi_k)$, connected to ψ and a proof rule r_i such, that ψ is a consequence of r_i and $\psi_1, ..., \psi_k$ are its premises (label of ψ is calculated using f_{r_i}).

We say, that a labeled formula ψ is a *syntactic consequence* of a set of labeled formulas KB ($KB \vdash \psi$) if there exist a proof of ψ from KB.

3   Proof Searching Algorithm

Algorithm presented below (algorithm 1) is based on AUTOLOGIC system developed by Morgan [5]. To limit the number of nodes and to generate optimal proofs, algorithm A* [4,6] is used.

Input: φ – formula, KB – finite set of labeled formulas
Output: If $\exists l_\varphi \in A : KB \vdash \varphi : l_\varphi$, then success, P – proof of $\varphi : l_\varphi$ from KB,
       else: failure.
$T :=$ tree with one node (root) s; label s by $[\varphi]$; $OPEN := [s]$
while $OPEN$ *is not empty* **do**
     $n :=$ first element from $OPEN$; remove n from $OPEN$
     if n *has empty label* **then**
         Generate proof P using path from s to n; exit with success.
     $R :=$ rules, which consequence can be unified with first formula of n.
     $E :=$ nodes generated by replacing first formula of n
         by premises of rules from R and applying substitutions
         from unifier generated in the previous step
     if *first formula from n can be unified with element of KB* **then**
         Add to W node obtained from n by removing first formula and applying
         substitutions from unifier.
     end
     Remove from E nodes generating loops.
     Append E to T connecting nodes to n.
     Insert nodes from E into $OPEN$.
end
Exit with failure.

Algorithm 1: LPR proof algorithm

Input data is a set of labeled formulas KB – a knowledge base and a formula φ, which should be proved from KB. If there exist a label $l_\varphi \in A$ such, that $KB \vdash \varphi : l_\varphi$, then proof of $\varphi : l_\varphi$ from KB is returned, else procedure exits with failure.

The algorithm generates a tree T, which nodes (N) are labeled by sequences of formulas. For the purpose of algorithm, the LPR language is extended by adding countable set of variables, which cen be used instead of constant symbols in these sequences. Every edge of T is labeled by a rule, which consequence can be unified with the first formula of a parent node or is labeled by the term $kb(l)$ if the first formula of a parent node can be unified with ψ, such that $\psi : l \in KB$. Node s is the root of T. It is labeled by $[\varphi]$. The goal is to generate a node labeled by empty set of formulas.

$OPEN$ is an ordered sequence of nodes which are not expanded, initially it consist of a root of T. If $OPEN$ is empty, procedure exits with failure – φ is not a syntactic consequence of KB.

If first node of $OPEN - n$ has empty sequence of formulas, procedure exits with success and returns a proof tree P generated from T, starting from n.

If n has not empty sequence of formulas, it is expanded. $R = \{r_i\}$ is a set of rules, which consequence can be unified with the first formula of n using most general unifier θ_i. Expansion E of n is a set of nodes generated by replacing the first formula of n by premises of a rule $r_i \in R$ and applying θ_i generated in the previous step to resulting sequence of formulas. If the first formula of n can be unified with element of KB, a node with removed the first formula and applied substitutions from unifier is added to E.

It is not allowed, that the same formula with no variables appears more then once on any path of the proof tree constructed from T. Therefore nodes violating this condition are removed from E.

Nodes from E are added to T by connecting them with node n. Edges are labeled by rule names or by the term $kb(l)$ if the first formula was generated from a knowledge base with a label l.

Next, nodes from expansion are inserted into $OPEN$ sequence. Ordering of the sequence is kept. Nodes are ordered according to values of evaluation function $f : N \rightarrow \Re$, which is defined as follows:

$$f(n) = g(n) + h(n), \tag{1}$$

where $g : N \rightarrow \Re$ represents the actual cost of the proof and $h : N \rightarrow \Re$ is a heuristic function which estimates the cost from n to the goal node.

Function g is defined below:

$$g(n) = edgescost(n) + 1 - labelvalue(n). \tag{2}$$

Every rule application or accessing knowledge base has a cost assigned (the minimal cost is $\varepsilon > 0$). Function $edgescost(n) : n \in N \rightarrow \Re$ is a sum of costs on the path from the root to n. Function $labelvalue : N \rightarrow [0, 1]$ gives a value of a proof by considering a label l_φ of φ (higher values are preferred). Label l_φ is generated from path $n, ..., s$ by using plausible functions of rules corresponding to edges or using l if edge is labeled by $kb(l)$ and assuming that formulas in the sequence of n have the best possible label – such that $labelvalue(n)$ is highest (so ordering of labels with maximum element for every subset of labels corresponding to every type of formula is needed).

To define heuristic function we can use the length of the sequence of formulas $\text{len}(n)$ multiplied by minimal cost of edge (ε):

$$h(n) = \varepsilon \text{len}(n). \tag{3}$$

By assigning small values to edge costs, label of proved formula becomes more significant. It is also possible to use edge costs which grow when the node depth grows. In such case on the beginning algorithm is looking for short proofs with the best label of the φ, but later it is searching for any proof.

Algorithm defined above has several important properties. Because of the lack of space, proof is omitted.

Proposition 3.01 *Algorithm 1 has the following properties:*

1. *Stop property (it terminates for any KB and φ).*
2. *Correctness (if algorithm exits with success and returns P, then P is the proof of φ from KB).*
3. *Completeness (if $KB \vdash \varphi$, algorithm exits with success).*
4. *Optimality (if algorithm exits with success and returns P, it has the lowest possible cost – g value of the goal node is minimal).*
5. *Optimal efficiency (there is no other optimal algorithm, which expand fewer nodes).*

4   Conclusions and Further Works

In the paper a proof searching algorithm for the LPR has been presented, and its properties has been showed. It seems, that it would be a good choice to use it in intelligent information systems based on LPR.

Further works will concern building information system using LPR and creating real world knowledge base from a chosen domain. It will allow to test the proof algorithm and the formalism. Comparison to other proof (graph) search techniques (such as iterative deepening strategy or AO*) and testing other heuristic functions is also planned.

References

1. D. Boehm-Davis, K. Dontas, and R. S. Michalski. Plausible reasoning: An outline of theory and the validation of its structural properties. In *Intelligent Systems: State of the Art and Future Directions*. North Holland, 1990.
2. A. Collins and R. S. Michalski. The logic of plausible reasoning: A core theory. *Cognitive Science*, 13:1–49, 1989.
3. D. M. Gabbay. *LDS – Labeled Deductive Systems*. Oxford University Press, 1991.
4. P. Hart, N. J. Nilsson, and B. Raphael. A formal basis for the heuristic determination of minimum cost path. *IEEE Trans. Syst. Science and Cybernetics*, 4 (2):100–107, 1968.
5. C. G. Morgan. Autologic. *Logique et Analyse*, 28 (110-111):257–282, 1985.
6. S. Russell and P. Norvig. *Artificial Intelligence – A Modern Approach*. Prentice-Hall, 1995.
7. B. Śnieżyński. Verification of the logic of plausible reasoning. In M. Kłopotek et al., editor, *Intelligent Information Systems 2001*, Advances in Soft Computing. Physica-Verlag, Springer, 2001.
8. B. Śnieżyński. Probabilistic label algebra for the logic of plausible reasoning. In M. Kłopotek et al., editor, *Intelligent Information Systems 2002*, Advances in Soft Computing. Physica-Verlag, Springer, 2002.

Part VI

Time Dimension in Data Mining

Collective Intelligence from a Population of Evolving Neural Networks

Aleksander Byrski and Marek Kisiel-Dorohinicki

Department of Computer Science
University of Mining and Metallurgy, Kraków, Poland
e-mail: { olekb, doroh} @agh.edu.pl

Abstract. In the paper an agent-based system of evolving neural networks dedicated to time-series prediction is presented. First the idea of a multi-agent predicting system is introduced and selected methods of management of collective intelligence of agent population(s) are considered. Then the problems and possibilities of evolutionary design of a predicting neural network are shortly discussed. Finally a hybrid solution combining both approaches in an evolutionary multi-agent system (**EMAS**) is proposed. General considerations are illustrated by the particular realisation of such a system and selected experimental results conclude the work.

1 Introduction

Numerous computationally difficult problems may be solved utilising analogies to processes observed in nature. In particular this concerns various *soft computing* techniques like artificial neural networks or evolutionary algorithms. While the former allow for modelling of very complex functions and non-linear structures with large number of variables, the latter are successfully used to solve difficult search and optimisation problems. Even though these have completely different application areas, one may notice that it often happens that combinations of completely different techniques by the effect of synergy exhibit some kind of intelligent behaviour [?]. This is sometimes called *computational intelligence* as opposed to rather symbolic artificial intelligence.

This is the case with evolutionary neural networks, which combine neural computing with evolutionary computation paradigm (see e.g. [?]). For example the architecture of a neural network may be encoded in a chromosome and thus be a subject of evolutionary optimisation. Still the use of classical evolutionary computation may bring some shortcomings. In the considered case to evaluate a chromosome the network must be trained – and this must be repeated for each network in the whole population before creating the next generation.

A possible serious improvement to such problems comes with the use of intelligent agents and agent-based systems. These provide concepts and tools for development of intelligent decentralised systems. They may also be used as a means for analysis and realisation of hybrid systems, in which different techniques cooperate to fulfil specified demands. This cooperation, complied with a specific management algorithms, may be a source of *collective intelligence.*

Combining evolutionary computation with multi-agent systems (MAS) leads to the decentralisation of the evolutionary optimisation process. Training of a neural network may be entrusted to an autonomously acting agent, which can also perform actions of reproduction and death. This way various agents may train their networks, simultaneously with the process of evolution occurring in the whole population. Such defined *evolutionary multi-agent system* (EMAS) may help in search for the optimal configuration of a neural network for the given problem, or at least help to establish a starting point for further network structure development.

This paper presents the concept of such a hybrid system, in which different techniques (neural networks, evolutionary computation, agent systems) cooperate helping one another in attaining its own goals. The main task of the system is time-series prediction, but at the same time it tries to find the best neural network architecture to perform this task, as well as the values of many parameters describing its work (*meta-evolution*).

2   MAS for time-series prediction

From the outside, a time-series predicting system may be considered as a black box with some input sequences, and predictions of successive values of (some of) these sequences as output. Some mechanism inside that box should be able to discover hidden regularities and relationships in and between the input sequences. Assuming that the characteristics of the signal(s) may change in time, this mechanism should be also able to dynamically adapt to these changes ignoring different kinds of distortion and noise.

2.1   Predicting MAS

When the signal to be predicted is much complicated (e.g. when different trends change over time or temporal variations in relationships between particular sequences are present) the idea of a *multi-agent predicting system* may be introduced [?]. The predicting MAS may be viewed as the box (as above) with a group of intelligent agents inside (fig. 1). Subsequent elements of input sequence(s) are supplied to the environment, where they become available for all agents. Each agent may perform analysis of incoming data and give predictions of (a subset of) the next-to-come elements of input. Specialisation in function or time of particular agents allow for obtaining better results by cooperation or competition in the common environment. Based on predictions of all agents, prediction of the whole system may be generated.

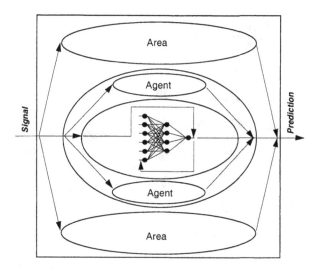

Fig. 1. Predicting neural MAS

2.2   Managing collective intelligence of a population of agents

One may notice that a group of autonomous yet cooperating intelligent agents can exhibit a kind of *collective intelligence*. While dealing with such a system an important problem arises: how to determine the answer of the whole system to given problem. For mentioned time-series prediction problem, it should be decided how to select an agent which output is the best or most representative and thus may be presented as the output of the whole system, or how to combine more than one agent's answers to produce the desired output value.

In literature various techniques of management of collective intelligence arising from cooperation of autonomous beings solving the same problem are proposed. This includes *voting* (linear combination of answers with fixed weights), *mixture of experts* (like voting but a *gating expert* assigns specific weights to particular answers), or *stacked generalisation* (gating expert can be designed as any non-linear model) [?]. In the particular system, the way of combining multiple individuals' answers can be based on probability analysis as shown in section 4.2.

3   Evolutionary neural networks for time-series prediction

The main advantage of using artificial neural networks is probably their ability to learn from examples and generalise acquired knowledge to new cases

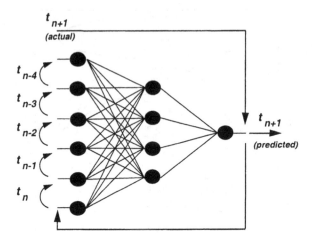

Fig. 2. Predicting neural network

in such a way that no explicit problem-dependent knowledge is needed. That is why they are often used in control problems, i.e. for management of some process or device, as well as in approximation problems, e.g. classification and prediction. The last is of our interest and will be discussed further.

3.1 Prediction with neural networks

A neural network may be used by an agent as a particular mechanism to model the characteristics of a signal in a system for time-series prediction [?]. Usually the next value of a series is predicted based on a fixed number of previous ones (fig. 2). Thus the number of input neurons correspond to the number of values the prediction is based on, and the output neuron(s) give prediction(s) of the next-to-come value(s) of the series.

The choice of a particular architecture of the network is in great measure determined by the particular problem. The feed-forward network on (fig. 2) should predict t_{n+1} value of the series, based on a fixed number of previous values, which are given on the inputs of the first layer. When t_{n+1} value is predicted, the inputs are shifted, and the value t_{n+1} is given as the input to the last neuron of the first layer.

3.2 Evolution of neural networks

Even though the use of neural networks allows to perform various difficult tasks with a little or even without problem-dependent knowledge, one still needs to define the suitable network configuration — its topology (number of

layers, number and type of neurons in particular layer, structure of connections), initial state and training algorithm. The network should have enough
complex structure to be able to solve the problem, at the same time this
structure should not be too complex to avoid overfitting the data.

Due to a huge number of possible network variants, manual process of
refining the neural network architecture requires many experiments and may
not necessarily lead to the best solution. That is why many techniques were
invented that allow for automatic design of a neural network that fulfils the
requirements of the problem. In evolutionary neural networks, the search for
a desirable neural network is made by an evolutionary algorithm [?,?].

In order to encode the structure of the network into a chromosome, several questions should be answered. Is the exact image of structure to be
encoded in this genotype? Or maybe it will be enough if some general outline
is encoded? Answering these questions leads to two main kinds of genotypic
representation of a network:

- in *direct encoding* (strong specification scheme) a chromosome contains
  complete information about the structure of the network and requires
  little effort do decode it – an example of such an encoding is a connection matrix that precisely and directly specifies the topology of the
  corresponding neural network,
- in *indirect encoding* (weak specification scheme) there are some rules
  of creating a neural network encoded in the chromosome and thus the
  construction of the network requires a considerable effort – an example
  of such an encoding is one that uses rewrite rules to specify a set of
  construction rules that are recursively applied to yield the whole network
  [?].

One should remember that in direct encoding longer chromosomes are needed,
and indirect encoding suffers from noisy fitness evaluation [?].

Of course during the evolution process not only connections of the network
can be evolved. The same process may operate on weights between neurons
(evolutionary learning). Yet because of high computational complexity it is
often much easier to evolve the structure of connections, leaving the search
for the values of weights to the training algorithm.

3.3   Radial basis function networks

Radial basis function networks consist usually of three layers. The input one
is used only for normalisation purposes. The hidden layer consists of neurons
with radial activation function:

$$F(X) = -\frac{||X - T||^2}{\sigma^2}$$

where $||X - T||$ denotes an Euclidean norm, σ describes the shape of the
radial function, X is an input vector and T is the centre of the function.

Third layer usually consists of one neuron used to combine hidden neurons' output values, and often uses linear activation function [?].

Training of RBF network can consist of two phases. First, the centres of the radial functions are set at random or may be a subject of evolutionary search. Second, the output layer of the network may be supervisory trained, using simple steepest gradient descent method, based on the comparison between values predicted and received as an error measure [?].

4   Neural EMAS for time-series prediction

The configuration of the agents in a predicting MAS (kind of specialisation or method of cooperation) is often difficult to specify. What is more, when dynamic changes of the characteristics of the signal are possible, the configuration of the agents should reflect these changes, automatically adapting to the new characteristics. The mechanisms of evolution may help to transform the whole population of agents (by means of mutation and/or recombination) so as it fits best current profile of the input signal (proper selection/reproduction) – this evolutionary development of predicting MAS meets the general idea of an evolutionary agent system (EMAS).

4.1   Evolutionary Multi-Agent Systems

The key idea of EMAS is the incorporation of evolutionary processes into a multi-agent system (MAS) at a population level [?,?]. It means that besides interaction mechanisms typical for MAS (such as communication) agents are able to reproduce (generate new agents) and may die (be eliminated from the system). A decisive factor of an agent's activity is its fitness, expressed by the amount of possessed non-renewable resource called life energy. Selection is realised in such a way that agents with high energy are more likely to reproduce, while low energy increases the possibility of death.

In EMAS training of a neural network may be entrusted to an agent while the search for a suitable network architecture may be realised as the process of evolution occurring in the whole population. To achieve this goal, each agent simply possesses some vector of parameters, which describes the configuration of its neural network. This vector plays role of agent's genotype, and as such may be modified by genetic operators when inherited by its offspring. The evaluation of agents is based on the quality of prediction obtained from a trained network by means of gained/lost life energy.

One may notice that such system performs not only search for the optimal neural network structure, but also exhibits collective intelligence at agent population level since agents are able to cooperate providing even better solutions to the given problem.

4.2   A population of agents as a dynamic modular neural network

In the above-described system every agent contains a neural network, which acts as a computational model for the given (prediction) problem. Entrusting the task of solving the problem to the complex system, one may expect to obtain more accurate answers. This is similar to the approach of modular neural networks such as the model of PREdictive MOdular Neural Network. PREMONN is a group (team) of neural networks, which solve the same problem, and their responses are combined together to yield the final result [?].

Applying PREMONN algorithm, every prediction of the given time-series may be assigned a certain probability, which can be used to determine the answer of the whole group of predicting individuals. After every prediction step, every individual based on its predictions and errors:

$$y_t^k = f_K(y_{t-1}, y_{t-2}, \ldots, y_{t-M})$$

$$e_t^k = y_t - \hat{y}_t^k$$

computes its credit function:

$$p_t^k = \frac{p_{t-1}^k \cdot e^{-\frac{|e_t^k|^2}{2\sigma^2}}}{\sum_{n=1}^{K} p_{t-1}^n \cdot e^{-\frac{|e_t^k|^2}{2\sigma^2}}}$$

Based on this function the response of the group of individuals can be a weighted combination of the answers or even can be the result of the winner-take-all combination:

1. Weighted combination:

$$\hat{y}_t = \sum_{k=1}^{K} p_{t-1}^k y_t^k$$

2. Winner-take-all combination:

$$\hat{y}_t = y^{\hat{z}_t}, \text{ where } \hat{z}_t = \arg\max_{k=1,2,\ldots,K} p_{t-1}^k$$

Using the above-mentioned approach a group of agents can produce the answer, which will be more accurate than the prediction of one arbitrarily chosen agent, as it comes from the group of agents using at least simple voting, or a more sophisticated bayesian stochastic scheme.

The system constructed in this way is also adaptive – its adaptation abilities base on its stochastic features. The probabilities of correct answers of agents are dynamically changed by the gating expert, so in the group of the agents, the answer of the whole group as the (somehow) weighted answer of every agent, is reliable, and the agents which produce worse answers should be replaced with new agents, in this way, globally, the system can adapt to the new features of the environment.

4.3 Evolving collective intelligence of agent populations

In such a complex organisation of agents, which was described above as a dynamically changing predictive modular neural network, it will be very difficult to determine correct values of additional parameters, which describe particular mechanisms of interaction. In order to improve the process of refining of such an organisation evolutionary processes may be also used (*meta-evolution*).

In an **EMAS** evolution usually is realised at the level of individual agent actions [?] (as each of them contains a chromosome which can be recombined, mutated, inherited etc.). At the same time perceiving the whole system as a group of modular neural networks (groups of agents, every group can be perceived as a single being), it seems natural to propose the way of evolving such ,,complex beings". Every modular network must have its own parameters, characterising its behaviour, such as parameters of credit function, and the parameters defining the way of evolving individual agents (amount of rewards and punishments [?]), which can be subject of the evolution process.

Such two-level evolution may lead to automatic determination of the system parameters, making it more reliable and adaptive to the changes of the work conditions (e.g. in time-series prediction, to the changes of the time-series to be predicted).

5 Experimental results

The neural networks used in the experiments were radial basis function networks with three layers. The number of neurons in the input and hidden layers and the centres of the RBF activation functions were the subject of evolution. The synaptic weights of the output neurons were changed with use of the simple steepest descent algorithm.

The results described below were obtained for simple sinusoidal time series, which period was 10 steps of the system's work and the signal range was from $0,1$ to $0,9$.

In the first graph (fig. 3a) a percent prediction error of a selected agent is presented. The process is naturally faster at the beginning and slows down after few hundreds of steps. It means that the networks used to predict a given time-series were properly constructed. In the second graph (fig. 3b) an average percent prediction error of the whole population of agents is presented. This error changes very fast at the beginning of the evolution process, and then, after several hundreds of steps it begins to stabilise, because many agents with better abilities than their parents are created.

A crucial task in these kind of systems is to maintain more or less the same number of agents in the population, so the system becomes stable and constant exchange of the individuals is assured (still we do not know how much agents we need for this particular task). In the conducted experiments the population of the agents seems to be stable, as it one may observe in the

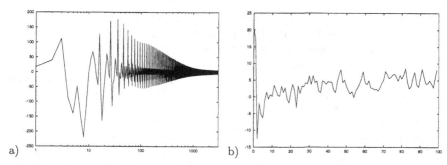

Fig. 3. Prediction error of a single agent (a) and average prediction error of the population (b)

third graph (fig. 4a), where the number of agents in the system is presented. It is to notify that (similar to average prediction error) the number of agents at the beginning of the evolution process changes very fast, then begins to stabilise.

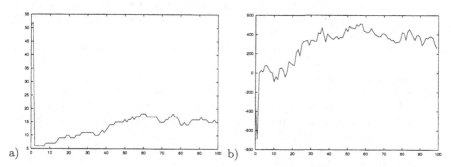

Fig. 4. Number of agents (a) and average agent energy (b) in the population

As the processes of evolution and death are based on the life energy of agents, it can be seen in the fourth graph (fig. 4b), that amount of this resource is also stable during the operation of the system, which proves that mechanisms of resource distribution agents (prices and penalties) were satisfactory.

6   Conclusions

Evolutionary design of neural networks may lead to obtaining faster and better results than these designed by a man from scratch. At the same time a population of cooperating agents may produce much more reliable results than even the best (as it may seem) agent at a time, which may be perceived as an example of collective intelligence. Two-level evolution conducted in **EMAS** can lead to new methods of solving complex problems, changing not only characteristics of the individuals, but also the features of the environment, where the evolution is performed.

Further research should concern more sophisticated mechanisms of management of collective intelligence in a population(s) of agents. Also more difficult problems and other application areas will be considered.

References

1. E. Alpaydin. Techniques for combining multiple learners. In *International ICSC Symposium on Engineering of Intelligent Systems (EIS'98)*. ICSC Academic Press, 1998.
2. P. Bonissone. Soft computing: the convergence of emerging reasoning technologies. *Soft Computing*, 1(1):6–18, 1997.
3. A. Byrski, M. Kisiel-Dorohinicki, and E. Nawarecki. Agent-based evolution of neural network architecture. In M. Hamza, editor, *Proc. of the IASTED Int. Symp.: Applied Informatics*. IASTED/ACTA Press, 2002.
4. K. Cetnarowicz, M. Kisiel-Dorohinicki, and E. Nawarecki. The application of evolution process in multi-agent world (MAW) to the prediction system. In M. Tokoro, editor, *Proc. of the 2nd Int. Conf. on Multi-Agent Systems (IC-MAS'96)*, AAAI Press, 1996.
5. J. Gonzalez, I. Rojas, H. Pomares, and J. Ortega. Rbf neural networks, multiobjective optimization and time series forecasting. In J. Mira and A. Prieto, editors, *Connectionist Models of Neurons, Learning Processes and Artificial Intelligence*, volume 2084 of *Lecture Notes in Computer Science*, 2001.
6. S. Haykin. *Neural networks: a comprehensive foundation*. Prentice Hall, 1999.
7. H. Kitano. Designing neural network using genetic algorithm with graph generation system. *Complex Systems*, pages 461–476, 1990.
8. T. Masters. *Neural, Novel and Hybrid Algorithms for Time Series Prediction*. John Wiley and Sons, 1995.
9. M. Mitchell. *An Introduction to Genetic Algorithms*. MIT Press, 1998.
10. V. Petridis and A. Kehagias. *Predictive Modular Neural Networks – Application to Time Series*. Kluwer Academic Publishers, 1998.
11. J. Sjoberg, H. Hjalmarsson, and L. Ljung. Neural networks in system identification. In M. Blanke and T. Soderstrom, editors, *Proc. of 10th IFAC Symposium on System Identification (SYSID'94)*, volume 2, 1994.
12. X. Yao and Y. Liu. Evolving artificial neural networks through evolutionary programming. In L. J. Fogel, P. J. Angeline, and T. Bäck, editors, *Evolutionary Programming V: Proc. of the 5th Annual Conf. on Evolutionary Programming*. MIT Press, 1996.

Taming Surprises

Zbigniew R. Struzik

CWI
Kruislaan 413, 1098 SJ Amsterdam
The Netherlands

Abstract. A methodological trajectory has been described dealing with the 'novelty' or 'surprise' issue in time series records arising from real world complex systems. It is based on extracting regularity (or scaling) characteristics of non-differentiable time series with the wavelet transform, on modelling the complex system using multi-fractal properties and on investigating novelty in the context of the possible non-stationarity of such a model.

1 Introduction: Novelty in Complex Systems

This paper has been motivated by the recent trend in automated data analysis suggesting techniques for novelty detection and knowledge discovery [1–3]. Such techniques are often applied to real life data known to be subject to complex dynamics inaccessible for modelling by deterministic dynamical systems. The local reconstruction of the phase space trajectories (or other deterministic characteristics) of such systems may not be possible or may pose substantial difficulty. These difficulties may be linked to critical behaviour characterising such systems and may be misguiding the capturing of the evolution of time series arising from such systems. This in turn may result in false novelty or surprise knowledge assessment.

Let us take an example. In the top panel of figure 1 we present a record of heartbeat from a healthy patient. To the untrained eye, this record is either plain noise or it is full of most interesting features, spikes etc. [1] Some degree of knowledge, suggesting that stable, cyclic, regular heartbeat rate should become a straight line in this plot, would prompt novelty discoveries at almost every single point. Indeed, the heartbeat rate shown is full of such 'surprises', each subsequent record seems to depart from the previous one in a most wild fashion. The methods of novelty (or discovery or surprise) detection suggested to date would also most likely detect surprises in every single point of the record.

Let us take an imaginary record of heartbeat with a few short beats followed by a long one. The novelty of detection of a long pause between beats of

---

[1] This duality of 'novelty' interpretation can be appreciated in a record as 'simple' as a white noise time series. On the one hand, it contains no correlations at all, no coherent structures and thus no novelty. On the other, it represents perfectly coded information - each and every new sample carries new surprising information, independent and orthogonal.

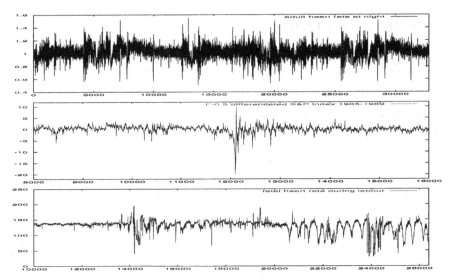

Fig. 1. Top: A healthy heart rate recording, about five hours long. Centre: A fractionally differentiated S&P500 record F(-0.5) about four years long centred at the '87 crash. Bottom: a fetal heartbeat record during labour, about ten hours long.

the heart after one or a series of shorter beats is, however, very questionable, if one realises that this is the most likely behaviour of the healthy heartbeat. Indeed, it has been established in the study of complex systems that heartbeat records show anti-persistence characterised by a high probability of long intervals after short intervals and vice versa. In fact, healthy heartbeat is at the ultimum of the range of anti-persistence measured by the Hurst exponent of anti-persistence - it reaches levels near $H = 0.0$ (on the scale 0-1) [4–6]. By comparison, Brownian walk, in which every next jump is independent of the previous one, is characterised by $H = 0.5$, central on the range of H.

Such seemingly 'anomalous' behaviour of the healthy heartbeat is not unique in nature. In fact many complex systems show characteristics which make it impossible to use low degree oscillatory models for modelling them. Such models would, of course, be relatively easy to use for discovery and fitting, including determination of instances when they do not perform. The essence of highly developed criticality of systems like heartbeat is that making and fitting models of such a smooth oscillatory character does not seem to make sense - they would fail at every instance and point.

The degree of difficulty of novelty detection in real life complex systems can be further appreciated in the following example, shown in the second (middle) panel of figure 1. The record of *S&P* index shows the crash of '87 centrally. It perhaps appears as an 'outlier' or novel feature to the observer. In fact it is probably neither of these. The (multi)-fractal structure [7–9] of the financial time series guarantees that there are crashes at any scale of observation. Smaller crashes will happen at smaller scales; their magnitude follows a law which is the subject of discussion and study. Large events like the '87 crash are 'rare' and may appear as outliers [10,11]. They are, however, possibly just the effect of the internal complex dynamics of the system - they are determined by the history of the system to the date of the crash. [12,13]

The last example again shows heart rate, but this time it is the heart rate of a baby being born. Fetal heartbeat is different in characteristics from adult heartbeat. Here, additionally, the record shows the dynamic evolution of the fetal heartbeat during the final hours of labour, ending in birth. This time it was a successful birth with positive fetal outcome. The baby born was healthy. Its heartbeat was used to make decisions about its status during labour. Such decisions are routinely taken upon observation of heartbeat, as it is the only indicator of the well-being of the fetus. The obstretician observes the heartbeat and judges it for development of characteristics which would prompt (or better not) intervention in the case of hypoxia (lack of oxygen). The case shown did not require intervention - the heartbeat, albeit extremely complex, did not deviate from normal.

Developed around the examples given above, the structure of this paper is as follows. First a method of analysis of regularity properties of time series (wavelet transform) is introduced in section 2. It is followed, in section 3, by its use for the characterisation of the so-called multi-fractal properties of the time series, originating from real life complex systems. Analysis of the possible non-stationarity of the multi-fractal properties is suggested in section 4, together with accommodating it in a heuristic model (section 5). For notes on model discovery from non-stationarity by the use of automated Bayesian net reasoning, the reader is referred to [14].

2   Estimating Regularity of Rough Time Series

Suppose we can *locally* approximate the time series (function f) with some polynomial P_n, but the approximation fails for P_{n+1}. One can think of this kind of approximation as the Taylor series decomposition:[2]

$$f(x)_{x_0} = c_0 + c_1(x - x_0) + \ldots + c_n(x - x_0)^n + C|x - x_0|^{h(x_0)} =$$
$$= P_n(x - x_0) + C|x - x_0|^{h(x_0)} .$$

It is traditionally considered to be important in data mining of time series to capture trend behaviour P_n. It is, however, widely recognised in other fields, as discussed in the previous section, that it is not necessarily the regular polynomial background but quite often the transient singular behaviour

---

[2] In fact the arguments to be given are true even if such a Taylor series decomposition does not exist, but it can serve as an illustration [15].

which can carry important information about the phenomena and the underlying system 'producing' the time series.

The exponent $h(x_0)$ characterises such local singular behaviour by capturing what 'remains' after approximating with P_n and what does not yet 'fit' into an approximation with P_{n+1}. Thus, our function or time series $f(x)$ is locally described by the polynomial component P_n and the so-called Hölder exponent $h(x_0)$.

$$|f(x) - P_n(x - x_0)| \leq C|x - x_0|^{h(x_0)}. \tag{1}$$

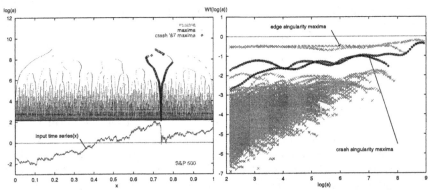

Fig. 2. Left: the input time series with the WT maxima above it in the same figure. The strongest maxima correspond to the crash of '87. The input time series is de-biased and L1 normalised. Right: we show the same crash related maxima highlighted in the projection showing the logarithmic scaling of all the maxima.

The advent of multi-scale techniques (like WT), capable of locally assessing the singular behaviour, greatly contributed to the advance of analysis of 'strange' signals, including (multi)fractal functions and distributions. The wavelet transform is a convolution product of the signal with the scaled and translated kernel - the wavelet $\psi(x)$. [16,17] The scaling and translation actions are performed by two parameters; the scale parameter s 'adapts' the width of the wavelet kernel to the resolution required and the location of the analysing wavelet is determined by the parameter b:

$$(Wf)(s,b) = \frac{1}{s} \int dx \, f(x) \, \psi(\frac{x-b}{s}) \tag{2}$$

where $s, b \in \mathbf{R}$ and $s > 0$ for the continuous version.

For analysis purposes, one is not so much concerned with numerical or transmission efficiency or representation compactness, but rather with accuracy and adaptive properties of the analysing tool. Therefore, in analysis tasks, continuous wavelet decomposition is mostly used. The space of scale

s and position b, is then sampled semi-continuously, using the finest data resolution available. [3]

The only *admissibility* requirement for the wavelet ψ is that it has zero mean - it is a wave function, hence the name *wavelet*. However, in practice, wavelets are often constructed with orthogonality to a polynomial of some degree n.

$$\int_{-\infty}^{\infty} x^n \, \psi(x) \, dx = 0 \qquad (3)$$

Indeed, if the number of the vanishing moments of the wavelet is at least as high as the degree of P_n, the wavelet coefficients will capture the local scaling behaviour of the time series as described by $h(x_0)$. Thus, what wavelets provide in a unique way is the possibility to tame and manage singularities and trends in a local fashion, through localised wavelets components [15-17,21].

$$W^{(n)} f(s, x_0) = \frac{1}{s} \int C |x - x_0|^{h(x_0)} \, \psi(\frac{x - x_0}{s}) \, dx$$

$$= C |s|^{h(x_0)} \int |x'|^{h(x_0)} \, \psi(x') \, dx' \, .$$

Therefore, we have the following power law proportionality for the wavelet transform (WT) of the (Hölder) singularity of $f(x_0)$:

$$W^{(n)} f(s, x_0) \sim |s|^{h(x_0)} \, .$$

From the functional form of the equation, one can attempt to extract the value of the local Hölder exponent from the scaling of the wavelet transform coefficients in the vicinity of the singular point x_0. A common approach to trace such singularities and to reveal the scaling of the corresponding wavelet coefficients is to follow the so-called maxima lines of the WT, converging towards the analysed singularity. This approach was first suggested by Mallat et al [21] (it resembled edge detection in image processing) and was later used and further developed among others in Refs [15,22,23]. However, any line convergent to the singularity can be used (to estimate the singularity exponent). Moreover, estimating local regularity at any point is possible by following the evolution (decay/increase) of the wavelet transform. This includes the smooth polynomial-like components of the time series. [4]

---

[3] The numerical cost of evaluating the continuous wavelet decomposition is not as high as it may seem. Algorithms have been proposed which (per scale) have complexity of the order n, the number of input samples, at a relatively low constant cost factor. [18]. Additionally, computationally cheap, discretised, semi-continuous versions of the decomposition are possible [19,20].

[4] Note that the interestingness of the time series is relative to application. The maxima of the WT can often be very well used since they converge to singular

In figure 2, we plot the input time series which is a part of the S&P index containing the crash of '87. In the same figure, we plot corresponding maxima derived from the WT decomposition with the Mexican hat wavelet. The maxima corresponding to the crash stand out both in the top view (they are the longest ones) and in the side log-log projection of all maxima (they have a value and slope different from the remaining bulk of maxima). The only maxima higher in value are the end of the sample finite size effect maxima. These observations suggest that the crash of '87 can be viewed as an isolated singularity in the analysed record of the S&P index for practically the entire wavelet range used.

The size, as reflected in maxima scale span, and the strength h of the crash related singularity, may suggest 'novelty' and 'surprise' to be associated with the event. For the strongest crashes observed, obviously due to their economic impact, there is a great interest and an ongoing debate as to whether they can be classified as outliers or whether they actually belong to the dynamics of the economic system [10–13]. In the case of the crash of '87, there are indications that it resulted from the past history of the development of the index [12], in particular as it lacked any evident external reason for occurring.

3   Multifractal Description of Complex Systems

The time series, the examples of which have been given in the previous sections belong to a class of systems recently characterised as multifractal (MF) [24–26,7–9]. Several models of multifractality have been suggested, starting at the early extensions of fractality and classical examples [27], to sophisticated wavelet cascade based models recently suggested [28,29]. Let us briefly hint at the main characteristic of multifractals.

For the stationary fractional Brownian noise, we would expect that any local estimate of the Hölder exponent h would conform to the mean or global Hurst exponent H. Of course, for finite length samples and single realisations, we will have fluctuations in the local h exponent, but they should prove to be marginal and diminish with increasing statistics. This will not be the case with a multifractal. The local h will show a wide range of exponents regardless of the resolution and sample size [26,30]. What we would expect to remain unchanged (or stationary) for the multifractal (cascade) is the multifractal spectrum of h, i.e. $D(h)$.

In figure 3, an example time series with the local Hurst exponent indicated in colour are shown. We have chosen the record of healthy (adult) heartbeat intervals and white noise for comparison. The background colour indicates

---

structures which they help in detecting and estimating. However, if one is interested in smooth components, this may not necessarily be the best alternative. Similarly, for localisation of oscillatory components, phase detection using complex wavelets may prove a more appropriate alternative than the maxima of the real valued WT.

the Hölder exponent h - the local counterpart to the Hurst exponent H. It is centred at the mean value corresponding with the Hurst exponent at green. The colour goes towards blue for higher h and towards red for lower h. In the same figure 3, we show corresponding log-histograms of the local Hölder exponent [5]. Each h measures a so-called singularity strength, and thus a histogram provides a way to evaluate the 'singularity spectrum'. In other words, the local h measures the local contribution to the multifractal spectra [30].

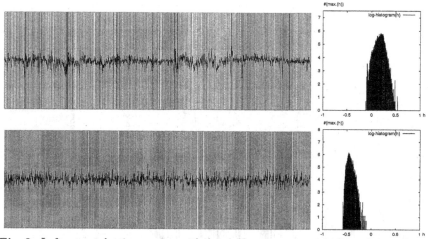

Fig. 3. Left: example time series with local Hurst exponent indicated in colour: the record of healthy heartbeat intervals and white noise. The background colour indicates the Hölder exponent locally, centred at the Hurst exponent at green; the colour goes towards blue for higher h and towards red for lower h. Right: the corresponding log-histograms of the local Hölder exponent.

4 'Novelty Hints' from the 'Failure' of the MF Model

A true multifractal process would share the same parameters (like MF spectrum) for any sub-part of the record. Thus, for an ideal multifractal system, each new data recorded would not affect the spectrum already estimated. Testing for this *stationarity* property can be done for our example records. In particular, new sample information can be simulated by running a simple

---

[5] They are made by taking the logarithm of the measure in each histogram bin. This conserves the monotonicity of the original histogram, but allows us to compare the log-histograms with the spectrum of singularities $D(h)$. By following the evolution of the log-histograms along scale, one can extract the spectrum of the singularities $D(h)$ (multifractal spectrum).

moving average (MA) filter, which may capture collective behaviour of the local h characteristic. A n-MA filtering of n base is defined as follows:

$$h_{MA_n}(i) = \frac{1}{n} \sum_{i=1}^{i=n} h_i(f(x)) \,, \tag{4}$$

where $h_i(f)$ are the subsequent values of the effective Hölder exponent of the time series f. Standard deviation from the $h_{MA_n}(i)$ mean exponent can also be calculated and is closely linked to the instantaneous MF spectrum width:

$$SDh_{MA_n}(i) = \frac{1}{n} \sqrt{\sum_{i=1}^{i=n} (h_i(f(x)) - h_{MA_n}(i))^2} \,. \tag{5}$$

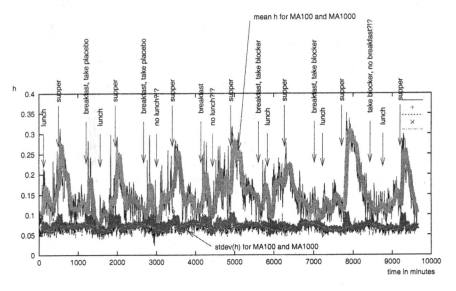

Fig. 4. The variability plot from a long run of experiments where the test persons were given placebo or beta-blocker. Two runs of MA filter were performed with 100 and 1000 maxima long window. An interesting pattern of response to food is evident.

An interesting pattern of 'surprising' features can be identified in the example (7 days long) record of the heartbeat. Upon verification, it confirms a pattern of response to activity, suggesting novel links to external information. Without going into much detail of the record given, there is a particularly strong response of the person in question to food. The observed shift towards higher values as the result of eating (it is almost possible to estimate the volume of the meal!) may indicate some nearly pathologic response in this individual case [31].

5 Exploiting Suspect 'Novelty' in a Heuristic Model

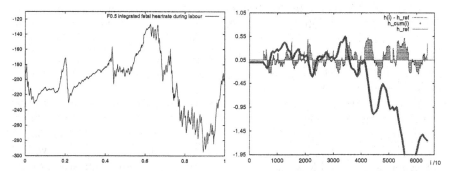

Fig. 5. Left: a fractionally $F0.5$ integrated 'normal' record contains no diagnostic surprises. Right: the cumulative indicator h_{cum} extracted from a bad fetal outcome record plunges after a period of homeostasis, suggesting diagnostic 'novelty'. Both the cumulative Hölder exponent $h_{cum}(V_i)$ (red line) and the deviation of the Hölder from the reference value $h_{ref} = 0.05$ (blue filled curve) are plotted in the right plot.

In the case of fetal heartbeat during labour, there is no reason why the local Hölder exponent $h(V_i)$ of the variability V_i component of fetal heartbeat intervals should be stationary. It reflects dynamic changes in the condition of the fetus and the degree of stress to which it is subjected. Despite the fact that stress has a rapid effect on heartbeat, the effects on the state of the fetus are not always immediate. This is why short dynamic changes in heartbeat characteristics (which determine the multifractal picture) may not be relevant and not representative of the state of the fetus. Rather than expanding the observation window, we have suggested [32] using a cumulative indicator, designed to capture the non-stationarity of the local h of the variability component of fetal heartbeat V_i [32]. [6] The cumulative h is defined from the beginning of the observation and with respect to some normal reference level h_{ref}:

$$h_{cum}(V_i) = -\sum_{l=1}^{i}(h_{eff}(V_l) - h_{ref}) \, . \qquad (6)$$

The minus sign is introduced to give the h_{cum} indicator increasing direction when the level of local correlations is lower than h_{ref}. This corresponds with a healthy condition. The case of higher correlations is associated with problems and, therefore, the accumulation of a positive difference

---

[6] To use the linear integral of the non-stationarity component of the variability of fetal heartbeat is simply a heuristic. Of course, other functional dependence than linear integral is possible. The discovery of a more suitable model of such possible functional dependence is the subject of our future research.

$(h_{eff}(V_l) - h_{ref})$ will lead to decreasing cumulative h. The cumulative indicator steadily increasing or remaining within some margin of fluctuations indicates no problems and a good prediction. When the indicator plunges down, it calls for intervention. This can, of course, happen at any moment during labour. The nature of this process is dramatically non-stationary, and a period of positive evaluation can be interrupted at any stage (for example by the occlusion of the umbilical cord due to movement). One of the examples given (figure 5 right) shows the cumulative indicator plunging after a prolonged homeostasis.

This record thus shows a 'novelty' or 'surprise' in the dynamic evolution of its multifractal properties as judged by the results of the related (bad) fetal outcome. For comparison, we show in the same figure 5 in the left panel a record of the raw, fractionally integrated fetal heartbeat time series which corresponds to a good fetal outcome. This time there is no diagnostic novelty or surprise in the time series - the corresponding fetal outcome has been pronounced OK. The apparent similarity of the record evolution would, however, likely prompt a different opinion, that in both cases a novelty occurred! The difference is that the panel on the right (in Fig. 5) has been drawn from a diagnostic carrying information derived from the time series using a heuristic model. The panel on the left has been prepared using the raw time series transformed linearly by (fractional F0.5) integration.

6 Conclusions

We have elaborated on the difficulties which may arise in the (novelty oriented) analysis of time series arising from real life complex systems. We have also presented a number of example approaches aiming at 1) tempering the singular behaviour by means of wavelet transforming, and 2) capturing complex multifractal model characteristics. Additionally, we have suggested evaluating novelty by monitoring the instances of failure of such a model, as is pronounced in non-stationarities of its characteristics. Such novelty can be linked to external variables, can be further incorporated in a heuristic model or investigated by probabilistic methods (Bayesian net) for evidence of a higher dimensional model. The methodology, however strongly it opposes attempting to model time series by the reconstruction of the dynamic properties of the system (e.g. phase space reconstruction, feature (based) linguistic), does not preclude it. In fact some aspects of tempering (taming) the singular time series involved and reasoning using multifractal (or other appropriate stochastic model) characteristics may prove to be practical in such an alternative approach.

As an additional concluding observation, it is possibly the lack of multiresolution capabilities which limits the possibilities of the techniques for novelty discovery [1–3]. They operate with one fixed resolution which makes it difficult to derive scaling laws. The wavelet transform makes multiresolution analysis possible and, in the context of novelty discovery, it has been shown to make possible the derivation of 'rules and laws from data' [33–35].

Acknowledgments

The author would like to express thanks to Plamen Ch. Ivanov, Eugene H. Stanley, Shomo Havlin, Arno Siebes, Robert Castelo and Willem van Wijngaarden, who in various ways contributed to the work referred to within this paper.

References

1. C. Shahabi, X. Tian, W. Zhao, TSA-tree: a Wavelet Based Approach to Improve the Efficieny of Multi-level Surprise and Trend Queries. In *Proc. of the 12th Int'l Conf. on Scientific and Statistical Database Management*, 55-68, Berlin, Germany, July 26-28, (2000).
2. D. Dasgupta, S. Forrest, Novelty Detection in Time Series Data Using Ideas from Immunology, In *Proceedings of the 4th International Conference of Knowledge Discovery and Data Mining*, 16-22, (AAAI Press 1998).
3. E. Keogh, S. Leonardi, B.Y. Chiu, Finding Surprising Patterns in a Time Series Database in Linear Time and Space, Proc. ACM Knowledge Discovery and Data Mining, pp 550-556, (2002).
4. M. Kobayashi, T. Musha, 1/f Fluctuation of Heartbeat Period, *IEEE Trans Biomed. Eng.*, **29**, 456-457 (1981).
5. C.-K. Peng, J. Mietus, J.M. Hausdorff, S. Havlin, H.E. Stanley, A.L. Goldberger, Long-Range Anticorrelations and Non-Gaussian Bahavior of the Heartbeat *Phys. Rev. Lett.*, **70**, 1343-1346, (1993).
6. J. B. Bassingthwaighte, L. S. Liebovitch and B. J. West. *Fractal Physiology*, (Oxford University Press, 1994).
7. A. Fisher, L. Calvet, B.B. Mandelbrot, Multifractality of the Deutschmark/US Dollar Exchange Rate, Cowles Foundation Discussion Paper, (1997).
8. M.E. Brachet, E. Taflin, J.M. Tchéou, Scaling Transformations and Probablity Distributions for Financial Time Series, preprint cond-mat/9905169, (1999).
9. F. Schmitt, D. Schwertzer, S. Levejoy, Multifractal Analysis of Foreign Exchange Data, *Appl. Stochasic Models Data Anal.* **15**, 29-53, (1999).
10. A. Johansen, D. Sornette, Large Stock Market Price Drawdowns Are Outliers arXiv:cond-mat/0010050, 3 Oct 2000, rev. 25 Jul 2001.
11. Z. R. Struzik. Wavelet Methods in (Financial) Time-series Processing. *Physica A: Statistical Mechanics and its Applications*, **296**, No. (1-2), 307-319, (2001).
12. D. Sornette, Y. Malevergne, J.F. Muzy, Volatility Fingerprints of Large Shocks: Endogeneous Versus Exogeneous, arXiv:cond-mat/0204626, (2002).
13. X. Gabaix, P. Gopikrishnan, V. Plerou, H.E. Stanley, Understanding Large Movements in Stock Market Activity, (2002), preprint available from http://econ-www.mit.edu/faculty/xgabaix
14. Z. R. Struzik, W. J. van Wijngaarden, R. Castelo. Reasoning from Nonstationarity. *Physica A: Statistical Mechanics and its Applications*, **314** No. (1-4), 245-254, (2002).

15. S. Jaffard, Multifractal Formalism for Functions: I. Results Valid for all Functions, II. Self-Similar Functions, *SIAM J. Math. Anal.*, **28**, 944-998, (1997).

16. I. Daubechies, *Ten Lectures on Wavelets*, (S.I.A.M., 1992).

17. M. Holschneider, *Wavelets - An Analysis Tool*, (Oxford Science, 1995).

18. A. Muñoz Barrutia, R. Ertlé, M. Unser, "Continuous Wavelet Transform with Arbitrary Scales and O(N) Complexity," *Signal Processing* **82**, 749-757, (2002).

19. M. Unser, A. Aldroubi, S.J. Schiff, Fast Implementation of the Continuous Wavelet Transform with Integer Scales, *IEEE Trans. on Signal Processing* **42**, 3519-3523, (1994).

20. Z. R. Struzik. Oversampling the Haar Wavelet Transform. Technical Report INS-R0102, CWI, Amsterdam, The Netherlands, March 2001.

21. S.G. Mallat and W.L. Hwang, Singularity Detection and Processing with Wavelets. *IEEE Trans. on Inform. Theory* **38**, 617 (1992). S.G. Mallat and S. Zhong Complete Signal Representation with Multiscale Edges. *IEEE Trans. PAMI* **14**, 710 (1992).

22. J.F. Muzy, E. Bacry and A. Arneodo, The Multifractal Formalism Revisited with Wavelets. *Int. J. of Bifurcation and Chaos* **4**, No 2, 245 (1994).

23. R.Carmona, W.H. Hwang, B. Torrésani, Characterisation of Signals by the Ridges of their Wavelet Transform, *IEEE Trans. Signal Processing* **45**, 10, 480-492, (1997).

24. H.E. Stanley, P. Meakin, Multifractal Phenomena in Physics and Chemistry, *Nature*, **335**, 405-409, (1988).

25. A. Arneodo, E. Bacry, J.F. Muzy, Wavelets and Multifractal Formalism for Singular Signals: Application to Turbulence Data, *PRL*, **67**, No 25, 3515-3518, (1991).

26. P.Ch. Ivanov, M.G. Rosenblum, L.A. Nunes Amaral, Z.R. Struzik, S. Havlin, A.L. Goldberger and H.E. Stanley, Multifractality in Human Heartbeat Dynamics, *Nature* **399**, 461-465, (1999).

27. K. Falconer, Fractal Geometry: Mathematical Foundations and Applications, (John Wiley, 1990; paperback 1997).

28. E. Bacry, J. Delour, J.F. Muzy, A Multifractal Random Walk, arXiv:cond-mat/0005405, (2000).

29. M.J. Wainweight, E.P. Simoncelli, A.S. Willsky, Random Cascades on Wavelet Trees and Their use in Analysing and Modeling Natural Images, *Applied and Computational Harmonic Analysis* **11**, 89-123 (2001).

30. Z. R. Struzik, Determining Local Singularity Strengths and their Spectra with the Wavelet Transform, *Fractals*, **8**, No 2, 163-179, (2000).

31. Z. R. Struzik. Revealing Local Variablity Properties of Human Heartbeat Intervals with the Local Effective Hölder Exponent. *Fractals* **9**, No 1, 77-93 (2001).

32. Z. R. Struzik, W. J. van Wijngaarden, Cumulative Effective Hölder Exponent Based Indicator for Real Time Fetal Heartbeat Analysis during Labour. In *Emergent Nature: Fractals 2002*, M. M. Novak, Ed., (World Scientific, 2002).

33. A. Arneodo, J.F. Muzy, D. Sornette, Causal Cascade in the Stock Market from the "Infrared" to the "Ultraviolet", *Eur. Phys J. B* **2**, 277 (1998).

34. A. Arneodo, E. Bacry and J.F. Muzy, Solving the Inverse Fractal Problem from Wavelet Analysis, *Europhysics Letters* **25**, No 7, 479-484, (1994).

35. Z.R. Struzik The Wavelet Transform in the Solution to the Inverse Fractal Problem. *Fractals* **3** No. 2, 329 (1995).

Problems with Automatic Classification of Musical Sounds

Alicja A. Wieczorkowska[1], Jakub Wróblewski[1],
Dominik Ślęzak[2,1], and Piotr Synak[1]

[1] Polish-Japanese Institute of Information Technology
   ul. Koszykowa 86, 02-008 Warsaw, Poland
[2] Department of Computer Science University of Regina
   Regina, SK, S4S 0A2, Canada

Abstract. Convenient searching of multimedia databases requires well annotated data. Labeling sound data with information like pitch or timbre must be done through sound analysis. In this paper, we deal with the problem of automatic classification of musical instrument on the basis of its sound. Although there are algorithms for basic sound descriptors extraction, correct identification of instrument still poses a problem. We describe difficulties encountered when classifying woodwinds, brass, and strings of contemporary orchestra. We discuss most difficult cases and explain why these sounds cause problems. The conclusions are drawn and presented in brief summary closing the paper.

1 Introduction

With the increasing popularity of multimedia databases, a need arises to perform efficient searching of multimedia contents. For instance, the user can be interested in finding specific tune, played by the guitar. Such searching cannot be performed efficiently on raw sound or image data. Multimedia data should be first annotated with descriptors that facilitate such search. Standardization of multimedia content description is a scope of MPEG-7 standard [9,13]. However, algorithms of descriptors extraction or database searching are not within a scope of this standard, so they are still object of research. This is why we decided to investigate labeling of mono sounds with the names of musical instruments that play these sounds.

In this paper, we deal with problems that arise when automatic classification of musical instrument sounds is performed. It is far from perfect, and we especially focus on these sounds (and instruments) that are misclassified.

2 Classification of Musical Instruments

There exist numerous musical instruments all over the world. One can group them into classes according to various criteria. Widely used Sachs-Hornbostel system [8] classifies musical instruments into the categories, which can be seen in Table 1.

Table 1. Categories and subcategories of musical instruments

Category	Criteria for subclasses	Subclasses	Instruments
idiophones	material	struck together	castanets
	whether pitch is important	struck	gongs
	no. of idiophones in instrument	rubbed	saw
	no. of resonators	scraped	washboards
		stamped	floors
		shaken	rattles
		plucked	Jew's harp
membrano-	whether has 1 or 2 heads	drums: cylindrical,	
phones	if there are snares, sticky balls	conical, barrel,	
	how skin is fixed on drum	hourglass, long,	
	whether drum is tuned	goblet,	darabukke
	how it is tuned	kettle, footed,	
	how it is played	frame drum	tambourine
	position of drum when played	friction drum	
	body material	mirliton/kazoo	kazoo
chordophones	number of strings	zither	piano
	how they are played	lute plucked	guitar
	tuning	lute bowed	violin
	presence of frets	harp	harps
	presence of movable bridges	lyre, bow	
aerophones	kind of mouthpiece:	flutes: side-blown,	
	blow hole	end-blown, nose,	
	whistle	multiple, panpipes	
	single reed	globular flute	ocarina
	double reed	whistle mouthpiece	recorder
	lip vibrated	single reed	clarinet
		double reed	oboe
		air chamber	accordion
		lip vibrated	brass
		free aerophone	bullroarers
electrophones			keyboards

Membranophones and idiophones are together called percussion. Contemporary classification also adds electrophones to this set. The categories are further divided into subcategories [17]. As one can see, the variety of instruments complicates the process of classifications, especially in case of percussion, when classification depends on the shape of the instrument. The sound parameterization for the classification purposes is often based on harmonic properties of sound, so dealing with definite pitch (fundamental frequency) sounds is much more common and convenient. This is why we decided to limit ourselves to instruments of definite pitch.

In our paper, we deal with chordophones and aerophones only. The instruments we analize include bowed lutes (violin, viola, cello, and double bass), side-blown flute, single reed, double reed, and lip vibrated. All of them produce sounds of definite pitch. We use fundamental frequency of musical instrument sounds as a basis of sound parameterization, as well as the envelope of the waveform. Further parameters are described in the next section.

3 Sound Parameterization

In our research, we dealt with 667 singular sounds of instruments, recorded from MUMS CDs with 44.1kHz frequency and 16bit resolution [15]. MUMS library is commonly used in experiments with musical instrument sounds [4–6,10,14,20], so we can consider them to be a standard.

There exist many descriptors that can be applied for the instrument classification purposes. Recently elaborated MPEG-7 standard for Multimedia Content Description Interface provides for 17 audio descriptors [9], but many other sound features have been applied by the researchers so far, including various spectral and temporal features, autocorrelation, cepstral coefficients, wavelet-based descriptors, and so on [1,2,5–7,10,11,14,16,18–20]. The sound parameterization we applied starts with extraction of the following temporal, spectral, and envelope descriptors [18]:

Temporal descriptors:

- *Length*: Signal length
- *Attack*, *Steady* and *Decay*: Relative length of the attack (till reaching 75% of maximal amplitude), quasi-steady (after the end of attack, till the final fall under 75% of maximal amplitude) and decay time (the rest of the signal), respectively
- *Maximum*: Moment of reaching maximal amplitude

Spectral descriptors:

- *EvenHarm* and *OddHarm*: contents of even and odd harmonics in spectrum
- *Brightness* and *Irregularity* [12]:

$$Br = \frac{\sum_{n=1}^{N} nA_n}{\sum_{n=1}^{N} A_n} \quad Ir = \log \sum_{n=2}^{N-1} |20 \log A_n - \frac{20 \log(A_{n+1}A_nA_{n-1})}{3}| \quad (1)$$

  where A_N is the amplitude of nth partial (harmonic) and N is number of available partials

- *Tristimulus* 1, 2, 3 [16]:

$$Tr_1 = \frac{A_1^2}{\sum_{n=1}^{N} A_n^2} \quad Tr_2 = \frac{\sum_{n=2,3,4} A_n^2}{\sum_{n=1}^{N} A_n^2} \quad Tr_3 = \frac{\sum_{n=5}^{N} A_n^2}{\sum_{n=1}^{N} A_n^2} \quad (2)$$

- *Frequency*: Fundamental frequency

Envelope descriptors:

- $Val_{Amp}1, \ldots, Val_{Amp}7$: Average values of amplitudes within 7 intervals of equal width for a given sound
- *EnvFill*: Area under the curve of envelope, approximated by means of values $Val_{Amp}1, \ldots, Val_{Amp}7$
- *Cluster*: Number of the closest of 6 representative envelope curves (obtained via clustering) shown in Figure 1 [18].

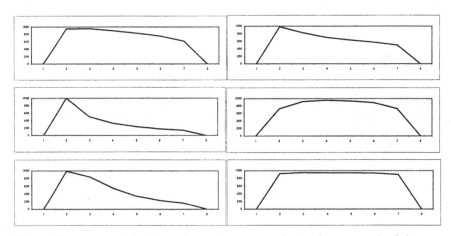

Fig. 1. The most typical shapes of sound envelopes, obtained as centroids of clusters

Some of the descriptors we calculated for the whole sound, whereas other where calculated for frames of length equal to 4 periods of sound and rectangular window through the whole sound (table WINDOW). The obtained basic data are represented as relational database. The structure of this database is shown in Figure 2 [18]. Objects of the database are classified according to both the instrument and articulation (how the sound is played) to the following classes: violin vibrato (denoted vln), violin pizzicato (vp), viola vibrato (vla), viola pizzicato (vap), cello vibrato (clv), cello pizzicato (clp), double bass vibrato (cbv), double bass pizzicato (cbp), flute (flt), oboe (obo), b-flat clarinet (cl), trumpet (tpt), trumpet muted (tpm), trombone (tbn), trombone muted (tbm), French horn (fhr), French horn muted (fhm) and tuba (tub).

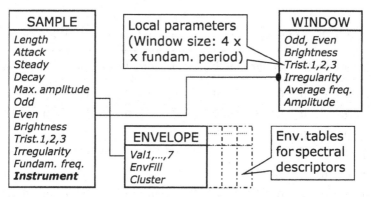

Fig. 2. Relational database of musical instrument sound descriptors

Apart from the data described above, we used new parameters, calculated for the basic ones. Namely, we extended the database by a number of new attributes defined as linear combinations of the existing ones. Additionally, the WINDOW table was used to search for the temporal templates, in order to use the found frequent episodes as new sound features [18].

4   Experiments

The data described in the previous section were used in experiments with automatic recognition of musical instrument sounds. The most common methods used in such experiments include k-nearest neighbor classifier (also with genetic algorithm to seek the optimal set of weights for the features), Bayes decision rules, decision trees, rough set based algorithms, neural networks, hidden Markov models and other classifiers [1,3,5,6,10,11,19,20]. Sometimes classification is performed in 2 stages:first, the sound is classified into a group (according to instrument category or articulation), and then the instrument is recognized. Extensive review of research in this domain is presented in [7].

Results of experiments vary, but apart from small data sets (for instance, 4 classes only), they are far from perfect, generally around 70-80% for instruments and about 90% for groups. The results are usually presented in form of correctness tables for various settings of the experiment method, and only some papers cover confusion matrices for the investigated instruments.

In the research for four woodwind instruments: oboe, sax, clarinet, and flute [3], confusions are presented in percentage for every pair. No overall pattern was observed for these data. In experiments for larger set of instruments (19), detailed confusion matrices are presented [10]. For these data, sax was frequently mistaken for clarinet (10 out of 37 samples for the best combined feature classifier) and trombone for French horn (7 out of 28 samples for the same classifier). Confusion matrix for our research is presented in Table 2.

Table 2. Confusion matrix for all (spectral and temporal) attributes

	cl	cbv	cbp	tpt	tpm	fhr	fhm	flt	obo	tbn	tbm	tub	vla	vap	clv	clp	vln	vp	%
cl	37	0	0	0	0	0	0	0	0	0	0	0	0	0	0	0	0	0	100
cbv	0	36	0	0	0	2	0	0	2	1	1	0	2	0	0	0	0	0	81.82
cbp	0	0	25	0	0	0	0	0	0	0	0	0	0	1	0	2	0	2	83.33
tpt	4	0	0	24	3	0	1	0	0	0	0	1	1	0	0	0	0	0	70.59
tpm	3	0	0	2	22	0	1	0	0	0	0	0	0	0	0	0	3	0	70.97
fhr	0	2	0	0	0	33	2	0	0	0	0	0	0	0	0	0	0	0	89.19
fhm	0	0	0	0	0	0	32	0	2	2	0	0	0	0	0	0	1	0	86.49
flt	3	0	0	0	0	0	0	33	0	1	0	0	0	0	0	0	0	0	89.19
obo	0	1	0	0	0	0	3	1	22	1	1	0	0	0	1	0	2	0	68.75
tbn	0	0	0	0	0	5	3	5	1	19	3	0	0	0	0	0	0	0	52.78
tbm	0	2	0	0	0	3	1	0	0	3	22	0	0	0	2	0	0	0	66.67
tub	0	1	0	0	0	0	0	1	0	0	0	29	0	0	1	0	0	0	90.63
vla	6	7	1	0	0	1	3	0	0	0	0	0	13	0	2	0	9	0	30.95
vap	0	0	0	0	0	0	0	0	0	0	0	0	0	21	0	4	0	9	61.76
clv	0	7	0	0	1	1	0	0	5	0	3	0	2	0	27	0	1	0	57.45
clp	0	0	6	0	0	0	0	0	0	1	0	0	0	13	0	18	0	1	46.15
vln	1	0	0	0	0	0	0	1	5	0	0	0	4	0	1	0	33	0	73.33
vp	0	0	1	0	0	0	0	0	0	0	0	0	0	14	0	0	0	25	62.50

As we can see, the most difficult instruments to classify in our case were violin and cello pizzicato, in 14 and 13 cases respectively, misclassified for viola pizzicato. Since pizzicato sounds are very short and these instruments belong to the same family, this is not surprising. Clarinet was perfectly classified (we have not investigated sax that was problematic in [10]). Other instruments did not show such distinct patterns, but generally strings yielded lower results, with viola being the most difficult to classify correctly. Average recognition rate was 70.61%. These results are comparable with other research, and also with human achievements in musical instrument sound classification [3].

5   Conclusions

Classification of musical instrument sounds must take into account various articulation methods and categorization of instruments. In our research, we investigated sounds of non-percussion instruments of contemporary orchestra, including strings, woodwind, and brass. The most difficult were string sounds, especially when the investigated sounds are very short and change dramatically in time, i.e. played pizzicato (string plucked with finger). Such sounds are also difficult to parameterize, since analyzing frame must be also very short, and the sound features change very quickly. String sound played vibrato are also quite similar, so it is understandable that they can be mistaken. Generally, instruments belonging to the same category, or, even worse, to the same subcategory, are more difficult to discern. Additionally, vibration introduces fluent changes of sound features and also makes recognition more challenging. However, we hope that investigation of various sound parameterization techniques, combined with testing of various classification algorithms, may move forward the research on automatic indexing of musical sounds.

6   Acknowledgements

This research was partially supported by Polish National Committee for Scientific Research (KBN) in form of PJIIT Project No. $ST/MUL/01/2002$.

References

1. Batlle, E. and Cano, P. (2000) Automatic Segmentation for Music Classification using Competitive Hidden Markov Models. Proceedings of International Symposium on Music Information Retrieval. Plymouth, MA.
2. Brown, J. C. (1999) Computer identification of musical instruments using pattern recognition with cepstral coefficients as features. J. Acoust. Soc. of America, 105, 1933–1941
3. Brown, J. C., Houix, O., and McAdams, S. (2001) Feature dependence in the automatic identification of musical woodwind instruments. J. Acoust. Soc. of America, 109, 1064–1072
4. Cosi, P., De Poli, G., and Lauzzana, G. (1994) Auditory Modelling and Self-Organizing Neural Networks for Timbre Classification. Journal of New Music Research, 23, 71–98

5. Eronen, A. and Klapuri, A. (2000) Musical Instrument Recognition Using Cepstral Coefficients and Temporal Features. Proceedings of the IEEE International Conference on Acoustics, Speech and Signal Processing ICASSP 2000. Plymouth, MA. 753–756

6. Fujinaga, I. and McMillan, K. (2000) Realtime recognition of orchestral instruments. Proceedings of the International Computer Music Conference. 141–143

7. Herrera, P., Amatriain, X., Batlle, E., and Serra X. (2001) Towards instrument segmentation for music content description: a critical review of instrument classification techniques. In: Proc. of ISMIR 2000, Plymouth, MA

8. Hornbostel, Erich M. v. and Sachs, C. (1914) Systematik der Musikinstrumente. Ein Versuch. Zeitschrift für Ethnologie, **46**, (4-5):553–90. Available at http://www.uni-bamberg.de/ppp/ethnomusikologie/HS-Systematik/HS-Systematik

9. ISO/IEC JTC1/SC29/WG11 (2002) MPEG-7 Overview. Available at http://mpeg.telecomitalialab.com/standards/mpeg-7/mpeg-7.htm

10. Kaminskyj, I. (2000) Multi-feature Musical Instrument Classifier. MikroPolyphonie **6** (online journal at http://farben.latrobe.edu.au/)

11. Kostek, B. and Czyzewski, A. (2001) Representing Musical Instrument Sounds for Their Automatic Classification. J. Audio Eng. Soc., **49(9)**, 768–785

12. Krimphoff, J., Mcadams, S., and Winsberg, S. (1994) Caractérisation du Timbre des Sons Complexes. II. Analyses acoustiques et quantification psychophysique. Journal de Physique IV, Colloque C5, J. de Physique III, 4, 3ème Congrès Français d'Acoustique, I, 625–628

13. Lindsay, A. T. and Herre, J. (2001) MPEG-7 and MPEG-7 Audio – An Overview. J. Audio Eng. Soc., **49(7/8)**, 589–594

14. Martin, K. D. and Kim, Y. E. (1998) 2pMU9. Musical instrument identification: A pattern-recognition approach. 136-th meeting of the Acoustical Soc. of America, Norfolk, VA

15. Opolko, F. and Wapnick, J. (1987) MUMS – McGill University Master Samples. CDs

16. Pollard, H. F. and Jansson, E. V. (1982) A Tristimulus Method for the Specification of Musical Timbre. Acustica, **51**, 162–171

17. SIL International (1999) 534 Musical Instruments subcategories. http://www.sil.org/LinguaLinks/Anthropology/ExpnddEthnmsclgyCtgrCltrlMtrls/MusicalInstrumentsSubcategorie.htm

18. Ślęzak, D., Synak, P., Wieczorkowska, A., and Wróblewski, J. (2002) KDD-based approach to musical instrument sound recognition. In Hacid M.-S., Raś Z., Zighed D. A., Kodratoff Y. (Eds.), Foundations of Intelligent Systems. Proc. 13th International Symposium ISMIS 2002. LNAI 2366, Springer, 29–37

19. Wieczorkowska, A. A. (1999) The recognition efficiency of musical instrument sounds depending on parameterization and type of a classifier (in Polish), Ph.D. Dissertation, Technical University of Gdańsk, Gdańsk.

20. Wieczorkowska, A. (1999) Rough Sets as a Tool for Audio Signal Classification. In Z. W. Ras, A. Skowron (Eds.), Foundations of Intelligent Systems, LNCS/LNAI 1609, Springer, 367–375

Discovering Dependencies in Sound Descriptors

Alicja A. Wieczorkowska[1] and Jan M. Żytkow[2]

[1] Polish-Japanese Institute of Information Technology, ul. Koszykowa 86, 02-008
Warsaw, Poland
[2] The University of North Carolina at Charlotte, 9201 University City Blvd,
Charlotte, NC 28223

Abstract. Multimedia and sound databases require special attention to perform
automatic searching for any musical data. To enable automatic search, sound processing and parameterization is needed. This paper investigates dependencies between most popular sound attributes used for sound description purposes. Apart
from experiments and results, we present considerations on possible further research, including industry applications.

1 Introduction

Automatic classification of sounds is practically impossible to perform on the
basis of row audio data, and preprocessing that parameterizes the sounds is
necessary. The produced set of parameters (descriptors) should match the
recognition task. In this paper, the task is to identify musical instruments
that produced the analyzed sounds.

Investigations on automatic musical instrument sound classification has
been progressing recently, and various parameterization techniques have been
applied to sound data as preprocessing. Extracted features are based on various sound analysis methods, like Fourier transform, wavelet analysis, and
so on. Features used in the research worldwide include spectral parameters,
i.e. spectral moments [4], statistical properties of spectrum [1], contents of
the selected groups of partials in sound spectrum [8], [11], inharmonicity,
spectral envelope [9]; time-related parameters, i.e. onset duration [9], amplitude envelope [7]; wavelet-based parameters [12]; cepstral coefficients based
on constant Q transform [2], and other parameters [5], [12]. Classification algorithms applied in the above mentioned research include k-nearest neighbor
[4], [7], [9], statistical methods [9], decision trees, rough set based algorithms,
neural networks [12], and others [5].

Musical instrument sounds can be classified on various levels. First of all,
sounds are classified at instrument level. However, the method of sound production (articulation) or families of instruments are also considered [5], [9],
[12]. In this paper, we classify musical instrument sounds at various levels, additionally looking into inside structure of parameterization using Forty-Niner

system approach [14], see Section 2. Starting with popular sound attributes based on Fourier analysis, we test dependencies between the attributes. We believe it is both interesting and useful in the research on automatic musical sound classification.

Presented work comes from the research performed while preparing Ph.D. dissertation under the direction of A. Czyżewski from Sound and Vision Engineering Department of the Gdańsk University of Technology.

2   Searching for Regularities

Forty-Niner (49er) system used in the described research is a general-purpose database mining system [14]. 49er conducts large-scale search for regularities in subsets of the investigated data. Such regularities (or patterns) can be found in form of contingency tables, equations, or logical equivalences [13]. Firstly, contingency tables are found, and if the data indicate a functional relationship, then more costly search for equations is conducted.

The searching procedure can be applied to any relational table (data matrix). Initially, 49er examines the contingency table for each pair of attributes, and then search in the space of equation is invoked for the discovered regularities. A contingency table shows actual distribution for a pair of attributes. Each entry in the table equals to the number of records that have the corresponding combination of values of both attributes. An example of contingency table for strongly correlated attributes x and y is shown in Table 1.

Table 1. Contingency table for highly correlated attributes x and y, with attribute domains $\{1, 2, 3\}$ and $\{a, b, c, d\}$ respectively

Attribute x				
1	65	0	0	1
2	0	1	0	26
3	2	0	73	0
	a	b	c	d Attribute y

In 49er system, regularities are approved based on their significance, measured by the probability that they are random fluctuations. This probability is determined on the basis of χ^2 test, which measures the distance between tables of actual and expected counts as follows:

$$\chi^2 = \sum_{i,j} \frac{(A_{ij} - E_{ij})^2}{E_{ij}} \qquad \text{where}$$

$E_{ij} = \frac{n_{x_i} \cdot n_{y_j}}{N}$ - expected number of records with $x = x_i$ and $y = y_j$, where x, y - parameters, x_i, y_j - parameter values, N - total number of records

A_{ij} - actual frequency distribution

Since χ^2 depends on the size of the data set, Cramer's V coefficient is also calculated, in order to abstract from the number of data:

$$V = \sqrt{\frac{\chi^2}{N \cdot \min(M_{row} - 1, M_{col} - 1)}}$$

where $M_{row} \times M_{col}$ - size of the contingency table

For ideal correlation, Cramer's $V = 1$, and $V = 0$ for perfectly non-correlated data.

3   The Database

Experiments described in this paper are based on musical instrument sound parameterization. The sounds used for database creation come from a collection of CDs, prepared at McGill University (MUMS) for sampling purposes [10]. MUMS features stereo samples of musical instrument sounds, recorded chromatically within the standard playing range of instruments, with timbral variations. 16-bit stereo recording with 44.1 kHz sampling frequency were made on the basis of these CDs. In the described experiments, 679 sounds of the following instruments have been chosen:

- strings: violin, viola, cello, and double bass, played vibrato and pizzicato;
- woodwinds: flute, oboe, and b-flat clarinet;
- brass: trumpet, trombone and French horn, played with and without muting, and tuba without muting.

Each record in the investigated database corresponds to a singular sound of one instrument. The records include 69 attributes (parameters), 62 of them are numerical. These 62 attributes describe general properties of the whole sound, as well as specific properties of 2 most important parts of the sound: the initial phase, called the attack (parameterized at the beginning, in the middle, and at the end), and the main phase, called quasi-steady state (parameterized for the maximal and minimal amplitude). The parameters calculated for the attack and for the quasi-steady state in the selected points are:

- n_{fv} - number of partial $n_{fv} \in \{1, \ldots, 5\}$ of greatest frequency variation,
- $\overline{fv}$ - weighted mean frequency variation for 5 lowest partials

$$\overline{fv} = \frac{\sum_{k=1}^{5} A_k \cdot \frac{\Delta f_k}{k f_1}}{\sum_{k=1}^{5} A_k} \qquad \text{where}$$

A_k - amplitude of k-th partial

f_k - frequency of k-th partial

- Tr_1 - modified first Tristimulus parameter [11]

$$Tr_1 = \frac{A_1^2}{\sum_{n=1}^{N} A_n^2}$$

- A_{1-2} - difference of amplitudes between the first partial (fundamental) and the second one

$$A_{1-2} = 20 \log \frac{A_1}{A_2}$$

- $H_{3,4}, H_{5,6,7}, H_{8,9,10}, H_{rest}$ - contents of the selected groups of harmonics in spectrum

$$H_{3,4} = \frac{\sum_{n=3,4} A_n^2}{\sum_{n=1}^{N} A_n^2}$$

$$H_{5,6,7} = \frac{\sum_{n=5,6,7} A_n^2}{\sum_{n=1}^{N} A_n^2}$$

$$H_{8,9,10} = \frac{\sum_{n=8,9,10} A_n^2}{\sum_{n=1}^{N} A_n^2}$$

$$H_{rest} = \frac{\sum_{n=11}^{N} A_n^2}{\sum_{n=1}^{N} A_n^2}$$

- Od - contents of odd harmonics in spectrum, excluding fundamental

$$Od = \frac{\sqrt{\sum_{k=2}^{\lfloor N/2+1 \rfloor} A_{2k-1}^2}}{\sqrt{\sum_{n=1}^{N} A_n^2}}$$

- Ev - contents of even harmonics in spectrum

$$Ev = \frac{\sqrt{\sum_{k=1}^{\lfloor N/2 \rfloor} A_{2k}^2}}{\sqrt{\sum_{n=1}^{N} A_n^2}}$$

- Br - brightness of sound

$$Br = \frac{\sum_{n=1}^{N} n \cdot A_n}{\sum_{n=1}^{N} A_n}$$

General properties of the whole sound are described by the following seven parameters:

- Vb - depth of vibrato

$$Vb = |f_{1max} - f_{1min}|$$

where f_{1max}, f_{1min} - frequency of fundamental for maximal and minimal amplitude in the quasi-steady state respectively
- f_1 - fundamental in the middle of the sound [Hz]
- d_{fr} - approximate fractal dimension of graph of spectrum amplitude envelope in decibel scale

$$d_{fr} = -\frac{\log N(\Delta s)}{\log \Delta s} \qquad \text{where}$$

Δs - mesh length of grid that covers the plane where the graph is drawn (here $\Delta s = 10^{-10}$)
$N(\Delta s)$ - number of nonempty mesh; for $f_s = 44.1kHz$, analysis frame 5520 samples was used, in order to cover at least 2 periods of the analyzed sound;
- $f_{1/2}$ - contents of subharmonics in the spectrum (overblow)

$$f_{1/2} = \frac{\sum_{m=1}^{M}[A(m \cdot f_1 - f_1/2)]^2}{\sum_{n=1}^{N}[A(n \cdot f_1)]^2}$$

- Qt - duration of the quasi-steady state in proportion to the total sound time, $Qt \in [0,1]$
- Et - duration of the ending transient of the sound in proportion to the total sound time, $Et \in [0,1]$
- Rl - velocity of fading of ending transient [dB/s]

$$Rl = \frac{10 \log \frac{\sum_{n=S-l+1}^{S}[A(nT_s)]^2}{\sum_{m=R-l+1}^{R}[A(mT_s)]^2}}{(R-S) \cdot T_s}, \qquad \text{where}$$

T_s - sampling period
R - time moment of the end of the sound
S - time moment of the beginning of the ending transient
l - number of samples in the sampling period
$A(t)$ - amplitude for the time instant t

All conditional attributes are presented in Table 2.
Attributes 63-69 represent various ways of classification of sounds in the database, grouping all objects from the database into 2-18 classes:

- attribute 63: 18 classes, each one contains records representing sounds of the same instrument, played with the same technique (articulation)
  - flute, oboe, clarinet, trumpet, trumpet muted, trombone, trombone muted, French horn, French horn muted, tuba, violin vibrato, violin pizzicato, viola vibrato, viola pizzicato, cello vibrato, cello pizzicato, double bass vibrato, double bass pizzicato,
- attribute 64: 2 classes - groups of instruments
  - strings, winds

Table 2. Conditional attributes of musical instrument sounds in the database

Attack:			Steady state:		General
Beginning	Middle	End	Maximum	Minimum	
1. n_{fv}	12. n_{fv}	23. n_{fv}	34. n_{fv}	45. n_{fv}	56. Vb
2. $\overline{fv}$	13. $\overline{fv}$	24. $\overline{fv}$	35. $\overline{fv}$	46. $\overline{fv}$	57. f_1
3. Tr_1	14. Tr_1	25. Tr_1	36. Tr_1	47. Tr_1	58. d_{fr}
4. A_{1-2}	15. A_{1-2}	26. A_{1-2}	37. A_{1-2}	48. A_{1-2}	59. $f_{1/2}$
5. $H_{3,4}$	16. $H_{3,4}$	27. $H_{3,4}$	38. $H_{3,4}$	49. $H_{3,4}$	60. Qt
6. $H_{5,6,7}$	17. $H_{5,6,7}$	28. $H_{5,6,7}$	39. $H_{5,6,7}$	50. $H_{5,6,7}$	61. Et
7. $H_{8,9,10}$	18. $H_{8,9,10}$	29. $H_{8,9,10}$	40. $H_{8,9,10}$	51. $H_{8,9,10}$	62. Rl
8. H_{rest}	19. H_{rest}	30. H_{rest}	41. H_{rest}	52. H_{rest}	
9. Od	20. Od	31. Od	42. Od	53. Od	
10. Ev	21. Ev	32. Ev	43. Ev	54. Ev	
11. Br	22. Br	33. Br	44. Br	55. Br	

- attribute 65: 3 classes - groups of instruments
  - strings, woodwinds, brass
- attribute 66: 2 classes, representing articulation techniques
  - vibrato, non-vibrato (including pizzicato)
- attribute 67: 5 classes, representing groups of instruments, played with the same articulation
  - strings vibrato, strings pizzicato, winds vibrato not muted, winds non-vibrato not muted, winds muted
- attribute 68: 5 classes, representing groups of instruments, played with the same articulation
  - strings vibrato, strings pizzicato, woodwinds, brass not muted, brass muted
- attribute 69: 11 classes - instruments
  - flute, oboe, clarinet, trumpet, trombone, French horn, tuba, violin, viola, cello, double bass.

These groups are based on general classification of musical instruments [3], and on methods of sound production, i.e. articulation. Such grouping also reflects research in this domain, since musical instrument sounds are classified on various levels, including instrument, family, and articulation level.

4   Experiments

In our experiments, we performed search for dependencies within the investigated data. As we expected, some of the dependencies within one column in Table 2 also appeared in adjacent columns (especially representing the same stage of the sound). For instance, attributes no. 33 and 55, and also 44 and 55, representing brightness, show strong functional dependencies (i.e. Cramer's V close to 1). Real-value attribute domain was divided into 6 intervals of equal width in our case. We can conclude that brightness of sound is predictable since the attack is finished. Generally, most of attributes from the same row for the last three columns of Table 2 are strongly connected, apart from the first 2 rows, i.e. number of a partial with the greatest frequency deviation (for 5 low partials) and mean frequency deviation for low 5 partials. For instance, $H_{3,4}$ (content of 3rd and 4th partial), i.e. attributes no. 27, 38 and 49, are strongly connected with respect to their values. It lets us conclude that sounds basically stabilize at the end of the attack, but still there happen changes in inharmonicity of low partials.

There are also dependencies within columns of Table 2. Attributes Tr_1, A_{1-2}, i.e. values of energy of fundamental in proportion to the whole spectrum, amplitude difference between 1st and 2nd partial, and contents of even partials in the spectrum, seem to be almost equivalent for all columns. Starting from the end of the attack, all these attributes appear to be almost equivalent. Probably predominant fundamental flattens the results obtained for higher partials. We can expect that calculation of all attributes in logarithmic scale may lessen these dependencies, and also division of the domain onto more sub-intervals.

Strong dependencies also appear between H_{rest}, i.e. for energy of the higher partials in the spectrum and Br, brightness of the sound, starting with the end of the attack (attributes no. 30, 33, 41, 44, 52, and 55). Therefore, we can conclude that brightness of sound is then based mainly on higher partials, and these parameters evolve similarly after the sound has stabilized.

For the beginning and the middle of the attack (the first two columns of Table 2, there are no significant dependencies between attributes from the same rows. Second and third column, apart from attributes no. 18 and 29 (energy of harmonics no. 8, 9, and 9 in the spectrum), also do not show significant dependencies, so we can see that musical instrument sounds evolve dramatically in their initial phase.

All the results commented above illustrate significant changes that the sound undergoes during the attack. This also reflects attention that human experts pay to the onset of the sound - it is necessary for them to classify musical instruments correctly. Sounds basically stabilize in the quasi-steady state, but still there are changes in its spectrum.

5   Conclusions

The investigations described in our paper focus on searching for the best representation of musical instrument sounds for timbre classification (instrument classification) purposes. This research can be generalized to searching for similarities and taxonomy of any sounds. In this paper, we focused on discovering regularities and knowledge in the investigated database. We found strong dependencies between some commonly used sound attributes, like dependency between brightness of sound and contents of higher harmonics in spectrum, and dependencies between attributes describing contents of selected groups of partials in spectrum. We also observed partial stabilization of sound properties after the end of the attack, although during the attack sound features evolve quite dramatically, and there are still changes during the quasi-steady state of the sound.

Finding dependencies in sound description and hidden hierarchies in audio data may help in exploration of sound databases in the future. We can expect further research on discovering knowledge from audio data, especially after development of MPEG-7 standard for multimedia content description. However, the research on audio databases is still a little bit neglected in comparison with image and video databases. Therefore, it is very important to investigate audio domain and focus more attention of scientists in this field.

References

1. Ando, S. and Yamaguchi, K. (1993) Statistical Study of Spectral Parameters in Musical Instrument Tones. J. Acoust. Soc. of America, **94(1)**, 37–45
2. Brown, J. C. (1999) Computer identification of musical instruments using pattern recognition with cepstral coefficients as features. J. Acoust. Soc. of America, **105**, 1933–194
3. Fletcher, N. H. and Rossing, T. D. (1991) The Physics of Musical Instruments. Springer
4. Fujinaga, I. and McMillan, K. (2000) Realtime recognition of orchestral instruments. Proceedings of the International Computer Music Conference. 141–143
5. Herrera, P., Amatriain, X., Batlle, E., and Serra X. (2001) Towards instrument segmentation for music content description: a critical review of instrument classification techniques. In: Proc. of ISMIR 2000, Plymouth, MA
6. ISO/IEC JTC1/SC29/WG11 (2002) MPEG-7 Overview. Available at http://mpeg.telecomitalialab.com/standards/mpeg-7/mpeg-7.htm
7. Kaminskyj, I. (2000) Multi-feature Musical Instrument Classifier. MikroPolyphonie **6** (online journal at http://farben.latrobe.edu.au/)
8. Kostek, B. and Wieczorkowska, A. (1997) Parametric representation of musical sounds. Archives of Acoustics, **22**, 3–26
9. Martin, K. D. and Kim, Y. E. (1998) 2pMU9. Musical instrument identification: A pattern-recognition approach. 136-th meeting of the Acoustical Soc. of America, Norfolk, VA
10. Opolko, F. and Wapnick, J. (1987) MUMS – McGill University Master Samples. CDs
11. Pollard, H. F. and Jansson, E. V. (1982) A Tristimulus Method for the Specification of Musical Timbre. Acustica, **51**, 162–171
12. Wieczorkowska, A. A. (1999) The recognition efficiency of musical instrument sounds depending on parameterization and type of a classifier (in Polish), Ph.D. Dissertation, Technical University of Gdańsk, Gdańsk.
13. Zembowicz, R. and Żytkow, J. M. (1996) From Contingency Tables to Various Forms of Knowledge in Databases. In: Kobsa, U. M., Piatetsky-Shapiro, G., Smyth, P., and Uthurusamy, R. (Eds.) Advances in Knowledge Discovery and Data Mining. AAAI Press, 329–349
14. Żytkow, J. M. and Zembowicz, R. (1993) Database Exploration in Search of Regularities. Journal of Intellingent Information Systems, **2**, 39–81

Part VII

Invited Session: Information Extraction and Web Mining by Machine

Intelligent Systems and Information Extraction - True and Applicative

Matjaz Gams

Department of intelligent systems, Jozef Stefan Institute, Jamova 39, 1000
Ljubljana, Slovenia
matjaz.gams@ijs.si, http://ai.ijs.si/mezi/matjaz.html

Abstract. First we analyze true intelligent systems with emphasis on supercomputing mechanisms, in particular the principle of multiple knowledge. The first part of the paper lays some background for further AI research towards true intelligence, and provides some practical clues for innovative applications. In the second part of the paper we present successful applications mimicking intelligence, especially in the information extraction area. These systems offer several advantages over classical computer systems, and currently represent the best intelligence that humans can implement on digital computers. We present a couple of advanced applications developed or being developed in our group: an employment agent, a speech agent, a semantic web speaking system, a still-mill application, an agent-based group server, and an e-commerce agent. The major conclusion is that the field is ripe for innovative applications, and that in this way we can significantly enrich information society in any country.

1 Intelligent Systems and True Intelligence

Intelligent systems simulate intelligence so that a normal user perceives them as intelligent [8,19]. The level of intelligence is not necessarily close to the human lever; rather, is it kind-of-intelligent for computers, which are essentially fast data-processing machines.

It is very hard to implement true intelligence or consciousness on digital computers, which are theoretically defined by the Turing machine. In recent years it is more or less becoming evident that the mathematical model of the universal Turing machine is not sufficient to encapsulate the human mind and the brain [15].

Yes, there are several very smart and successful AI researchers, who still say that thermostats have minds. So, computers are intelligent and conscious. But this position is becoming more and more denoted as a version of strong AI [37]. Yes, the Church-Turing thesis still indicates that all solvable functions can be mechanically solved by the Turing machine and humans [10], and furthermore, that there is no possible procedural counterexample or counterproof. But one of the best indicators that there is something with old AI is a simple empirical test. Science is first of all empirical, and if we can't find positive instances of some theory over an extended time and efforts, then the theory is quite possibly questionable.

Comparing humans and computers, one first notices the enormous growth rate of computers. Their performances grow exponentially [29,20], doubling roughly each 1-2 years. Compared to humans, and put on a logarithmic scale, performance of humans stays practically the same during the last century. Computer performance, on the other hand, grows pretty linearly on a logarithmic scale, and in more and more areas sooner or later outperforms humans. It happened practically immediately after the introduction of computers in calculating, and several other areas followed. Now it is happening in chess or mass memory (See Fig. 1).

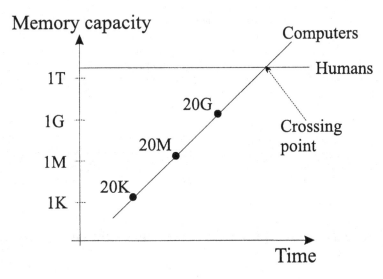

Fig. 1. Computers are just surpassing humans in mass memory capacity, like they did in several other areas before.

But if we put together on the same graph top human mental properties like intelligence and consciousness, we get a very different picture. Human performances still remain unchanged in the last century, but computer performances remain indistinguishable from zero over the same period. So there must be something wrong with the strong AI thesis that digital computers will soon become truly intelligent and conscious [41].

The discussion about the Turing machine sometimes resembled discussions between scientists and mentalists [3,2,14], but mostly it remained inside two scientific disciplines [30]: strong and weak AI. Other related disciplines are cognitive sciences and studies of Turing machines.

While strong AI relies on the Church-Turing thesis, claiming that digital computers will sooner or later achieve human performance in all areas including intelligence, consciousness and feelings, weak AI basically says that the universal Turing machine is not strong enough to encapsulate top human

performances. In other words - humans are either in principle computationally stronger than humans or in reality so different that computers cannot
copy humans efficiently.

2   Supercomputing

In a recent special issue [12], over a dozen authors propose their mathematical
versions of advanced Turing machines. Terms like "hypercomputation" (some
other authors use "superminds" [7]) are extensively used. These terms denote
computing mechanisms that are in principle stronger than those attained by
the Universal Turing machine. Copeland presents a quick overview of this
discipline:

Turing himself proposed a Turing machine with an oracle, capable of answering any question with always correct Yes/No. Scarpellini [35] suggested
that nonrecursive functions (demanding stronger mechanisms than the universal Turing machine) are abundant in real life. Komar [25] proposed that
an appropriate quantum system might be hypercomputational. This is unlike
Penrose who proposed that only the undefined transition between quantum
and macroscopic is nonrecursive. Putnam and Gold [33] described a trial-
and error Turing machine, which can also compute the Turing-uncomputable
functions like the halting problem. Abramson [1] introduced the Extended
Turing machine, capable of storing real numbers on its tape. Since not all
numbers are Turing-computable, Turing machines cannot compute with those
numbers. Boolos and Jeffrey [5] introduced the Zeus machine, a Turing machine capable of surveying its own indefinitely long computations. Karp and
Lipton [24] introduced McCulloch-Pitts neurons, which can be described by
Turing machines, but not if growing at will. Rubel [34] proposed that brains
are analog and cannot be modeled in digital ways. These basic ideas were
later advanced in several ways. Here we mention only some of them.

One of the best-known authors in this field is Roger Penrose. He claimed
[30,31] that the Goedelian argument shows that humans are in principle
stronger than computers. Namely, each formal system can be Goedelized - a
statement can be constructed that can not be proven right or wrong although
humans immediately "see" that it is true. This is the problem of all formal
systems, but humans are not formal systems, they are stronger than formal
systems (computers). Furthermore, since humans "see" the truth and cannot
describe it formally, the solution is not procedural, therefore, humans use
nonrecursive mechanisms. The other major idea by Penrose is related to supercomputing mechanisms in the nerve tissue. This is the Penrose-Hameroff
theory [21].

The basic idea that the mind is stronger than the Turing machine was
indicated already by Turing himself [38], although disputed by various researchers. The idea that humans introduce any new mechanism necessary
whenever the current computing mechanism (or computing system) gets

Goedelized is essential also in our principle of multiple knowledge [15]. Namely, if there are lots of multiple thinking processing going on, then only each of them can be Goedelized, and no integrated system can be Goedelized unless frozen in time. But this is not the way our minds and our world works. Similar performance is achieved by growing communities of final number of mathematicians. What we argue and later present in the principle of multiple knowledge, is that one human brain and mind performs like a community of humans. The idea is similar to the Society of minds [27,28], where Minsky presented the mind as a society of agents, but presented no claim that this computing mechanism is stronger than a universal Turing machine. Minsky also introduced the idea that multiple knowledge representations are in reality more useful than single knowledge representations. However, Minsky seems inclined to the validity of the Church-Turing idea, otherwise he would not propose hard disks for additional human storage capacities.

Wegner in 1997 [40] presented an idea that interaction is more powerful than algorithms. In a practical example, intelligent agents on the Internet [6] are in principle stronger than stand-alone universal Turing machines. The major reason for superior performance is in an open truly interactive environment, which not only cannot be formalized, it enables solving tasks better than with the Turing machine. Therefore, a stronger version is needed, the "interaction machine".

Copeland and Sylvan [13] describe a "coupled Turing machine", a version of the Turing machine very similar to the interaction machine. This Turing machine is coupled to its environment through an input channel, which not only enables undefined input, it also makes it impossible for the universal Turing machine to perform it.

A similar idea of introducing mechanisms that cannot be performed by the universal Turing machines comes from partially random machines [39,11] - these Turing machines get random inputs and therefore can not be modeled by a Turing-computable function as shown already by Chruch [9].

The problem with these ideas is that it is hard to find physical evidence for these mechanisms in the real life around us. For example, which computing mechanism in reality would correspond to a Turing mechanism with an oracle? There is no oracle in real world that would always correctly reply to a Yes/No question. On the other hand, nearly all practical, e.g. mathematical and physical tasks are computable (meaning Turing computable). There is no task we are not able to reproduce by a Turing machine. Copeland [12] objects to this idea citing Turing and Church. Referring to the original statements presented by the founders of computer science, one observes discrepancies between mathematical definitions and interpretations later declared by the strong AI community. Original mathematical definitions are indeed much stricter, e.g. referring to mathematical definitions. Yet, to fully accept Copeland's claims, humans should produce one simple task that can be solved by humans and not by computers.

The example we propose is simply the human mind and lack of intelligence and consciousness in computers. There is a clear distinction between the physical world and the mental world. The Turing machine might well be sufficient to perform practically all meaningful practical tasks in practical life. But the mental world is a challenge that is beyond the Turing machine. Every moment in our heads a new computing mechanism rumbles on. In computing terms, our mind performs like an unconstrained growing community of at any time final number of humans.

The principle of multiple knowledge is in a way similar to the Heisenberg principle, which discriminates the physical world of small (i.e. atomic) particles from the physical world of big particles. Our principle differentiates between universal Turing computing systems and multiple interaction computing systems like human minds.

Confirmations of the principle of multiple knowledge are presented in machine learning where several hundreds of publications show that multiple learning mechanisms achieve better results; simulations of multiple models and formal worst-case analyses which all show that reasonable multiple processes can outperform best single-ones; fitting the formal model to real-life domains shows that it can be done sufficiently good; human multiple reasoning currently and through the historic progress also strongly indicate that multiple thinking is essential for the progress of the human race (just consider the left-right brain asymmetry); studies of cognitive sciences and common sense. Some consequences are analyzed, for example, Occam's razor and Bayesian classifiers. Finally, many-worlds theory and quantum computing enable nice analogy with a little over-general interpretation of the quantum theory. For further details see [15] or http://ai.ijs.si/mezi/weakAI/weakStrongAI.htm.

3   Intelligent Information Processing - Applications

In Sections 1 and 2 we presented the supercomputing idea claiming that current digital computers cannot perform top human functions like true intelligence and consciousness. But we can copy (not perform on its own!) several human properties on digital computers quite well. With AI methods we can implement more or less preprogrammed flexible patterns of human behavior, which resembles simple intelligence [18,23]. With these extensions over classical systems, intelligent systems aim at improving the applicability of computers and providing a technological basis for new and improved information services. In addition, we can copy some of the super-computing mechanisms, e.g. those introduced by the principle of multiple knowledge.

In real-life application, the pure computing power and quick exact storing of information of computers enable several advantages over humans. The speed of communication and calculation enables a single computer to communicate with hundreds of human users at the same time 24 hours per day.

The major advantage of intelligent systems over classical ones is as mentioned limited machine intelligence [8,22].

Another argument in favor of intelligent processing is the information overload. Huge amounts of data are processed by computers several orders of magnitude faster than by humans [26]. Without intelligent assistants, humans are unable to cope quickly with huge amounts of information.

At the same time, computer capabilities are still growing exponentially, thus opening a huge and ever-growing space of possibilities for new applications. In other words, technical possibilities grow much faster than they are exploited. This is quite different compared to other technologies, such as car industry. Only a couple of world-wide car companies survived in these global times, and no group of - say - 10 persons can start car industry on its own. On the other hand, small groups of software developers with new ideas can create successful new companies. In fact, several now major SW companies started with a couple of staff a decade or two ago. This idea is further elaborated in each country since each country has its own specifics. While classical industry is restricted only to the richest and strongest countries in the world at the global level, practically every qualified information technology (IT) group can in principle develop world-class computer programs especially in their own country.

In summary, information extraction is one of the most relevant fields for implementing artificial intelligence - it is desperately needed by overloaded humans, it offers several advantages already due to the brute force of incredibly fast improving computers, the task itself (finding, preprocessing and using information) is well suited for intelligent computer systems, and technical possibilities enable a huge space for possible applications in each country.

In the next section we present a couple of successful applications that were developed or are being developed in our Department of Intelligent Systems. The developed systems have already been used by several tens of thousands users monthly, they controlled the production of several hundred thousands tons of products and overall had a considerable impact on our society.

3.1   Information Gathering in Employment Tasks

The basic task of each computer employment system is to provide employment information through the Internet thus replacing human agents with an agent system. We developed the EMA employment agent [16,18].

The system consists of several tens of modules, including job-description ontologies, natural speech modules in English and Slovenian, and an information-gathering module - a global automatic database wrapper agent. When a user declares a specific request for a specific profession, the system searches on its own through the Internet, through employment bases etc. and extracts relevant information. Also, the system is able to report in a user-predefined time intervals, e.g. weakly, when and where it encountered relevant information. There are several principal advantages over search engines: a uniform output gathered from heterogeneous sites; gathering information from HTML and databases; specialized knowledge in a specific area (ontologies) enables additional functions. EMA uses Slovenian speech modules [36], which are freely available on the Internet (http://ai.ijs.si/mezi/govor/govor.html). The system was implemented for Slovenian language based on the concatenation of basic speech units, diphones, using TD-PSOLA technique improved with a variable length linear interpolation process.

The system became operational eight years ago. It offered over 90% of all nationally available vacant jobs. At that time, no other country provided similar percentage of all jobs on the Internet. In one year, the system was on average used by every citizen of Slovenia. Of course, several citizens even now don't have access to the Internet, but others used the system a lot thus creating the high average use. Due to 2 million inhabitants, absolute numbers of users of the system were even at that time small compared to large countries.

Overall, the system was the most often used intelligent system in Slovenia so far, and the percentage of nationally Internet-available jobs temporarily put it at top world level.

3.2   Four Applications

Here we briefly mention four interesting intelligent-processing applications designed in cooperation with our Department: an e-shopping agent, an Internet speech system, a still-mill emulsion application, and a group-server.

The domain-dependent universal e-commerce agent [32] is capable of creating dynamic uniform databases for specific products by dynamic information gathering. For example, a user wishes to get information about tennis shoes in Slovenia. The system creates information about tennis shoes dynamically from the Internet sites in a similar way as does Amazon for books from its own databases. The system experimentally works for 50 e-stores.

The GIVE project enables speech communication through a telephone instead of through a computer. The idea is that due to several circumstances users need e.g. Internet functions through mobile or stationary phones. For example, one drives a car and can only use a phone.

There are two types of speech: prerecorded human speech and dynamically generated program speech. The speech system enables voice output from any URL address by parsing HTML pages and speaking it in Slovenian. More relevant, the system enables any user to put text information through the GIVE system into GIVE's local databases. In this way, any user can customize not only his Internet presentation but also provides automatic speech secretary available through the Internet and telephones.

The most common access is through a specific input telephone number and the telephone number of a particular institution or a person. In this way, for instance, anybody can get information about governmental institutions or services even though no human is present there late at night. The GIVE project is still in progress. The system is operational and is going to be released to public soon.

The still-mill application [17] has been our most relevant industrial application of intelligent systems, since for over seven years our system yearly controlled the quality of oil emulsion in the rolling mill, and thus quality of the surface of iron products. Each year the company produced over 100.000 tons of iron rods and belts. The hybrid system integrated two machine learning systems, one knowledge acquisition system, and one expert system into one system estimating the quality of the oil emulsion. From the estimated quality, a direct match was made into proposed actions. The system replaced one person/expert thus indicating that an intelligent system can replace a human by copying his/her knowledge through knowledge acquisition, an expert system, and machine learning. Here we already introduced practical application of the principle of multiple knowledge, however simplified [15].

Finally, we propose agent-based programming as an improvement over object-based programming. As a practical application, an Internet group-server system was developed for an American company [4]. It is based on agents communicating in KQML. The major advantages are more robust structure and easier upgrading since each computer/task/agent represents a robust self-contained entity cooperating with other agents. The system is operational.

4   Conclusion

We argue that current digital computers might be inadequate to perform truly intelligent and conscious information processing. Digital computers are far the best mechanism enabling copying of nearly all mental performance, but we need significantly improvements to perform (!) top human performances. Currently, it is not clear which of the possible improvements over the universal Turing machine is the correct one. There are several super-computing or hypercomputing mechanisms proposed. We have proposed our principle of multiple knowledge differentiating between universal Turing machines and human brains and minds. Thinking processes in our heads perform like a group of actors, using different computing mechanisms interacting with each other and always being able to include new ones. At the moment we can partially implement the principle of multiple knowledge and get better practical results; however, the key aspect of the Principle is missing, and currently we see no way to fully implement it on current digital computers.

Although not being able to perform like humans on their own, computer intelligent systems copy several human functions quite well. We have managed to implement several successful applications of intelligent systems that were used in real-life, especially for information-extraction tasks. These applications were innovative in the sense of introducing new computer approaches.

Our thesis is that intelligent systems offer several advantages. There are great opportunities and great challenges in the direction of new applications of intelligent systems and in the search for true artificial intelligence.

References

1. Abramson, F.G. (1971), Effective computation over the real numbers, Twelfth annual symposium on switching and automata theory, Nortridge, CA.
2. Abrahamson, J.R. (1994) Mind, Evolution, and Computers, AI magazine, 19–22.
3. Angell, I.O. (1993) Intelligence: Logical of Biological, Communications of the ACM 36, 15–16.
4. Bezek, A. (2002) An agent-based Internet group server, M.Sc. thesis (in Slovene).
5. Boolos, G.S., Jeffrey, R.C. (1974) Computability and logic, Cambridge Computability Press.
6. Bradshaw, M. (ed.) (1997) Software Agents, AAAI Press/The MIT Press.
7. Bringsjord, S., Zenzen, M. J. (2003) Superminds, Kluwer.
8. Buchanan, B., Uthurusamy, S.(eds.) (1999) Innovative applications of artificial intelligence, AI Magazine.
9. Church, A. (1940) On the concept of a random sequence, American Math. Soc. Bulletin 46, 130-135.
10. Copeland, B.J. (1997) The Church-Turing thesis, in E. Zalta (ed.), Stanford Encyclopedia of Phylosophy.
11. Copeland, B.J. (2000) Narrow versus wide mechanisms, Journal of philosophy 96, 5–32.
12. Copeland, B.J. (2002) Hypercomputation, Minds and Machines, Vol. 12, No. 4, 461–502
13. Copeland, B.J., Sylvan, R. (1999) Beyond the Universal Turing Machine, Australian journal of philosophy 77, 46–66.
14. Dreyfus, H.L. (1979) What Computers Can't Do, Harper and Row.
15. Gams, M. (2001) Weak intelligence: Through the principle and paradox of multiple knowledge, Advances in computation: Theory and practice, Volume 6, Nova science publishers, inc., NY.
16. Gams, M. (2001) A Uniform Internet-communicative Agent, Electronic Commerce Research 1, Kluwer Academic Publishers, 69–84.
17. Gams, M., Drobnic, M., Karba, N. (1996) Average-Case Improvements when Integrating ML and KA, Applied Intelligence 6, No. 2, 1996, 87–99.
18. Gams, M., Karalic, A., Drobnic, M., Krizman, V. (1998) EMA - an intelligent employment agent, Proc. of the Forth World Congress on Expert Systems, Mexico, 57–64.
19. Goonatilake, S., Treleaven, P. (1996) Intelligent systems for Finance and Business, John Wiley & Sons Ltd.
20. Hamilton, S. (1999) Taking Moore's Law into the Next Century, IEEE Computer, 43–48.
21. Hameroff, Kaszniak, Scott, Lukes (eds.), (1998) Consciousness Research Abstracts, Towards a science of consciousness, Tucson, USA.

450   Matjaz Gams

22. Hedberg, S.R. (1998)Is AI going mainstream at least? A look inside Microsoft research, IEEE Intelligent Systems, March/April, 21–25.

23. Hopgood, A.A. (2001) Intelligent Systems for Engineers and Scientists, CRC Press.

24. Karp, R.M., Lipton R.J. (1982) Turing machines that take advice, in Engeler et al. (eds.) Logic and algorithmic, L Enseignement Marhematique.

25. Komar A. (1964) Undicedability of macroscopically distinguishable states in quantum field theory, Physical Review, 133B, 542–544.

26. Lewis, T. (1999) Microsoft rising, IEEE Computer Society.

27. Minsky, M. (1987) The Society of Mind, Simon and Schuster, New York.

28. Minsky, M. (1991) Society of mind: a response to four reviews, Artificial Intelligence 48, 371–396.

29. Moore, G.E. (1975) Progress in Digital Integrated Electronics, Technical Digest of 1975, International Electronic Devices Meeting 11.

30. Penrose, R. (1989) The Emperor's New Mind: Concerning computers, minds, and the laws of physics, Oxford University Press.

31. Penrose, R. (1994) Shadows of the Mind, A Search for the Missing Science of Consciousness, Oxford University Press.

32. Pivk, A., Gams M. (2002) Domain-dependent information gathering agent. Expert Systems Applilactions, 23, 207–218

33. Putnam, H. (1965) Trial and error predicates and the solution of a problem of Mostowski, Journal of Symbolic Logic 30, 49–57.

34. Rubel, L.A. (1985) The brain as an analog computer, Journal of theoretical neurobiology 4, 73–81.

35. Scarpellini, B. (1963) Zwei Unentscheitbare Probleme der Analysis, Zeitschrift fuer Mathematische Logik und Grundlagen der Mathematik 0, 265–354.

36. Sef, T., Dobnikar, A., Gams, M. (1998) Improvements in Slovene Text-to-Speech Synthesis, Proceedings ICSLP'98, 2027–2030.

37. Sloman, A. (1992) The emperor's real mind: review of the Roger Penrose's The Emperor's New Mind: Concerning Computers, Minds and the Laws of Physics, Artificial Intelligence, 56, 335–396.

38. Turing, A.M. (1947) Lecture to the London Mathematical Society on 20 February 1947, in Carpenter, Doran (eds.) A.M. Turing's ACE Report of 1946 and other papers, MIT Press.

39. Turing, A.M. (1948), Intelligent Machinery, in B. Meltzer, D. Michie (Eds.), Machine Intelligence 5, Edinburgh University Press.

40. Wegner, P. (1997) Why Interaction is More Powerful than Computing, Communications of the ACM, Vol. 40, No. 5, 81–91.

41. Wilkes, M. W. (1992) Artificial Intelligence as the Year 2000 Approaches, Communications of the ACM, 35, 8, 17–20.

Ontology-based Text Document Clustering (Extended Abstract of Invited Talk)

Steffen Staab[1,2] and Andreas Hotho[1]

[1] Institute AIFB, University of Karlsruhe, Germany
   http://www.aifb.uni-karlsruhe.de/WBS
[2] Ontoprise GmbH, Karlsruhe, Germany
   http://www.ontoprise.de

Text document clustering plays an important role in providing intuitive navigation and browsing mechanisms by organizing large amounts of information into a small number of meaningful clusters. Standard partitional or agglomerative clustering methods efficiently compute results to this end.

However, the bag of words representation used for these clustering methods is often unsatisfactory as it ignores relationships between important term that do not co-occur literally. Also, it is mostly left to the user to find out why a particular partitioning has been achieved, because it is only specified extensionally. In order to deal with the two problems, we integrate background knowledge in the form of a core ontology into the process of clustering text documents.

First, we preprocess the texts, enriching their representations by background knowledge — provided in a core ontology, *viz.* Wordnet. Then, we cluster the documents by a partitional algorithm. Our experimental evaluation on Reuters newsfeeds compares clustering results with pre-categorizations of news. In the experiments, improvements of results by background knowledge compared to the baseline can be shown for many interesting tasks (cf. [1]).

Second, the clustering partitions the large number of documents to a relatively small number of clusters, which may then be analyzed by *conceptual clustering*. Conceptual clustering techniques are known to be too slow for directly clustering several hundreds of documents, but they give an intensional account of cluster results. They allow for a concise description of commonalities and distinctions of different clusters. With background knowledge they even find abstractions like "food" (vs. specializations like "beef ∨ corn"). Thus, partitional clustering reduces the size of the problem such that it becomes tractable for conceptual clustering, which then facilitates the understanding of the clustering result produced from partitional clustering (cf. [2]).

Furthermore, partitional and conceptual clustering in the manner just described is language-independent to a large extent. Therefore, we may apply this method also to documents stemming from multiple languages. In the talk, we will show the improvements that arise compared to corresponding naive baselines (cf. [3]).

References

1. Hotho A., Staab S. (2003). Clustering Text Documents using Background Knowledge. Submitted for Publication.
2. Hotho A., Staab S., Stumme G. (2003). Integrating Background Knowledge and Conceptual Clustering For Clustering Text Documents. Submitted for Publication.
3. Hotho A., Peters W., Staab S. (2003). On Multi-lingual Text Document Clustering. Submitted for Publication.

Ontology Learning from Text:
Tasks and Challenges for Machine Learning
(Extended Abstract)

Jörg-Uwe Kietz[1]

kdlabs AG, Zürich, Switzerland, http://www.kietz.ch/

The Semantic Web heavily relies on formal ontologies to enable machine-interpretation, reasoning and communication about the underlying data. However, it turned out that ontology engineering is an as difficult and time consuming task as knowledge engineering in general. Therefore there is an increasing amount of research concerned with the support of manual ontology engineering by automatic methods called ontology learning [11].

As for the support of knowledge acquisition by machine learning it is unrealistic to assume fully-automated ontology acquisition through machine learning. Therefore, the balanced-cooperative-modeling approach [13] could be adapted as well. In this approach manual knowledge-engineering is supported by an integrated environment, in which the knowledge engineer and the system work together to build an adequate knowledge-base. The system is supporting the knowledge engineer by doing reasoning, reason maintenance, revision, restructuring and automatic acquisition of theory parts by learning methods from test cases. For rule-based knowledge-bases this paradigm has been implemented with the MOBAL system [15] and used for real-world applications. But so far such a system supporting the acquisition of formal ontologies from texts does not exist. This is mainly due to three problems.

First, the linguistic text processing needed to transform unstructured texts into a structured representations, as e.g. used in [4] to learn from texts with ILP methods, is a quite difficult task. So far, it only works reliably in quite restricted domains where a sophisticated semantic lexicon (i.e an ontology of word meaning) about the domain of discourse already exists.

Second, the reasoning support for formal ontologies is much less investigated than that for rule-based systems so far. This is even true if the formal ontologies (e.g. OIL) used for the Semantic Web are strongly related to description logics [7]. There exist a lot of research about consistency checking for different description logics (see e.g. [1] for an overview), and based on that for subsumtion checking and instance classification given a consistent ontology, but nearly nothing about reasoning with inconsistencies, knowledge revision (only [16]) or reason maintenance.

Third, the situation for machine learning is quite similar to that of reasoning. There is a lot of research concerned with the optimization of learning classification rules mainly in attribute-value representation and to a lesser extend in structured first-order representation, but nearly nothing about learning an ontology.

In this talk we mainly address the third problem. We review the results of existing approaches to descriptions logic concept and rule learning [14,3,5,10,17,2] and relate them to the tasks of ontology learning. We will show how the recent result in [8], which reduces $\mathcal{ALN}$-description logic learning to inductive logic programming could be used to make further results from inductive logic programming namely association rule learning and clustering available to ontology learning. We finally point to some of the still open issues in ontology learning.

References

1. BAADER, F., AND SATTLER, U. Tableau algorithms for description logics. In *Proceedings of the International Conference on Automated Reasoning with Tableaux and Related Methods (Tableaux 2000)* (2000), R. Dyckhoff, Ed., LNCS, Springer-Verlag, pp. 1–18.

2. BADEA, LIVIU ANDNIENHUYS-CHENG, S.-H. Refinement operator for description logics. In *Proc. Tenth International Conference on Inductive Logic Programming, ILP'2000* (2000), J. Cussens and A. M. Frisch, Eds., Springer.

3. COHEN, W. W., AND HIRSH, H. The learnability of description logics with equality constraints. *Machine Learning 17* (1994), 169–199.

4. ESPOSITO, F., FERILLI, S., FANIZZI, N., AND SEMERARO, G. Learning from language with inthelex. In *Proc. of the 2nd learning language in logic (LLL) workshop* (http://www.lri.fr/ cn/LLL-2000/programme.html, 2000).

5. FRAZIER, M., AND PITT, L. Classic learning. In *Proc. of the 7th Annual ACM Conference on Computational Learning Theory* (1994), pp. 23–34.

6. GENNARI, J. H., LANGLEY, P., AND FISHER, D. H. Models of incremental concept formation. *Artificial Intelligence 40* (1989), 11 – 61.

7. HORROCKS, I., FENSEL, D., BROEKSTRA, J., DECKER, S., ERDMANN, M., GOBLE, C., VAN HARMELEN, F., KLEIN, M., STAAB, S., AND STUDER, R. The ontology inference layer oil, on-to-knowledge eu-ist-10132 project deliverable no. otk-d1. Tech. rep., Free University Amsterdam, Division of Mathematics and Computer Science, Amsterdam, NL, 2000.

8. KIETZ, J.-U. Learnability of description logic programs. In *Proc of the 12th Int. Conf. on Inductive Logic Programming, ILP-2002* (2002), S. Matwin and C. Sammut, Eds., Springer Verlag.

9. KIETZ, J.-U., MAEDCHE, A., AND VOLZ, R. A method for semi-automatic ontology acquisition from a corporate intranet. In *Workshop "Ontologies and Text", co-located with the 12th International Workshop on Knowledge Engineering and Knowledge Management (EKAW'2000)* (2000).

10. KIETZ, J.-U., AND MORIK, K. A polynomial approach to the constructive induction of structural knowledge. *Machine Learning 14*, 2 (1994), 193–217.

11. MAEDCHE, A., AND STAAB, S. Ontology learning for the semantic web. *IEEE Intelligent Systems 16*, 2 (2001).

12. MICHALSKI, R. S. Inferential learning theory as a basis for multistrategy task-adaptive learning. In *Proceedings of the first International Workshop on Multistrategy Learning* (1991), George Mason University, pp. 3 – 18.

13. MORIK, K. Balanced cooperative modeling. *Machine Learning 11*, 2/3 (1993), 217–235. Also appeared in *Proc. Workshop on Multi-Strategy Learning*, pp. 65 – 80, ed. R. S. Michalski and G. Tecuci, 1991.

14. MORIK, K., AND KIETZ, J.-U. A bootstrapping approach to conceptual clustering. In *Proc. Sixth Intern. Workshop on Machine Learning* (1989).

15. MORIK, K., WROBEL, S., KIETZ, J.-U., AND EMDE, W. *Knowledge Acquisition and Machine Learning: Theory, Methods and Applications*. Academic Press, London, 1993.

16. NEBEL, B.   *Reasoning and Revision in Hybrid Representation Systems*. Springer, New York, 1990.

17. ROUVEIROL, C., AND VENTOS, V. Towards learning in CARIN-⊣↕\. In *Proc. Tenth International Conference on Inductive Logic Programming, ILP'2000* (Berlin, 2000), J. Cussens and A. M. Frisch, Eds., Springer Verlag.

Part VIII

Invited Session: Web Services and Ontologies

Caching Dynamic Data for E-Business Applications*

Mehregan Mahdavi[1], Boualem Benatallah[1], and Fethi Rabhi[2]

[1] School of Computer Science and Engineering
The University of New South Wales, Sydney, NSW 2052, Australia
[2] School of Information Systems, Technology and Management
The University of New South Wales, Sydney, NSW 2052, Australia

Abstract. This paper is concerned with business portals; one of the rapidly growing web applications. It investigates the problem of providing a fast response time in such applications, particularly through the use of caching techniques. It discusses issues related to caching providers' response messages at business portals and proposes a caching strategy based on the collaboration between the portal and providers. .

1 Introduction

It is now common for many businesses to offer a web site through which customers can search and buy products or services on-line. Such businesses are referred to as product or service *providers*. Due to the large number of existing providers, *business portals* have emerged as Internet-based applications which enable access to different providers through a single web interface. The idea is to save time and effort for customers who only need to access the portal's web interface instead of having to navigate through many provider web sites.

Fig. 1 shows the general architecture of a business portal. It shows that each provider may have a membership relationship with a number of portals. Moreover, each provider may have a number of sub-providers. Each provider stores its own catalog and the *integrated catalog* represents the aggregation of all providers' catalogs. The portal deals with a request from the customer by sending requests to the appropriate providers. Responses from providers are sent back to the portal, processed and a final response is returned to the customer.

Emerging technologies such as web services promise to take portal-enabled applications a step further [3]. However, providing fast response time is one of the most critical issues in such applications. Network traffic between the portal and individual providers, server workload, or failure at provider sites are some contributing factors for slow response time. Previous research has shown that abandonment of web sites dramatically increases with the increase in the response time [20], resulting in loss of revenue by businesses. In general, providing a fast response time is one of the critical issues that today's e-business applications must deal with.

Caching is one of the key techniques which promises to overcome some of the performance issues. In particular, caching response messages (which we also refer

---

* This research has been partly supported by an Australian Research Council (ARC) Discovery grant number DP0211207.

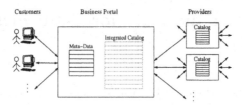

Fig. 1. The Architecture of Business Portals

to as *dynamic objects* or *objects*[1], for short) gives portals the ability to respond to some customer requests locally. As a result, response time to the customer is improved, customer satisfaction is increased and better revenue for the portal and the providers is generated. In addition, network traffic and the workload on the providers' servers are considerably reduced. This in turn improves scalability and reduces hardware costs.

The best candidates for caching are objects which are requested frequently and not changed very often. Products such as *Oracle Web Cache*, *IBM WebSphere Edge Server*, and *Dynamai* from *Persistence Software* enable system administrators to specify caching policies. Server logs (i.e., access log, and database update log) are also used to identify objects to be cached.

Caching dynamic objects at business portals introduces new problems. Since the portal may be dealing with a large number of providers, determining such objects by an administrator or by processing logs is impractical. On one hand, an administrator cannot identify candidate objects in a highly dynamic environment where providers may join and leave the portal frequently. On the other hand, keeping and processing access logs in the portal is impractical due to high storage space and processing time requirements. Moreover, database update logs are not normally accessible by the portal.

We propose a caching strategy based on the collaboration between portal and providers. Providers trace their logs, extract information to identify good candidates for caching and notify the portal. Providers associate a score with each response message which represents the usefulness of caching this object. Our strategy deals with the problem of inconsistencies between scores from different providers. The portal can trace the performance of the cache and dynamically regulate the scores from different providers. The major contributions of our work include:

(i) a collaborative scheme for caching dynamic objects in remote servers (i.e., business portals) based on a score given by providers, and

(ii) a regulation technique for monitoring and adjusting caching scores.

The remainder of this paper is organised as follows. Section 2 presents our caching scheme. Section 3 discusses implementation aspects. Related work along with conclusions are presented in Section 4.

---

[1] A dynamic object is a data item requested by the portal, such as the result of a database query, the result of a JSP page, an XML or SOAP response message.

2   Caching at Web Portals

Caching a particular object at the portal depends on the available storage space, response time (QoS) requirements, access and update frequency of objects[11]. As mentioned earlier, identifying caching objects by the portal is not realistic. As owners of the objects, providers are more eligible and capable of deciding which objects should be selected. In accordance with this approach, our strategy allocates to each object a caching score determined mainly from a score (called cache worthiness) given by providers to response messages. The caching score is also determined using other parameters such as recency of objects, importance of providers, and correlation between objects. However, these parameters are not the focus of this paper.

The cache worthiness is a score assigned by the content provider to each object. Its value, which is in the range $[0, 1]$, represents the usefulness of caching this object at a portal. At both extremes, a value equal to zero indicates that the object cannot be cached at the portal while a value equal to 1 indicates that the object must be cached. We assume that each provider defines these scores independently based on its own policy and priority.

The rest of this section explains the caching strategy in more detail. The first sub-section describes the meta-data being used to support the caching strategy and the next sub-section explains how the providers can use logs to determine object scores.

2.1   Meta-Data Support

The caching strategy is supported by two major tables; The *cache look-up table* used by portal to keep track of the cached objects, and the *cache validation table* used by provider to validate the cached objects at portal(s).

An entry in the cache look-up table mainly consists of a *Request-Instance* (RI) and a *Generation Time-Stamp* (GTS). An RI represents the request based on which the provider generates and returns a response. It contains the name of the requested service operation plus the values for the input parameters. For example, a web service operation can be invoked through a URI which includes the name of a Servlet and input values. In this case the RI is represented by a URI. A GTS contains the time the response was generated by the content provider. This is used for validation when a cache hit occurs.

The cache validation table keeps track of all the responses sent to the portal and is used to validate the freshness of the cached objects. For each cached object, it contains the RI and GTS which correspond to previous requests and response times respectively.

When a hit is detected at the portal using the cache look-up table, a validation request message is sent to the relevant provider. The message includes the corresponding RI and GTS. The provider checks the freshness of the object by probing the cache validation table to find the relevant entry for the request instance. If the cache validation table does not contain an entry for the request instance, it means that the corresponding object is not fresh anymore due to changes in the database. It is also possible that entries are removed for other reasons such as space limitations. After the object is sent back, the portal responds to the customer request and a copy of the object may be cached at the portal for future requests. If an entry is found, the GTS in the message is compared with the corresponding field in the cache validation table. If it is equal to the GTS in the table, the provider will confirm validation of the cache.

Changes in the back-end database invalidate entries in the cache validation table. If changing the content of the database affects the freshness of any object, then the appropriate entry in the provider cache validation table will be removed. Solutions for detecting changes in the back-end database and invalidating the relevant entry/object are provided in [2,17,6].

The cache look-up table reflects the content of the portal's cache. When an object is cached, the relevant entry is created in the table. Similarly, if an object is removed from the cache, the relevant entry is removed from the table. However, the cache validation table may become inconsistent with the cache in the following two cases:

(i) An object is cached at a portal but there is no entry for the object in the cache validation table at any provider. In this case, when the portal sends a validation request message to the provider, it can not check the freshness of the object and it has to generate the object, even though the object in the cache may be still fresh.

(ii) There is an entry in the cache validation table but the relevant object is not cached anywhere. In this case, storage space is being wasted by the provider to keep such entries in cache validation table.

To provide an effective caching strategy, the cache validation tables should closely reflect the content of the caches at the portal(s). For this reason, we assume that an entry is created in the cache validation table only when the cached object has a cache worthiness above a certain threshold (τ). We use an LRU replacement strategy to create free space in the table when it becomes full. More efficient methods are under investigation.

2.2 Heterogeneous Cache Policy Support

This section discusses inconsistencies between scores given by different providers and describes a mechanism to regulate these scores. First, an object scoring technique that uses server logs is described. Then, the issue of how inconsistencies arise, even though all providers may use the same scoring strategy is discussed. Finally, our solution for regulating the scores is described.

Calculating Cache Worthiness. The best candidates for caching are objects (i) requested frequently, (ii) not changed frequently, and (iii) expensive to compute or deliver [19]. For other objects, the caching overheads may outweigh the caching benefits. Server logs at provider sites can be used to calculate a score for cache worthiness. In the rest of the paper we use $O_{i,m}$ to denote an object i from a provider m. We identify four important parameters:

- The **access frequency** is represented by $A(O_{i,m}, k)$. It indicates the access frequency of $O_{i,m}$ through a portal k in a specified time period. $AR(O_{i,m}, k) = \frac{A(O_{i,m},k)}{\sum_r A(O_{r,m},k)}$ represents the popularity (i.e., access rate). Values close to 1 represent more popular objects and values close to 0 represent less popular objects. $AR(O_{i,m}, k)$ is calculated by processing web application server access log.

- The **update frequency** is represented by $U(O_{i,m})$. It indicates the number of changes on the back-end database which invalidate $O_{i,m}$. The update rate is represented by $UR(O_{i,m}) = \frac{U(O_{i,m})}{\sum_r U(O_{r,m})}$. Values close to 1 represent more frequently changed objects and values close to 0 represent less frequently changed objects. $UR(O_{i,m})$ is calculated by processing database update log.

- The **computation cost** is represented by $C(O_{i,m})$. It indicates the cost of generating $O_{i,m}$ in terms of database access. It is calculated by processing the database request/delivery log and calculating the time elapsed between the request and delivery of the result from the database. $CR(O_{i,m}) = \frac{C(O_{i,m})}{K_1 + C(O_{i,m})}$ represents the calculation cost rate. Values close to 1 represent more expensive and values close to 0 represent less expensive objects to generate from the database. $K_1 \in R^+$ is a constant value required by the transform function and is set by the system administrator. It can be simply set to $\overline{C}(O_{i,m})$.

- The **delivery cost** is represented by the size of the object. Larger objects are more expensive to deliver in terms of time and network bandwidth consumption. $SR(O_{i,m}) = \frac{Size(O_{i,m})}{K_2 + Size(O_{i,m})}$ results in a rate in $[0,1]$ from the size of $O_{i,m}$. Values close to 1 represent larger and values close to 0 represent smaller objects. $K_2 \in R^+$ is a constant value required by the transform function and is set by the system administrator. It can be simply set to $\overline{Size}(O_{i,m})$.

The score for cache worthiness can be calculated based on the above parameters as follows:

$$W(O_{i,m}, k) = \nu_1.AR(O_{i,m}, k) + \nu_2.(1 - UR(O_{i,m})) + \nu_3.CR(O_{i,m}) + \nu_4.SR(O_{i,m})$$
$$0 \leq \nu_1, \nu_2, \nu_3, \nu_4 \leq 1 , \quad \sum_{i=1}^{4} \nu_i = 1$$

The term $W(O_{i,m}, k) \in [0,1]$ represents the cache worthiness and indicates how useful caching $O_{i,m}$ at portal k is. Tuning the effect of each term in calculating the score is enabled by ν_1, ν_2, ν_3, and ν_4.

Regulating Cache Worthiness. Although, all providers may use the same strategy to score their objects, the scores may not be consistent. This is mainly due to the fact that: (i) each provider uses a limited amount of logs to extract required information which varies from one to another, (ii) each provider may use different weight for each term in the above formula (i.e., ν_1, ν_2, ν_3, and ν_4), (iii) the value of $CR(O_{i,m})$ depends on the provider hardware and software platform, workload and etc., and (iv) providers may use other mechanisms to score the objects.

To achieve an effective caching strategy, the portal should detect these inconsistencies and regulate the scores given by different providers. For this purpose, the portal uses a regulating factor $\lambda(m)$ for each provider. When the portal receives a cache worthiness score, it multiplies it by $\lambda(m)$ and uses the result in the calculation of the overall caching score. This factor can be set by an administrator at the beginning. It is adapted dynamically based on the performance of the cache.

Before explaining the process of dynamically adjusting $\lambda(m)$, two terms must be defined: (i) *false hit*, and (ii) *real hit*. A hit which occurs at the portal while the object is already invalidated is called a false hit. In contrast, a hit which occurs when the object is still fresh and can be served by the cache is called a real hit. False hits degrade the performance and increase the overheads both at portal and provider sites, without any outcome. These overheads include probing the cache validation table and generating validation request messages by portal, and probing cache look-up table by provider.

Adapting $\lambda(m)$ dynamically is done through the following process carried out by the portal:

- When the ratio of false hits (i.e., the number of false hits divided by the total number of hits) for a provider exceeds a threshold (τ_1), then the portal decreases $\lambda(m)$ for the provider. The new value of $\lambda(m)$ will be: $\lambda(m) \leftarrow \frac{\lambda(m)}{\theta}$.
- When the ratio of real hits (i.e., the number of real hits divided by the total number of hits) for a provider exceeds a threshold (τ_2), then the portal increases $\lambda(m)$. The new value of $\lambda(m)$ will be: $\lambda(m) \leftarrow \rho.\lambda(m)$.

3   Implementation Aspects

In order to evaluate the performance of our caching scheme, we have implemented a business portal in a case study involving providers offering accommodation and car rental services. Fig. 2 shows the architecture of the system.

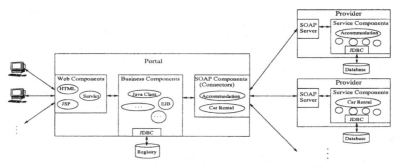

Fig. 2. The Architecture of the Implemented System

Each provider offers the following operations: getHotels, getRoomDetails, bookRoom, and unBookRoom for accommodation providers and getPickUpLocations, getVehicleDetails, bookVehicle, and unBookVehicle for car rental providers. To support caching, these providers also supply an implicit operation which calculates cache worthiness for a response message. Once the provider registers its service with the portal an entry for the provider is created in the registry at the portal. When the portal receives a request (e.g., search request) the portal first accesses the registry to find appropriate providers to process the request (e.g., based on the type of request for accommodation or car rental, or location). For each provider, a *connector object* is created to handle the request/response. Connectors create SOAP requests and send them to providers and respectively receive SOAP responses from providers.

In this implementation, *BEA WebLogic* application server and *Oracle 8i* DBMS are used to implement the portal service. We have used *Jakarta-Tomcat* web server and *MySQL* DBMS to implement provider services. To create, send, and receive messages, *Apache Axis* SOAP server is used.

Ongoing work includes evaluating the performance of the proposed caching scheme. The evaluation uses the following two metrics: (i) number of web interactions per second , and (ii) consumed network bandwidth. We are also studying the accuracy of the regulation process (i.e., choosing the appropriate values of θ and ρ). Finally, we are examining the implications of divergence between cache look-up and cache validation tables on the overall performance.

4   Related Work and Conclusions

Proxy server is still the most popular caching mechanism. However, it only deals with static objects and does little or nothing about dynamic objects. In [13], caching dynamic objects at the proxy server level is enabled by transferring some programs to the proxy server which generate the dynamic part of the objects. *ICP* [18] was developed to enable querying of other proxies in order to find requested web objects. In *Summary Cache* [10], each cache server keeps a summary table representing the content of caches in other servers to minimise the number of ICP messages. *CARP* [14] can be considered as a routing protocol which uses a hash function to determine the owner of a requested object in an array of proxy servers.

Companies such as *Akamai*, and *Digital Island* have been providing Content Delivery/Distribution Network (CDN) services for several years. CDN services are designed to deploy cache/replication servers at different geographical locations called *edge servers*. The first generation of these services was designed to cache static objects such as HTML pages, image, audio and video files. Nowadays, *Edge Side Includes* (ESI) enables the definition of different cachability for different fragments of an object. Processing ESI at these servers enables dynamic assembly of objects at edge servers which otherwise may be done at server accelerator, proxy server or browser.

Caching policies for web objects are studied in [1,4,7]. Weave [19] is a web site management system which provides a language to specify a customized cache management strategy. Maintaining cache consistency has been studied in [8,9,15,5,12]. Triggers can be deployed to detect changes on back-end database and invalidate cached objects. *Oracle Web Cache* [16] uses a time-based invalidation mechanism [2]. CachePortal [17] intercepts and analyses system logs to detect changes on database and invalidate the corresponding object(s). *Data Update Propagation (DUP)* algorithm [6] uses an object dependence graph for determining the dependences between cached objects and the underlying data. It provides API for application programs to explicitly manage caches to add, delete, and update cache objects. *Dynamai* from *Persistence Software* uses an event-based invalidation technique. It includes a request-based invalidation mechanism through the application, or external events such as a system administrator.

In summary, we have discussed the limitations of current approaches for providing an effective caching strategy in business portals. We proposed a caching scheme for business portals based on providers' collaboration. The required metadata to support the strategy consists of cache look-up and cache validation tables. We have provided a mechanism for providers to score objects for caching. We have also provided a mechanism to regulate the scores at portals.

References

1. Aggrawal C., Wolf J L., Yu P. S (1999) Caching on the World Wide Web. IEEE TKDE
2. Anton J., Jacobs L., Liu X., Parker L., Zeng Z., Zhong T. (2002) Web Caching for Database Applications with Oracle Web Cache. ACM SIGMOD
3. Benatallah B., Casati F., editors. (2002) Special Issue on Web Services. Distributed and Parallel Databases, An International Journal, **12**, 2–3
4. Cao P., Irani S. (1997) Cost-Aware WWW Proxy Caching Algorithms. The USENIX Symposium on Internet and Systems
5. Cao Y., Ozsu M. T. (2002) Evaluation of Strong Consistency Web Caching Techniques. WWW Journal
6. Challenger J., Iyengar A., Dantzig P. (1999) A Scalable System for Consistently Caching Dynamic Web Data. IEEE INFOCOM
7. Cheng K., Kambayashi Y. (2000) LRU-SP: A Size-Adjusted and Popularity-Aware LRU Replacement Algorithm for Web Caching. IEEE Compsac
8. Deolasee P., Katkar A.,Panchbudhe A., Ramamaritham K., Shenoy P. (2000) Adaptive Push-Pull: Disseminating Dynamic Web Data. The Tenth World Wide Web Conference (WWW-10)
9. Duvuri V., Shenoy P., Tewari R. (2000) Adaptive Leases: A Strong Consistency Mechanism for the World Wide Web. IEEE INFOCOM
10. Fan L., Cao P., Broder A. (2000) Summary Cache: A Scalable Wide-Area Web Cache Sharing Protocol. IEEE/ACM Transactions on Networking
11. Kossmann D., Franklin M. J. (2000) Cache Investment: Integrating Query Optimization and Distributed Data Placement. ACM TODS
12. Liu C., Cao P. (1998) Maintaining Strong Cache Consistency in the World-Wide Web. International Conference on Distributed Computing Systems
13. Luo Q., Naughton J. F. (2001) Form-Based Proxy Caching for Database-Backed Web Sites. VLDB Conference
14. Microsoft Corporation (1997) Cache Array Routing Protocol and Microsoft Proxy Server 2.0. White Paper. http://www.mcoecn.org/WhitePapers/Mscarp.pdf
15. Olston C., Widom J. (2001) Best-Effort Synchronization with Source Cooperation. ACM SIGMOD Conference
16. Oracle Corporation (2001) Oracle9iAS Web Cache. White Paper
17. Selcuk K., Li W. S., Luoand Q., Hsiung W. P., Agrawal D. (2001) Enabling Dynamic Content Caching for Database-Driven Web Sites. ACM SIGMOD Conference
18. Wessels D., Claffy K. (1997) Application of Internet Cache Protocol (ICP), Version 2. Network Working Group
19. Yagoub K., Florescu D., Valduriez P., Issarny V. (2000) Caching Strategies for Data-Intensive Web Sites. VLDB Conference
20. Zona Research Inc. (2001) Zona Research Releases Need for Speed II. http://www.zonaresearch.com/info/press/01-may03.htm

A Contextual Language Approach for Multirepresentation Ontologies

Djamal Benslimane[1], Ahmed Arara[2], Christelle Vangenot[2], and Kokou Yetongnon[3]

[1] LIRIS, University of Claude Bernard, Lyon1 69622 Villeurbanne, France
{djamal.benslimane,ahmed.arara}@iuta.univ-lyon1.fr
[2] Swiss Federal Institute of Technology, Database Laboratory, Lausanne,
Switzerland
christelle.vangenot@epfl.ch
[3] LE2I UMR-CNRS, University of Burgandy, BP 47870 - 21078 Dijon Cedex,
France
kokou.yetongnon@u-bourgogne.fr

Abstract. Taking a step forward from syntactic interoperability to the semantic one has become of a great challenge to researchers in the discipline of computer networking and Internet technology. Ontologies have proved to be the most efficient in providing common understanding among machines and/or humans. Several ontology languages have been proposed. These languages, however, are not capable of defining the multirepresentation ontology - that is: an ontology that characterizes a concept by a variable set of properties in several contexts. Multirepresentation ontologies are becoming very essential to diversified user communities that need to share and exchange information and applications in a domain. The objective of this paper is to define a contextual ontology language to support multiple representations of ontologies. In our research, we focus on the logic-based ontology languages. As a matter of fact, we will consider only languages that are based on description logics (DLs). At first, we propose a sub-language of DL as an ontology language. Furthermore we achieve multiple representations of ontological concepts by extending such sub-language through the use of stamping mechanism proposed in the context of multiple representation of spatial databases. The proposed $C-DL$ language should offer a modest solution to the problem of multirepresentation ontologies.

1 Introduction

1.1 Motivation

The recent emerging of computer networking and the internet technologies has customized the monopoly of information in knowledge dissemination and availability. An extensive research activities in networking technology have been going on in the last decade. The most prominent and challenging problem, however, is the semantic aspect of information. The semantic web has evolved from the existing web technology to satisfy the formal semantic need for information exchange and sharing. Ontologies, as an explicit specification of conceptualized terminologies, is a key element for the semantic web. Consequently, ontology languages and tools are essentially needed to create and represent sementically enriched web contents. Several languages are commonly used such as Ontolingua [5], KARL [1], OIL [6], OCML [7].

Domain ontologies are constructed by capturing a set of concepts and their links according to a given context. A context can be viewed by various criteria such as the abstraction paradigm, the granularity scale, interest of user communities, and the perception of ontology developer. So, the same domain can have more than one ontology, where each one of them is described in a particular context. We call each one of these ontologies a **MonoRepresentation ontology (MoRO)**. Thus, concepts in MoRO are defined with one and only one representation. Our motivation is to see how we can describe ontology according to several contexts at the same time. We shall call such ontology a **MultiRepresentation Ontologies (MuRO)**. A MuRO is an ontology that characterizes an ontological concept by a variable set of properties or attributes in several contexts. So, in a MuRO, a concept is defined once with several representations, such that a single representation is available for one context.

Existing ontology languages are not capable of defining a single MuRO ontology. For instance, there isn't any possibility to define a unique road concept with different sets of properties at the same time. So, our motivation is to deal with the problem of multirepresentation in the context of ontology languages. In our research, we focus on the logic-based ontology languages. As a matter of fact, we will consider only languages that are based on description logics (DLs).

1.2   Multirepresentation needs

The multirepresentation ontology extends classical, monorepresentation ontology by offering new facets of several advantages as for example:

- The same concept is represented at different levels of granularity. A concept "ROAD" in urban domain can be seen as of more or less properties depending on the subdomain or context (traffic control, transportation, land use management, etc).
- Two distinct semantic domains can share concepts and describe them with some differences. In the same way, concepts of the same sub-domain can have more or less information depending on the granularity sought. Granulation of concepts allows partitionning and packaging of concepts in a specific domain.

1.3   Objectives and outline of the paper

In this paper, we are mainly concerned about the multirepresentation of ontological concepts in different contexts. The underlying key idea of our work is
to adapt the stamping mechanism proposed in [8] for the needs of multirepresentation in spatial databases to the needs of ontology languages. To achieve
this goal, we first propose a sub-language of DL as an ontology language and
then we extend it by introducing stamping mechanism to the constructs of the
DL language to allow multiple representations of concepts. The DL language
obtained is called **Contextual Description Logic** ($C - DL$). The outline
of the rest of the paper is as it follows. Section 2 gives a general background of
the multirepresentation solutions. Section 3 presents the essential elements,
namely DLs formalisms and stamping technique, participating in the definition of the proposed language. Section 4 is devoted to informal presentation
of the key principles of our $C - DL$ language. The formal presentation of
$C - DL$ language is given in section 5. Finally, section 5 will conclude the
paper.

2   Basic elements of the the proposed solution

We present in this section the description logics formalism and the stamping
mechanism which are the basic elements for the $C - DL$ language proposed
in this paper.

2.1   Presentation of a sub-language of DL for ontologies

Description logics are a family of logics designed to represent the taxonomical
and conceptual knowledge of a particular application domain in an abstract
and logical level. DLs are equipped with well-defined and set-theoretic semantics. Furthermore, the interesting reasoning problems such as subsumption
and satisfiability are, for most description logics, decidable (see, for example, [4]). Based on atomic concepts and roles, complex concepts (and roles)
are built by using a set of constructors. For example, from atomic concepts
Human and Female and the atomic role child we can build the expression
Human ⊓ ∀child.Female which denotes the set of all Human whose children
are (all) instances of Female. Here, the symbol ⊓ denotes conjunction of concepts, while ∀ denotes (universal) value restriction.

In our work, ontological concepts and relations between them are defined
by applying the constructors summarized in table 1. In this table, C and D
designate a concept name, R a role name and n a natural number in $\mathcal{N}$.

Let A be a concept name and let D be a concept term. Then $A = D$ is a
terminological axiom (also called defined-as axiom). A terminology ($TBox$) is
a finite set T of terminological axioms with the additional restriction that no

Table 1. Concept constructors

Constructor name	Notation
Conjunction	⊓
Disjunction	⊔
Complement	¬
Value restriction	∀
Existential quantification	∃
At most number restriction	≤
At least number restriction	≥

concept name appears more than once in the left hand side of the definition. An interpretation $\mathcal{I}$ is a model for a $TBox$ $\mathcal{T}$ if and only if $\mathcal{I}$ satisfies all the assertions in $\mathcal{T}$.

In this paper, we consider that ontologies are specified by means of a set of terminological axioms. Hence, each ontology is a terminology.

Example: A very simple ontology concerning a "Driving School" context is shown in table 2. It consists of four concepts: *Road, Vehicle, Student* and *Instructor* and two subconcepts namely *SchoolVehicle* and *TestingRoad*.

2.2   Stamping mechanism

A stamping mechanism to characterize database elements (in spatial databases and GIS applications) was proposed in [8] to support several representations of data. Stamping is a technique used to distinguish one representation of the same phenomenon from other representations. Hence, a concept can be used in one or more representation but each representation of a concept is stamped or labeled differently.

Example: The concept AUTOMOBILE can be viewed in many different contexts. If we consider two different facets of the same concept namely traffic control and environment, then stamping mechanism will be applied to distinguish the attributes and instances of the two contexts as shown below: Stamp s1 to designate ("traffic-control"), and Stamp s2 to mark ("environment"). Thus, the concept AUTOMOBILE is defined in Pascal-like code in the two contexts (s1,s2) as:

```
Def-concept AUTOMOBILE ( STAMP s1,s2)
Begin
Insurance-no (s1): string;
Registration-number (s1, s2): string;
Engine-type (s2): string;
END
```

Table 2. Monorepresentation Ontology example for Driving School Context

---

Concepts defintion

Student $= (\leq 1Name) \sqcap (\geq 1Name) \sqcap$
$(\forall Name.String) \sqcap (\leq 1Birthdate) \sqcap$
$(\geq 1Birthdate) \sqcap (\forall Birthdate.Date)$

Instructor $= (\leq 1Name) \sqcap (\geq 1Name) \sqcap$
$(\forall Name.String)$

Vehicle $= (\leq 1CarModel) \sqcap (\geq 1CarModel) \sqcap$
$(\forall CarModel.String) \sqcap (\leq 1LicencePlate) \sqcap$
$(\geq 1LicensePlate) \sqcap (\forall LicensePlate.String) \sqcap$
$(\leq 1InsurancePolicy) \sqcap (\geq 1InsurancePolicy) \sqcap$
$(\forall InsurancePolicy.String)$

Road $= (\leq 1Name) \sqcap (\geq 1Name) \sqcap$
$(\forall Name.string) \sqcap (\leq 1RoadType) \sqcap$
$(\geq 1RoadType) \sqcap (\forall RoadType.String)$

SchoolVehicle $= Vehicle \sqcap (\leq 2SteeringWheel) \sqcap (\geq 2SteeringWheel) \sqcap$
$(\forall SchoolSign.String)$

TestingRoad $= Road \sqcap (\leq 4Stopsign) \sqcap (\geq 1Stopsign) \sqcap$
$(\leq 2TrafficLight) \sqcap (\geq 1TrafficLight)$

. . .

The two contexts share the attribute Registration-number for record-keeping purposes but the two other attributes are specific to each context. Attribute Engine-type is very relevant to the pollution aspect of the environment whereas Insurance-number is obligatory in the context of traffic control.

3 Towards a multirepresentation language ontology: Informal presentation

3.1 Generality

In DL, we have the notions of concepts (that we denote classical concepts for more clarity) as a set of individuals (unary predicates) and roles (binary predicates) as attributes or relationships. Complex concepts are derived from atomic concepts by applying DL constructors (also will be denoted as classical constructors). For our requirements of multi-representation ontologies, we propose the notion of contextual concepts that may describe concepts associated with different contexts.

Contextual concepts are derived basically from atomic concepts by using a set of classical and/or contextual constructors. Contextual constructors are defined by specializing the classical ones to allow the construction of a concept that is partially or completely available in some contexts. Table 3 presents the new contextual constructors proposed.

It is important to note the following remarks concerning the proposed $C - DL$ langauge:

1. The definition of classical concepts remains always possible. Such concepts will exist in all contexts with a single representation.
2. The concepts and constructors used in deriving a contextual concept could be either contextual or a combination of classical and contextual.

Table 3. Contextual constructors

Constructor name	Notation
contextual value restriction	$\forall_{s_1,\cdots,s_n}$
contextual existential quantification	$\exists_{s_1,\cdots,s_n}$
contextual at most number restriction	$\leq_{s_1,\cdots,s_n}$
contextual at least number restriction	$\geq_{s_1,\cdots,s_n}$

3.2   Description of contextual constructors

In the following, we briefly describe contextual constructors accompanied by simple examples.

The contextual value restriction constructor $(\forall_{s_1,\cdots,s_n} R.C)$: It will define a new concept all of whose instances are related via the role R only to the individuals of class C and in the contexts s_1 to s_n. For example, in two contexts s_1 and s_2, the concept Employee is defined as a person with an attribute Birth-Date that can be of a type DATE in context s_1 or STRING in context s_2. The concept is expressed in $C - DL$ as it follows:

$$Employee = Person \sqcap \forall_{s_1} BirthDate.DATE \sqcap$$
$$\forall_{s_2} BirthDate.STRING \sqcap \geq_{s_1,s_2} 1BirthDate \sqcap$$
$$\leq_{s_1,s_2} 1BirthDate$$

The contextual existential quantification constructor $(\exists_{s_1,\cdots,s_n} R.C)$ will construct a new concept all of whose instances are related via the role R to at least one individual of type C and only in contexts s_1 to s_n. For example, the following expression describes that student is a person that has at least one graduation diploma in context s_1 and participates in at least one sport in context s_2.

$$\text{Student} = Person \sqcap \exists_{s_1} Diploma.Graduate \sqcap \exists_{s_2} Play.Sport$$

The contextual number (at most, at least) restriction constructors
($\leq_{s_1,\cdots,s_n} nR, \geq_{s_1,\cdots,s_n} nR$): They specify the number of role-fillers. The $\leq_{s_1,\cdots,s_n}$ nR is used to indicate the maximum cardinality whereas, the expression $\geq_{s_1,\cdots,s_n}$ nR indicates the minimum cardinality. The following example illustrates two cardinalities in different contexts: context s_1 where a man is allowed for one and only one wife at one period of time, and context s_2 (some cultures) where he is allowed to have 4 wives at the same time.

$$\text{Man} = Person \sqcap \forall Sex.Male \sqcap$$
$$\leq_{s_1} 1NumberOfWife \sqcap \leq_{s_2} 4NumberOfWife \sqcap$$
$$\forall_{s_1,s_2} NumberOfWife.Number$$

3.3 Inheritance with conceptual concepts

With the presence of contextual concepts, the property of inheritance could be adapted to obtain the contextual inheritance. This is to allow context restrictions in the contextual subconcept definition. For instance, we would like to define B in context $s1$ as a subconcept of contextual concept A defined in contexts $s1$ and $s2$. Thus, in order to restrict the inheritance mechanism to only some contexts, we propose the **contextual defined-as** axiom that specializes the existing classical defined-as axiom.

The contextual defined-as axiom, noted as $(= s_1, \cdots, s_n)$, is labelled with several contexts to indicate that the concept is described only in those contexts and not others. For example, the following expression defines the concept RetiredPerson to be a person that has a varying range of retirement age depending on the contexts s_1 or s_2. The subconcept RetiredFrenchPerson restricts the definition of Retired-Person to only context s_1. So, the range of retirement age is 55-60.

$$\text{RetiredPerson} = Person \sqcap \leq_{s_1} 60Age \sqcap \geq_{s_1} 55Age \sqcap \leq_{s_2} 65Age \sqcap \geq_{s_2} 61Age$$
$$\text{RetiredFrenchPerson} =_{s_1} RetiredPerson \sqcap \geq_{s_1} 30workyears$$

Note that with the classical defined-as axiom, as it is used in the following expression, both ranges of retirement age (55-60 in s_1 and 61-65 in s_2) will be preserved for the ReiredFrenchPerson concept.

$$\text{RetiredFrenchPerson} = RetiredPerson \sqcap \geq_{s_1} 30workyears$$

4 Formal presentation of $C - DL$ language

4.1 Syntactical aspect

Definition 1. (Syntax of concept terms) Let $s_1, \cdots, s_n$ be a set of context names. Classical and Contextual concept terms C and D can be formed by means of the following syntax:

$$C, D \longrightarrow A \mid$$

$A \mid$	(atomic concept)
$\top \mid \bot \mid$	(top, bottom)
$C \sqcap D \mid$	(conjunction)
$C \sqcup D \mid$	(disjunction)
$\neg C \mid$	(complement)
$\forall R.C \mid \forall_{s_1,\cdots,s_n} R.C \mid$	(classical and contextual value restriction)
$\exists R.C \mid \exists_{s_1,\cdots,s_n} R.C \mid$	(classical and contextual existential quantification)
$(\leq nR) \mid (\leq_{s_1,\cdots,s_n} nR)$	(classical and contextual at most number restriction)
$(\geq nR) \mid (\geq_{s_1,\cdots,s_n} nR)$	(classical and contextual at least number restriction)

Definition 2. (Syntax of contextual terminology axioms) Let $\mathcal{A}$ be a concept name and let $\mathcal{D}$ be a classical/contextual concept term. Also, let $s_1, \cdots, s_n$ be a set of context names. A contextual terminology axioms $C - TBox$ is a finite set $\mathcal{T}$ of the two following terminology axioms:

1. $\mathcal{A} = \mathcal{D}$ (Classical defined-as axiom)
2. $\mathcal{A} =_{s_1,\cdots,s_n} \mathcal{D}$ (Contextual defined-as axiom)

The $C - TBox$ constitutes the so called multirepresentation ontologies.

4.2   Decidability of $C - DL$

The Contextual constructors proposed in $C - DL$ language are not really new concept constructors. In fact, they are adaptive constructors that specializes the existing ones. By the same manner, contextual terminology axiom is derived from the existing terminology axiom. Thus, the subsumption for our $C - DL$ language will be decidable since it constitutes a subset of the description logic $\mathcal{ALCNR}$ [3].

4.3   Example of multirepresentation ontology

Table 4 presents a very simple multirepresentation ontology described in $C - DL$ Language. Two contexts are considered: the traffic control and driving school designated by s_1 and s_2 respectively.

5   Conclusion

Ontologies play a crucial role in providing common understanding for the semantic Web contents. Recent expressive ontology languages (such as OIL, DAML+OIL, OWL) are designed to be used on the Web. DAML+OIL, however, does not take in consideration the aspects of the multirepresentation ontologies. Multirepresentation ontologies will be essentially needed in many diversified applications and heterogeneous information in a domain. In this article, we modestly proposed a contextual multirepresentation ontologies language ($C - DL$) based on description logics.

For future work, we aim at completing the set of constructs of the $C - DL$ language by augmenting more of contextual axioms, constructors and

Table 4. Multirepresentation Ontology Example for Road Traffic and Driving School Contexts

---

Concepts defintion

Road $= (\leq_{s_1} 1StartPoint) \sqcap (\geq_{s_1} 1StartPoint) \sqcap (\forall_{s_1} StartPoint.Point) \sqcap$
$(\leq_{s_1} 1EndPoint) \sqcap (\geq_{s_1} 1EndPoint) \sqcap (\forall_{s_1} EndPoint.Point) \sqcap$
$(\leq_{s_1,s_2} 1Name) \sqcap (\geq_{s_1,s_2} 1Name) \sqcap (\forall_{s_1,s_2} Name.String) \sqcap$
$(\leq 1With_{s_1}) \sqcap (\geq_{s_1} 1With) \sqcap (\forall_{s_1} With.Number) \sqcap$
$(\leq_{s_1} 1Lanes) \sqcap (\geq_{s_1} nLanes) \sqcap (\forall_{s_1} Lanes.Number) \sqcap$
$(\leq_{s_1} 1SpeedLimitWith) \sqcap (\geq_{s_1} 2SpeedLimit) \sqcap$
$(\forall_{s_1} SpeedLimit.Number) \sqcap (\leq_{s_2} 1RoadType) \sqcap$
$(\geq_{s_2} 1RoadType) \sqcap (\forall_{s_2} RoadType.String)$

Vehicle $= (\leq_{s_1} 1Speed) \sqcap (\geq_{s_1} 1Speed) \sqcap (\forall_{s_1} Speed.Number) \sqcap$
$(\leq_{s_1} 1VehicleType) \sqcap (\geq_{s_1} 1VehicleType) \sqcap$
$(\forall_{s_1} VehicleType.String) \sqcap (\leq_{s_1,s_2} 1LicencePlate) \sqcap$
$(\geq_{s_1,s_2} 1LicensePlate) \sqcap (\forall_{s_1,s_2} LicensePlate.String) \sqcap$
$(\leq_{s_1,s_2} 1InsurancePolicy) \sqcap (\geq_{s_1,s_2} 1InsurancePolicy) \sqcap$
$(\forall_{s_1,s_2} InsurancePolicy.String) \sqcap (\leq_{s_2} 1CarModel) \sqcap$
$(\geq_{s_2} 1CarModel) \sqcap (\forall_{s_2} CarModel.String)$

RoadSign $= (\leq_{s_1} 1Type) \sqcap (\geq_{s_1} 1Type) \sqcap (\forall_{s_1} Type.string) \sqcap$
$(\leq_{s_1} 1Location) \sqcap (\geq_{s_1} 1Location) \sqcap (\forall_{s_1} Location.String) \sqcap$
$(\leq_{s_1} 1Purpose) \sqcap (\geq_{s_1} nPurpose) \sqcap (\forall_{s_1} Purpose.String)$

Student $= (\leq_{s_2} 1Name) \sqcap (\geq_{s_2} 1Name) \sqcap (\forall_{s_2} Name.String) \sqcap$
$(\leq_{s_2} 1Birthdate) \sqcap (\geq_{s_2} 1Birthdate) \sqcap (\forall_{s_2} Birthdate.Date)$

Instructor $= (\leq_{s_2} 1Name) \sqcap (\geq_{s_2} 1Name) \sqcap (\forall_{s_2} Name.String)$

. . .

---

operations. We, also, intend to verify and validate the C-DL language in an urban applications domain. Moreover, The proposed C-DL language should be compatible with the existing ontology languages. Hence, an interfacing to the OIL kernel becomes necessary. Finally, an adaptation of the class-based logic query language [2] will be considered to allow for browsing and searching concepts in very large multirepresentation ontologies.

References

1. J. Angel, D. Fensel, and R.Studer. The model of expertise in karl. In *Proceedings of the 2nd World Congress on Expert Systems, Lisbon,/Estoril,Portugal*, January 1994.
2. Djamal Benslimane, Mohand-Said Hacid, Evimaria Terzi, and Farouk Toumani. A class-based logic language for ontologies. In *In the proceedings of the Fifth*

International Conference on Flexible Query Answering Systems. October 27 - 29, 2002, Copenhagen, Denmark. Proceedings. Lecture Notes in Computer Science 1864 Springer 2000, ISBN 3-540-67839-5, pages 56–70, October 2002.

3. M. Buchheit, F.M. Donini, and A. Scharaef. Decidable reasoning in terminological knowledge representation systems. *Journal of Artificial Intelligence Research*, 1:109–138, 1993.

4. F.M. Donini, M. Lenzirini, D. Nardi, and w. Nutt. The complexity of concept language. Technical report, Report RR-95-07, Deutshes Forschunggszentrum fur kunstliche intelligenz (DFKI), Kaiserslautern, 1995.

5. A. Farquhar, R. Fikes, and J.P. Rice. The ontolingua server: A tool for collaborative ontology construction. In *Journal of Human-Computer Studies, 46:*, pages 707–728, 1997.

6. D. Fensel, I. Horrocks, F. Van Harmelen, S. Decker, M. Erdmann, and M. Klein. Oil in a nutshell. In *In Proceedings of the ECAI-2000 Workshop on Applocations of Ontologies and Problem-Solving Methods, Berlin (Allemagne)*, 2000.

7. Enrico Motta. An overview of the ocml modelling language. In *In Proceedings of the 8th Workshop on Knowledge Engineering Methods and Languages (KEML'98), Karlsruhe, Germany*, pages 125–155, January 1998.

8. Stefano Spaccapietra, Christine Parent, and Christelle Vangenot. Gis databases: From multiscale to multirepresentation. In *Abstraction, Reformulation, and Approximation, 4th International Symposium, SARA 2000, Horseshoe Bay, Texas, USA. Proceedings. Lecture Notes in Computer Science 1864 Springer 2000, ISBN 3-540-67839-5*, pages 57–70, July 2000.

An Ontology-based Mediation Architecture for E-commerce Applications

O. Corcho[1], A. Gomez-Perez[1], A. Leger[2], C. Rey[3], and F. Toumani[3]

[1] Universidad Politécnica de Madrid (ocorcho@fi.upm.es, asun@fi.upm.es)
[2] France Telecom R&D (alain.leger@rd.francetelecom.com)
[3] LIMOS, Université Blaise Pascal, France (rey@isima.fr, ftoumani@isima.fr)

Abstract. As part of the MKBEEM project, we present an ontology based mediation framework for electronic commerce applications. The framework is based on a mediator/wrapper approach that supports an integrated view over multiple heterogeneous sources. The MKBEEM mediation system allows to fill the gap between customers queries (possibly expressed in a natural language) and diverse specific providers offers. In contrast with many existing mediator based systems, our approach rests on a three-layer knowledge representation architecture which includes an *electronic services* ontology besides the usual domain ontology and sources descriptions layers. At the reasoning level, we propose a new mechanism, namely *dynamic discovery of e-services*, that acts in collaboration with the Picsel mediator system to effectively achieve the MKBEEM mediation tasks.

1 Introduction

The recent progress and wider dissemination of electronic commerce via the world wide web has dramatically increased the diversity and heterogeneities in corresponding systems. To avoid this progress to be hampered by closed markets with incompatible applications that cannot use each other services, there is a need for efficient and flexible mechanisms to handle the interoperability problem. Today, we expect that emerging standards will solve most of the syntactic infrastructure problems. However, dealing with sources from different domains and their semantics differences requires to handle the interoperability problem at the semantic level.

The notion of mediation has been proposed as the principal means to resolve problems of semantic interoperation [13]. Mediation architectures are generally based on the mediator/wrapper paradigm where information sources are "wrapped" to logical views so that their interfaces to the outside world are uniform [13,8]. These logical views are "glued" together through an integrated global schema usually called the domain ontology. In such architectures, a mediator acts as an interface between user queries or application programs and existing information sources: queries against the domain ontology are reformulated in terms of logical views and then sent to the remote sources.

In this paper, we present a **three-layer ontology-based mediation framework** for electronic commerce applications. In contrast with many existing mediator based systems, the proposed architecture includes an *electronic services ontology* besides the usual domain ontology and sources descriptions layers. Roughly speaking, an electronic service (e-service)[1] can be seen as an application made available via Internet and accessible by humans and/or other applications [2,3,6]. Examples of e-services currently available range from weather forecast, on-line travel reservation or banking services to entire business functions of an organization. Whereas in many e-commerce platforms, e-services are associated to providers, we propose to define an e-service as an *integrated provider-independent offer* available on a given e-commerce platform. Then, to effectively handle the mediation tasks, we rely on two reasoning mechanisms:

- the first, called **dynamic discovery of e-services**, allows to reformulate users queries against the domain ontology in terms of e-services. The aim here is to allow the users and applications to automatically discover the available e-services that best meet their needs at any given time,
- the second, called *query plan generation*, allows to reformulate a user query, expressed as a combination of e-services, in terms of providers views. The aim of this second issue is to allow the identification of the views that are able to answer to the query (knowing that, afterwards, the query plans will be translated into databases queries via the corresponding wrappers).

While the second reasoning mechanism, known as *query rewriting using views*, has already been addressed in the literature [11,5,8], the first is a quite new problem for which we propose a solution in this paper.

The rest of this paper is organized as follows. The context of this study, namely the MKBEEM[2] project, is introduced in Section 2. The architecture of the MKBEEM ontology is presented in Section 3. In Section 4, the dynamic identification of e-services is detailed. We conclude in Section 5.

2   The MKBEEM mediation system

The global aim of the MKBEEM project is to extend current electronic commerce platforms to reach a truly pan European and culturally open electronic commerce market. The main technical aim of MKBEEM is to create an *intelligent knowledge based multilingual* mediation service which displays the following features:

- Natural language interfaces for both the end user and the system's content providers/service providers.
- Automatic multilingual cataloguing of products by service providers.

---

[1] Also called web services.

[2] MKBEEM stands for Multilingual Knowledge Based European Electronic Marketplace (IST-1999-10589, 1st Feb. 2000 - 1st Aug. 2002).

- On-line e-commerce contractual negotiation mechanisms in the language of the user, which guarantee safety and freedom.

A detailed description of the MKBEEM project and relevant references related to this project can be found in [1].

Let us consider an example of a simplified *natural language request* scenario trained on the generic MKBEEM architecture to sketch the roles of the different components of the MKBEEM system: an end user submits to the MKBEEM system a natural language query. The query is received by the User Agent component which recognizes the user language and then forwards it to the corresponding Human Language Processing Server (HLP Server). The HLP Server is in charge of *meaning extraction*: he analyzes the input string and converts the query to an *Ontological Formula* (OF) which is a language-independent formula containing the semantic information of the corresponding phrase of human language. The OF is sent to the Rational Agent (RA) which is responsible of the dialogue management between the different components of the MKBEEM system. The RA forwards the OF to the Domain Ontology Server (DOS). The DOS module is responsible of storing, accessing and maintaining the ontologies used by the MKBEEM system. It also provides the core reasoning mechanisms needed to support the mediation services. Continuing with the example, the DOS achieves a *contextual interpretation of the formula* using its knowledge about the application domain. This task consists mainly in the identification of the offers (e-services) delivered by the MKBEEM platform that *"best match"* the ontological formula. The set of solutions computed by the DOS is sent back to the RA which, in collaboration with the *User Agent*, will ask the user to choose one solution and to complete, if any, the parameters that are missing. After this dialogue phase, the retained solution is sent back to the *DOS* to generate the *query plans*. A query plan contains information about the Content Providers Agents (CPA) that are able to answer to the user query. Then, thanks to specific wrappers belonging to CPA a query plan is translated into databases queries which are executed on the remote providers sources.

This example is a typical mediation instance: the user poses queries in terms of the integrated schema (e-services and domain ontology) rather than directly querying specific provider information sources. This enables users to focus on *what* they want, rather than worrying about *how* and *from where* to obtain the answers. Apart from the natural language processing and the wrapping steps, which are no further considered in this paper (cf. [1] for more details), the MKBEEM mediation relies on two rewriting steps achieved by the DOS:

1. A contextual interpretation step: the OF is rewritten into a few E-Services Formula (ESF) which have been dynamically identified as the combinations of e-services that "best match" the OF. Here are mainly used the e-services ontology and the domain ontology layers. In the sequel, we refer to this task as *dynamic discovery of e-services*.

2. A query plan generation step: the ESF are rewritten into a few query plans where the Content Providers Agents that are able to answer the user query are identified. Here is mainly used the sources descriptions layer. In the sequel, we refer to this task as *query plan generation*.

The problem of query plan generation, known as *query rewriting using views*, has already been studied in the research area of information integration [5,8]. That is why, in the MKBEEM mediation system, the DOS relies on the Picsel [8] mediator to handle this task. Picsel is a description logic based information integration system. It uses the $\mathcal{ALN}$-CARIN[3] language as the core logical formalism to represent both the domain ontology and the contents of information sources. The query rewriting algorithm implemented in the Picsel system is proved to be sound and complete. On the contrary, the dynamic discovery of e-services is a new research problem. This is our proposal to solve it, that is detailed in section 4, after the presentation of the three-layer ontology.

3   Architecture of the MKBEEM ontology

The MKBEEM Ontologies are defined as a knowledge structure to enable sharing and reuse of information. They provide a consensual representation of the electronic commerce field in two typical Domains (Tourism and Mail order) allowing the exchanges independently of the language of the end user, the service, or the content provider. Ontologies are used for classifying and indexing catalogues, for filtering user's query, for facilitating man-machine dialogues between users and software agents, and for inferring information that is relevant to the user's request.

The $\mathcal{ALN}$-CARIN language has been chosen as the ontology implementation language due to its inferential capabilities. Its description logic part, the $\mathcal{ALN}$ language, contains the following constructors:

- concept conjunction ($\sqcap$), e.g., the concept description *parent* $\sqcap$ *male* denotes the class of fathers (i.e., male parents),
- the universal role quantification ($\forall R\,C$), e.g., the description $\forall child\,male$ denotes the set of individuals whose children are all male,
- the number restriction constructors $\geq nR$ and $\leq nR$, e.g., the description ($\geq 1\ child$) denotes the class of parents (i.e., individuals having at least one children), while the description ($\leq 1\ leader$) denotes the class of individuals that cannot have more than one leader.
- the negation restricted to atomic concepts ($\neg A$), e.g., the description $\neg male$ denotes the class of individuals which are not males (females).

---

[3] $\mathcal{ALN}$-CARIN is a logical language combining Description Logics and Datalog rules [12].

As all the description logics, $\mathcal{ALN}$ comes equipped with well-defined, set-theoretic semantics. Furthermore, the interesting reasoning problems such as subsumption and satisfiability are decidable for $\mathcal{ALN}$ [7].

The MKBEEM ontologies are structured in three layers, as shown in Figure 1.

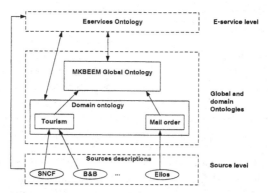

Fig. 1. Knowledge representation in the MKBEEM system.

These three layers are described below.

1. Global and domain Ontologies.
   The global ontology describes the common terms used in the whole MK-BEEM platform. This ontology represents knowledge reusable on different domains while each domain ontology contains concepts corresponding to one of the domains of the MKBEEM partners (e.g;, tourism, mail orders, etc.). This decomposition in global and domain modular ontologies has allowed us to reuse many definitions from existing ontologies and more easily create definitions of the domain ontologies.

   Example 1. Table 1 provides the definitions of the concept Date and Time of the global ontology. The travel domain ontology contains, among others, the concept trip which has a departure place (the role depPlace), an arrival place (the role arrPlace) and may have a transport mean (the role transportMean), and the concept accommodation which is located in a place (role placedIn) and has a starting date (the role startDate).

2. E-services ontology.
   All the offers available in the MKBEEM platform are integrated and described in the e-services ontology. An e-service can be seen as a provider-independent predefined query corresponding to an existing offer in the MKBEEM platform.

   Example 2. According to e-services ontology given in Table 1, the MK-BEEM platform delivers four offers:

- hotel, which allows to consult a list of hotels.
- apartment, which allows to consult a list of apartments.
- timetable1, which allows to consult a trip given the departure place, the arrival place, the departure date and the departure time.
- timetable2, which allows to consult a trip given the departure place, the arrival place, the arrival date and the arrival time.
  It is worth noting that these e-services are specified independently from a given provider.

3. Sources descriptions.
   At the lower knowledge level, sources descriptions specify the providers competencies, i.e., the description of the contents of the providers information sources. Each content provider agent is identified by its name (e.g., SNCF[4]) and can provide many views, each of which corresponds to a possible query (offer) on the provider information source.

   Example 3. In the example of the MKBEEM ontologies given in Table 1, four providers are defined: SNCF, cheapJet, Expedia and GitesDeFrance. There are two possibilities for querying the SNCF source to have information about train trips: given information about the departure place and the arrival place and either the departure date and the departure time (view timetableD) or the arrival date and the arrival time (view timetableA). The cheapJet source allows to consult a list of plane trips given the departure place, the arrival place, the departure date and the departure time (view planeTrip). The Expedia provider delivers information about hotels (view ExpediaHotel) while GitesDeFrance delivers information about apartments (view Renting)

The following section introduces the e-services dynamic discovery mechanism and illustrates how the proposed modular architecture is used to handle the DOS mediation tasks (i.e., contextual interpretation of ontological formulas and query plan generation).

4   E-services discovery

In the MKBEEM project, the *dynamic discovery* mechanism is used in association with the Picsel system to achieve the reasoning tasks in the DOS. The complementary roles of these two complex logical reasoning constitutes the description logic core for query processing in the MKBEEM system. They are in fact two different instances of the problem of rewriting concepts using terminologies [4].

In the following we illustrate the interest of *e-services discovery* using the MKBEEM ontologies provided in table 1.

---

[4] SNCF is the french railway company.

Table 1. Example of MKBEEM ontologies.

Global ontology

time $\doteq$ ($\leq$ 1 hour) $\sqcap$ ($\geq$ 1 hour) $\sqcap$ ($\forall$ hour integer) $\sqcap$ ($\leq$ 1 minute)
$\sqcap$ ($\geq$ 1 minute) $\sqcap$ ($\forall$ minute integer)

date $\doteq$ ($\leq$ 1 day) $\sqcap$ ($\geq$ 1 day) $\sqcap$ ($\forall$ day integer) $\sqcap$ ($\leq$ 1 month)
$\sqcap$ ($\geq$ 1 month) $\sqcap$ ($\forall$ month string) $\sqcap$ ($\leq$ 1 year)
$\sqcap$ ($\geq$ 1 year) $\sqcap$ ($\forall$ year integer) $\sqcap$ ($\leq$ 1 weekday)
$\sqcap$ ($\geq$ 1 weekday) $\sqcap$ ($\forall$ weekday string)

Travel Domain ontology

trip $\doteq$ ($\leq$ 1 depPlace) $\sqcap$ ($\geq$ 1 depPlace) $\sqcap$ ($\forall$ depPlace string) $\sqcap$
($\leq$ 1 arrPlace) $\sqcap$ ($\geq$ 1 arrPlace) $\sqcap$ ($\forall$ arrPlace string) $\sqcap$
($\geq$ 1 transportMean) $\sqcap$ ($\forall$ transportMean transportMeanType)

accommodation $\doteq$ ($\leq$ 1 placedIn) $\sqcap$ ($\geq$ 1 placedIn) $\sqcap$ ($\forall$ placedIn string) $\sqcap$
($\leq$ 1 startDate) $\sqcap$ ($\geq$ 1 startDate) $\sqcap$ ($\forall$ startDate date)

E-services ontology

hotel $\doteq$ accommodation $\sqcap$ ($\leq$ 1 numberOfBeds) $\sqcap$ ($\geq$ 1 numberOfBeds) $\sqcap$
($\forall$ numberOfBeds number) $\sqcap$ ($\leq$ 1 hotelCategory) $\sqcap$
($\geq$ 1 hotelCategory) $\sqcap$ ($\forall$ hotelCategory string)

apartment $\doteq$ accommodation $\sqcap$ ($\leq$ 1 numberOfRooms) $\sqcap$ ($\geq$ 1 numberOfRooms) $\sqcap$
($\forall$ numberOfRooms number) ($\leq$ 1 apartmentCategory) $\sqcap$
($\geq$ 1 apartmentCategory) $\sqcap$ ($\forall$ apartmentCategory string)

timetable1 $\doteq$ trip $\sqcap$ ($\forall$ depTime time) $\sqcap$ ($\geq$ 1 depTime) $\sqcap$ ($\leq$ 1 depTime)
$\sqcap$ ($\leq$ 1 depDate) $\sqcap$ ($\geq$ 1 depDate) $\sqcap$ ($\forall$ depDate date)

timetable2 $\doteq$ trip $\sqcap$ ($\forall$ arrTime time) $\sqcap$ ($\geq$ 1 arrTime) $\sqcap$ ($\leq$ 1 arrTime)
$\sqcap$ ($\leq$ 1 arrDate) $\sqcap$ ($\geq$ 1 arrDate) $\sqcap$ ($\forall$ arrDate date)

Sources Descriptions

Provider	View Name	View Description
SNCF	timetableD	timetable1 $\sqcap$ ($\forall$ transportMean train)
SNCF	timetableA	timetable2 $\sqcap$ ($\forall$ transportMean train)
cheapJet	planeTrip	timetable2 $\sqcap$ ($\forall$ transportMean plane)
Expedia	ExpediaHotel	hotel
GitesDeFrance	Renting	apartment

Example 4. Let us assume that a given user submits to the MKBEEM system the following request expressed in a human language:

Q1: "I'll arrive in Paris on Monday and I look for an accommodation with swimming pool".

The query Q1 is first processed by the HLP Server and converted to the following *ontological formula*:

OF1 : "(trip)(V8), (accommodation)(V7), (arrDate)(V8,C9), (date)(C9),
(day)(C9,15), (weekday)(C9,monday), (month)(C9,april),
(year)(C9,2002), (arrPlace)(V8,paris), (leisure)(V7,swimmingPool)".

Then, given such an ontological formula, the *e-services discovery* is used by the DOS to identify the corresponding relevant service(s) in the ontology of e-services. This task is achieved in two steps:

1. Converting an ontological formula F into a concept description Q_F
   This task depends on the structure of the ontological formula and on the expressive power of the target language. In the context of the MK-BEEM project, the current ontological formulas generated by the HLP Server have relatively simple structures that can be described using the small description logic $\mathcal{FL}_0 \cup \{(\geq nR)\}$. This logic contains the concept conjunction constructor ($\sqcap$), the universal role quantification constructor ($\forall R.C$) and the minimal number restriction constructor ($\geq nR$). In this case, we can achieve this task by computing the so-called *most specific concept* [7] corresponding to the ontological formula.

   Example 5. The concept description Q_{OF1} corresponding to the ontological formula $OF1$ given in the previous example is:
   $$Q_{OF1} \doteq \text{trip} \sqcap \text{accommodation} \sqcap (\geq 1 \text{ arrPlace}) \sqcap (\forall \text{ arrPlace string}) \sqcap$$
   $$(\geq 1 \text{ arrDate}) \sqcap (\forall \text{ arrDate (date} \sqcap (\geq 1 \text{ day}) \sqcap (\forall \text{ day integer})$$
   $$\sqcap (\geq 1 \text{ year}) \sqcap (\forall \text{ year integer}) \sqcap (\geq 1 \text{ month}) \sqcap$$
   $$(\forall \text{ month string}) \sqcap (\geq 1 \text{ weekday}) \sqcap (\forall \text{ weekday string}))) \sqcap$$
   $$(\geq 1 \text{ leisure}) \sqcap (\forall \text{ leisure string})$$

2. Selecting the relevant e-services.
   This problem can be stated as follows: given a user query Q_F and an ontology of e-services T, find a description E, built using (some) of the names defined in T, such that E contains as much as possible of common information with Q_F and as less as possible of extra information with respect to Q_F. We call such a rewriting E a *best cover* of Q_F using T. Therefore, our goal is to rewrite a description Q_F into the closest description expressed as a conjunction of (some) concept names in T.

A best cover E of a concept Q using T is defined as being any conjunction of concept names occurring in T which shares some common information with Q, is consistent with Q and minimizes, in this order, the extra information in Q and not in E and the extra information in E and not in Q. Once the notion of a best cover has been formally defined, the second issue to be addressed is how to find a set of e-services that best covers a given query. This problem, called *best covering problem*, can be stated as follows: given an ontology T and a query description Q, find all the best covers of Q using T.

More technical details about the best covering problem can be found in [9,10]. To sum up, the main results that have been reached are:

- The precise formalization of the best covering problem in the framework of languages where the difference operation is semantically unique (e.g., the description logic $\mathcal{FL}_0 \cup \{(\geq nR)\}$).

- A study of complexity showed that this problem is NP-Hard.
- A reduction of the best covering problem to the problem of computing the minimal transversals with minimum cost of a weighted hypergraph.
- Based on hypergraph theory, a sound and complete algorithm that solves the best covering problem was designed and implemented.

Example 6. Continuing with the example, we expect the following result to be returned by the DOS:

	Identified services	Rest	Missed information
Solution 1	timetable2, apartment	leisure	arrTime.hour, arrTime.minute numberOfRooms, apartmentCategory
Solution 2	timetable2, hotel	leisure	arrTime.hour, arrTime.minute numberOfBeds, hotelCategory

These solutions correspond to the combinations of e-services that best match the ontological formula OF1. For each solution, the DOS computes the extra information (column *Missed Information*) brought by the e-services but not contained in the user query. The column Rest contains the extra information (leisure) contained in the user query and not provided by any e-services. This means that, in the proposed solutions the requirement concerning the leisure is not taken into account.

To continue with the example, assume that the user chooses the first solution (timetable1, apartment). Then, he is asked to complete the missed information: the arrival time (hour and minutes), the apartment category and the number of rooms in the apartment. The result is a global query Q, expressed as an E-service Formula (ESF), that will be sent to the Picsel system to identify the Content Provider Agent (CPA) that are able to answer to this query. In our example, the execution of the query Q using the ontologies depicted in Table 1 leads to the unique query plan: $[SNCF\_timetableA, GitesDeFrance\_Renting]$.

Implementation and validation The *e-services discovery* algorithm has been implemented as an integrated component in the MKBEEM prototype. This prototype is built as a set of Enterprise Java Beans (EJB) components that interact with each other. There are also some components dedicated to the interaction with the user interface built with Java Server Pages (JSPs) or Servlets. Finally, some of these components have the functionality of interacting with the remote (or locally duplicated) databases in the provider information systems. The MKBEEM prototype has been validated on a pan-European scale (France and Finland), with three basic languages (Finnish, English and French) and one optional language (Spanish), in two distinct end-user fields: 1) Business to consumer on-line sales, and 2) Web based travel/tourism services.

5   Conclusion

In this paper, we have presented an ontology-based mediation architecture structured in three layers: global/domain ontology, e-services ontology and provider sources descriptions. In this context, a new reasoning mechanism, called *dynamic discovery of e-services*, is proposed to allow the users to automatically discover the available e-services that best meet their needs. This reasoning mechanism is used in association with the Picsel mediator system to effectively handle the mediation tasks. The modularity of this architecture together with the associated reasoning mechanisms allow to make the whole system provider-independent and more capable to face the great instability and the little lifetime of e-commerce offers and e-services. The proposed architecture has been successfully experienced in the context of the MKBEEM project.

References

1. MKBEEM web site : http://www.mkbeem.com.
2. Data Engineering Bulletin: Special Issue on Infrastructure for Advanced E-Services. 24(1), IEEE Computer Society, 2001.
3. The VLDB Journal: Special Issue on E-Services. 10(1), Springer-Verlag Berlin Heidelberg, 2001.
4. F. Baader, R. Küsters, and R. Molitor. Rewriting Concepts Using Terminologies. In *KR'2000, Colorado, USA*, pages 297–308, Apr. 2000.
5. C. Beeri, A.Y. Levy, and M-C. Rousset. Rewriting Queries Using Views in Description Logics. In *ACM PODS , New York, USA*, pages 99–108, Apr. 1997.
6. F. Casati and M-C. Shan. Models and Languages for Describing and Discovering E-Services. In *ACM SIGMOD, Santa Barbara, USA*, May 2001.
7. F. Donini and A. Schaerf M. Lenzerini, D. Nardi. Reasoning in description logics. In *Gerhard Brewka, editor, Foundation of Knowledge Representation*, pages 191–236. CSLI-Publications, 1996.
8. F. Goasdoué and M-C Rousset V. Lattès. The Use of CARIN Language and Algorithms for Information Integration: The PICSEL System. *IJICIS*, 9(4):383–401, 2000.
9. M.S. Hacid, A. Leger, C. Rey, and F. Toumani. Computing concept covers: A preliminary report. In *Workshop on Description Logics. Toulouse, France*, Apr. 2002.
10. M.S. Hacid, A. Leger, C. Rey, and F. Toumani. Dynamic discovery of e-services: A description logics based approach. In *Proceedings of the 18th French conference on advanced databases (BDA), Paris*, pages 21–25, Oct. 2002.
11. A.Y.Levy, A .Rajaraman, and J.J. Ordille. Querying Heterogeneous Information Sources Using Source Descriptions. In *VLDB'96, Mumbai (Bombay), India*, pages 251–262. Morgan Kaufmann, Sep. 1996.
12. A.Y. Levy and M-C. Rousset. Combining Horn Rules and Description Logics in CARIN. *Artificial Intelligence*, 104(1-2):165–209, 1998.
13. G. Wiederhold and Michael R. Genesereth. The Basis for Mediation. In *CoopIS, Vienna, Austria*, pages 140–157, May 1995.

Part IX

Invited Session: Reasoning in AI

On generalized quantifiers, finite sets and data mining

Petr Hájek[*][1]

Institute of Computer Sciences, Academy of Sciences
182 07 Prague, Czech Republic

Abstract. Logical foundations of data mining are described using the apparatus of finite model theory and generalized quantifiers and in the spirit of the theory of the GUHA method of hypothesis formation. Several results on coputational coplexity are obtained.

1 Introduction

Data mining is a modern spontaneously developing domain. See e.g. [1]. It should be stressed that it can have and in fact does have logical foundations: the monograph [13] (from 1978, now fully available on internet) presents serious logical (and statistical) treatment of the GUHA method of "automatic generation of hypotheses from data" and its formalism is by far not restricted to GUHA (whose "hypotheses" are just rather general "associations" found in data) but for many methods of what we now called "data mining". Here we restrict our attention just to *observational languages,* languages speaking on data. *Data* will be understood as a rectangular matrix of zeros and ones (objects – attributes), even if much more general structure can be considered. The starting point is to understand data as *finite models* (interpretations) of a monadic predicate language having unary predicates $P_1, \ldots, P_n$; they can be combined by logical connectives to define composed attributes. Now the basic observation is that patterns (association rules,...) are defined by *generalized quantifiers* saying (in some well defined sense) "Many φ's are ψ's", "φ makes ψ more likely (than $\neg\varphi$ does)" etc. etc. Given a data matrix $\mathbf{M}$, any two (composed) attributes φ, ψ determine their *four-fold table* (a, b, c, d) of frequencies of $\varphi\&\psi, \varphi\&\neg\psi, \neg\varphi\&\psi, \neg\varphi\&\neg\psi$ in $\mathbf{M}$. The semantics of a quantifier $\sim$ is given by a function $tr_\sim$ assigning to each quadruple (four-fold table) (a, b, c, d) the truth value $tr_\sim(a, b, c, d) \in \{0, 1\}$ (1-valid, 0-non valid).

Two kinds of quantifiers are particularly investigated in [13] and relevant for data mining:

- implicational, i.e. such that whenever $a' \geq a$, $b' \leq b$ and $tr_\sim(a, b, c, d) = 1$ then $tr_\sim(a', b', c, d) = 1$
- associational: whenever $a' \geq a$, $b' \leq b$, $c' \leq c$, $d' \geq d$ and $tr_\sim(a, b, c, d) = 1$ then $tr_\sim(a', b', c', d') = 1$.

[*] Acknowledgement: This research is a part of the COST-Action 274 (TARSKI)

Example 1 As an example of an implicational quantifier let us take the quantifier FIMP of founded implication (denoted $\Rightarrow_{p,s}$, where $0 < p \leq 1$, s natural);

$$tr_{\Rightarrow_{p,s}}(a, b, c, d) = 1 \text{ iff } a \geq s \text{ and } a/(a+b) \geq p.$$

Mining associational rules as iniciated by Agrawal [2] (in ignorance of the literature of GUHA method) uses the quantifier which is an unessential variant of FIMPL. (You get Agrawal's quantifier by replacing the condition $a \geq s$ by $a/(a+b+c+d) \geq t$ for $0 < t \leq 1$; he calls p *confidence* and t *support.*)

As an example of an asasociational quantifier which is not implicational take the quantifier SIMPLE (of simple association) denoted $\sim_h^0$ where $h > 1$:

$$tr_{\sim_h^0}(a, b, c, d) = 1 \text{ iff } ad > h.bc.$$

Note that GUHA implementations use several particular associational and implicational quantifiers, cf. [13,14,16,17,19,20]. Rauch has introduced several subclasses of associational quantifiers. From the purely logical point of view, investigation of our quantifiers belongs to the domain of *finite model theory* iniciated by Fagin [8] and in relation to generalized quantifiers by the present author [9]. A recent paper on this domain is Gottlob's [4]; for finite model theory in general see e.g. [7] and an extensive bibliography [3].

Thus let us agree that we shall consider a first-order language with n unary predicates $P_1, \ldots, P_n$ and one quantifier symbol $\sim$ binding one variable in a pair of formulas (if φ, ψ are formulas and x a variable then $(\sim x)(\varphi, \psi)$ is a formula written also $\varphi \sim_x \psi$ or, if x is clear from context, just $\varphi \sim \psi$). The truth function $tr_\sim$ can vary over all functions from some class (associational, implicational, other). Let Q be such a class; and let φ, ψ be open formulas containing no other variable than x. We call $\varphi \sim \psi$ a Q-*tautology* if for each $tr_\sim \in Q$ and each data matrix **M** (interpreting $P_1, \ldots, P_n$), $\varphi \sim \psi$ is true in **M**.

Example 2 (1) It is easy to verify that the formula

$$((\varphi \& \delta) \Rightarrow^* \psi) \to (\varphi \Rightarrow^* (\psi \vee \neg\delta))$$

is an implicational tautology (but not an associational tautology).
(2) Also verify that the following is an associational tautology:

$$(\varphi \sim \psi) \to [((\varphi \& \psi) \vee \delta) \vee (\varphi \& \neg\psi \& \neg\delta)] \sim [((\varphi \& \psi) \vee \delta) \vee (\neg\varphi \& \psi \& \neg\delta)].$$

For δ being φ you get the associational tautology

$$(\varphi \sim \psi) \to (\varphi \sim (\varphi \vee \psi)).$$

It is natural to ask what is the computational complexity of the set of all Q-tautologies. In [11] I show that the class of all implicational tautologies (in normal form, see below) is co-NP-complete. In Sect. 2 of the present paper this result is extended to associative tautologies and some other classes of quantifiers. In Sect. 3 we extend the language by *partialization* (relativization of quantifiers) and show its undefinability in the original language; we also extend the complexity results to formulas with relativization. In Sect. 4 we comment on a quantifier of conviction.

2 Generalized quantifiers and computational complexity

Recall that a closed formula in in *normal form* if it is a boolean combination of formulas of the form $\varphi \sim_x \psi$ where φ, ψ are open. (Each closed formula is equivalent to a formula in normal form, see [13] 3.1.30.) In [11], the set of formulas in normal form that are implicational tautologies is proved to be co-NP-complete by showing that the problem of deciding if a formula in normal form is satisfiable in a data matrix (finite model) $\mathbf{M}$ for an implicational quantifier is NP-complete. We first prove an analogous theorem for associational quantifiers, in full analogy to the proof in [11].

Theorem 1. The set of all associational tautologies in normal form is co-NP-complete.

Proof. We prove that the set of all formulas in normal form satisfiable by an association quantifier (in a model $\mathbf{U}$) is NP-complete. A given formula Φ in normal form has the form $A(\Phi_1, \ldots, \Phi_n)$ where $A(p_1, \ldots, p_n)$ is a propositional formula and Φ results from it by substituting Φ_i (a formula of the form $\varphi_i \sim \psi_i$) for the propositional variable p_i ($i = 1, \ldots, n$). Guess an evaluation e of $p_1, \ldots, p_n$ by zeros and ones and test if e makes $A(p_i, \ldots p_n)$ true. (If not, fail.) Call Φ_i *positive* if $e(p_i) = 1$. Each Φ_i determines four critical formulas:

$$\varphi_i \& \psi_i, \varphi_i \& \neg \psi_i, \neg \varphi \& \psi_i, \neg \varphi_i \& \neg \psi_i.$$

An arbitrary linear preorder of all the critical formulas is *acceptable* if for each i, j the following holds:

If $\varphi_i \& \psi_i \leq \varphi_j \& \psi_j$, $\varphi_i \& \psi_i \geq \varphi_j \& \neg \psi_j$, $\neg \varphi_i \& \psi_i \geq \neg \varphi_j \& \psi_j$, $\neg \varphi_i \& \neg \psi_i \leq \neg \varphi_j \& \psi_j$ and Φ_i is positive then Φ_j is also positive.

The preorder $\leq$ is *realizable* if for some finite model $\mathbf{U}$, $\leq$ is given by frequencies of the formulas in $\mathbf{U}$ (i.e. for each pair γ, δ of critical formulas, $\gamma \leq \delta$ iff the number of objects in U satisfying γ is $\leq$ the number of objects in U satisfying δ).

Now guess a preorder $\leq$ and test acceptability (easy). Then decide realizability; the fact that this is possible by an NP-algorithm is proved in detail in [11] using results on fuzzy probability logics from [10] and [15]. If $\leq$ is realizable then define

$$tr_\sim(a, b, c, d) = 1 \text{ iff for some positive } \Phi_i, a \geq a_i, b \leq b_i, c \leq c_i, d \geq d_i$$

where (a_i, b_i, c_i, d_i) is the four-fold table of (φ_i, ψ_i) in $\mathbf{U}$. Clearly, this makes $\sim$ to an associational quantifier and, for $\sim$, all positive Φ_i's are true in $\mathbf{U}$ and all non-positive Φ_i's are false in $\mathbf{U}$. This shows that associationally satisfiable formulas are in NP. (Up to now the only difference from [11] was a different notion of critical formulas and acceptability.)

Finally show NP-hardness exactly as in [11] by showing, for each open φ, that $\varphi \sim \varphi \,\&\, \neg(\varphi \sim true)$ is satisfiable by an associational quantifier iff $\neg\varphi$ is propositionally satisfiable: if φ is a Boolean tautology then for each $\mathbf{U}$ the four-fold table of (φ, φ) is the same as that of $(\varphi, true)$; and if $\neg\varphi$ is satisfiable then e.g. for the quantifier $\Leftrightarrow$ of logical equivalence and any $\mathbf{U}$ in which $\neg\varphi$ is satisfiable, $\varphi \Leftrightarrow \varphi$ is true but $\varphi \Leftrightarrow true$ is not.

Now let us investigate an important subclass of associational quantifiers called *saturable*. (Cf. [13] 3.2.23; the present definition is slightly modified.)

Definition 1 A quantifier $\sim$ *saturable* if for each a, b, c, d (all positive) there are $a' \geq a$, $b' \geq b$, $c' \geq c$, $d' \geq d$ such that

$$tr_\sim(a', b, c, d) = tr_\sim(a, b, c, d') = 1, \; tr_\sim(a, b', c, d) = tr_\sim(a, b, c', d) = 0.$$

In words, each four-fold table (a, b, c, d) with all entries positive can be made to a table with the value 1 of the quantifier by sufficiently increasing a or sufficiently increasing d, and can be made to a table with the quantifier value 0 by sufficiently increasing b or c.

The *simple* associational quantifier $\sim_0$ with $tr_{\sim_0}(a, b, c, d) = 1$ iff $ad > bc$ is evidently saturable. It is proved in [13] that also the statistical quantifiers FISHER and CHISQUARE are saturable.

Theorem 2. A closed formula Φ is a saturable associational tautology (i.e. tautology for each saturable associational quantifier) iff it is an associational tautology.

Proof. Clearly each associational tautology is a saturable associational tautology. Conversely, let us show that each Φ satisfiable by an associational quantifier is satisfiable (in the same model $\mathbf{U}$) by a saturable associational quantifier. Assume $\Phi = A(\Phi_i, \ldots, \Phi_n)$ in normal form and let Φ be true in $\mathbf{U}$ for a given associational quantifier $\sim$. Define positive and negative Φ_i's (as true/false in $\mathbf{U}$) and let (a_i, b_i, c_i, d_i) be the four-fold table of Φ_i in $\mathbf{U}$. Let q be bigger than all the numbers a_i, b_i, c_i, d_i $(i = 0, \ldots, n)$. Let $tr_{\underset{\sim}{1}}(a, b, c, d) = 1$ iff

- for some positive Φ_i, $a \geq a_i$, $b \leq b_i$, $c \leq c_i$, $d \geq d_i$, or
- $\max(a, d) > q$ and $\max(a, d) > \max(b, c)$.

Then $\underset{\sim}{1}$ is saturable associational, makes all positive Φ_i true and all non-positive Φ_i false in $\mathbf{U}$.

Corollary 1 The set of all saturable associational tautologies is co-NP-complete.

Remark 1 One may investigate the set of tautologies (dually: set of satisfiable formulas) of a fixed quantifier. For example it is easy show that the set of formulas satisfiable for the logical implication quantifier $(tr_\Rightarrow(a, b, c, d) = 0$ iff $b = 0)$ is NP-complete. One can use techniques of this paper (and of [11] together with the result of [15] to show that the set of tautologies of the SIMPLE quantifier (see above) is in PSPACE. Statistically motivated quantifiers (FISHER, LIMPL,...) should be also analyzed.

3   Relativization

In classical logic, relativized quantifiers are definable: $(\forall x/\varphi(x))\psi(x)$ ("for all x satisfying $\varphi(x), \psi(x'')$ is defined as $(\forall x)(\varphi(x) \to \psi(x))$ and $(\exists x/\varphi(x))\psi(x)$ "there is an x satisfying $\varphi(x)$ such that $\psi(x)$") is $\exists x)(\varphi(x)\&\psi(x))$. They say that something holds in the submodel of **U** consisting of objects satisfying $\varphi(x)$. Call this submodel $U \upharpoonright \varphi(x)$. Clearly, $(\forall x/\varphi(x))\psi(x)$ is true in U iff $(\forall x)\psi(x)$ is true in $U \upharpoonright \varphi(x)$, similarly for $\exists$. Formally, restricted quantifier $\forall/$ binds a variable x in a pair of formulas φ, ψ.

It is often useful to work also with relativized versions of associational quantifiers (and other quantifiers applying to a pair of formulas), cf. [18]; we get formulas $(\varphi \sim \psi)/\chi$ (pedantically, $(\sim x)(\varphi, \psi, \chi)$). For simplicity, we continue to restrict ourselves to formulas containing just one object variable x. Assuming the semantics of $\sim$ given, we define $(\varphi \sim \psi)/\chi$ to be true in a model **U** iff $\varphi \sim \psi$ is true in $U \upharpoonright \chi$. (Caution: $U \upharpoonright \chi$ may be empty; but $tr_\sim(0, 0, 0, 0)$ is defined somehow.)

The natural question arises, if the formula $(\varphi \sim \psi)/\chi$ is definable by some formula containing only non-relativized quantifier $\sim$. We present one positive and one negative result.

Theorem 3. For each implicational quantifier $\Rightarrow^*$, the formula $(\varphi \Rightarrow^* \psi)/\chi$ is logically equivalent to $(\varphi\&\chi) \Rightarrow^* \psi$.

Proof. Clearly the four-fold table for φ, ψ in $U \upharpoonright \chi$ and the four-fold table for $\varphi\&\chi, \psi$ in **U** have the same first row (frequency of $\varphi\&\psi\&\chi, \varphi\&\psi\&\neg\chi$ respectively). And only the first row decides on validity of $\Rightarrow^*$.

Theorem 4. Assume $\sim$ is a saturable associational quantifier and φ, ψ, χ are factual open (neither propositional tautologies nor propositionally identically false). Then there are no boolean combinations $\alpha(\varphi, \psi, \chi), \beta(\varphi, \psi, \chi)$ of φ, ψ, χ such that $(\varphi \sim \psi)/\chi$ would be equivalent to $\alpha(\varphi, \psi, \chi) \sim \beta(\varphi, \psi, \chi)$.

Proof. Assume we have α, β such that $(\varphi \sim \psi)/\chi$ is logically equivalent to $(\varphi \sim \psi)/\chi$. Since $\sim$ is saturable, there is a model $\mathbf{U}_1$ in which $\alpha \sim \beta$ is true and another $\mathbf{U}_2$ in which $\alpha \sim \beta$ is false.

Case 1. If $\neg\chi$ is consistent with $\alpha\&\beta$ start with $\mathbf{U}_2$ and expand it by so many objects satisfying $\alpha\&\beta\&\neg\chi$ that $\alpha \sim \beta$ is true in this expansion $\mathbf{U}_2'$: if (a, b, c, d) is the four-fold table of α, β is $\mathbf{U}_2$ then the four-fold table of α, β in U_2' is $(a + q, b, c, d)$ where q is the number of added objects. But clearly, $\mathbf{U}_2 \upharpoonright \chi$ is the same as $\mathbf{U}_2' \upharpoonright \chi$, hence $(\varphi \sim \psi)/\chi$ in false in $\mathbf{U}_2'$ since it is false in $\mathbf{U}_2$ (being equivalent to $\alpha \sim \beta$).

Case 2. Similarly if $\neg\chi$ is consistent with $\neg\alpha\&\neg\beta$.

Case 3. $\neg\chi$ implies $(\alpha\&\neg\beta) \vee (\neg\alpha \& \beta)$. Then start with $\mathbf{U}_1$ in which $\alpha \sim \beta$ as well as $(\varphi \sim \psi)/\chi$ is true and add so many copies of objects satisfying $(\alpha\&\neg\beta\&\neg\chi)$ or $(\neg\alpha\&\beta\&\neg\chi)$ that $\alpha \sim \beta$ is false in the expansion (whereas $(\varphi \sim \psi)/\chi$ remains true).

Let us now discuss computational complexity of our monadic language with a generalized quantifier $\sim$ enriched by relativization. Since evidently, for any semantics of $\sim$, the formula $\varphi \sim \psi$ is logically equivalent to $\varphi \sim \psi/true$, let us define formulas in normal form as having the form $A(\Phi_1, \ldots, \Phi_n)$ where $A(p_1, \ldots, p_n)$ is a propositional formula and Φ_i is $(\varphi_i \sim \psi_i)/\chi_i$.

Theorem 5. For the present notion of formulas in normal form, both the set of implicational tautologies and the set of associational tautologies is co-NP-complete.

Proof. The above proof is modified as follows (we discuss the case of associational quantifiers): given a formula $A(\Phi_1, \ldots, \Phi_n)$ as above, the *critical formulas* of Φ_i are $\alpha_i, \beta_i, \gamma_i, \delta_i$ where α_i is $\varphi_i\&\psi_i\&\chi_i$, β_i is $\varphi_i\&\neg\psi_i\&\chi_i$, γ_i is $\neg\varphi_i\&\psi_i\&\chi_i$ and δ_i is $\neg\varphi_i\&\neg\psi_i\&\chi_i$. Given an evaluation e of $p_1, \ldots, p_n$ (dividing Φ_i's to positive and negative), a linear preorder $\leq$ of critical formulas is *acceptable* if it satisfies the following: If Φ_i is positive, $\alpha_i \leq \alpha_j$, $\beta_i \geq \beta_j, \gamma_i \geq \gamma_j, \delta_i \leq \delta_j$, then Φ_j is positive. Guess an acceptable preorder $\leq$ and test if it is *realizable*, i.e. if there is a model $\mathbf{U}$ such that $\leq$ is given by numbers of objects satisfying formulas, in symbols: for open σ, φ, $\sigma \leq \varphi$ iff $fr_\mathbf{U}(\sigma) \leq fr_\mathbf{U}(\varphi)$, $fr_\mathbf{U}(\sigma)$ being the number of objects in $\mathbf{U}$ satisfying σ. Everything else as before (for implicational quantifiers delete the conditions on γ's and δ's).

4   The quantifier of conviction

Recent works on mining associational rules suggest some notions of rules alternative to that of Agrawal et al.(i.e. other quantifiers in our terminology); we mention just one.

Definition 2 The quantifier $\sim_h^{conv}$ of conviction [6,1,5] is defined as follows: $tr_{\sim_h^{conv}}(a, b, c, d) = 1$ if $conv(a, b, c, d) \geq h$ where

$$conv(a, b, c, d) = \frac{(a + b)(b + d)}{b(a + b + c + d)} = \frac{rl}{bm} = \frac{bm + ad - bc}{bm}$$

and $r = a+b$, $l = b+d$, $m = a+b+c+d$ – marginals; for $b = 0$ and $a+c > 0$, $conv(a, b, c, d) = +\infty)$.

[1] develops a nice theory of this quantifier, including statistical analysis.

Theorem 6. For $h \geq 1$, the conviction quantifier is associational.

Proof. First show $conv(a + 1, b, c, d) \geq conv(a, b, c, d)$, similarly for $d, d + 1$. We show the first inequality. Clearly,

$$\frac{(a + 1 + b)(b + d)}{b(a + 1 + b + c + d)} = \frac{(a + b)(b + d) + b + d}{b(a + b + c + d) + b} \geq \frac{(a + b)(b + d)}{b(a + b + c + d)}.$$

Thus $conv$ is non-decreasing in a, d. Now observe that $conv(a, b, c, d) \geq 1$ iff $ad \geq bc$; hence if $tr_{\sim_h^{conv}}(a, b, c, d) = 1$ for $h \geq 1$ then $ad \geq bc$. Assuming this we show $conv(a, b-1, c, d) \leq conv(a, b, c, d)$. Use the formula $conv(a, b, c, d) = \frac{bm+ad-bc}{bm}$. Then we claim

$$\frac{(b - 1)(m - 1) + ad - (b - 1)c}{(b - 1)(m - 1)} \geq \frac{bm + ad - bc}{bm}; \text{ i.e.}$$

$$bm(ad - (b - 1)c) \geq (b - 1)(m - 1)(ad - bc),$$

$$bm(ad - bc) + bmc \geq bm(ad - bc) + (1 - b - m)(ad - bc),$$

$$bmc \geq (1 - b - m)(ad - bc),$$

which is true since $1 - b - m \leq 0$ (for $m \geq 1$, i.e. non-empty model) and $ad - bc \geq 0$ by the above assumption about h. A similar computation gives the result for c.

Remark 2 Since $conv(a, b, c, d) \geq h \geq 1$ implies $ad \geq bc$, it is immediate that for each (a, b, c, d) with $a, b, c, d > 0$ there is a $b' \geq b$ and $c' \geq c$ such that $ad < b'c$, hence $conv(a, b', c, d) < 1$ and $conv(a, b, c', d) < 1$. This is the "negative" part of saturability. On the other hand, clearly

$$\lim_{a \to +\infty} conv(a, b, c, d) = \frac{b + d}{b},$$

$$\lim_{a \to +\infty} conv(a, b, c, d) = \frac{a + b}{b},$$

which shows that the "positive" part of saturability does not hold; the conviction quantifier is *not* saturable. And clearly it is not implicational since it does depend on d. It would be nice to define a natural class of quantifiers that are "conviction-like" in an analogy to implicational quantifiers that are "FIMPL-like" and saturable associational quantifiers that are "SIMPLE-like". Finally let us mention that the set of tautologies of the conviction quantifier (for all $h > 1$ or, alternatively, for a given rational $h > 1$) can be shown to be in PSPACE by the methods presented above.

References

1. Adamo J. M.: Data mining for associational rules and sequential patterns, Sequential and parallel algorithms, Springer 2001.
2. Agrawal R., Imielinski T., Swami A.: Mining associations between sets of items in massive databases, Proc. ACM-SIGMOD Int. Conf. on Data, 1993, pp. 207-216.
3. Arratia A.: References for finite model theory and related subjects, www.ldc.usb.ve/~arratia/refs/all.ps
4. Gottlob G.: Relativized logspace and generalized quantifiers over finite structures, Journal Symb. Logic, 1997, pp. 545-574.
5. Bayardo R. J., jr., Agrawal R., Gunopulos D.: Constraint-based rule mining in large, dense databases, Proc. 15th Int. Conf. on Data Engineering, 1999.
6. Brin S., Motwani R., Ullman J., Tsur S.: Dynamic itemset countion and implication rules for market basket data, Proc. ACM-SIGMOD Int. Conf. of Management of Data, 1997, pp. 255-264.
7. Fagin R.: Finite model theory – a personal perspective, Theor. Comp. Science, 116, 1993, pp. 3-31.
8. Fagin R.: Contributions to the model theory of finite structures. PhD thesis, Univ. of California, Berkley, 1973.
9. Hájek P.: Generalized quantifiers and finite sets, In Prace Naukowe Inst. Mat. Politech. Wroclaw, 14, 1977, pp. 91-104.
10. Hájek P.: Metamathematics of fuzzy logic, Kluwer, 1998.
11. Hájek P.: Relations in GUHA style data mining, Proc. Relmics 6, Tilburg, Netherlands, 2001.
12. Hájek P.: The GUHA method and mining associational rules, Proc. CIMA'2001, Bangor, U. K., ICSC Acad. Press 2001, pp. 533-539.
13. Hájek P., Havránek T.: Mechanizing hypothesis formation – Mathematical foundations for a general theory, Springer-Verlag 1978. Internet version (free): www.cs.cas.cz/~hajek/guhabook
14. Hájek P., Sochorová A., Zvárová J.: GUHA for personal computers, Comp. Statistics and Data Analysis, 19, 1995, pp. 149-153.
15. Hájek P., Tulipani S.: Complexity of fuzzy probability logic, Fundamenta Informaticae, 45, 2001, pp. 207-213.
16. Rauch J.: classes of four-fold table quantifiers. In (Zytkow et al., ed.) Principls of data mining and knowledge discovery. Springer verlag 1998, 203-211
17. Rauch J.,Šimůnek M.:Mining for 4ft association rules by 4ft-miner. In: INAP 2001 Tokyo, 285-294
18. Rauch J.: Interesting association rules and multi-relational associational rules. In Communications of Institute of Information and Computing Machinery, Taiwan. Vol. 5, No. 2, May 2002. pp. 7782
19. www.cs.cas.cz – research – software
20. http://lispminer.vse.cz/overview/4ftminer.html

Part X

Invited Session: AI Applications in Medicine

MLEM2—Discretization During Rule Induction

Jerzy W. Grzymala-Busse

Department of Electrical Engineering and Computer Science, University of
Kansas, Lawrence, KS 66045, USA

Abstract. LEM2 algorithm, a rule induction algorithm used by LERS, accepts
input data sets only with symbolic attributes. MLEM2, a new algorithm, extends
LEM2 capabilities by inducing rules from data with both symbolic and numerical attributes including data with missing attribute values. MLEM2 accuracy is
comparable with accuracy of LEM2 inducing rules from pre-discretized data sets.
However, compared with other members of the LEM2 family, MLEM2 produces
the smallest number of rules from the same data. In the current implementation of
MLEM2 reduction of the number of rule conditions is not included, thus another
member of the LEM2 family, namely MODLEM based on entropy, induces smaller
number of conditions than MLEM2.

1 Introduction

The algorithm MLEM2 (Modified LEM2) is a new option in the data mining
system LERS (Learning from Examples based on Rough Set theory) inducing
rules from raw data [6], [7]. Other data mining systems based on rough set
theory were described in [15]. The first version of LERS was implemented
at the University of Kansas in 1988. Initially, LERS tests the input data for
consistency. Consistent data has no conflicting cases, i.e., cases having the
same values for all attributes but different decision values. LERS uses an
approach to handle inconsistent data based on *rough set theory* [13], [14]. For
inconsistent data the system computes *lower* and *upper approximations* for
each concept. Rules induced from the lower approximations are called *certain*,
while rules induced from the upper approximations are called *possible*.

The algorithm MLEM2 is used to process either lower or upper approximation of a concept. Thus, the input data of MLEM2 are always consistent.
The algorithm MLEM2 induces a *discriminant* rule set, i.e., the smallest
set of minimal rules describing each concept in the input data. The original
version of the algorithm, LEM2 (Learning from Examples Module, version
2), was presented in [3], [4], [6], though the algorithm was implemented for
the first time in 1990 [7]. The algorithm LEM2 was re-implemented and
modified, under different names, in a number of places, some examples are
MODLEM implemented in the Poznan Technology University, Poland [8],
[9], and ELEM2 from the University of Waterloo, Canada [1]. All of these
implementations are based on the same idea of using the set of all relevant
attribute-value pairs as a search space.

2   LEM2

In algorithm LEM2 a rule set is induced by exploring the search space of blocks of attribute-value pairs. LEM2 induces a local covering and then converts it into the rule set. To define a local covering a few auxiliary definitions will be quoted. For a variable (attribute or decision) x and its value v, a block $[(x, v)]$ of a variable-value pair (x, v) is the set of all cases for which variable x has value v.

Let B be a nonempty lower or upper approximation of a concept represented by a decision-value pair (d, w). Set B *depends* on a set T of attribute-value pairs t if and only if

$$\emptyset \neq [T] = \bigcap_{t \in T} [t] \subseteq B.$$

Set T is a *minimal complex* of B if and only if B depends on T and no proper subset T' of T exists such that B depends on T'. Let $\mathcal{T}$ be a nonempty collection of nonempty sets of attribute-value pairs. Then $\mathcal{T}$ is a *local covering* of B if and only if the following conditions are satisfied:

- each member T of $\mathcal{T}$ is a minimal complex of B,
- $\bigcap_{t \in \mathcal{T}} [T] = B$, and
- $\mathcal{T}$ is minimal, i.e., $\mathcal{T}$ has the smallest possible number of members.

The user may select an option of LEM2 with or without taking into account attribute priorities. The procedure LEM2 with attribute priorities is presented below. The option without taking into account priorities differs from the one presented below in the selection of a pair $t \in T(G)$ in the inner loop WHILE. When LEM2 is not to take attribute priorities into account, the first criterion is ignored. In our experiments all attribute priorities were equal to each other. The original algorithm is presented below.

Procedure LEM2
(**input**: a set B,
output: a single local covering $\mathcal{T}$ of set B);
begin
      $G := B$;
      $\mathcal{T} := \emptyset$;
      while $G \neq \emptyset$
            begin
            $T := \emptyset$;
            $T(G) := \{t | [t] \cap G \neq \emptyset\}$;
            while $T = \emptyset$ **or** $[T] \not\subseteq B$
                begin
                      select a pair $t \in T(G)$ with the highest
                      attribute priority, if a tie occurs, select a pair

$t \in T(G)$ such that $|[t] \cap G|$ is maximum;
if another tie occurs, select a pair $t \in T(G)$
with the smallest cardinality of $[t]$;
if a further tie occurs, select first pair;
$T := T \cup \{t\}$;
$G := [t] \cap G$;
$T(G) := \{t|[t] \cap G \neq \emptyset\}$;
$T(G) := T(G) - T$;
end {while}
for each $t \in T$ **do**
    if $[T - \{t\}] \subseteq$ B **then** $T := T - \{t\}$;
$\mathcal{T} := \mathcal{T} \cup \{T\}$;
$G := B - \bigcup_{T \in \mathcal{T}} [T]$;
end {while};
for each $T \in \mathcal{T}$ **do**
    if $\bigcup_{S \in \mathcal{T} - \{T\}} [S] = B$ **then** $\mathcal{T} := \mathcal{T} - \{T\}$;
end {procedure}.

For a set X, $|X|$ denotes the cardinality of X.

3   MLEM2

MLEM2 is a modified version of the algorithm LEM2. The original algorithm LEM2 needs discretization, a preprocessing, to deal with numerical attributes. Discretization is a process of converting numerical attributes into symbolic attributes, with intervals as values. LEM2 treats all attributes as symbolic, thus producing too specific rules when input data are not discretized. Also, the original LEM2 algorithm considers missing attribute values as special values. An approach to extend LEM2 to induce rules from data with missing attributes was presented in [10]. We will use the same approach to missing attribute values in MLEM2.

First we will describe how MLEM2 induces rules from data with numerical attributes. MLEM2 has an ability to recognize integer and real numbers as values of attributes, and labels such attributes as numerical. For numerical attributes MLEM2 computes blocks in a different way than for symbolic attributes. First, it sorts all values of a numerical attribute. Then it computes cutpoints as averages for any two consecutive values of the sorted list. For each cutpoint c MLEM2 creates two blocks, the first block contains all cases for which values of the numerical attribute are smaller than c, the second block contains remaining cases, i.e., all cases for which values of the numerical attribute are larger than c. The search space of MLEM2 is the set of all blocks computed this way, together with blocks defined by symbolic attributes. Starting from that point, rule induction in MLEM2 is conducted the same way as in LEM2.

In addition, MLEM2 handles missing attribute values during rule induction. For any attribute with missing values, blocks are computed only from the existing attribute-value pairs. Thus, during the computation of attribute-value blocks, cases that correspond to missing attribute values are ignored. As a result, it is possible that some cases from the training data are not classified by any rule at all. This is an expected effect of missing attribute values—rules are induced only from existing attribute-value pairs.

Example. To illustrate the idea of MLEM2, let us induce a rule set from the data presented in Table 1.

	Attributess		Decision
Case	Height	Hair	Attractiveness
1	60	blond	−
2	70	blond	+
3	60	red	+
4	80	dark	−
5	60	dark	−
6	?	dark	−

Table 1. An example of the data set

Table 1 contains one numerical attribute (*Height*). The sorted list of values of *Height* is 60, 70, 80. Thus, MLEM2 computes two cutpoints: 65 and 75.

The set of all blocks on which MLEM2 is going to operate, i.e., the search space for MLEM2, consists of the following blocks: $[(Height, 60..65)]$ = {1, 3, 5}, $[(Height, 65..80)]$ = {2, 4}, $[(Height, 60..75)]$ = {1, 2, 3, 5}, $[(Height, 75..80)]$ = {4}, $[(Hair, blond)]$ = {1, 2}, $[(Hair, red)]$ = {3}, and $[(Hair, dark)]$ = {4, 5, 6}.

Our first concept $G = B = [(Attractiveness, -)] = \{1, 4, 5, 6\}$. The attribute-value pairs relevant with G, i.e., attribute-value pairs (A, v) such that

$$[(A, v)] \cap [(Attractiveness, -)] \neq \emptyset,$$

are $(Height, 60..65)$, $(Height, 65..80)$, $(Height, 60..75)$, $(Height, 75..80)$, $(Hair, blond)$ and $(Hair, dark)$. The most relevant attribute-value pair is $(Hair, dark)$, since for this attribute-value pair

$$|[(A, v)] \cap [(Attractiveness, -)]|$$

is the largest.

The set consisting of the pair $(Hair, dark)$ is a minimal complex since

$$[(Hair, dark)] \subseteq [(Attractiveness, -)].$$

Our new goal G is $B = [(Attractiveness, -)] - [(Hair, dark)] = \{1, 4, 5, 6\} - \{4, 5, 6\} = \{1\}$.

Among attribute-value pairs relevant with $G = \{1\}$, i.e., $(Height, 60..65)$, $(Height, 60..75)$, $(Hair, blond)$, the attribute-value pair $(Hair, blond)$ is selected since $\|[(Hair, blond)]\|$ is the smallest. However,

$$[(Hair, blond)] \not\subseteq [(Attractiveness, -)],$$

so the set $\{(Hair, blond)\}$ is not a minimal complex and we need to select additional attribute-value pairs. The next candidate is $(Height, 60..65)$, since $\|[(Height, 60..65)]\| = 3 < \|[(Height, 60..75)]\| = 4$.

Now

$$[(Hair, blond)] \cap [Height, 60..65] \subseteq [(Attractiveness, -)],$$

so $\{(Hair, blond), (Height, 60..65)\}$ is the second minimal complex. Furthermore, it is not difficult to see that

$$\{\{(Hair, dark)\}, \{(Hair, blond), (Height, 60..65)\}\}$$

is a local covering of $[(Attractiveness, -)] = \{1, 4, 5, 6\}$.

The remaining local covering, for $B = [(Attractiveness, +)] = \{2, 3\}$, may be computed in the similar way. This local covering is

$$\{\{(Hair, red)\}, \{(Hair, blond), (Height, 65..80)\}\}.$$

Therefore, the rule set induced by MLEM2 from the data presented in Table 1 is

(Hair, red) -> (Attractiveness, +)
(Hair, blond) & (Height, 65..80) -> (Attractiveness, +)
(Hair, dark) -> (Attractiveness, -)
(Hair, blond) & (Height, 60..65) -> (Attractiveness, -)

4   Classification system

The classification system of LERS is a modification of the *bucket brigade algorithm* [2], [12]. The decision to which concept a case belongs is made on the basis of three factors: strength, specificity, and support. They are defined as follows: *strength* is the total number of cases correctly classified by the rule during training. *Specificity* is the total number of attribute-value pairs on the left-hand side of the rule. The matching rules with a larger number of attribute-value pairs are considered more specific. The third factor, *support*, is defined as the sum of scores of all matching rules from the concept. The concept C for which the support, i.e., the following expression

$$\sum_{\text{matching rules } R \text{ describing } C} Strength\_factor(R) * Specificity\_factor(R)$$

| | Number of | | |
	cases	attributes	concepts
Bank	66	5	2
Bricks	216	10	2
Bupa	345	6	2
Buses	76	8	2
German	1000	24	2
Glass	214	9	6
HSV	122	11	2
Iris	150	4	3
Pima	768	8	2
Segmentation	210	19	7

Table 2. Data sets

is the largest is the winner and the case is classified as being a member of C.

In the classification system of LERS, if complete matching is impossible, all partially matching rules are identified. These are rules with at least one attribute-value pair matching the corresponding attribute-value pair of a case. For any partially matching rule R, the additional factor, called *Matching _ factor* (R), is computed. Matching_factor (R) is defined as the ratio of the number of matched attribute-value pairs of R with a case to the total number of attribute-value pairs of R. In partial matching, the concept C for which the following expression is the largest

$$\sum_{\substack{partially\ matching \\ rules\ R\ describing\ C}} Matching\_factor(R) * Strength\_factor(R)$$
$$* Specificity\_factor(R)$$

is the winner and the case is classified as being a member of C.

Every rule induced by LERS is preceded by three numbers: specificity, strength, and the total number of training cases matching the left-hand side of the rule.

5   Experiments

For experiments we used the same ten data sets that were used for experiments reported in [8], [9], see Table 2. Complexity of rule sets induced by the two versions of MODLEM, based on Laplacian accuracy and entropy and on data preprocessed by discretization based on entropy and then processed by LEM2 [5] are presented in Tables 3 and 4. These results, previously reported in [8], [9], are compared with new results of our experiments using MLEM2. Both versions of MODLEM were described, e.g., in [8], [9]. Discretization based on entropy was presented in [5]. Accuracy for the ten data sets, presented in Table 5, was computed using ten-fold-cross validation.

Data set	MODLEM Laplace	MODLEM Entropy	Discretization based on entropy and LEM2	MLEM2
Bank	6	3	10	3
Bricks	22	16	25	12
Bupa	101	79	169	73
Buses	5	4	3	2
German	253	182	290	160
Glass	80	43	111	33
HSV	54	35	62	23
Iris	12	10	14	9
Pima	188	125	252	113
Segmentation	47	22	108	14

Table 3. Number of rules

Data set	MODLEM Laplace	MODLEM Entropy	Discretization based on entropy and LEM2	MLEM2
Bank	7	16	13	5
Bricks	38	33	61	40
Bupa	228	219	501	345
Buses	5	5	4	4
German (numeric)	774	751	1226	1044
Glass	139	111	262	137
HSV	96	92	206	101
Iris	24	20	33	23
Pima	400	426	895	599
Segmentation	80	45	322	48

Table 4. Number of conditions

An example of the rule set induced by MLEM2, from the well-known data *Iris*, is presented below. The current version of MLEM2 does not simplify rule conditions involved in numerical attributes. For example, the second rule, presented below, has two conditions (petal_length, 3.15..6.9) and (petal_length, 1..4.95). These two conditions may be combined together into one condition (petal_length, 3.15..4.95), thus reducing the total number of conditions.

1, 50, 50

(petal_length, 1..2.45) -> (class, Iris-setosa)

3, 46, 46

(petal_length, 3.15..6.9) & (petal_width, 0.1..1.65)
& (petal_length, 1..4.95) -> (class, Iris-versicolor)

5, 2, 2

(petal_length, 1..5.15) & (petal_width, 0.1..1.85) & (petal_width, 1.55..2.5)
& (petal_length, 4.95..6.9)
& (sepal_length, 5.95..7.9) -> (class, Iris-versicolor)
    4, 23, 23
(sepal_length, 4.3..5.95) & (petal_length, 1..4.85) & (sepal_width, 2..3.25)
& (sepal_length, 5.05..7.9) -> (class, Iris-versicolor)
    2, 43, 43
(petal_width, 1.75..2.5) & (petal_length, 4.85..6.9) -> (class, Iris-viginica)
    2, 39, 39
(sepal_length, 5.95..7.9) & (petal_width, 1.75..2.5) -> (class, Iris-viginica)
    2, 6, 6
(petal_length, 4.95..6.9) & (sepal_width, 2..2.65) -> (class, Iris-viginica)
    2, 16, 16
(sepal_width, 2..2.85) & (petal_width, 1.65..2.5) -> (class, Iris-viginica)
    2, 35, 35
(petal_length, 5.05..6.9) & (sepal_length, 6.25..7.9) -> (class, Iris-viginica)

Data set	MODLEM Laplace	MODLEM Entropy	Discretization based on entropy and LEM2	MLEM2
Bank	94	94	97	95
Bricks	91	91	92	92
Bupa	68	66	66	65
Buses	97	97	99	96
German (numeric)	73	73	74	70
Glass	58	72	67	72
HSV	63	57	56	60
Iris	91	94	97	95
Pima	74	74	74	71
Segmentation	72	85	64	80

Table 5. Accuracy

The rule set with simplified conditions is presented below. The total number of conditions of this rule set is 19. As it is clear from Table 4, none of the four algorithms induce so few conditions for Iris. Likely, another version of MLEM2 that will simplify conditions in the above way, combining conditions with the same numerical attributes, will produce the smallest number of conditions as well.

    1, 50, 50
(petal_length, 1..2.45) -> (class, Iris-setosa)
    2, 46, 46
(petal_length, 3.15..4.95) & (petal_width, 0.1..1.65)
-> (class, Iris-versicolor)

3, 2, 2

(petal_length, 4.95..5.15) & (petal_width, 1.55..1.85)
& (sepal_length, 5.95..7.9) -> (class, Iris-versi color)

3, 23, 23

(sepal_length, 5.05..5.95) & (petal_length, 1..4.85) & (sepal_width, 2..3.25)
-> (class, Iris-versicolor)

2, 43, 43

(petal_width, 1.75..2.5) & (petal_length, 4.85..6.9) -> (class, Iris-viginica)

2, 39, 39

(sepal_length, 5.95..7.9) & (petal_width, 1.75..2.5) -> (class, Iris-viginica)

2, 6, 6

(petal_length, 4.95..6.9) & (sepal_width, 2..2.65) -> (class, Iris-viginica)

2, 16, 16

(sepal_width, 2..2.85) & (petal_width, 1.65..2.5) -> (class, Iris-viginica)

2, 35, 35

(petal_length, 5.05..6.9) & (sepal_length, 6.25..7.9) -> (class, Iris-viginica)

6   Conclusions

Rule sets induced from data sets by data mining systems may be used for classification of new, unseen cases (usually such rule sets are incorporated in expert systems) or for interpretation by the users. In the latter, the user will see some regularities hidden in the user's data and may come to some conclusions on this basis. In both cases simplicity of rules is important. Too numerous rule sets in expert systems slow down the systems. In the interpretation of rule sets, or direct rule analysis by the users, simplicity is crucial.

Results of experiments for the four rule induction algorithms: two versions of MODLEM, a preliminary discretization based on entropy and then LEM2, and MLEM2 were compared using two-tailed Wilcoxon matching-pair signed rank test with the 5% significance level [11].

Among members of the LEM2 family of rule induction algorithms, the algorithm MLEM2 produces the rule sets with the smallest number of rules. On the other hand, MLEM2 needs an additional tool to simplify conditions using numerical attributes. For time being, MODLEM based on entropy is the algorithm producing the smallest number of conditions, MLEM2 and MODLEM based on Laplacian accuracy trail behind the version of MODLEM based on entropy.

Finally, all four algorithms do not differ significantly in the most important area: accuracy of induced rules.

Acknowledgment. The author would like to thank Dr. Jerzy Stefanowski for his kind consent to quote results of his experiments on both versions of MODLEM in this paper.

References

1. An, A. and Cercone, N. (1999) An empirical study on rule quality measures. Proc of the 7th Int. Workshop, RSFDGrC'99, Yamaguchi, Japan, November 9–11, 1999. Lecture Notes in AI 1711, Springer Verlag, Heidelberg, 482–491.

2. Booker, L. B., Goldberg, D. E., and Holland, J. F. (1990) Classifier systems and genetic algorithms. In Machine Learning. Paradigms and Methods. Carbonell, J. G. (ed.), The MIT Press, Menlo Park, 235–282.

3. Chan, C. C. and Grzymala-Busse, J. W. (1991) On the attribute redundancy and the learning programs ID3, PRISM, and LEM2. Department of Computer Science, University of Kansas, TR-91-14.

4. Chan, C. C. and Grzymala-Busse, J. W. (1994) On the two local inductive algorithms: PRISM and LEM2. Foundations of Computing and Decision Sciences **19**, 185–203.

5. Chmielewski, M. R. and Grzymala-Busse, J. W. (1996) Global discretization of continuous attributes as preprocessing for machine learning. Int. Journal of Approximate Reasoning **15**, 319–331.

6. Grzymala-Busse, J. W. (1992) LERS—A System for Learning from Examples Based on Rough Sets. In: Slowinski, R. (ed.): Intelligent Decision Support. Handbook of Applications and Advances of the Rough Sets Theory. Kluwer Academic Publishers, Boston, MA, 3–18.

7. Grzymala-Busse, J. W. (1997) A new version of the rule induction system LERS. Fundamenta Informaticae **31**, 27–39.

8. Grzymala-Busse, J. W. and Stefanowski, J. (1999) Two approaches to numerical attribute discretization for rule induction. Proc. of the 5th International Conference of the Decision Sciences Institute, Athens, Greece, July 4–7, 1999, 1377–1379.

9. Grzymala-Busse, J. W. and Stefanowski, J. (2001) Three discretization methods for rule induction. International Journal of Intelligent Systems **16**, 29–38.

10. Grzymala-Busse, J. W. and Wang, A. Y. (1997) Modified algorithms LEM1 and LEM2 for rule induction from data with missing attribute values. Proc. of the Fifth International Workshop on Rough Sets and Soft Computing (RSSC'97) at the Third Joint Conference on Information Sciences (JCIS'97), Research Triangle Park, NC, March 2–5, 1997, 69–72.

11. Hamburg, M. (1983) Statistical Analysis for Decision Making. New York, NY: Harcourt Brace Jovanovich, Inc., New York, 546–550 and 721.

12. Holland, J. H., Holyoak, K. J., and Nisbett, R. E. (1986) Induction. Processes of Inference, Learning, and Discovery. The MIT Press, Menlo Park, Cambridge MA, London.

13. Pawlak, Z. (1982) Rough Sets. International Journal of Computer and Information Sciences **11**, 341–356.

14. Z. Pawlak, (1991) Rough Sets. Theoretical Aspects of Reasoning about Data. Kluwer Academic Publishers, Boston, MA.

15. Polkowski, L. and Skowron, A. (eds.) (1998) Rough Sets in Knowledge Discovery, 2, Applications, Case Studies and Software Systems, Appendix 2: Software Systems. Physica Verlag, Heidelberg, New York, 551–601.

Part XI

Short Contributions

A Hybrid Model Approach to Artificial Intelligence

Kevin Deeb and Ricardo Jimenez

Department of Information Technology, Barry University, 11415 NE 2nd Ave,
Miami Shores, Florida, 33161,USA
Email: {kdeeb,rjimenez}mail.barry.edu

Abstract. This paper establishes a learning paradigm for natural language. The paradigm is then effectively modeled by using a hybrid approach to Artificial Intelligence (AI). A synergy of established AI paradigms was found by examining various methodologies, and choosing aspects of each that can replicate the desired learning paradigm. The hybrid model postulated is a combination of object-oriented deductive database with elements of fuzzy logic, neural networks, and natural language understanding methodologies. A three-word, transitive verb-phrase, natural language understanding model was created, based on fuzzy values, and inference mechanisms that endow the model with deductive capabilities. The resulting model creates new phrases based on input of data, while queries return information that is relevant to the universe of the models matrix. The resulting output is based on a dynamic relationship between the static objects in the matrix and the fed input. The results showed that the mechanisms involved in this model can be effective, and realistically implemented.

1 Introduction

This paper explorers the possibilities of approaching artificial intelligence (AI) by infusing elements of different established AI methodologies in order to produce a hybrid model that can be used in natural language understanding. The hybrid model postulated is a combination of object-oriented deductive database with elements of fuzzy logic, neural networks, and natural language understanding paradigms. This approach allowed the creation of simple three-word, transitive verb-phrase, natural language understanding model. Connection-weight algorithms, based on fuzzy values, trigger inference mechanisms that endow the model with deductive capabilities with minimal initial framework and no supervised training. The resulting model creates new phrases based on input of data, while queries return information that is relevant to the universe of the models matrix. The resulting output is based on the relationship between the static objects in the matrix and the fed input. Weights are adjusted for all objects in the matrix with each new input thus changing the truth-value of each object. Therefore, the model can be viewed as being event driven. Frequency of repeated relationships increases the validity of each object phrase. In essence, the model learns from experience and its truth responses are based on the universal value of each object, as it relates to all other objects in the matrix. The approach to the proposed model sprang from the vantage point of establishing the learning paradigm itself; and then, develops heuristics that may simulate that paradigm. Our hypothesized learning paradigm stems from our observations of early child development; for example, children first learn simple noun-verb phrases and then begin to develop grammar. It is evident that syntactic knowledge is derived secondarily after semantic and discourse knowledge. Indeed, syntax is needed for complex concepts such as stories; but if humans are used to model natural language understanding, then a model, which starts with no predefined predicates or background knowledge, should begin with simple noun-verb phrases. Furthermore, children often make what seem as logical conclusions from their limited knowledge base; these conclusions either are reinforced or suppressed through further event exposure. In the knowledge matrix of a child, a talking purple dinosaur has a high truth-value, simply because the event is repeatedly reinforced in relation to all other events in the childs knowledge matrix. Only when the event is later suppressed, through the lack of relational magnitude, does the truth-value decrease. Hence, the truth-value for the talking purple dinosaur decreases from the set of Real Characters while proportionally increases in the set of Fictional Characters. Once a learning paradigm for the proposed model was conceived and to determine the best approach, it was first postulated that a method based on a synergy of established AI paradigms should be found by examining various methodologies that replicate the desired learning paradigm. The goal is to infuse strengths of one AI methodology with other ones, in an effort to arrive at a model that can satisfy our approach. Once the elements are identified, a simple model is proposed to illustrate the viability of taking a hybrid approach to AI. The assumption that cognizers should mirror the real world accurately, by background predicate-definitions, will yield limited AI models. One of our initial premises is that true AI accumulates knowledge from novel events, and should ultimately produce predicates dynamically. Moreover, modeling artificial intelligence should not be approached from a uniquely deterministic viewpoint [7]. This would indicate that a hybrid model is better suited, if the presumption is to model human intelligence. Inductive Logic Programming (ILP), though shown to be effective in natural language understanding, requires extensible revisions of hierarchal predicates, if the model is to evolve [11]. Hence, only certain aspects of ILP would serve in our model. In addition, ILP has not been applied much into existing database technologies, because of the background knowledge necessary before the model can become intelligent [15]. Furthermore, ambiguity must be incorporated into any true cognizer. Indeed, natural language incorporates these ambiguities with modifiers like, somewhat, somewhere, likely, could be, and so forth. Thus, elements of fuzzy logic seem ideal by allowing the definition of set memberships, which can simulate ambiguity. It is our view, that association through repetitive reinforcement and fuzzyfication of relevance values between all data in a matrix is the key to learning. Neural networks that use back-propagation and biasing have been shown to be particularly effective with front-end fuzzyfication [6]. Recurrent neural networks have been shown to be computationally as powerful as Turing machines [13], and neural networks have been used effectively in natural language understanding [9]. Lofti A. Zadehs fuzzy logic methodology, which he termed Computing with Words (CW), has had a profound influence on the development of the proposed model. CW provides a mechanism to facilitate synergism between natural language and computation with fuzzy variables. Fuzzy logic permits the prototyping of ambiguity; and thus, addresses a key issue in our model. The most important technique in CW is fuzzy constraint propagation; which can be illustrated in the phrase *John is tall*, the fuzzy set tall is a constraint on the height of John [16]. Fuzzy logic is easily adapted to other methodologies and is well suited for the proposed model. In addition, it has already been effectively used in recurrent neural nets [13]. Fuzzy logic interpolation is also used in the proposed model, which has previously been successfully implemented in feed-forward neural nets [13]. Fuzzy neural nets based on Takagi-Sugeno-Kang (TSK) model and have also been shown to be capable of effectively modeling a self-constructing inference network [8]. An additional mechanism explored for learning can be found in structural word association, it has been hypothesized that groups of words tend to evoke each other as word associates, and has been shown in word response experiments that associated word clusters exhibit faster response latency than non associative clusters [14]. At the core of the model is an object-oriented database (OOD). The model is defined using ODMG 2.0 standard for Object Definition Language (ODL) and Object Query Language (OQL) that should be easily adapted to O2. An OOD was chosen for the proposed model because of its relative strengths in describing object behavior over conventional relational database. In addition to object-oriented databases like O2, the proposed model incorporates some aspects of conventional deductive databases like Datalog. Consequently, the proposed model may be viewed as quasi-Deductive Object Oriented Database (DOOD), which has been shown to be, computationally, as powerful as Turing machines [12]. DOOD strength lies in its declarative queries, while for OOD lies in modeling. While it would seem ideal, attempts to integrate databases with machine learning to construct Intelligent Learning Database Systems (ILDB), has been limited [15]. Aspects of WHIRL (Word-based Heterogeneous Information Representation Language), a subset of Datalog, were incorporated into this model. In particular, a variation of the use of term-weights in WHIRL is fused with the concept of connection weights in feed-forward neural nets. It was felt that the methods used in WHIRL are an effective model of semantic content [2], and fit well with the idea of structural-word association and the even-driven paradigm postulated in this model. WHIRLs representation of simple text as vectors, and the representation of connection weights between the input and output ($W_{ij} = X_i \rightarrow Y_j$) in neural nets seemed to indicate room for a possible synergy between to the two. In particular, aspects of single-layer feed-forward neural nets using the Hebb rule [6]. Finally, our viewpoint is that AI should be approached as syntactic objects, which are event driven, and structural word association is accomplished through dynamically changing truth-values based on fuzzy relationships between the objects. The proposed models approach is not syntactic parsing; but, rather an association of words in predefined syntactic objects. Therefore, the proposed model obtains semantic knowledge through the objects relationships. This concept is coupled with an event driven paradigm, which brings novel events into the matrix while fuzzy values are used to represent the frequency of these relationships.

2 Proposed Model

The model created is to be implemented in an OOD DBMS like O2. The Interface definitions state the necessary functions and data types of the model.

Step 1 The model is composed of objects whose atomic attributes are three strings. The combined attributes would form a phrase based on natural language syntactic rules. The three strings must follow a natural language logical form [1], and comply with the sentence structure Subject? Verb ? Object. In addition, the verb must be of the transitive form. The meaning of a transitive verb is incomplete without a direct object, for example The boy goes is incomplete, and must contain a direct object, The boy goes home. This would be entered into our model as boy, goes, and home. The object attribute is referred to as target in the proposed model.

Call create new object function, Phrase:Object new() with elements (S,V,T) Insert elements (S=Subject,V=Verb,T=Target)
// (S=boy, V=goes, T=home)
// O:= phrase object.
// O_i.S or O_i.V or O_i.T

Step 2 A connection weight algorithm compares atomic attribute of all objects in the matrix. $O_i.S \rightarrow O_j.S$, $O_i.V \rightarrow O_j.V$, $O_i.T \rightarrow O_j.T$. Each compared object, which contains a matching string, is given a predetermined connection-weight fuzzy value of .33 for each matching attribute. Therefore, Connection weight states can be seen as:

0 Null, .33 Low .66 Medium .99 High Where a null state would indicate phrases with no common attributes, a low state indicates phrases with one matching attribute, medium state with two matching attributes, and a high state as duplicate attributes. These values are stored in vectored arrays (Wi,j) that indicate the connection state between every object in the matrix.

Call Connection Weight Function,ConnectionWeight:Object ConWeight() Loop compares object elements S,V,T
// O_1(boy, goes, home) O_2 (girl, goes, school)
// O_1(boy, goes, home) O_7(boy, goes, park)

// O_7(boy, goes, park) O_{15}(boy, goes, park)
// O_{18}(man, crosses, road) O_{19}(girl, jumps, rope) Int array Wi,j same as function assigns like elements a fuzzy value of .33 and placed in array Wi,j.
// $W_{1,2}$=.33 fuzzy state low
// $W_{1,7}$=.66 fuzzy state med
// $W_{7,15}$=.99 fuzzy state high (duplicate object)
//$W_{18,19}$=0 fuzzy state null

Step 3 Once the states are computed new phrases can be generated by the system based on a subject swap algorithm. In essence, two new objects are created and back propagated into the matrix for each low weight connection that has been previously generated. The subject swap algorithm finds low connection weights and creates two new objects composed of those attributes. This process continues until all objects with low states have swapped subjects. Newly created objects that are in high state (duplicate) are destroyed upon construction.

Call Subject Swap Function, Object SubSwap
Loop checks for connection weights in low state.
If $W_{i,j}$ is low and O_i.S not the same as O_j.S
Call new()
Copy subject of O_i to new object
Copy verb of O_j to new object
Copy Target of O_j to new object
// O_1(boy, goes, home) O_2 (girl, goes, school) $\rightarrow$ O_3(boy, goes, school)
Call Conweight()
// re-calculate conection weights with newly created object.
If new object connection weight is high
destroy object // high connection is duplicate object
rollback // maintain OID continuaty.
// create second object of swap
Call new();
Copy subject of O_j to new object;
Copy verb of O_i to new object;
Copy Target of O_i to new object;
// O_1(boy, goes, home) O_2 (girl, goes, school)$\rightarrow$ O_4(girl, goes, home)
Call Conweight() // Recalculate connection weights with newly created object.
If new object connection weight is high
destroy object // high connection is duplicate object
rollback // maintain OID continuity.
Return

Step 4 Upon completion of the subject swap, each object is given a universal fuzzy value that indicates its relational magnitude to all other objects in the matrix. Summing all $W_{mathrmi,j}$ of $Object_n$ and dividing by the total number of objects in the matrix calculates the universal weight. The universal weight is calculated by summing all $W_{mathrmi,j}$ of $Object_n$ and dividing by the total number of objects in the matrix.

$UW_n = \sum_{i \to j} W_{(n,i)_{j-1}}$

Where n= object number, $i \neq n$

Call Universal Connection Weight function, UniversalWeight: Object Uni-Weight()

Initialize i

Call Iterator()

get_element(O_n)

inner loop increments i

get_element $(W_{n,i+1})$

sum all connection weights

Universal connection weight = average of all Objects $W_{i,j}$

Returns values

Step 5 The universal truth weight expresses a fuzzy truth-value, which is a mean derived from all universal connection weights in the matrix, and recalculated following each iteration. The truth-value is calculated from one standard deviation from the dynamic mean. Therefore the universal truth-value is as follows: Threshold <=0.1 Saturation =1 True >= (Dynamic mean) (one standard deviation) False <= (Dynamic mean) (one standard deviation) The truth-value is an essential part of the query process and can later be used in an expanded model for set membership.

Call Universal Truth Value, UniversalTruthValue: Object UniTruth()

Call Iterator()

get_element $(O.UCW_n)$

Call Mean () // Calculates mean of all current universal connections in the matrix

Call StdDev () Calculates standard deviation of all current UC in the matrix.

Returns values

The following section shows an ODMG definition [5] of our proposed model as well as the algorithms previously illustrated in steps one through five.

3   Connection Weight Algorithm

Algorithm pseudo code loosely based on ODMG 2.0 [5] and C++ [4]. The connection weight algorithm compares the elements of O_i.SVT with O_j.SVT and assigns a fuzzy value of .33 for each match. Thus if O_i.SVT is same as O_j.SVT, a total fuzzy connection value of .999 would be assigned to that relation. The connection weight, $W_{i,j} \to O_i$, O_j. Thus $W_{i,j}$ represents the connection weight between Objecti and Objectj. Connection weight fuzzy values are:

$W_{i,j}$=0 Null $W_{i,j}$=.33 low $W_{i,j}$=.66 medium $W_{i,j}$=.999 High

ConWeight() double W [I] [j] = 0, 0;//An array is created of size I and J
While (I < j, I++)
{
If O_i.S same_as O_j.S &&
O_i.V same_as O_j.V &&
O_i.T same_as O_j.T;
Then c.$W_{i,j}$=.999;
    Else If O_i.S !same_as O_j.S && O_i.V !same_as O_j.V && O_i.T !same_as
O_j.T; Then c.$W_{i,j}$=0;
      Else If O_iS same_as O_jS && O_iV same_as O_jV &&; Then c.$W_{i,j}$=.66;
      Else If O_iS same_as O_jS && O_iT same_as O_jT &&; Then c.$W_{i,j}$=.66;
      Else If O_iV same_as O_jV && O_iT same_as O_jT &&; Then c.$W_{i,j}$=.66;
      Else c.$W_{i,j}$=.33; Return (c.$W_{i,j}$);
}

4   Subject Swap Algorithm

Algorithm pseudo code loosely based on ODMG 2.0 [5] *and C++* [4]. The
subject swap algorithm finds low connection weights and creates two new
objects composed of elements, which swaps subjects of those objects with
such a relation. An additional predicate specifies that subjects may not be
the same.
    SubSwap()
{
int I=1;
While (I < j I++) {
If c.$W_{i,j}$=.33 && O_i.S !same as O_j.S;
New (); // create new object O_{j+1}
copy (O_{j+1}.S O_i.S);
copy (O_{j+1}.V= O_j.V);
copy (O_{j+1}.T= O_j.T);
ConWeight() ;// Recalculate new weights with newly created objects
GetElement (c.$W_{i,j+1}$);
If c.$W_{i,j+1}$=.999 ;
Destroy (O_{j+1}); //destroy any duplicate objects created.
Rollback ();// maintain object number sequence
New (); // create new object
copy (O_{j+1}.S, O_jS);
copy (O_{j+1}.V, O_i.V);
copy (O_{j+1}.T, O_i.T);
Else
copy (O_{j+2}.S, O_j.S);
copy (O_{j+2}.V, O_i.V);
copy (O_{j+2}.T, O_i.T);.

ConWeight(); // Recalculate new weights with newly created objects
GetElement $(W_{i,j+2})$;
If $W_{i,j+2}$=.999;
Destroy (O_{j+2}); //destroy any duplicate objects created.
Rollback () ;// maintain object number sequence
}
Return
}

5   Universal Weight Algorithm

UniWeight()
{
Int i;
i=0;
Iterator ();
get_element();
u.UW_n=0;
While (i < j, i ++){
u.UW_n= u.UW_n +(c.$W_{n,i+1}$);
}
O.UCW_n= (u.UW_n) /j-1;
return (O.UCWn)
next_element();
}

6   Universal Truth Value

The universal truth weight expresses a fuzzy truth-value. The threshold and
saturation values are as follows: The truth-value is calculated from one standard deviation from the dynamic mean. Therefore the universal truth-value
is as follows: Threshold Θ <=0 Saturation Ω=1 Queries return truth and
false values True >= (Dynamic mean) (one standard deviation) False <=
(Dynamic mean) (one standard deviation)

**Calculation of Dynamic Mean and Standard Deviation Algo-
rithm:** UniTruth ()
{
Iterator ();
get_element (O.UCW_n);
T.MCW= Mean(O.$UCW_{1 \text{ to } n}$);
T.UTV= StdDev (O.$UCW_{1 \text{ to } n}$);
next_element;
return (T.STD, T.MCW)
}

Queries: *Queries are loosely based on OQL []*

Retrieve all True values in matrix:

<?, all_true> select O*, T.USTV, T.MCW from O in Phrase and T in UniversalTruthValue Where O.UCW$_n$ is_greater T.MCW-T.USTV Order by O.UCW desc;

Retrieve all False values in matrix:

<?, all_false> select O*, T.USTV, T.MCW from O in Phrase and T in UniversalTruthValue Where O.UCW$_n$ less_than T.MCW-T.USTV Order by O.UCW asc;

Retrieve Highest Truth Value: <?, highest_true> select O* max_O.UCW max(Select O.UCW from O in Phrase) from O in Phrase and T in UniversalTruthValue

Retrieve Highest FalseValue: <?, highest_False> select O* min_O.UCW min(Select O.UCW from O in Phrase) from O in Phrase and T in UniversalTruthValue

Phrase Queries: These are a few examples of the queried data. The samples return all true and false or by subject. Naturally many ad-hoc variations to these queries may be performed.

All True Phrase Query: All true phrase query returns elements that satisfy query in descending order of truth value. ? can be placed in any of the elements S,V,T

<?,all_true, verb, target> select O*, T.USTV, T.MCW, O.? from O in Phrase and T in UniversalTruthValue Where O.UCW$_n$ is_greater T.MCW-T.USTV and O.S same_as O.? Order by O.UCW desc;

All False Phrase Query: <?, all_false, verb target> select O*, T.USTV, T.MCW, O.? from O in Phrase and T in UniversalTruthValue Where O.UCW$_n$ less_than T.MCW-T.USTV and O.S same_as O.? Order by O.UCW asc;

Highest True Phrase Query: <?, highest_true, verb, target> select O*, O.?, max_O.UCW max(Select O.UCW from O in Phrase) from O in Phrase and T in UniversalTruthValue Where O.? same_as O.S

Highest False Phrase Query: <?, highest_false, verb, target> select O* min_O.UCW min(Select O.UCW from O in Phrase) from O in Phrase and T in UniversalTruthValue Where O.? same_as O.S

7   Conclusion

The hybrid model that is proposed is a combination of object-oriented deductive database, fuzzy logic, neural networks, and natural language understanding paradigms. This approach allowed the creation of simple three-word, transitive verb-phrase, natural language understanding model. Algorithms based on fuzzy values, trigger inference mechanisms that endow the model with deductive capabilities. We first hypothesized a learning paradigm from our observations of early child development; and adapted a hybrid methodology to it. The sample data clearly shows the mechanisms involved in this model can be effective, and realistically implemented. In particular, it is felt that the approach of establishing a learning paradigm, prior to heuristics, is quite advantageous Therefore, this hybrid methodology, if properly expanded can produce significant results.

References

1. J. Allen, Natural language understanding, 2nd ed., The Benjamin/Cummings Publishing Company Inc., Redwood City, CA, 1995.
2. W. Cohen, WHIRL: A word based information representation language, Artificial Intelligence, 118, (2000) 163-196
3. M. Delgado, A. F., Gomez-Skarmeta, and F. Martin, A Fuzzy clustering-based rapid prototyping for fuzzy rule-based modeling, IEEE Transaction on fuzzy systems, 5 (2), (1997), 223-233.
4. H. M. Deitl and P. J. Deitl, C++ how to program, 3rd ed., Prentice Hall, New Jersey, 2001.
5. R. Elmasri and S. B. Navathe, Fundamentals of database systems, 3rd ed., Addison-Wesley, Reading, MA, 2000.
6. L. Fausett, Fundamentals of neural networks, 1st ed., Prentice Hall, New Jersey, 1994
7. R. C. Johnson and C. Brown, Cognizers, 1st ed., John Wiley & Sons Inc., New York, 1988.
8. C-F Juang and C-T Lin, An on-line self-constructing neural fuzzy inference network and its applications, IEEE Transaction on fuzzy systems, 6 (1), (1998) 12-32.
9. C-T Lin and M. C. Kan, Adaptive fuzzy command acquisition with reinforcement learning IEEE Transaction on fuzzy systems, 6 (1), (1998), 102-121.
10. Y. Lin, G. A. Cunningham III, S. V. Coggeshall, Using fuzzy partitions to create fuzzy systems from input-output data and set the initial weights in a fuzzy neural network, IEEE Transaction on fuzzy systems, 5 (4), (1997), 614-621
11. S. Muggleton, Inductive logic programming: issues, results and the challenge of learning language in logic, Artificial Intelligence, 114, (1999), 283-296
12. T. Niemi, M. Christensen, K. Jarvelin, Query language approach based on the deductive object oriented database paradigm, Information and Software Technology, 42, (2000), 777-792.
13. C. W. Omlin, K. K. Thornber and C. L. Giles, Fuzzy finite-state automata can be deterministically encoded into recurrent neural networks, IEEE Transaction on fuzzy systems, 6 (1), (1998), 76-89.
14. H. R. Pollio, The structural basis of word association behavior, 1st ed., Mouton & Co., The Hague, The Netherlands, 1966.
15. X. Wu, Building intelligent learning database systems, AI magazine, (2000), 61-67.
16. L. A. Zadeh, Fuzzy logic = computing with words, IEEE transactions on Fuzzy Systems, 4 (2), (1996), 103-111.

Applying Transition Networks in Translating Polish E-Mails

Krzysztof Jassem, Filip Graliński, and Tomasz Kowalski

Adam Mickiewicz University, Faculty of Mathematics and Computer Science,
Poznań, Poland
jassem@amu.edu.pl, filipg@amu.edu.pl, chiro@interia.pl

Abstract. The paper presents an adoption of transition networks to recognizing and translating fragments of texts characteristic of e-mails written in Polish. An extension of XTND (XML Transition Networks Definition) — abbreviated as PTND — is introduced in order to provide a convenient way to describe these networks. A new tool for graphical representation, modification and validation of network descriptions is proposed.

1 Introduction

POLENG is a rule-based Polish-to-English machine translation system, which has been developed since 1995. In the years 1995–2001 POLENG had a status of strictly academic project. Since 2002 the research has been directed towards commercial use. The first commercial partner of the project was the Allied Irish Bank (AIB) who envisaged the system as a communication tool between its centre in Ireland and its branches in Poland. One of the most urgent needs of AIB was the translation of e-mails.

Examination of an e-mail corpus delivered by AIB has shown that most e-mails contain texts that would not be translated correctly by a rule-based translation algorithm. This concerned mainly such parts of e-mails as greetings, farewells, adding attachments or expressions of gratitude or respect.

A few potential solutions to this problem have been considered. Example-based translation was rejected because of scarcity of bilingual e-mail corpora. Lexicon-based approach, i.e listing of all possible phrases characteristic of e-mails seemed to burden the system with heavy data of minimal use.

The regularity of these expressions made transition networks a convenient tool for their description and translation.

2 The origin of the idea

Transition networks (or transducers) have long been hoped to be a useful tool for translation. The reality has revised these hopes but the tool is still used for recognizing simple patterns prior to the translation process. SYSTRAN researchers report the use of translation networks for recognizing Hungarian temporal expressions in Hungarian-English translation [1] and for translation of support files in the AUTODESK system [2].

The approach presented here is different from the one suggested by SYS-TRAN. We do not construct target (i.e. English) networks, and thus we do not make any alignment between networks of two languages. Our experiments on networks composed for both languages have shown that rules of alignment are of almost the same complexity as those of rule-based transfer.

We have decided that transition networks should generate equivalent English text in the process of recognition of a Polish pattern. In order to overcome well-known restrictions of transducers (e.g. discrepancies between orderings of words) we have augmented each network with a history stack, which stores the information about visited states, traversed transitions (and their pre-conditions) as well as executed actions.

3   PTND — POLENG Transition Network Definition

The format of our description for transition networks - that will be further on referred to as PTND (POLENG Transition Network Definition) - is based on XTND, the XML Transition Network Definition ([3]). The aim of the authors of XTND was to create a standard format for transition networks in a universally used language.

3.1   XTND — XML Transition Network Definition

The note [3] realizes two objectives: it gives a general and universally applicable definition of a transition network and provides a tool for the formal description of all elements of a network. This description is given in the XML language and forms a DTD (Document Type Definition). For example, the state is defined as an object with the following type:

```
state = object {
    name          = string;
    preconditions = set of predicates;
    prelude       = ordered set of actions;
    postconditions = set of predicates;
    postlude      = ordered set of actions;
};
```

A state is represented by its name, a set of *preconditions*, i.e. conditions that must be satisfied to enter the state, a set of actions that should be executed before entering the state, a set of *postconditions*, i.e. conditions that must be satisfied to exit the state and a set of actions that should be executed on exiting the state. The above definition as mirrored in the DTD for the XTND in the following way:

```
<!ELEMENT   state   {properties?,
                     preconditions?,
                     prelude?,
                     postlude?,
                     postconditions?) >
<!ATTLIST   state   id     ID     #REQUIRED
                    name   CDATA  #REQUIRED >
```

The transition is defined as an object which consists of the following elements: *from, to, preconditions, actions*. The element definition of a transition in DTD is analogous to that of a state.

XTND suggests also a language for marking actions, called XEXPR, an example of which follows:

```
<set name="AA" value="AA Value"/>
<print newline="true"><get name="AA"/></print>
```

3.2   Applying XTND to PTND

In PTND the conditions are assigned to transitions only (and not to states). Actions are executed only during transitions. The notion of 'event' is not used in PTND.

3.3   Conditions in PTND

Transition condition is a "condition that guards the transition from one state to another" [3]. In PTND a condition may have one of the following types:

- Empty
- Raw — string of letters
- Netref — reference to a subnetwork
- Code

The syntax of *Code* is the same as the syntax for 'condition' in the C language.

The need for using *Code* appears for *dictionary conditions*, i.e. conditions which have to be verified by looking up the dictionary of the POLENG system (see section 5. for Examples).

3.4   Actions

Action is an operation performed when a transition is traversed. In PTND actions are expressed with a sublanguage of C — limited to logical operations, comparisons and assignments. The code for actions is left verbatim (except for variables) when transition networks are compiled into a C++ code (see section 6).

3.5   Variables

The language used in conditions and actions allows for using variables. Two types of variables are used:

- local read-only variables, marked in the examples with a character '@'
- global read-write variables, marked in the examples with a character '$'.

3.6   References to stacked arguments

In order to be able to cope with variations of order of Polish expressions, each network is augmented with a history stack. The stack allows for referring to the values of local variables of previous transitions. For example, GEN[2] refers to the value of the local variable GEN that was set two transitions back.

3.7   Example

Below a part of description of a network that recognizes and translates a type of e-mail greetings, is presented:

```
<transition from="s1" id="t1" name="t1" to="s11">
  <conditions>
    <![CDATA[@LEX=="mój" && @CASE=="Nom";]]>
  </conditions>
  <actions>
    <![CDATA[$E += "my";]]>
  </actions>
</transition>
<transition from="s11" id="t2" name="t2" to="s21">
  <conditions>
    <![CDATA[@LEX=="drogi" && @CASE=="Nom" && @GEN==@GEN[1];]]>
  </conditions>
  <actions>
    <![CDATA[$E += "dear";]]>
  </actions>
</transition>
<transition from="s21" id="t3" name="t3" to="s31">
  <conditions>
    <![CDATA[@CASE=="Voc" && @GEN==@GEN[1];]]>
  </conditions>
  <actions>
    <![CDATA[$E += @E;]]>
  </actions>
</transition>
```

The network recognizes exemplary headers: "Moja droga Julio" and "Mój drogi chłopcze" and produces "My dear Julia" and "My dear boy" respectively".

The first transition recognizes all nominative forms of the lexeme "mój" (i.e. singular, plural forms of all genders), the second transition recognizes nominative forms of the lexeme "drogi" and verifies consistency of the gender with the gender of the word recognized in the previous transition. The third transition recognizes vocative forms of Polish nouns and adds the equivalent found in the POLENG dictionary (@E) to the value of the global variable $E.

Picture 1. shows the representation of the above network produced by *Gichon*.

4 The backtracking algorithm

A typical backtracking algorithm for transition networks goes back following the trace in order to find a state which still has some untraversed transitions exiting from it. The only information that has to be stored in such an algorithm is the list of visited states and the list — for each state — of untraversed transitions that exit from the state.

The backtracking algorithm that we use stores the information on transitions. If a transition is based on a network, then the success path is remembered in order for the algorithm to be able to look for another path in the backtracking process (if needed). If a transition is of a dictionary type and a dictionary entry is ambiguous, then the backtracking algorithm needs to know which interpretations of the dictionary entry have already been verified.

It is worth noting that in our approach backtracking is called not only upon failure but also in the reference to historical values of local variables.

5 The drawing tool

Experience has shown that even a universal standard of storing networks (like XML) raises difficulties in creating networks that are compatible both with linguistic data, and the formal description of the network. We have designed a drawing tool, called *Gichon*, which:

- allows for creation and modification of a network consistent with PTND
- reads a PTND description of a transition network and displays its graphical representation
- validates the description of a network against PTND
- stores a graphically designed (or modified) network in .xml file consistent with PTND
- checks the syntax of actions and conditions.

We have considered using some public domain software for drawing networks. One of the user-friendly public domain drawing tools is DIA. It was formerly designed for LINUX systems (see [Toga, 2001] for the description of the tool). Nowadays it is available also for Win32 systems (downloadable from http://hans.breuer.org/dia).

Eventually we have made the decision to design a new tool. The advantages of having a tool dedicated especially for our needs are following:

- *Gichon*, apart from drawing capabilities, supports the PTND formalism. It also means that the application can be easily re-designed in order to support XTND.
- *Gichon* is 100% Java code, which makes it platform-independent. The same application may be used both for LINUX and Win32.
- *Gichon* allows for checking the syntax of the Action language.

The next section presents examples of PTND networks as well as their graphical representations produced by Gichon. The application is downloadable from the POLENG web site: http://ceti.pl/~poleng.

6   Using transition networks in the translation process

Similarly to SYSTRAN's approach, the POLENG mechanism of transition networks is used only for recognizing regular patterns prior to the translation process. The networks are applied to a tokenized text, i.e. a list of tokens, which represents the input text. The output of the recognition process is a list of tokens, based on the input list, with all segments of texts recognized by networks being marked appropriately. The recognized segments form units, which are described by the same data structures as 'regular' tokens. This means that the input and output list of recognition are of the same format. The output list of network recognition is transferred to the proper translation process. The translation procedures 'are unaware' of the fact whether any transition networks have been actually applied to the text. In other words, the network recognition stage is transparent in the whole translation process as far as data structures are concerned.

It would be ineffective to call all transition networks for any type of texts and at any position. Recognition according to a specific network takes place only when the POLENG system is in a definite state. The state comprises information about the domain of the text (e.g. *information technology, banking, science*), the type of the text (e.g. *e-mail, WWW page*), the current position in the text (e.g. *the beginning of the text*). For example, the network for recognizing greetings is checked only at the beginning of an e-mail text. Naturally, the networks themselves are allowed to change the state of the system.

Network recognition procedures are implemented in a C++ code. For convenience sake, the XML representations of networks are converted into

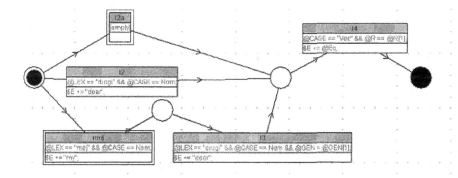

Fig. 1. Picture 1. Gichon's visualization of an exemplary TN

pure C++ code. Transition conditions are converted into statements of conditions in `if` instructions. Actions are embraced by curly braces in order to create C++ statements. Network variables (attributes) inside conditions and actions are detected and converted into valid C expressions.

7   Conclusions

The research has led to the following conclusions:

- Transitions networks may still find their applications in describing some restricted phenomena of natural language.
- It has been a good idea to create the XTND specification. The document can be easily adopted for various implementations of transition networks.
- In order to deal with free-order languages transition networks must be augmented with stacks, which store the history of transitions.

We believe that the tool presented here may prove helpful in easy and convenient describing of quite complex patterns of natural language.

References

1. Senellart J., Dienes P. Varadi T., *New Generation Systran Translation System*, MT Summit VIII, Santiago de Compostela 2001
2. Senellart J., Plitt M. Bailly C, Cardoso F., *Resource Alignment and Implicit Transfer*, MT Summit VIII, Santiago de Compostela 2001
3. XTND - *XML Transition Network Definition, W3C Note 21 November 2000*, available at: http://www.w3.org/TR/xtnd/
4. Toga K. DIA. Creating Charts and Diagrams, available at: http://www.togaware.com/linuxbook/dia.pdf

Link Recommendation Method Based on Web Content and Usage Mining

Przemysław Kazienko and Maciej Kiewra

Wrocław University of Technology, Wyb. Wyspiańskiego 27, Wrocław, Poland,
kazienko@pwr.wroc.pl, makie@eui.upv.es

Abstract. Hyperlink recommendation overcomes the problem of quick and easy access to information in web systems. A method that integrates web usage and content mining was proposed and examined in this paper. Potentially interesting documents are prompted to the user on the basis of usage patterns and conceptual spaces matched against the active user session. Automatic term selections and web usage distinction according to the time of visit were introduced to enhance method effectiveness.

1 Introduction

Since WWW is more and more competitive, the creation of well-designed web site is not sufficient to attract users. Therefore personalization is more and more meaningful. One of the personalization techniques is hyperlink recommendation often utilizing information about navigation activity and site content. This information is not known explicitly and should be obtained using web mining techniques, which may be divided into two groups: *web usage mining* (analyses of data related to users' activity, e.g. navigation patterns [4,6]) and *content mining* (processing of documents' content [1,2,8]). We propose a hyperlink recommendation method based on the integration of both these approaches.

2 Method Overview

Our method improves and extends works from [5]. In respect of content mining we introduced the original method of term selection for clustering and the new conception of document weight calculation. In web usage mining, the time factor was added. Additionally, the new integration algorithm was presented.

The whole process (Fig. 1) can be divided into two independent tasks. The former extracts text features — terms from site documents in order to discover thematic areas from the site content — *conceptual spaces*. The latter is based on recognition of typical *navigation patterns*.

Each technique uses N-dimensional vectors, $N = \text{card}(D)$ and D is the set of all documents (site pages). Each vector refers to one term in the content mining issue and to one user session in the web usage mining and its coordinates correspond to particular documents in both cases.

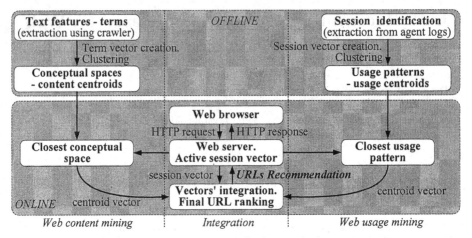

Fig. 1. The link recommendation method overview

3   Content Mining — Term Selection and Clustering

The first step of term clustering is the text feature extraction. As documents are very often generated dynamically, it is recommendable to use a crawler. Normally, the term frequency is calculated for each indexed document. We propose to use other text features (such as HTML title, keywords, etc.) and the search engine queries. Many extracted terms are poor descriptors, thus we suggest selecting only the terms t_i for which clustering usefulness function $f_{cu}(t_i)$ value is the highest:

$$f_{cu}(t_i) = \frac{n^{t_i} - k_1^t}{n^{t_i}} \cdot \exp\left(-\left(\frac{2\left(n^{t_i} - k_2^t\right)}{n^{t_i} + k_3^t}\right)^4\right) + \frac{tf_i^q}{tf_{\max}^q}$$

where: n^{t_i} — the number of documents from D in which term t_i occurs, k_1^t, k_2^t, k_3^t — constants, tf_i^q — the frequency of the term t_i in all search engine's queries, $tf_{\max}^q$ — max. value of tf_i^q. The last component denotes how often the term t_i is used in queries by users in comparison with other terms. The function f_{cu} is used to eliminate terms, which occur rarely (more seldom then k_1^t) or too often. The most emphasized are terms, which are in k_2^t documents. The factor k_3^t determines "flatness" of the function. According to our experiments the parameter k_3^t should have the value of $4 \div 6$. Values of the last two depend on the web site's size and according to experiments: $k_2^t = k_3^t = N/10$. Only maximum n terms are selected.

For every selected term t_i a N-dimensional vector $c_i^t = \langle w_{i1}^c, w_{i2}^c, \ldots, w_{iN}^c \rangle$ is created, where w_{ij}^c denotes the weight of the term t_i in the document d_j:

$$w_{ij}^c = (tf_{ij}^b + \alpha tf_{ij}^t + \beta tf_{ij}^d + \gamma tf_{ij}^k) \cdot \log_2 \left(\frac{N}{n^{t_i}} \right),$$

where: tf_{ij}^b, tf_{ij}^t, tf_{ij}^d, tf_{ij}^k — term frequency of term t_i respectively in the body, title, description and keywords of the d_j page; α, β, γ stress place of term occurrence. Concerning the experiments from [2] they can be set as follows: $\alpha = 10$, $\beta = 5$, $\gamma = 5$.

The set of N-dimensional vectors can be clustered to discover groups of terms that are close to each other. These terms describe *conceptual spaces* existing within the web site. Once we have clusters we can calculate essences of the *conceptual spaces* — vectors c^c called centroids:

$$c_i^c = \frac{1}{\max_i} \sum_{j=1}^{m_i} c_{ij}^t, \tag{1}$$

where: c_i^c — centroid of the ith cluster; c_{ij}^t — j^{th} term vector belonging to the ith cluster; m_i — number of terms in the ith cluster, $\max_i$ — the maximal value of coordinate from the ith cluster used for normalization.

4   Usage Mining — Historical Session and Active Session Processing

The first step of usage mining is acquisition of HTTP requests and creation of user's sessions. A user session, in this context, is a series of pages requested by the user during one visit. Since web server logs do not provide any easy method to group these requests into sessions, each request coming to the web server should be captured and assigned to a particular session using a unique identifier passed to a client's browser (i.e. by means of cookies [3]).

Many users visit only few pages and abandon the site. Such insignificant sessions should not be used in recommendation. Thus, we omit all session in which less than n^s documents where visited. In the implementation $n^s = 4$ has been assumed.

The next step is to form N-dimensional session vectors $s_i = \langle w_{i1}^s, \ldots, w_{iN}^s \rangle$, one for each separate i^{th} session. We used geometric sequence in coordinates w_{ij}^s of the vectors to weaken influence of the old session in the following way:

$$w_{ij}^s = \begin{cases} (tc)^{n_i^{tp}}, & \text{when document } d_j \text{ was visited during the } i^{\text{th}} \text{ session}, \\ 0, & \text{when document } d_j \text{ was not visited during the } i^{\text{th}} \text{ session}, \end{cases}$$

where: tc — constant time coefficient from the interval $[0, 1]$; n_i^{tp} — number of time periods since beginning of the ith session until vector creation moment.

Time period length (a unit of measure for n_i^{tp}) depends on how often users enter the web site. Time coefficient tc denotes changeability of links between pages and the site content. The more often site changes, the smaller should be the tc value. In that way older sessions have less impact on clustering results.

Session vectors s_i are clustered in the same way like term vectors c_i^t, using (1). Finally, for every i^{th} cluster, we obtain one session centroid s_i^c, that describes one typical navigational path throughout the web site. Each its coordinate indicates whether a corresponding document is strongly represented in the navigational path or not. Based on historical session information clustering discovers standard user behaviours — *usage patterns*.

An active session describes pages visited by the user during the current session. N-dimensional active session vector $a = \langle w_1^a, w_2^a, \ldots, w_N^a \rangle$ is formed to facilitate processing with above obtained vectors. w_j^a is the weight of the jth document in the active session. Similarly to creation of historical session vectors we propose geometric sequence to strengthen last visited documents:

$$w_j^a = \begin{cases} (\lambda)^{n_j^a}, & \text{when document } d_j \text{ was visited during the active session,} \\ 0, & \text{when document } d_j \text{ was not visited during the active session,} \end{cases}$$

where: λ — constant parameter for the interval $[0, 1]$, determined experimentally, in implementation $\lambda = 0{,}95$ was assumed; n_j^a — consecutive number of document d_j in active session in reverse order. For the just viewed document $n_j^a = 0$ ($w_j^a = 1$), for the previous document $n_j^a = 1$ ($w_j^a = \lambda$), etc. If the document was visited more than once, the least value is assumed to n_j^a.

5   Document Ranking and Link Recommendation

The described process results in: the set of content centroids c_i^c, the set of historical session centroids s_i^c and the active session vector a. Normalized cosine vector similarity formula [7] is used in order to find the centroid c_i^c closest to active user vector a. It denotes to *conceptual* space most similar to the active session. The vector c_i^c is multiplied by the cosine value:

$$c_i^{c'} = c_i^c \cdot \cos(c_i^c, a)$$

The most suitable session centroid s_i^c (a *usage pattern* the closest to the active session) is found in the same way. Thus, we obtain the centroid transformation $s_i^{c'} = s_i^c \cdot \cos(s_i^c, a)$.

Integration of the content mining with usage mining is done by the *rank'* function — sum of transformed centroids vectors:

$$rank' = c_i^{c'} + s_i^{c'} = \langle w_1^r, w_2^r, \ldots, w_N^r \rangle.$$

The vector rank is multiplied by the modified active session vector to not recommend the documents that have been just seen:

$$rank = rank' \cdot (1 - a) = \langle w_1^r \cdot (1 - w_1^a), w_2^r \cdot (1 - w_2^a), \ldots, w_N^r \cdot (1 - w_N^a) \rangle$$

For link recommendation first n^r document corresponding to the vector *rank* coordinates with the highest value are selected. The coordinate adequate to active document has the value of 0 ($w^a = 1$). Link to this document will not be suggested.

6   Conclusions and Future Works

As the set of extracted terms contains weak descriptors, an automatic selection of the terms for clustering was proposed. The time factor was introduced in order to weaken the influence of old user sessions (usage patterns) and to strengthen last visited pages (active session).

The method implementation (within the project ROSA — *Remote Object Site Agent*) — reveal some interesting facts. Firstly, the documents that contain a lot of relevant terms tend to appear in many content clusters on the highest position. Secondly, the documents that occur in many historical sessions appear in the almost all clusters with strong weights. The problem can be solved by setting to 0 all session vector coordinates corresponding to the documents that occur at least in n user sessions (n is about 80 %).

The future work will concentrate on introducing a special mechanism that will promote new site documents (for example by increasing new documents' weights in c^c centroids). Typical usage patterns and thematic *conceptual spaces* can be also used to propose the user advertising banners or special product offers.

References

1. Chakrabarti S., et al. (1999) Mining the Web's Link Structure. *IEEE Computer* **32** (8) 60–67.
2. Kazienko P. (2000) Hypertekst Clustering based on Flow Equivalent Trees. *Wrocław University of Technology, Dep. of Inf. Systems, Ph.D. Thesis* (in Polish)
3. Kiewra M. (2002) Web Management Using Users' Data and Activities. *Wrocław University of Technology, M.Sc. Thesis.*
4. Lin W., Alvarez S.A., Ruiz C. (2002) Efficient Adaptive-Support Association Rule Mining for Recommender Systems. *Data Mining and Knowledge Discovery* **6** (1) 83–105.
5. Mobasher B., Dai H., Luo T., Sun Y., Zhu J. (2000) Integrating Web Usage and Content Mining for More Effective Personalization. *Lecture Notes in Computer Science* **1875** Springer 156–176.
6. Mobasher B., Dai H., Luo T., Nakagawa M. (2002) Discovery and Evaluation of Aggregate Usage Profiles for Web Personalization. *Data Mining and Knowledge Discovery* **6** (1) 61–82.
7. Salton G., McGill M.J. (1983) Introduction to Modern Information Retrieval. *McGraw-Hill Book Co.*
8. Wulfekuher M.R., Punch W.F. (1997) Finding Salient Features for Web Page Categories. *Computer Networks and ISDN Systems* **29** (8–13) 1147–1156.

Conceptual Modeling of Concurrent Information Systems with General Morphisms of Petri Nets

Boleslaw Mikolajczak[1,2] and Zuyan Wang[1]

[1] Unversity of Massachusetts Dartmouth, North Dartmouth, MA, USA
[2] Polish-Japanese School of Information Technology, Warsaw, Poland

Abstract. Development of complex concurrent information systems is very often performed in top-down or bottom-up approach depending on design circumstances. Petri net morphisms have been proven to be useful in this process as long as certain desired structural and behavioral properties of such systems are preserved. In particular, for general morphisms of Petri nets, we study their structural and behavioral properties and we use Petri net model of Peterson's mutual exclusion algorithm to illustrate a step-wise process of bottom-up abstraction.

1 Motivation and Introduction

Petri net is a formal, graphical, and executable mathematical model that is appropriate for the development of concurrent, discrete-event dynamic systems. It can be used in design and analysis of concurrent and distributed systems, workflow management systems, requirement specifications in software engineering, specifications of communication networks, and so on.

Usually there are two different approaches in Petri net system modeling. One is the top-down approach and the other is the bottom-up approach. In top-down approach, one can start modeling a system from the highest conceptual level of abstraction, then refine the system using techniques such as rule-based refinement [6] and general refinement [3] until reaching a satisfying level of detail of the system. In bottom-up approach, one starts the modeling of a system at the lowest level of abstraction, and then abstracts the system model step by step until reaching the highest level of abstraction.

In formal object-oriented software engineering, rigorous software development requires continuous verification during all phases of the software development process. However, resources are often very restricted and a totally new verification at each step is usually too expensive and time consuming. Thus, vertical structuring techniques that can preserve desired properties for both top-down and bottom-up approaches will be very helpful.

There exist several different concepts of Petri net morphisms which have already been introduced in literature, such as vicinity respecting morphism [2], Winskel's morphism [8], Lakos's net morphism and system morphism [4], and general morphism [1]. These morphisms respect different types of Petri nets and preserve different structural and behavioral properties of Petri nets. The properties of Petri net morphisms make them useful in the refinement and abstraction of Petri net system modeling and analysis of the system being modeled.

In this paper, we will focus on the abstraction of Petri net model, and try to provide a solution to the following problem using general morphisms of Petri nets: Suppose we have a detailed system modeled using Petri net, how can we abstract the system model so that some desired properties of the system can be preserved. In section two, we will give some basic definitions about Petri nets and Petri net morphisms. Section three presents detailed example of conceptual modeling of concurrent system using general morphisms of Petri nets.

2   Petri Nets and Their Morphisms

[Def.2.1] A Petri net is a 4-tuple $N = (P, T, F, M_0)$ satisfying the following conditions:

1. P is a finite non-empty set of places,
2. T is a finite non-empty set of transitions satisfying $P \cap T = \emptyset$,
3. F is a multiset of $(P \times T) \cup (T \times P)$, called the flow relation,
4. M_0 is a non-empty multiset of places, called initial marking.

Here $X = P \cup T$ denotes the set of all elements of a Petri net. We call the set $\{y \in X | (y, x) \in F\}$ the *pre-set* of x, the set $\{y \in X | (x, y) \in F\}$ the *post-set* of x, denoted by ${}^\bullet x$ and $x^\bullet$, respectively. We will use the notation ${}^\circ x = {}^\bullet x \cup \{x\}$ to represent *pre-vicinity* and $x^\circ = x^\bullet \cup \{x\}$ to represent *post-vicinity*, too.

The abstraction of Petri nets is not new and can date back to Petri [6]. The method of relating a Petri net and its abstraction net is called Petri net morphism. Roughly speaking, Petri net morphism is a mapping from elements of a source net to elements of a target net that preserves some properties of the source net. There are two types of Petri net morphisms. One focuses on the structural relationship between source and destination net, i.e. how places and transitions are connected via arcs in both nets. The other one respects the behavioral relationship, i.e. the relationship between different markings caused by firing of transitions in Petri net.

[Def.2.2] Let $N = (P, T, F, M_0)$ and $N' = (P', T', F', M_0')$ be Petri nets. A morphism from N to N' is a partial function $\eta : T \to_\mu T'$ and a multirelation $\beta : P \to_\mu P'$ such that

$$\beta M_0 = M_0' \text{ and } \forall A \in \mu T, \; {}^\bullet(\eta A) = \beta({}^\bullet A) \text{ and } (\eta A)^\bullet = \beta(A^\bullet)$$

This definition is based on Def.2.1, and respects behavioral aspects of Petri net. Here $\eta : T \to_\mu T'$ refers to a partial function between T and T' and $\beta : P \to_\mu P'$ refers to a multirelation between P and P', respectively. "A" refers to a multiset of transitions, ${}^\bullet A$ refers to the pre-set of A, $A^\bullet$ refers to the post-set of A. From here, one can see that such Petri net morphism preserves initial marking and the environments of transitions, that is, the pre-set of the image of a transition set is equal to the image of the pre-set of that transition set, the post-set of the image of a transition set is equal to the image of the post-set of that transition set. Here η is specified as a partial function between transition of source and destination net, this eliminates the possibility that a transition in source net can be mapped to two different transitions using a single morphism and allows a source net to refine or shrink as needed.

[Def.2.3] Let $N = (P, T, F, M_0, \mathbf{M})$ and $N' = (P', T', F', M_0', \mathbf{M'})$ be two augmented Petri nets. A general morphism $f : N \to N'$ between N and N' consists of a partial function $\eta : T \to T'$ and a multirelation $\beta : P \to P'$ which together fulfill the following conditions:

1. $\beta M_0 = M_0'$
2. $M \in \mathbf{M}$ implies $\beta M \in \mathbf{M'}$
3. $^\bullet(\eta t) \leq \beta(^\bullet t)$
4. $\beta(t^\bullet) = \beta(^\bullet t) -^\bullet (\eta t) + (\eta t)^\bullet$

In this definition (P, T, F, M_0) is a classical Petri net, and $\mathbf{M} \subseteq \mu P$ is a set of markings of Petri Net satisfying the following conditions: $M_0 \in \mathbf{M}$ and $M \in \mathbf{M}$ and $M[e\rangle M'$ implies $M' \in \mathbf{M}$.

3   Conceptual modeling of concurrent systems with Petri net morphisms

General morphisms preserve many behavioral properties of Petri nets and can be useful means to refine and abstract concurrent systems modeled with Petri nets.

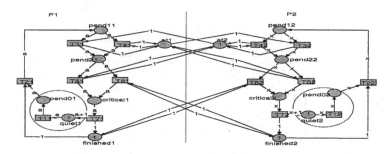

Fig. 1. Original Petri net model of Peterson's algorithm

We present an example of conceptual modeling of concurrent system with general morphisms of Petri nets - the Petri net implementation of Peterson's mutual exclusion algorithm as shown in Fig.1[7]. In the figures, *P1* and *P2* represent two concurrent processes.

By step-wise application of general morphisms, we can simplify the model to different levels of abstraction as shown in Fig.2.

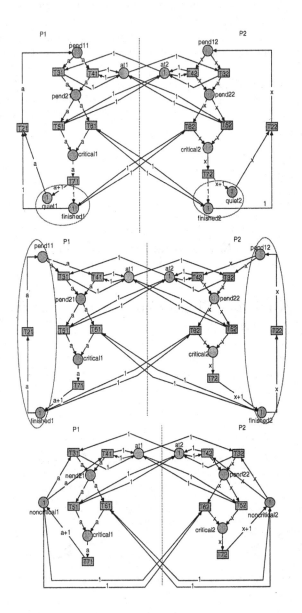

Fig. 2. Different level of abstraction of Peterson's algorithm using general morphisms

From Fig.1, we can see that process *P1* can enter place *critical1* only if there is a token in place *finished2* or place *at2*. If both *P1* and *P2* want to enter their critical section, then both place *pend21* and place *pend22* have one token, and there is no token in place *finished1* and place *finished2*. At any time, because there is only one token that can be in either place *at1* or *at2* but not in both place, So either *P1* can enter place *critical1* or *P2* can enter place *critical2*, mutual exclusion requirement is satisfied. If process *P1* is not in place *critical1* and *P1* does not want to enter place *critical1*, then there is one token in place *finished1*. If *P2* wants to enter place *critical2*, because there is a token in place *finished1*, *P2* can enter place *critical2* at any time it wants. The progress requirement is satisfied.

Process *P1* makes a request to enter place *critical1*, then there is a token in place *pend21* and place *at1*, separately. If there is a token in place *critical2* or place *pend22*, that is, there is no token in place *finished2*, *P1* needs to wait *P2* to finish its execution and put a token in place *finished2*, then *P1* can enter place *critical1*. If process *P2* is not in place *critical2* and *P2* does not want to enter place *critical2*, then there should have a token in place *finished2*, *P1* can enter place *critical1* without any wait. That means once *P1* makes a request to enter its critical section, *P1* needs to wait at most one entry by *P2*, so the bounded waiting requirement is satisfied. Thus the Petri Net implementation of Peterson's algorithm is correct.

References

1. Bednarczyk, M. A.; Borzyszkowski, A. M., General Morphisms of Petri nets, Lecture Notes in Computer Science, vol. 1644: Automata, Languages and Programming, Springer-Verlag 1999.
2. Desel, J.; Merceron, A., Vicinity Rrespecting Net Morphisms, Lecture Notes in Computer Science, vol. 483: Advances in Petri nets, pp. 165-185, Springer-Verlag 1991.
3. Devillers, R.; Klaudel, H.; Riemann, R. C., General Refinement for High Level Petri Nets, Lecture Notes in Computer Science, vol. 1346: Foundations of Software Technology and Computer Science pp. 297-311, Springer-Verlag 1997.
4. Lakos, C., Composing Abstraction of Colored Petri nets, Lecture Notes in Computer Science, vol. 1825: Application and Theory of Petri nets, pp. 323-345, Springer-Verlag 2000
5. Petri, C. A., Introduction to General Net Theory Net and Applications, Lecture Notes in Computer Science, W. Brauer (ed.), pp. 1-19, Springer-Verlag 1980.
6. Padberg, J.; Gajewsky, M.; Ermel, C., Rule-based Refinement of High-level Nets Preserving Safety Properties, Lecture Notes in Computer Science, vol. 1382: Fundamental Approaches to Software Engineering, pp. 221-238, Springer-Verlag 1998.
7. Reisig, W., Elements of Distributed Algorithms: Modeling and Analysis with Petri Nets, 1998.
8. Winskel, G., Petri Nets, Algebras, Morphisms and Compositionality, Information and Computation, 72: pp. 197-238, 1987.

Reliability of the Navigational Data

Marek Przyborski and Jerzy Pyrchla

The Naval University of Gdynia

Abstract. We present results of the experiment which was set up to give some view on the possible nature of the navigational data gathered during rescue mission. In particular we addresses the question - Is it possible that errors made during estimating navigational parameters like bearing and distance can have deterministic nature.

1 Introduction

This note investigates model of navigational observations presented in Ref. [9]. The model is based on fuzzy set theory with the idea that fuzzy sets can enhance the qualitative information usually extracted from the observations into quantitative data. In this way more precise description of uncertainty included in the visual observations can be achieved.

We addresses the question about the nature of the data consisted of the observations made by the ordinary observers who have to determine the bearing and distance from the shore to the given object at sea as described in Ref. [8–12].

If the data has deterministic elements then we can assume that the influence of the external factors on the accuracy of collected data can be considered as a deterministic one. Otherwise we cannot say anything about the possible influence on the reliability of navigational visual observations.

Our attention have been concentrated on one of the navigational parameters - bearing's error. We have taken a statistical approach to detect nonlinearity, by showing that a given linear model is unlikely to describe our data.

2 Measures of nonlinearity

We decided to apply three types of nonlinear statistics $t = t(\{x_n\})$:

1. As a first one, nonlinear prediction error with respect to a locally constant predictor F defined by

$$t^1(m, \tau, \epsilon) = \left(\sum [x_{n+1} - F(x_n)]^2\right)^{1/2}. \tag{1}$$

   The prediction is performed over one time step and it is done by averaging over the future values of all neighboring delay vectors closer than ϵ in m dimensions.

2. The second quantity it is cross-prediction errors, which can be expressed by the following formula

$$\sigma_{X,Y}^2 = \frac{1}{L'} \sum_{k=(m-1)\tau+1}^{L-1} \| \, \boldsymbol{y}_{k+1} - F_X(\boldsymbol{y}_k) \, \|^2, \tag{2}$$

where: $L' = L - (m-1)\tau - 1)$ - number of delay vectors, and F zeroth order model as it is proposed in the Ref. [3].

3. A simple quantity which is frequently used to detect deviations from time-reversibility is

$$t^3(\tau) = \langle (x_n - x_{n-\tau})^3 \rangle \tag{3}$$

Calculating those nonlinear observable requires using time delay embedding according to the following scheme, where embedding vectors in m dimensions are created by: $\boldsymbol{x}_n = (x_{n-(m-1)\tau}, \ldots, x_n)$, - τ is the delay time.

3   Results

The method of Schreiber and Schmitz Ref. [4] is based on the phase randomized surrogate series $S = \{s_n, n = 1, \ldots, N\}$ which has the same power spectrum as the time series $X = \{x_n, n = 1, \ldots, N\}$. The temporal correlations in the original data are not preserved in the surrogates. The surrogate is obtained by determine the Fourier transform of the original data X, randomizing the phases, and inverting the transform.

Conducted tests revealed that for one-sided test we achieved the 95% level of significance, results of one of the tests are presented on the Fig. 1, the same level was achieved when applying two-sided test. Time reversal asymmetry of the original data is found to be different from the surrogate data. Results of that test are presented on the Fig. 2.

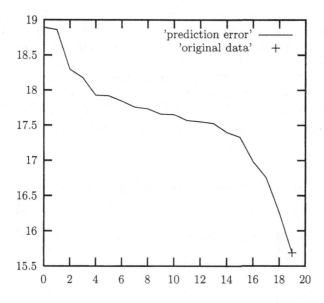

Fig. 1: Prediction error.

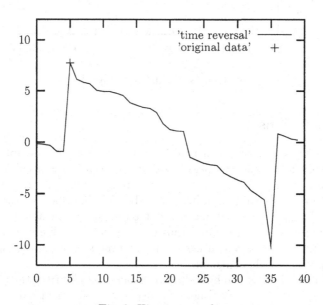

Fig. 2: Time reversal .

In the clustering algorithm, the cross-prediction errors have been used as a dissimilarity measure. For conducting the test we used 9 surrogates and the original data, thus the probability that the algorithm turned out the original data is $1/K = 0.1$, if it is true then we can reject the null hypothesis with the $(1 - (1/K)) \times 100\% = 90\%$ of significance. During the tests we have obtained the rejection of the null hypothesis with the 90% level of confidence. The answer of the clustering algorithm is presented on the Fig. 3. We may noticed there two clusters and one of them singled out the one element which contains the original data.

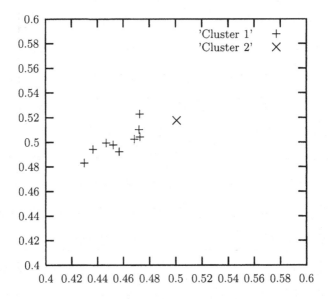

Fig. 3: Results of clustering algorithm .

4   Conclusions

We applied three nonlinear statistics in order to reveal the possible nature of visual observations made by people, who might report sea accidents. Taking into account results of conducted tests we can conclude that the null hypothesis can be rejected, original data cannot be well described by the Gaussian linear stochastic process. It means that our assumption about the nature of presented observation is false, thus even in this case due to deterministic nature of this process we can use nonlinear time series analysis methods to describe presented phenomenon.

Possibility to reject the hypothesis of stochastic nature of that signal create a new question what parameters constitutes that the visual observations have deterministic nature? Answer to this question allows us to classify those observations to different classes of navigational information according to the underlying nature.

Acknowledgments

We wish to thank Rainer Hegger, Holger Kantz and Thomas Schreiber for using their package TISEAN which is available as a free software from its home page [1]. This work is partially supported by the Polish State Committee for Scientific Research [2].

References

1. C. Diks, J. C. van Houwelingen, F. Takens, and J. DeGoede, Reversibility as a criterion for discriminating time series, Phys. Lett. A **201**, 221 (1995).
2. P. E. Rapp, A. M. Albano, T. I. Schmah, L. A. Farwell, "Filtered noise can mimic low-dimensional chaotic attractors" Phys. Rev. E **47**, 2289 (1993)
3. T. Schreiber, A. Schmitz, "Discrimination power of measures for nonlinearity in a time series" Phys. Rev. E **55**, 5443 (1997)
4. T. Schreiber and A. Schmitz, "Improved surrogate data for nonlinearity tests" Phys. Rev. Lett. **77**, 635 (1996)
5. T. Schreiber, "Constrained randomization of time series data" Phys. Rev. Lett. **80**, 2105 (1998)
6. J. Theiler, D. Prichard, Constrained-realization Monte-Carlo method for hypothesis testing, Physica D **94**, 221 (1996).
7. J. Theiler, D. Prichard, Generating surrogate data for time series with several simultaneously measured variables, Phys. Rev. Lett. **73**, 951 (1994).
8. M. Bednarczyk and J. Pyrchla, *Zbiory rozmyte w planowaniu poszukiwań morskich.* Proc. II Sympoozjum NAWIGACJA ZINTEGROWANA, Szczecin, 2000
9. J. Pyrchla, Usefulness of fuzzy set theory for locating sea accidents, Geodezja i Kartografia, **4**, 2001
10. M. Bednarczyk and J. Pyrchla and A. Stateczny, Location of an accident at sea in the SAR system - an attempt at formalizing the problem, Zeszyty naukowe WSM Szczecin, **55**, 2000,
11. M. Bednarczyk and M. Holec and J. Pyrchla, Problem określenia i hierarchizacji czynników wpływających na dokładność wizualnych obserwacji nawigacyjnych, Proc. XII Międzynarodowa Konferencja Naukowo-Techniczna *Rola Nawigacji w Zabezpieczeniu Działalności Ludzkiej na Morzu*, Gdynia, 200 0
12. J. Pyrchla, Zależność rozkładu błędu obserwacji wzrokowych od ukształtowania linii brzegowej, Spraw. VI Konferencja Naukowo-Techniczna *Bezpieczeństwo Morskie i Ochrona Naturalnego Środowiska*, Kołobrzeg, 2002,

---

[1] http://www.mpipks-dresden.mpg.de/~tisean
[2] Grant 0 T00A 030 19

Evaluation of Authorization with Delegation and Negation

Chun Ruan[1], Vijay Varadharajan[2], and Yan Zhang[1]

[1] School of Computing and Information Technology, University of Western Sydney, Penrith South DC, NSW 1797 Australia
[2] Department of Computing, Macquarie University, North Ryde, NSW 2109, Australia

Abstract. Access evaluation is a significant issue in any intelligent information system. In this paper, we develop a logic programming based approach for decentralized authorization delegations in which users can be delegated, granted or forbidden some access rights. A set of domain-independent rules are given to capture the features of delegation correctness, conflict resolution and authorization propagation along the hierarchies of subjects, objects and access rights. The basic idea is to combine these general rules with a set of domain-specific rules defined by user to derive the authorizations holding in the system.

1   Introduction

Logic based approaches have been developed by many researchers for the purpose of formalizing authorization specifications and evaluations. The advantage of this methodology is to separate policies from implementation mechanisms, give policies precise semantics, and provide a unified framework that can support multiple policies.

Abadi et al proposed a modal logic based approach for access control in distributed systems [1]. Their work focuses on how to believe that a principal (subject) is making a request, either on his/her own or on someone else's behalf. The delegation in their model mainly concerns on the access right itself. Jason Crampton et al's work is also based on modal logic [3], which investigates the ability of representing and reasoning about the implementation of a real-world access control mechanism. Woo and Lam proposed an expressive language to authorization in distributed systems [5]. In their method, they consider structural properties inherent in authorization and provide formal semantics evaluation which is based on extended logic program. Jajodia et al also proposed a logical language and illustrated how it can specify authorization, conflict resolution, access control and integrity constraint checking [4]. Bertino et al [2] also proposed a logic framework in which they considered hierarchically structured domain of subjects, objects and access rights for authorization, supported both negation as failure and classical negation, and provided a conflict resolution method.

One restriction of the above works is that the delegation of administrative privilege is not supported, which is a key issue for discretionary access control. This paper presents a logic based formulation which supports authorization delegations, authorization negations and authorization inheritance. A conflict resolution method based on the underlying delegation relation is presented which gives higher priorities to the predecessors to achieve the controlled delegation. In our method, a set of domain-independent rules are given to capture the features of delegation correctness, conflict resolution and authorization propagation along the hierarchies of subjects, objects and access rights. The basic idea is to combine these general rules with a set of domain-specific rules defined by user to derive the authorizations holding in the system.

2   Syntax

Our language $\mathcal{L}$ is a many-sorted first order language, with four disjoint *sorts* S, O, A, and T for subject, object, access right and authorization type respectively. Variables are denoted by strings starting with lower case letters, and constants by strings starting with upper case letters. Three partial orders $<_S, <_O, <_A$ are defined on sorts S, O, A respectively, which represent the inheritance hierarchical structures of subjects, objects and access rights. We use $\sharp$ in S to denote the security administrator, and it is not comparable to any subjects in S w.r.t. $<_S$. In the constant set of authorization types $T = \{-, +, *\}$, $-$ means *negative*, $+$ means *positive*, and $*$ means *delegatable*. A negative authorization specifies that the access must be forbidden, while a positive authorization specifies that the access must be granted. A delegatable authorization specifies that the access must be delegated as well as granted. That is, $*$ means $+$ plus administrative privilege on the access.

The predicate set P consists of a set of ordinary predicates defined by users, and one built-in predicate symbol for delegatable authorization, *grant*. *grant* is a 5-term predicate symbol with type $S \times O \times T \times A \times S$. Intuitively, $grant(s, o, t, a, g)$ means s is granted by g the access right a on object o with authorization type t. A *literal* is either an atom p or the negation of the atom $\neg p$, where the negation sign $\neg$ represents classical negation. Two literals *complementary* if they are of the form p and $\neg p$, for some atom p. A *rule* r is a statement of the form:

$b_0 \leftarrow b_1, ..., b_k, not\, b_{k+1}, ..., not\, b_m, m >= 0$

where $b_0, b_1, ..., b_m$ are literals, and not is the negation as failure symbol. A delegatable authorization program consists of a finite set of rules.

3   Authorization evaluation using domain-independent rules

3.1   Rules for delegation correctness

Definition 1. (delegation correct) an authorization set is *delegation correct* if it satisfies the following two conditions: (a) subject s can grant other subjects an access right a over object o if and only if s is the security administrator $\sharp$ or s has been granted a over o with a delegatable type $*$; (b) if subject s receives a delegatable authorization directly or indirectly from another subject s' on some object o and access right a, then s cannot grant s' any further authorization on the same o and a later on.

We will require the authorization set derived by a delegatable authorization program be delegation correct. To capture the feature of delegation correctness, we introduce several new auxiliary predicates and treat them as system reserved words. *delegate* has a type of $S \times S \times O \times A$. The arguments are *grantor, subject, object,*and *access right* respectively from left to right. Intuitively, $delegate(g, s, o, a)$ means subject g has directly or indirectly granted subject s access a on object o with type $*$. $existdelegate(g, s, o, a)$ has the same type with *delegate* which is used to avoid the existential quantifier to be used in a rule, since in extended logic programs all the variables in clauses are considered to be universally quantified. It is true if there is any delegation from g to s on o and a. *grant1* has the same type and meaning with *grant* except that *grant1* has passed through the delegation correctness check. The following five rules are used to deal with the feature of delegation correctness.

$(D1)$ $grant1(s, o, t, a, \sharp) \leftarrow grant(s, o, t, a, \sharp)$

$(D2)$ $grant1(s, o, t, a, g) \leftarrow grant(s, o, t, a, g), grant1(g, o, *, a, g'),$
$$g \neq s, not\, existdelegate(s, g, o, a)$$

$(D3)$ $delegate(g, s, o, a) \leftarrow grant1(s, o, *, a, g)$

$(D4)$ $delegate(s, s_1, o, a) \leftarrow delegate(s, s_2, o, a), delegate(s_2, s_1, o, a)$

$(D5)$ $existdelegate(s, g, o, a) \leftarrow grant(s, o, t, a, g), delegate(s, g, o, a)$

Rules $(D1)$ and $(D2)$ define *grant1* which is an authorization satisfying the delegation correctness requirement. Rule $(D3)$ and $(D4)$ derive the delegation relation. Rule $(D5)$ defines *existdelegate*.

3.2   Rules for authorization propogation

$(H1)$ $grant(s, o, t, a, g) \leftarrow grant(s', o, t, a, g), s' <_S s, s \neq g$

$(H2)$ $grant(s, o, t, a, g) \leftarrow grant(s, o', t, a, g), o' <_O o, s \neq g$

$(H3)$ $grant(s, o, t, a, g) \leftarrow grant(s, o, t, a', g), a' <_A a, t \neq -, s \neq g$

$(H4)$ $grant(s, o, t, a, g) \leftarrow grant(s, o, t, a', g), a <_A a', t = -, s \neq g$

Noted that, unlike other propagations that are downward along the hierarchies, when the grant type is $-$, the propagation is upward along the access right hierarchy as indicated by rule (H4). For example, if someone is forbidden to read, then he will be implicitly forbidden to write.

3.3   Rules for conflict resolution

The basic idea of our method of solving conflicts is outlined as follows:

- Policy 1: Solving conflicts using delegation relation. According to the delegation relation, if subject s delegate subject s' directly or indirectly an authorization on object o and access right a, then, when a conflict w.r.t o and a occurs, the authorization from s (i.e. s is the grantor) will always override the one from s'. Given a delegatable authorization program, the delegation relation for different subject and object w.r.t. it will be dynamically derived by the system.

- Policy 2: Solving conflicts according to the types of authorizations. If the grantors of two conflicting authorization are identical, we consider types of the authorizations next. To achieve maximum security, we solve the conflicts in terms of the negative-take-precedence principle by giving type $-$ the highest priority followed by $+$ which is again followed by $*$.

- Policy 3: Cascading overriding.

When a delegatable authorization is overridden, the authorizations granted by the grantee of that authorization should also be overridden. In other words, we support cascading overriding.

We will use a set of domain-independent rules to realise the above conflict resolution policies. First, we introduce four more system reserved predicates *overriden1, overriden2, grant2* and *hold* with the same type $S \times O \times T \times A \times G$. *overriden1*$(s,o,t,a,g)$ means the authorization *grant1*(s,o,t,a,g) has been overridden by some other authorizations according to Policy 1. *grant2*(s,o,t,a,g,ti) means the authorization *grant1*(s,o,t,a,g,i) is not overridden by any other authorization according to Policy 1 and 3. *overriden2*(s,o,t,a,g) means *grant2*(s,o,t,a,g) has been overridden by some other authorizations according to Policy 2. *hold*(s,o,t,a,g) means the authorizations that actually hold(not overridden by any other authorization). We define $- < + < *$, then the rules are as follows:

$(C1)$ *overriden1*$(s,o,t,a,g) \leftarrow grant1(s,o,t,a,g), grant1(s,o,t',a,g'),$
$\qquad\qquad delegate(g',g,o,a), t \neq t'$

$(C2)$ *grant2*$(s,o,t,a,\natural) \leftarrow grant1(s,o,t,a,\natural), not\ overriden1(s,o,t,a,\natural)$

$(C3)$ *grant2*$(s,o,t,a,g) \leftarrow grant1(s,o,t,a,g), not\ overriden1(s,o,t,a,g),$
$\qquad\qquad grant1(g,o,*,a,g')$

$(C4)$ *overriden2*$(s,o,t,a,g) \leftarrow grant2(s,o,t,a,g), grant2(s,o,t',a,g'), t' < t$

$(C5)$ *hold*$(s,o,t,a,\natural) \leftarrow grant2(s,o,t,a,\natural), not\ overriden2(s,o,t,a,\natural)$

$(C6)$ *hold*$(s,o,t,a,g) \leftarrow grant2(s,o,t,a,g), not\ overriden2(s,o,t,a,g),$
$\qquad\qquad hold(g,o,*,a,g')$

Rules $(C1)$ is corresponding to the conflict resolution policy 1. Rules $(C2)$ and $(C3)$ are used to derive the authorizations that are not overridden or cascading overridden by using policy 1. Rule $(C4)$ corresponds to conflict resolution policy 2. Rules $(C5)$ and $(C6)$ are used to derive the authorizations that are not overridden or cascading overridden by any other authorizations. Let R denotes all the general rules, i.e. $R = \{D1, ...D5, H1, ...H4, C1, ..., C6\}$.

4   The formal semantics

Let Π be a delegatable authorization program, the *Base* B_Π of Π is the set of all possible ground literals constructible from the predicates appearing in the rules of Π and the constants occurring in S, O, A, T. Two ground literals are *conflicting* on subject s, object o and access right a if they are of the form $hold(s, o, t, a, g)$ and $hold(s, o, t', a, g')$ and $t \neq t'$. Let $G(\Pi)$ denotes all ground instances of the rules occurring in Π. A subset of the Base of B_Π is *consistent* if no pair of complementary or conflicting literals is in it. An *interpretation* I is any consistent subset of the Base of B_Π.

Definition 2. Given a delegatable authorization program Π, an interpretation for Π is any interpretation of $\Pi \cup R$.

Definition 3. Let I be an interpretation for a delegatable authorization program Π, the reduction of Π w.r.t I, denoted by Π^I, is defined as the set of rules obtained from $G(\Pi \cup R)$ by deleting (1) each rule that has a formula not L in its body with $L \in I$, and (2) all formulas of the form not L in the bodies of the remaining rules.

Given a set Z of ground rules, we denote by $pos(Z)$ the positive version of Z, obtained from Z by considering each negative literal $\neg p(t_1, ..., t_n)$ as a positive one with predicate symbol $\neg p$.

Definition 4. Let M be an interpretation for Π. We say that M is an answer set for Π if M is a minimal model of the positive version $pos(\Pi^M)$.

5   Future work

We plan to implement a prototype of our framework based on logic programming technique.

References

1. M.Abadi, M.Burrows, B.Lampson, G.Plotkin (1993) A calculus for access control in distributed systems. *ACM Trans. on programming languages and systems*, **15**(4):706-734
2. E. Bertino, F.buccafurri, E.Ferrari, P.Rullo (1999) A logical framework for reasoning on data access control policies. *proceedings of the 12th IEEE Computer Society Foundations Workshop*, IEEE Computer Society Press, Los Alamitos, 175-189.
3. J.Crampton, G.Loizou, G.O'Shea (2001) A logic of access control. *The Computer Journal*, **44**, 54-66
4. S. Jajodia, P. Samarati, and V.S. Subrahmanian (1997) A logical language for expressing authorizations. In *Proceedings of the 1997 IEEE Symposium on Security and Privacy*, IEEE Computer Society Press, 31-42
5. T. Woo and S. Lam (1992) Authorization in distributed systems: a formal approach. *Proceedings of IEEE on Research in Security and Privacy*, 33-50

Decision Units as a Tool for Rule Base Modeling and Verification

Roman Simiński and Alicja Wakulicz-Deja

University of Silesia, Institute of Computer Science, Poland, 41-200 Sosnowiec, Będzińska 39, Phone (+48 32) 2 918 381 ext. 768, email: siminski@us.edu.pl, wakulicz@us.edu.pl

Abstract. Expert systems are problem solvers for specialized domains of competence in which effective problem solving normally requires human expertise. The transition of expert systems technology from research laboratories to software development centers highlighted the fact that the quality assurance for expert system is a very important issue for most real-word problems. Although the basic verification concepts are shared by software engineering and knowledge engineering, verification methods of conventional software are not directly applicable to expert systems and the new, specific methods of verification are required. The main aim of this work is to present our own rule base verification method. In our opinion the decision units conception allows us to consider different verification and validation issues together. Thanks to properties of the decision units we can perform different verification and validation actions during knowledge base development and realization.

1 Introduction

In recent years, expert systems technology has proven itself to be a valuable tool for solving hitherto intractable problems in domains such a telecommunication, aerospace, medicine and the computer industry itself. The transition of expert systems technology from research laboratories to software development centers highlighted the fact that the quality assurance for expert system is a very important issue for most real-word problems [1,4]. The quality covers verification, validation and evaluation of expert systems [6]. Expert systems are programs and programs must be validated. Therefore expert systems has to meet the same standards as other software [7].

An essential part of developing knowledge bases for most real-word problems is determining whether the knowledge base is adequate and can reliably solve the problem. After acquiring domain knowledge much of the effort in building a knowledge base goes into verifying that the knowledge is encoding correctly [2–4].

The main impediment to successful verification of knowledge bases is the nature of expert systems themselves. Expert systems are often employed for working with incomplete or uncertain information or ill-structured situations. The second impediment are implementation methods and tools [5]. Therefore, although the basic verification concepts are shared by software engineering and knowledge engineering, verification methods of conventional software are not directly applicable to expert systems and the new, specific methods of verification are required [8].

The main aim of this work is to present our own rule base verification method. The paper firstly presents the decision units conception needed to establish our verification approach. Then we present the decision unit as the tool for local rule base verification and modeling. Next we briefly present the usage of decision units net in global verification and modeling issues. A last we present kbBuilder system as an interactive tool for rule base building and verification. The last chapter draws the main conclusions and states some directions for further research.

2   Conception of decision units

We assume, that *decision units* are the main tool for our rule base verification method [8,9]. In the real-word rule knowledge bases literals are often coded using *attribute-value* pairs. Now we introduce conception of decision units for literals as attribute-value pairs. The paper [9] contains an example rule base with attribute-value pairs. All rules with the same attribute we can group together. Decision unit U contains the set of rules R with the same attribute in the decision part of each rule $r \in R$. Output entries of U are attribute-value pairs which appear in the conditional part of each rule $r \in R$, input entries I are attribute-value pairs appearing in the conditional part of each rule $r \in R$. Fig. 1 presents the structure of the decision unit U.

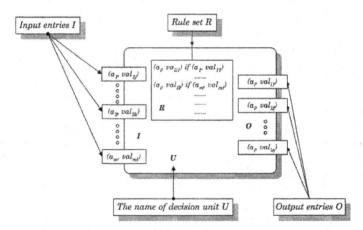

Fig. 1. The structure of the decision unit

3   The usage of decision units for modeling and verification

Decision unit may be considered as a model of elementary decision produced by the knowledge base. Each decision unit allows to confirm set of goals described by the output entries O. Knowledge engineer can work with a decision unit like a programmer works with a procedure or function. Therefore decision unit separately considered is a tool for modeling on local level - the level of elementary decision. Decision unit can be considered as a tool for verification on the local level too. Verification may be done using back box or glass box testing method. We can also apply static verification (classical anomaly detection) or dynamic techniques (using forward and backward chaining inferences).

The net of the decision units may be considered as a global model of decisions produced by the system. Expert systems often confirm global goal using subgoals we assume that the each subgoal is modeled by the appropriate decision unit. Therefore knowledge engineer may check if the current content of rule base is consistent with intended global decision model. It is specially useful in real-word problems if the particular knowledge representation language doesn't provide a solution for knowledge base partitioning.

The decision unit net allows us to formulate the global verification method similarly to local verification on the level of decision unit. Unchained output entries represent main goals of rule base, chained output entries represent subgoals. Unchained input entries represent the input data (facts) necessary to produce proper inference results. We can apply static and dynamic verification on the global level using black box and glass box techniques. For example knowledge engineer can observe the current inference path on the global level for selected goal, can check which input data are required. In the case of improper system behavior, knowledge engineer can apply selected verification method for detection the sources of anomalies.

We use a verification strategy base on decision units in the kbBuilder system. The *kbBuilder* system is a tool for interactive, incremental construction, validation and refinement of the rule knowledge bases. Current version of the system is dedicated for Sphinx knowledge representation language, next version will utilize the other ones. The kernel of kbBuilder is implemented in pure C++, user interface is implemented using Borland C++ Builder. System works under the control of Windows 9x/ME/2000. In the current version, system provides selected set of local verification techniques static and dynamic, in the glass box and black box mode. Current works concern on designing and implementing a new verification algorithms for verification on the global level. We are going to use the decision unit conception for modeling a rule base too actually decision units net is only use for presentation and verification issues.

4   Concluding remarks

In our opinion the decision units conception allows us to consider different verification and validation issues together. Thanks to properties of the decision units we can perform different verification and validation actions during knowledge base development and realization. We can divide anomalies into the two levels local and global anomalies and perform verification on those levels. Decision unit is a simple decision model, useful and efficient for knowledge base modeling and verification. The net of the decision units is a simple tool for modeling large real-word databases. Decision unit net allows us to perform global verification - i.e. circularity detection, dead end rules, auxiliary rules. Graphical representation of knowledge base in the form of the decision units net is user friendly and is an efficient and useful way of presentation of current knowledge base contents. We are going to extend the kbBuilder system to use the decision unit conception for modeling a rule base.

References

1. O'Keefe R.M., Balci O., Smith E.P. (1987) Validating Expert Systems Performance, IEEE Expert, Vol. **2**, No. 4, Winter, pp.8190.
2. Preece A.D. (1994), Foundation and Application of Knowledge Base Verification, International Journal of Intelligent Systems, **9** pp. 683701.
3. Preece A.D. (1996) Validating Dynamic Properties of RuleBased Systems, International Journal of Human-Computer Studies, **44**, pp. 146169.
4. Preece, A.D. (1990) Towards a Methodology for Evaluating Expert System, Expert Systems, **7** (4), pp. 215223.
5. Siminski R. (1998) Methods and Tools for Knowledge Bases Verification and Validation, Proceedings of the CAI'98, Colloquia in Artificial Intelligence, 2830.09.1998, d, Poland, pp. 273-291.
6. Siminski R., Wakulicz–Deja A. (1998) Principles and Practice in Knowledge Bases Verification, Proceedings of the IIS VII, Intelligent Information Systems, IIS'98, 1519.6.1998, Malbork, Poland, pp. 203211.
7. Siminski R., Wakulicz–Deja A. (1999) Errors And Anomalies In Rule Knowledge Bases, Proceedings of EUFIT'99, 1114.9.1999, Aachen, Germany.
8. Siminski R., Wakulicz–Deja A. (1999) Dynamic Verification Of Knowledge Bases, Proceedings of the IIS VIII, Intelligent Information Systems,VIII, IIS'99, 1418.6.1999, Ustro, Poland, pp 327-331.
9. Siminski R., Wakulicz–Deja A. (2000) Verification of Rule Knowledge Bases Using Decision Units, Advances in Soft Computing, Intelligent Information Systems, Physica-Verlag, Springer Verlag Company, 2000, ISBN 3-7908-1309-5, pp. 185-192.

Advantages of Deploying Web Technology in Warfare Simulation System

Zbigniew Świątnicki[1] and Radosław Semkło[2]

[1] Military University of Technology, Institute of Automated Command Systems and Logistics, Kaliskiego 2, 00-908 Warsaw 49, Poland
[2] Air Force Computer Science Centre, Radiowa 2, 00-908 Warsaw, Poland

Abstract The simple simulation system based on web technology, which enables CRC staff training is described in this paper. The advantages of using servlets, applets and Enterprise Java Beans have been presented. The influence on failover, scalability and reliability has been discussed too.

1 Introduction

The changes in structure and command methods in Polish Air Force, caused that new automated systems have been deployed. These command, control and communication (C3) systems support airspace management and the subjected forces command.

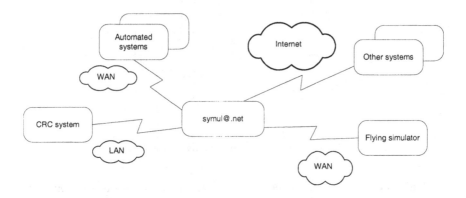

Figure1. The environment of symul@.net system

There are, among others, Mikoajek-RL and Dunaj systems. These systems are able to exchange the information between other automated systems operating in NATO defence system.

The deployment of new automated systems used by operational services (forces) requires creating new units to train and exercise these forces. Air Force Computer Science Centre was obliged to fulfil this task. As a result of undertaken activities, the new system that enables teaching and training the Command and Reporting Centre (CRC) staff has been created. This system allows CRC staff to practise procedures and reaction in situations that at peace happen very rarely or at all. These are for example, the procedures that are used in case of Air Policing operations or massed air attack.

2   Origin of the system

New applications are found for the web technologies every day. It takes place in a significant number of cases due to current fashion for modern technologies as well as undeniable benefits, which they bring. As a result of analysis conducted by Air Force Computer Science Centre, the concept has emerged to create the system that, first of all, would enable CRC staff training in new-deployed automated system (Dunaj).

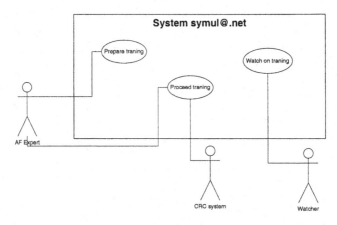

Figure2. Use cases for symul@.net system

Additionally this system should exchange the information between flight simulators as well as other automated systems used at different levels and by different kinds of forces. The conducted analysis has shown that it is possible to use web technologies for this purpose, especially application server compatible with J2EE specification. Thanks to the use of web technologies symul@.net system enables remote input and workout of planned scenarios. So the man ordering training is able to prepare or verify stored scenario individually. Besides, it is possible to present the Recognize Air Picture (RAP) in the window of Internet browser. It allows people, which relocation is far too expensive or inadvisable, to take part in exercises.

3   System architecture

The part of the symul@.net system designed for inputting data has been deigned and implemented in accordance with J2EE specification. Due to such approach the multilayered system in architecture called Model View Controller (MVC) has been built. The first separate layer is responsible for storage and data processing. The second layer that is separated from presentation layer, implements application logic. In this way it is easy to modify the system without the necessity of system stop or exchange some elements without altering others. The next advantage of using application servers is the possibility of adding new hardware units to increase the system efficiency without changing system source codes. The symul@.net system uses four kinds of computer nodes. The are computers designed as servers, computational nodes as well as computers working as interfaces and workstations. The database and application servers have been installed on server nodes.

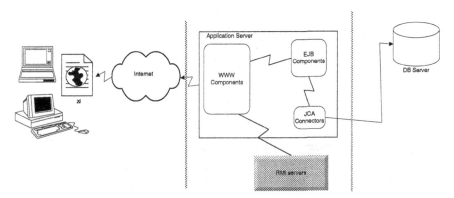

Figure3. Multilayers in symul@.net system

The separate group of nodes is computers designed for interface processes. This class of nodes is assigned for exchanging information between symul@.net and other systems. There are also workstations, which can start user applications or applets (depending on hardware resources). Because all processes have been written in JAVA language, they can be deployed on any hardware platform, where Java Virtual Machine is present. Thus, server processes (providing information) can be installed on large UNIX machine or on PC computer. Of course the efficiency differs in these two cases.

4   Deployed solutions

The server application platform together with its technologies made it possible to create system with open interfaces. Many solutions related to the

web technologies have been used in symul@.net system. Such approach resulted both from platforms applied and functionality, reliability and efficiency
requirements. Most of these requirements has been used due to modern technology (UML, [1]) applied to system design and implementation. Both HTML
pages and JSP have been used to create the presentation layer. Enterprise
Java Bean (EJB) have been used to data management and the session beans
have been used to implement application logic. To improve the system efficiency, the connectors in accordance with Java Connector Architecture (JCA)
have been created. The EJB have been created as Bean Manage Persistence
(BMP). Now the process of changing the model management from BMP to
Container Manage Persistence (CMP) is taking place. Due to separating the
application logic layer from data management layer, this process performs
imperceptible for the user and without consuming additional labour.

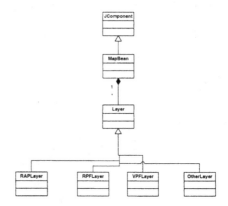

Figure4. Class diagram for layers in Map module

The processes inside the symul@.net system communicate with the RMI
mechanism. This mechanism for distributed computing (characteristic of JAVA
language) enables using accumulated resources located wherever the computer network with TCP/IP protocol exists. The implementation of one
datasource with RMI interface make meeting the external systems requirements easier. Therefore, the processes responsible for communication with
external systems are using resources stored in datasources and communicate
with their own system in a specific way. Furthermore, processes responsible
for communication with user can be started in two ways. One of the man-
ners is when the application starts on workstation machine. The other way
is starting the applet inside WWW page. The functionality of both solutions is identical since they use the same B-codes. They only differ about
the way they are evoked. Such solution makes software modification easier
and meets functional requirements apart from hardware resources. The special kind of applet/application is a Map module for displaying Recognize Air

Picture (RAP) in the Internet browser window. The topographical grounds recorded in vector format, for example VPF (Vector Product Size) can consist of forests, rivers, cities etc. However, it can be raster grounds recorded in Raster Product Format (RPF) as a map for low flights. The Map unit with the layer interface makes it possible to connect (with RMI) to datasources that provide information about tactical situation. That is why it is possible to observe the simulation from any place in the computer network.

5 Conclusions

The symul@.net is a relatively young system. The process of its creation has not been finished yet. The deployment stage of the test version has not reached the end. At present, the deployment of the system is taking place in National Air Operation Centre located in Pyry. Certainly current situation does not exclude capability of conducting the CRC staff training. Due to employed technologies the foundation for planned exercises can by remotely inserted into the system installed in Bemowo. Then generated course of simulation can be sent online to Pyry as well as Bemowo and used in exercise process there. The generated warfare picture can be used to execute the exercise without limits in functionality of the system. The symul@.net system has been introduced, the architecture and selected elements realized with the use of the web technology has been presented. The employment of these technologies has brought significant improvement of the system resistance to damages and increases the scalability. The better efficiency of hardware resources has also been achieved due to web technology. The web technology assigned exclusively for open systems (Internet) for selected, closed users group as an innovative approach has been showed. It is the only case known in Polish Army where the system based on web technologies and systems assigned for command have been combined. The solutions presented in this paper have showed that implementing modern technologies is possible without resigning from fundamental requirements formulated by users. The requirements concern the security, reliability, efficiency and functionality. The first part of trial training conducted with National Defense Academy cadets has showed that combination of automated system and system based on web technologies is possible

References

1. Fowler M., Scott K. (2000) UML Distilled, LTP, Warsaw
2. Semkło R. (2001) Supporting objects identification in Polish Air Forces Republic of Poland, National Defense Academy, Warsaw
3. Świątnicki Z., Semkło R. (1998) Computational Intelligence and Applications, Springer-Verlag, Berlin
4. AFCSC Designers (2000) The conception of training system for CRC staff, AFCSC, Warsaw

Knowledge Management and Data Classification in Pellucid

Dang T.-Tung, Hluchy L., Nguyen T. Giang, Budinska I., Balogh Z., and Laclavik M.

Institute of Informatics, SAS, Dubravska c. 9, Bratislava 84507, Slovakia

Abstract. The main aim of the Pellucid project is to develop a platform based on the multi-agent technology for assisting public employees in their organization. This paper deals with a problem of classification and identification of needed information for agents performance. This paper presents methods for encoding data and creating the database, so that agents can have an easy access to the required information. Furthermore, two methods applicable with every type of database for classification and selection of historical information are presented.

1 Introduction

Organizationally mobile employees belong to the class of large-dimensional applications related to many various domains, which require optimalization. Each employee can execute many different activities, with different results or effects, moreover each of them might have different capabilities, possess various kinds of information or knowledge; therefore it is really necessary to optimally re-organize their work, share knowledge or activities, so that the collective performance could be improved as much as possible. The overall objective of Pellucid is to develop an adaptable platform for assisting organizationally mobile employees, in effect re-engineering their work in the organization. Because of the short frame of this paper we will deal with only a problem, that is how the Pellucid agents are able to automatically identify and capture desired information that they need in the current situation. This task appears during realization of the Pellucid system and however it does not belong to one of the main declared tasks, but it has strong influences to the quality of the final product.

2 Problem formulation

The agents functionality in the Pellucid system is described as follows: on the basis of information from the system and from user, an agent has to calculate and choose the most optimal action for the users execution in the current situation and recommends it to the user. This process is repeated at every time when the user meets a new situation. In order to make a quality decision, the Pellucid agents use experiences provided by former users to assist the new ones. As a result, one of the tasks that agents have often to repeat is to search for data with certain specific attributes in the database. A problem how to find the desired data, which is the closest to the target, is just the main focus of this paper. Since all information, knowledge or experiences have different forms that cause Software agents a lot of difficulties. This paper presents a simple method for maintaining and recording data based on decomposition principle. Each real situation is described by a number of basic elements and operators that express relationships among them. Elements are classified to individual classes, in hierarchical structure, in order to make easier search and access to the desired data. In addition, we design a number of classification rules, which enable to deduce the desired solution. Creating such a set of rules depends on the characteristics of the application, which applies the Pellucid system. This paper discusses a general method for classification and identification of data based on the fuzzy theory, which is applicable with every set of rules. In the case with many types of basic elements, we propose a method for classification that is combination between the traditional fuzzy method and the weight setting method.

3   Data representation and database

3.1   Data representation

The database is consisted of two parts. The first part (let denote as $SI = \{si\}$) involves all historical situations, which happened in the past. For each situation si let denote sol(si) as a set of all information associated with it, and it is recorded in the second part of the database. When an agent searches for historical experiences, at first it has to look at the set SI and extract from it a situation that is the nearest to the current one. To simplify the classification process, let assume that each situation is consisted of a number of basic elements. Let denote a set of basic elements as EL (this set includes also an empty element marked as ϕ), and assume that each situation could be described as a unification of basic elements from EL.

$$si = (el_1 U el_2 U...U el_n) \text{ where } el_1,..,el_n \in EL$$

The set of basic elements EL needs to be defined clearly and should contain as many elements as possible. Therefore, in our experience, it is a need to contact or to cooperate with some experts, who have been working in this domain and could help to identify these basic elements.

3.2   Database organization

In order to improve a search, the database is organized in hierarchical structure. Each element that appears in every situation is classified to one of a number of predefined classes. For example, the first class contains elements describing the main purpose of a situation or tasks that the user has to do e.g. writing a report, analyzing data, etc. The second class contains elements that express time constraints e.g. deadline is Friday, etc. Elements that describe technical constraints: LaTeX, MS Word, etc might be stored in the next class. Each situation is encoded as an array like:

1 analyzing data 2-3 MS Exel 4-5-6 less than 10 pages, color pictures 7-8.

In this array, numbers represent a class that includes the elements appeared in the situation. A number of classes depend on a concrete kind of applications. Description of each class must be defined clearly, so that every user is able to assort these elements creating the current situation to accurately corresponding classes. All information associated with each situation (the second part of the database) are stored in one object according to the situations code, in order to identify them easily. The form in which these data are stored is not an interest of this paper but it should take a suitable form, so that the database will be small as possible (depending on a method of implementation yet).

4   Methods for capturing information

4.1   Relationship among individual elements and situations

At first, in order to extract a situation, it is necessary to specify relationships among elements.Since all the basic elements cannot be expressed precisely by any numerical variable, their comparison is only relative. For that reason, using fuzzy sets for evaluation seems as the most appropriate way to resolve this problem.

Let define $re(el_1, el_2)$ as a real number, which expresses the degree of dependence between two elements $el_1, el_2 \in EL$. This parameter expresses to which degree the associated data with one element could be applied for the second one. In following, there are some important properties of this variable:

- $re(el_1, el_2)$ does not have to be equal to $re(el_2, el_1)$ or $(1 - re(el_1, el_2))$ (means this relation is not symmetric or inverse).

- $re(el_1, el_2) = 1$, when a domain of el_1s effects is a subset of the domain where el_2 can affect,

- $re(el_1, el_2) = 0$, when both the elements have disjoint domains of effects,

- $\forall el \in EL : re(\phi, el) = 1$,and $re(el, \phi) \geq 0$

For other cases this parameter can have a value in an interval [0,1]. These variables will be used for calculating the degree of similarity between two situations. Detail about that will be presented in the next sections.

4.2   Selection of historical data on the basis of fuzzy classification

This section presents a method for extracting information by using fuzzy classification. Firstly, for comparing situations, we define a fuzzy relation SI x SI $\supseteq$ Sim, which is used to expresses a degree of similarity between two arbitrary situations. The property of such a relation are:

- Reflex: $Sim(si, si) = 1 \ \forall si \in SI$,
- Symmetric: $Sim(si_1, si_2) = Sim(si_2, si_1)$,
- Transitive: $Sim(si_1, si_2) \geq min(Sim(si_1, si), Sim(si, si_2))$.

An algorithm for selecting historical situations 1.
Input: target tg $= el_1$ U ... U el_n, EL, SI, starting with $i = 1$.
1. Select randomly $si_i \in SI$.
2. Calculate $Sim(tg, si_i).SI = SI - si_i$
3. Choose $si_{i+1}|Sim(si_{i+1}, si_i) \geq Sim(tg, si_i)$
4. If step 3 does not have a solution si_i is the desired solution and STOP,
5. Else, remove all examined situations
$si \in SI|Sim(tg, si) > Sim(si, si_i), i = i + 1$ and return to step 1.
6. Stop until SI is empty.

4.3   Selection of historical data based on algebraic classification

Another method for classifying and discovering data is an algebraic classification. Let c_0 be a number of classes that are used to classify individual elements appeared in each situation (introduced in part 3.2). The similarity is calculated as a linear combination of these partial ones.

Let denote $Simp_l(si_1, si_2)$ as a degree of partial similarity between si_1 and si_2 within class $l \in [1, c_0]$, which moreover is defined similarly as in the previous part by using fuzzy sets. The degree of similarity between these situations is defined as following: $Sim(si_1, si_2) = (\sum_{l=1}^{c_0} w_l * Simp_l(si_1, si_2)$ Where w_l is the weight on the l-class.

The algorithm for extracting a similar situation (presented below) differs from the previous in such a way.It does not examine all elements including in each situation at once, but according to the importance of each class, from the top to the least important class.

An algorithm for selecting historical situations 2.
Input: $targettg = el_1 \cup ... \cup el_n, EL, SI, w_l|l = 1, .., c_0$
(let assume $w_1 \geq .. \geq w_{c_0}$). Starting with $i = 1, l = 1$.
1. Select randomly $si_i \in SI$.
2. Calculate $Simp_l(tg, si_i).SI = SI - si_i$.
3. Choose $si_{i+1}|Simp_l(si_i, si_{i+1}) \geq Simp_l(tg, si_i)$.
4. If step 3 does not have any solution, then $l = l + 1$ and return to step 3 while $l < c_0$, either si_i is the desired solution and STOP,
5. Else, remove all examined situations
$si \in SI|Simp_l(tg, si_i) > Simp_l(si, si_i), i = i + 1$, and return to step 1.
6. Stop until SI is empty.

5   Application of the approach in the Pellucid system

The presented results have been implemented in the Pellucid prototype. One of experiments is to create and maintain the database for document processing related to the public transport system in a city. Both the methods presented in Sections 3.1 and 3.2 were used for encoding and creating the database. The database contains a number of situations and their associated data, which have been generated at random. The input is a description of the users current task that he has to perform (the main task what he has to do, constraints, information about external environment, etc.). The output of the Pellucid system is recommendations to the user what he should do in the current case.

Both the presented algorithms in Sections 4.2 and 4.3 were used and implemented. A set of rules and basic elements SI were created on the basis of real situations and by cooperation with experts. To assess relationships among individual elements and to specify variables re(SI,SI) we use an adaptive method, which updates and recalculates these when new results have been achieved. The test has been executed with some different databases and it has achieved promising results.

6   Conclusion and future work

This paper described the problem of classification and identification of data, which plays very important role in Artificial Intelligence domain. Two methods for classification and selecting information based on the fuzzification principleare presented. To make easier access and maintain the database we propose a method for encoding data according to a hierarchical scheme, with a number of different levels and basic elements. Such method of encoding moreover can make easier programming and reduce the database size. In the future, our aim is to create an automatic system without using experts, so that the Pellucid agents could work fully autonomously to assist users.

References

1. R. Allen: Workflow: An introduction. Workflow Handbook 2001. Edited by Layna Fischer. 15 38, 2001.
2. B. Habegger, M. Quafafou: Multi-Pattern Wrappers for Relation Extraction from the Web." ECAI-02, Lyon, France, pp. 395-399, July 2002.
3. C. X. Ling and H. Wang: "Computing Optimal Attribute Weight Settings for Nearest Neighbor Algorithms", Artificial Intelligence Review, Vol. 11, 255-272, 1997.
4. Dang T.-Tung and B. Frankovic: "Agent based scheduling in production systems". Int. Journal of Production Research, Vol. 40, No. 15, p. 3669-3679, 2002.
5. I. Rudas: "Evolutionary operators; new parametric type operator families." Int. Journal of Fuzzy Systems, Vol. 23, No. 2, pp. 147-166. 1999
6. D. Dubois, F. Esteva, P. Garcia, L. Godo, R. L. de M'antaras, and H. Prade: "Fuzzy set modelling in case-based reasoning", Int. Journal of Intelligent Systems, 13, 345373, 1998.
7. E. Melis, J. Lieber, and A. Napoli: "Reformulation in Case-Based Reasoning", EWCBR-98, LNAI 1488, 172183. Springer, 1998.

A Representation of Relational Systems

Jacek Waldmajer

Department of Psychology
University of Opole, Poland
jwaldmajer@uni.opole.pl

Abstract. In this paper elements of a theory of multistructures are formulated. The theory of multistructures is used to define a binary representation of relational systems.

Keywords: multistructures, relations and operations on multistructures, binary representation of relational systems.

1 Introduction

In both computer science and discrete mathematics, there are diverse methods of a binary representation of mathematical objects. They have been known for years and mean numbers in the binary system, Boolean algebras and Boolean functions, graphs and relations on finite sets defined by binary matrices, binary trees as systems of identification of sets, etc.

Bonikowski in his PhD dissertation (1996), proposed an original Boolean representation of sets, which allows representing of approximate sets by means of Boolean functions, in the way Pawlak (1991) meant it. In this paper such a generalisation of construction of a representation of sets which allows a binary representation of relational systems is presented. The research attempts to uncover a theoretical formal-logical basis which would serve the purpose of description and explanation of processes of identification of objects and subsytems of reality treated as relational systems by learning systems, and more precisely, by natural and artificial networks.

2 Reality

By reality, in which features, relations, operations, individuals define things, we understand such a reality that any things in it possess certain features or are in some relations, or are results of the application of some operations, or are some individuals.

Definition 1 A **reality** is called a relational system $Re = \langle U,\ F,\ R,\ O,\ I \rangle$, where U is a non-empty set of things called the universe, F is a finite family of distinguished nonempty subsets of the set U called the features, R is the finite set of relations defined on U, O is the finite set of operations defined on U, I is a distinguished non-empty subset of the universe U called the set *of individuals, and for any $x \in U$ at least one of the following conditions is satisfied: x is an element of a certain set of the family F, x is an argument of a certain relation of the set R, x is an argument or value of a certain operation of the set O, x is an element of the set I.*

Definition 2 *A **tuple** is any ordered system $\langle \alpha_1, \alpha_2, ..., \alpha_n \rangle$ $(n \geq 1)$ of elements $\alpha_1, \alpha_2, ..., \alpha_n$ of the universe U.*

A **segment of the tuple** $t = \langle \alpha_1, \alpha_2, ..., \alpha_n \rangle$ is its any subsequence $t_1 = \langle \alpha_{i_1}, \alpha_i, ..., \alpha_{i_2} \rangle$, such that: $1 \leq i_1 \leq i_2 \leq n$. We accept the convention that $\langle \alpha_i \rangle = \alpha_i$ for any $i = 1, ..., n$.

Definition 3 *The **structure of reality** Re is the set S^{Re} of all tuples such that:*

(C1) *for any operation $o \in O$, if $\langle \beta_1, \beta_2, ..., \beta_k, \alpha \rangle \in o$, then*
$\langle \beta_1, \beta_2, ..., \beta_k \rangle \in S^{Re}$ and $\alpha \in S^{Re}$, where $\alpha = o(\beta_1, \beta_2, ..., \beta_k)$,
(C2)
$1°$ $\langle \alpha_1 \rangle \in S^{Re}$ if $\alpha_1 \in I$, or α_1 belongs to a certain set of F, or α_1 is determined by the condition $(C1)$,
$2°$ $t = \langle \alpha_1, \alpha_2, ..., \alpha_n \rangle \in S^{Re}$ if every element α_i $(i = 1, ...,n)$ is an element of any at least two element segment of t either determined by the condition $(C1)$ or belonging to a relation of R or an operation of O,
(C3) *no other tuple except that which is satisfies the conditions $(C1)$-$(C2)$ belongs to S^{Re}.*

If a tuple belongs to the structure of reality, it is called a **state**.

3   Multistructures

Within the confines of reality Re we may consider diverse types of interconnection of things.

Definition 4 *Let $M \subseteq S^{Re}$. We denote by $M^{\#}$ the set of all states of S^{Re} which are segments of certain tuples of M. The set $M^{\#}$ is called the **set of all segments of states** of M.*

In the family $P(S^{Re})$ of all subsets of the structure of reality S^{Re} we define the binary relation $\sim$.

Definition 5 *$M_1 \sim M_2 \Leftrightarrow M_1^{\#} = M_2^{\#}$, for any sets $M_1, M_2 \subseteq S^{Re}$.*

Fact 1 *The relation $\sim$ is an equivalence relation in the family $P(S^{Re})$.*

Definition 6 *The equivalence class $[M]_{\sim}$ of the relation $\sim$ represented by the set $M \subseteq S^{Re}$ is called the **multistructure determined by** M.*

From Definition 6 and Definition 5 we may conclude the following facts:

Fact 2 *For any two multistructures $[M_1]_{\sim}$ and $[M_2]_{\sim}$ the following conditions are satisfied: (1) $[M_1]_{\sim} = [M_2]_{\sim}$ iff $M_1^{\#} = M_2^{\#}$, (2) $[M_1]_{\sim} = [M_1^{\#}]_{\sim}$.*

4   Operations on multistructures

In this section we define basic relations and operations on multistructures.

Definition 7 *The state $x \in S^{Re}$ **belongs** to the multistructure $[M]_\sim$ (symbolically: $x \in_m [M]_\sim$) iff $x \in M^\#$.*

Definition 8 *Let $[M_1]_\sim$, $[M_2]_\sim$ be any multistructures. $[M_1]_\sim$ is included in $[M_2]_\sim$ (symbolically: $[M_1]_\sim \subseteq_m [M_2]_\sim$) iff for any state x if $x \in_m [M_1]_\sim$ than $x \in_m [M_2]_\sim$.*

From Definition 8 and Definition 7 follows:

Fact 3 *For any two multistructures $[M_1]_\sim$ and $[M_2]_\sim$*
$$[M_1]_\sim \subseteq_m [M_2]_\sim \Leftrightarrow M_1^\# \subseteq M_2^\#.$$

Now we shall define the operation $\cap_m$ of the **multiplication** of multistructures.

Definition 9 *The **intersection** $[M_1]_\sim \cap_m [M_2]_\sim$ of the multistructures $[M_1]_\sim$ and $[M_2]_\sim$ is the multistructure $[M_1^\# \cap M_2^\#]_\sim$.*

Fact 4 *For any multistructures $[M_1]_\sim$ and $[M_2]_\sim$ and any state x*
$$(x \in_m [M_1]_\sim \cap_m [M_2]_\sim) \Leftrightarrow (x \in_m [M_1]_\sim \land x \in_m [M_2]_\sim).$$

The next definition is a definition of the operation $\cup_m$ of the **addition** of multistructures.

Definition 10 *The **union** $[M_1]_\sim \cup_m [M_2]_\sim$ of the multistructures $[M_1]_\sim$ and $[M_2]_\sim$ is the multistructure $[M_1^\# \cup M_2^\#]_\sim$.*

Fact 5 *For any multistructures $[M_1]_\sim$ and $[M_2]_\sim$ and any state x*
$$(x \in_m [M_1]_\sim \cup_m [M_2]_\sim) \Leftrightarrow (x \in_m [M_1]_\sim \lor x \in_m [M_2]_\sim).$$

The definition of the operation $\setminus_m$ of the **subtraction** of multistructures is the following.

Definition 11 *The **difference** $[M_1]_\sim \setminus_m [M_2]_\sim$ of the multistructures $[M_1]_\sim$ and $[M_2]_\sim$ is the multistructure $[M_1^\# \setminus M_2^\#]_\sim$.*

Fact 6 *For any multistructures $[M_1]_\sim$ and $[M_2]_\sim$ and any state x*
$$(x \in_m [M_1]_\sim \setminus_m [M_2]_\sim) \Leftrightarrow (x \in_m [M_1]_\sim \land x \notin_m [M_2]_\sim).$$

The operation $'^m$ of the **complementation** of multistructures is defined as follows:

Definition 12 *The **complement** of the multistructure $[M]_\sim$ is the multistructure $[S^{Re} \setminus M^\#]_\sim$, which is denoted by the symbol $[M]_\sim^{'m}$.*

Definition 13 *Let M be a family of multistructures and*
$M^\# = \{M^\#\colon M \subseteq S^{Re} \wedge [M]_\sim \in M\}$. *The **generalized union** $\bigcup_m M$*
of the family M of multistructures is the multistructure $[\bigcup M^\#]_\sim$.

Fact 7 *For any family M of multistructures*
$x \in_m \bigcup_m M$ *iff there is a multistructure $[M]_\sim \in M$, such that $x \in_m [M]_\sim$.*

Definition 14 *The multistructure $0_m = [\emptyset]_\sim$ is called the **empty multi-**
structure and the multistructure $1_m = [S^{Re}]_\sim$ is called the **full multi-**
structure.*

Theorem 1 *The family of all multistructures with the operations of the addition $\cup_m$, the multiplication $\cap_m$ and the complementation $'^m$ defined on this
family, with the empty multistructure 0_m and the full multistructure 1_m is a
Boolean algebra.*

The proof of Theorem 1 is based on an obserwation that the family of all
multistructures with these operations is isomorphic with a field of subsets of
the structure S^{Re} of reality Re (see Rasiowa and Sikorski 1963).

5   A binary representation of relational systems

In this section we present a model of a binary representation of relational
systems. We use a method that is a generalisation of a method of a representation of sets introduced by Bonikowski (1996).

Definition 15 *Let us denote successively features, relations, operations and
the set I of individuals of reality Re by M_1, M_2, ..., M_n, and the family of
multistructures determined by these sets by $C = \{[M_1]_\sim, [M_2]_\sim, ..., [M_n]_\sim\}$.*
Let us introduce additionally the following notation for any $[M]_\sim$:

$$[M]_\sim^\sigma = \begin{cases} [M]_\sim^{'^m}, & when \ \sigma = 0 \\ [M]_\sim, & when \ \sigma = 1 \end{cases}$$

*A **binary representation** of the multistructure $[M]_\sim$ means a set of finite
binary sequences $(x_1, x_2, ..., x_n) \in E^n$ (E^n is a n - Cartesian power of the
set $E = \{0, 1\}$) such that the so-called **components of the multistructure**
$[M]_\sim$ defined in the following way:*

$$[M]_\sim^{(x_1, x_2, ..., x_n)} = [M_1]_\sim^{x_1} \cap_m [M_2]_\sim^{x_2} \cap_m ... \cap_m [M_n]_\sim^{x_n} \cap_m [M]_\sim$$

are non-empty multistructures.

From Definition 15 follow:

Theorem 2 *The sum of all components of the multistructure $[M]_\sim$ is equals
to the multistructure $[M]_\sim$, i.e. $\bigcup_m [M]_\sim^{(x_1, x_2, ..., x_n)} = [M]_\sim$.*

Theorem 3 *If* $(x_1, x_2, ..., x_n) \neq (x'_1, x'_2, ..., x'_n)$ *then*
$$[M]_{\sim}^{(x_1,x_2,...,x_n)} \cap_m [M]_{\sim}^{(x'_1,x'_2,...,x'_n)} = [\emptyset]_{\sim}.$$

A **Boolean representation** of the multistructure $[M]_{\sim}$ is a map $f_M : E^n \rightarrow E$ such that

$$f_M(x_1, x_2, ..., x_n) = \begin{cases} 1, & \text{when } [M]_{\sim}^{(x_1,x_2,...,x_n)} \neq [\emptyset] \\ 0, & \text{when } [M]_{\sim}^{(x_1,x_2,...,x_n)} = [\emptyset] \end{cases}$$

Fact 8 *For any multistructure there is only one Boolean representation.*

Among Boolean representations we may distinguish: representations of multistructures determined by the structure of reality Re, by its features, relations, operations and the set of all individuals of reality Re. The set of all such Boolean representations is called a **representation of the relational system** Re. The binary representation of reality as a relational system is a model of perception for learning systems.

Perspectives of research

In this paper we presented a conceptual apparatus enabling, step by step, to generate a certain Boolean representation for any relational system. The procedure may serve as a basis for drawing an algorithm to create a binary representation of reality by machines (computers, neural networks, digital and analogue learning machines). In this sense, this representation is a model of a perception. The author hopes that this research will help solve the problem of adequacy for a binary representation of reality, i.e. to announce the characteristics of classes of relational systems possessing the same binary representation and to formulate necessary and sufficient conditions for the existence of a common representation of relational systems (c.f. Bryniarski and Waldmajer, 2002).

References

1. Bonikowski, Z.: *Zbiory aproksymowane przez reprezentacje,(Sets aproximated by representations)*, Manuscript IPIPAN, Warszawa 1996.
2. Bryniarski, E., Waldmajer, J.: 2002 *Adequacy Problem of Representation of Knowledge in Neural Networks*, Proceedings of The Sixth International Conference on Soft Computing and Distributed Processing, June 24-25, Rzeszow 2002, (ed. Z. Suraj), p. 86−88.
3. Pawlak, Z.: *Rough Sets: Theoretical Aspects of Reasonong about Data*, Kluwer Academic Publishers, Boston 1991.
4. Rasiowa, H., Sikorski, R.: *The Mathematics of Metamathematics*, PWN, Warszawa 1963.

Concept of the Knowledge Quality Management for Rule-Based Decision System

Michal Wozniak[1]

Chair of Systems and Computer Networks, Wroclaw University of Technology, Wybrzeze Wyspianskiego 27, 50-370 Wroclaw, Poland

Abstract. The paper deals with the knowledge acquisition process. Different experts formulate the set of rules for decision support systems. We assume they have different knowledge about the problem and therefore obtained rules have different qualities. The knowledge base for the system under consideration does not have logical interpretation but probabilistic one. We will formulate a proposition of the rule confidence measure and its application to the decision process.

1 Introduction

Machine learning is the attractive approach for building decision support systems [8]. For this type of software, the quality of the knowledge base plays the key-role. One can get rules from different experts with different qualities. This problems was partly described for induction learning [1,2,6] and statistical method in [4]. The following paper deals with the quality of rules for the probabilistic reasoning. The content of this work is as follows: Section 2 introduces necessary background and provides the probabilistic decision problem statement. Next section presents a form of the rule for the probabilistic expert systems and proposes the rule-based algorithm. Section 4 defines statistical confidence measure of the knowledge and shows how to modify the knowledge base according to the confidence measure of rules. Section 5 presents the interpretation of the proposed measure for the estimation process based on the typical statistical model. The last section concludes the paper.

2 Decision problem statement

Among different concepts and methods of using 'uncertain' information in pattern recognition, Bayes decision theory is efficient approach and attractive from the theoretical point of view.This approach consists of assumption [3] that the feature vector $x = (x^{(1)}, x^{(2)}, ..., x^{(d)})$ (describing the object being under recognition) and in the class number $j \in \{1, 2..., M\}$ (the object belongs to) are the realization of the pair of the random variables X, J. The random variable J is described by the *prior* probability p_j. For each class $j - X$ has conditional probability density function $f_j(x)$. These parameters can be used for enumerating *posterior* probability $p(i|x)$ according to Bayes

formulae. The formalisation of the recognition in the case under consideration implies the setting of an optimal Bayes decision algorithm $\Psi(x)$, which
minimizes probability of misclassification for 0-1 loss function:

$$\Psi(x) = i \quad \text{if} \quad p(i|x) = \max_{k \in \{1, ..., M\}} p(k|x). \tag{1}$$

In the real situation the *prior* probabilities and the conditional density
functions are usually unknown. Furthermore we often have no reason to decide that the *prior* probability is different for each decision. Instead of them
we can use the rules and/or the learning set for constructing decision algorithms [7].

3   Rule-based decision algorithm

Rules are the most popular model for the logical decision support systems.
For systems we consider the rules given by experts have more the statistical interpretation than logical one. The form of a rule for the probabilistic
decision support system [5] is usually as follows

if A then B with the probability β,

where β is interpreted as an estimator of the *posterior* probability $P(B|A)$.
More precisely, in the case of human knowledge acquisition process, experts
are not disposed to formulate the exact value of the β, but he (or she) rather
prefers to give the interval for its value $\underline{\beta} \leq \beta \leq \bar{\beta}$.

The analysis of different practical examples leads to the following general
forms of rule $r_i^{(k)}$ pointed at the class i for x belonging to the decision area
$D_i^{(k)}$:

IF $x \in D_i^{(k)}$ **THEN** state of object is i

WITH posterior probability $\beta_i^{(k)}$ greater than $\underline{\beta}_i^{(k)}$ and less than $\bar{\beta}_i^{(k)}$.
Where

$$\beta_i^{(k)} = \int_{D_i^{(k)}} p(i|x)\, dx. \tag{2}$$

For that form of knowledge we can formulate the decision algorithm $\Psi_R(x)$
which points at the class i if $\hat{p}(i|x)$ (the *posterior* probability estimator
obtained from the rule set) has the biggest value.

The knowledge about probabilities given by expert estimates the average
posterior probability for the whole decision area. As we see for decision making we are interested in the exact value of the *posterior* probability for given
observation.

A rule, for the logical knowledge representation, with the small decision
area can be overfitting the training data (especially if the training set is

small). For our proposition we respect this danger for the rule set obtained from learning data (it will be described in section 5).

For the estimation of the *posterior* probability from a rule we assume the constant value for the rule decision area. Therefore let us propose the relation 'more specific' between the probabilistic rules pointing at the same class.

Definition

Rule $r_i^{(k)}$ is 'more specific' than rule $r_i^{(l)}$ if

$$\left(\bar{\beta}_i^{(k)} - \underline{\beta}_i^{(k)}\right)\left(\int_{D_i^{(k)}} dx \bigg/ \int_X dx\right) < \left(\bar{\beta}_i^{(l)} - \underline{\beta}_i^{(l)}\right)\left(\int_{D_i^{(l)}} dx \bigg/ \int_X dx\right). \quad (3)$$

Hence the proposition of the *posterior* probability estimator $\hat{p}(i\,|x)$ is as follows: from a subset of rules $R_i(x) = \left\{ r_i^{(k)} : x \in D_i^{(k)} \right\}$ choose the "most specific" rule $r_i^{(m)}$

$$\hat{p}(i\,|x) = \left(\bar{\beta}_i^{(m)} - \underline{\beta}_i^{(m)}\right) \bigg/ \int_{D_i^{(m)}} dx. \quad (4)$$

4   Proposition of knowledge confidence measure

We consider decision under the assumption that the learning set is noise free (or experts tell us always true), i.e.

$$P\,(\text{If A then B with probability } \beta) = 1.$$

During the expert system designing process the rules are obtained from different sources which have the different confidence. For the knowledge given by experts we can not assume that they do not make any mistakes or/and if the rule set is generated on the base on the learning set we cannot assume that it is noise free. Therefore we postulate we cannot trust the information we get or we can believe on it only with the γ factor, proposed as the confidence measure which can be formulated as

$$P\,(\text{If A then B with probability } \beta) = \gamma \leq 1.$$

Let $\gamma_i^{(k)}$ denotes the value of the confidence measure of rule $r_i^{(k)}$. Let us show how it utilize for the rules modification. We propose after the acquisition process for each rule $r_i^{(k)}$

$$\bar{\beta}_i^{(k)} = 1 - \left(1 - \bar{\beta}_i^{(k)}\right)\gamma_i^{(k)} \quad \text{and} \quad \underline{\beta}_i^{(k)} = \underline{\beta}_i^{(k)}\,\gamma_i^{(k)}. \quad (5)$$

5   Confidence measure for the statistical estimation

The central problem of our proposition is how to calculate the confidence measure. For human experts the values for their rules are fixed arbitrarily according to the expert quality. We can also find the presented problem in typical statistical estimation of unknown parameter β, where we assume the significance level[10]. The significant level can be interpreted as the confidence measure. Each rule gives the index of the class. If the feature vector value belongs to the decision area given by the rule, the decision depends on the previous state and on the applied therapy. While constructing the rules set, we have to define somehow the decision areas for the new rule set. For example we can want to obtain *posterior* probability estimator for each rule, which is not less than a fixed value or in practice we can use the one of well known machine learning algorithms based on *sequential covering* procedure [8].

For each of the given intervals we have to obtain the estimator of the *posterior* probability. We use the following statistical model [9]:

- the learning set is selected randomly from a population and there exist two class of points: marked (point at the class $i \in \{1, ..., M\}$) and unmarked (point at the class l, where $l \in \{1, ..., M\}$ and $l \neq i$),
- the expected value for the population is p,
- the best estimator of p is $\hat{p} = m/n$,

where n means the sample size and m - the number of the marked elements.
For the fixed significance level α we get

$$P\left(\frac{m}{n} - \mu_\alpha \sqrt{\frac{\frac{m}{n}\left(1 - \frac{m}{n}\right)}{n}} < p < \frac{m}{n} + \mu_\alpha \sqrt{\frac{\frac{m}{n}\left(1 - \frac{m}{n}\right)}{n}}\right) \approx 1 - \alpha \quad (6)$$

The μ_α is the value of:

- the t-distribution for $n-1$ degrees of freedom and for the significance level α. (for small sample $n < 100$),
- the normal standardized $N(0, 1)$ distribution for the significance level α (for big sample $n \geq 100$).

In those cases we get rule $r_i^{(k)}$, for which confidence measure of rule $\gamma_i^{(k)} = 1 - \alpha$ and

$$\underline{\beta}_i^{(k)} = \frac{m}{n} - \mu_\alpha \sqrt{\frac{\frac{m}{n}\left(1 - \frac{m}{n}\right)}{n}} \text{ and } \bar{\beta}_i^{(k)} = \frac{m}{n} + \mu_\alpha \sqrt{\frac{\frac{m}{n}\left(1 - \frac{m}{n}\right)}{n}}. \quad (7)$$

6   Conclusion

The paper concerned probabilistic reasoning and the proposition of the quality measure for that formulated decision problems. We hope this idea of confidence management can be applied to any other form of rule. E.g. for the logical rules (where "if-then" means logical implication) acquisition process we can attribute the value of confidence to each rule. It could be used in case of the contradiction detected in the set of rules. Then we propose to remove rule by rule according to their value of confidence measure until contradiction is detected. Presented method needs the analytical and simulation research. Let us draw some future works under the concept of the information quality:

1. developing the method how to judge the expert quality and providing analytical researches into proposed method properties,
2. applying proposed method to the real medical decision problems,
3. performing simulation experiments on computer generated data to estimate the dependencies between the size of the decision area and the data quality versus correctness of classification.

References

1. Bruha I., Quality of Decision Rules: Definitions and Classification Schemes for Multiple Rules, [in:] Nakhaeizadeh G., Taylor C.C. [eds], *Machine Learning and Statistic*, John Wiley and Sons, New York 1997.
2. Dean P., Famili A., Comparative Performance of Rule Quality Measures in an Inductive Systems, *Applied Intelligence*, no 7, 1997.
3. Devijver P. A., Kittler J., *Pattern Recognition: A Statistical Approach*, Prentice Hall, London 1982.
4. Hand D.J., *Construction and Assesment of Classificaton Rules*, J.Willey and Sons, New York 1997.
5. Giakoumakis E., Papakonstantiou G., Skordalakis E., Rule-based systems and pattern recognition, *Pattern Recognition Letters*, No 5, 1987.
6. Gur-Ali O., Wallance W.A., Induction of rules subject to a quality constraint: probabilistic inductive learning, *IEEE Transaction on Knowledge and Data Engineering*, vol. 5, no 3, 1993.
7. Kurzynski M., Wozniak M.: Rule-based algorithms with learning for sequential recognition problem, *Proceedings of the Third International Conference on Information Fusion "Fusion 2000"*. Paris, July 10-13, 2000.
8. Mitchell T., *Machine Learning*, McGraw Hill, 1997.
9. Sachs L., *Applied Statistic. A Handbook of Techniques*, Springer-Verlag, New York Berlin Heideberg Tokyo, 1984.
10. Wozniak M., Blinowska A., Unification of the information as the way of recognition the controlled Markov chains, *Proc. of the Congress on Information Processing and Management of Uncertainty in Knowledge Based Systems*, Granada, Spain 1996.

Druck: Strauss GmbH, Mörlenbach
Verarbeitung: Schäffer, Grünstadt